S0-DXD-661
COLLEGE
LIBRARY
RECEIVED
DEC 8 1976
DALLAS, PENNSYLVANIA

RODD'S CHEMISTRY OF CARBON COMPOUNDS

ELSEVIER SCIENTIFIC PUBLISHING COMPANY
335 JAN VAN GALENSTRAAT
P.O.BOX 211, AMSTERDAM, THE NETHERLANDS

AMERICAN ELSEVIER PUBLISHING COMPANY, INC.
52 VANDERBILT AVENUE, NEW YORK, N.Y. 10017

ISBN: 0-444-41209-3

LIBRARY OF CONGRESS CARD CATALOG NUMBER 64-4605

WITH 27 TABLES

PRINTED IN THE NETHERLANDS

RODD'S CHEMISTRY OF CARBON COMPOUNDS

ADVISORS

RODD'S CHEMISTRY OF CARBON COMPOUNDS

VOLUME I

GENERAL INTRODUCTION

ALIPHATIC COMPOUNDS

*

VOLUME II

ALICYCLIC COMPOUNDS

*

VOLUME III

AROMATIC COMPOUNDS

*

VOLUME IV

HETEROCYCLIC COMPOUNDS

*

VOLUME V

MISCELLANEOUS

GENERAL INDEX

*

RODD'S CHEMISTRY OF CARBON COMPOUNDS

A modern comprehensive treatise

SECOND EDITION

Edited by

S. COFFEY

M.Sc. (London), D.Sc. (Leyden), F.R.I.C.

formerly of

I.C.I. Dyestuffs Division, Blackley, Manchester

VOLUME III PART D

AROMATIC COMPOUNDS

Monobenzenoid hydrocarbon derivatives with functional groups in an acyclic side chain; aralkylamines and their derivatives, aralkanols and their oxidation products, monobenzenoid carbaldehydes and monoketones; phenolic aralkanols, their oxidation products and derivatives including carboxylic acids; depsides, hydrolysable tannins, lignans, lignin and humic acid.

ELSEVIER SCIENTIFIC PUBLISHING COMPANY

AMSTERDAM OXFORD NEW YORK

1976

CONTRIBUTORS TO THIS VOLUME

G. W. H. CHEESEMAN, D.SC., F.R.I.C.
Department of Chemistry, Queen Elizabeth College, London W8

J. GRIMSHAW, PH.D., D.SC., F.R.I.C.
Department of Chemistry, Queen's University, Belfast BT9 5AG

P. F. G. PRAILL, PH.D., F.R.I.C.
Department of Chemistry, Queen Elizabeth College, London W8

R. E. FAIRBAIRN, B.SC., PH.D., F.R.I.C.
formerly of Research Department, Dyestuffs Division, I.C.I. Ltd., Manchester 9 *(Index)*

PREFACE TO VOLUME III D

The first three parts of Volume III are concerned with the chemistry of the monobenzenoid hydrocarbons and their nuclear substituted derivatives in which the substituent groups are attached to the benzene nucleus through an element other than carbon.

The present volume, III D, deals with compounds in which an acylic side-chain attached to the benzene nucleus carries the substituent groups, *i.e.* with aralkyl compounds. The first two chapters, 13 and 14, are a revision by Dr. G. W. H. CHEESEMAN and Dr. P. F. G. PRAILL of Chapter IX in the original edition and deal respectively with (a) aralkylamines, aralkanols and their derivatives, and (b) oxidation products of the aralkanols, namely, monobenzenoid carbaldehydes and monoketones and their derivatives. Chapters 15 and 16 are a revision by Dr. J. GRIMSHAW of Chapter XII in the original edition. The first is concerned with phenolic aralkylamines, aralkanols, carbaldehydes, ketones and carboxylic acids of the benzene series and their derivatives, the second with the closely related naturally occurring compounds, the depsides, hydrolysable tannins, lignans, lignin and humic acids, and so will be of particular botanical and phytochemical interest.

In the strictly logical sense followed in the original edition, Chapter 14 should have been followed by a revised Chapter X of the original edition on monobenzenoid carboxylic acids. Unfortunately, the revised script for this chapter is not yet available; it will appear eventually in Volume III G.

With the publication of this volume the editor takes great pleasure in thanking the three authors for their cooperation and patience during its preparation and Mr. E. B. ROBINSON for checking the corrected proofs.

March 1976 S. COFFEY

CONTENTS

VOLUME III D

Aromatic Compounds: Monobenzenoid hydrocarbon derivatives with functional groups in an acyclic side chain; aralkylamines and their derivatives, aralkanols and their oxidation products, monobenzenoid carbaldehydes and monoketones; phenolic aralkanols, their oxidation products and derivatives including carboxylic acids; depsides, hydrolysable tannins, lignans, lignin and humic acid

PREFACE VII

OFFICIAL PUBLICATIONS; SCIENTIFIC JOURNALS AND PERIODICALS XV

LIST OF ABBREVIATED NAMES OF CHEMICAL FIRMS MENTIONED IN PATENT REFERENCES XVI

LIST OF COMMON ABBREVIATIONS AND SYMBOLS USED XVII

Chapter 13. Aralkylamines and Aralkanols and their Oxidation Products: Aralkylamines and Aralkanols and their Derivatives
by G. W. H. CHEESEMAN AND P. F. G. PRAILL

1. Aralkylamines and their derivatives 1

   a. Aralkylamines; methods of formation and preparation 1

      (*i*) Amines by aralkylation, 1 – (*ii*) Amines by reduction, 3 – (*iii*) Degradative methods, 8 – (*iv*) Miscellaneous methods, 8

   b. Physical properties of aralkylamines 11

   c. Reactions of aralkylamines 11

      (*i*) Acylation, 11 – (*ii*) Nitrosation, 12 – (*iii*) Reductive cleavage, 12 – (*iv*) Reactions with oxidising agents, 13 – (*v*) Nitration, 14 – (*vi*) Sommelet–Hauser rearrangement, 14 – (*vii*) Stevens rearrangement, 14

   d. Individual amines 16

      (*i*) Aralkylamines, 16 – (*ii*) Carbonic acid derivatives of benzylamine and its homologues, 20 – (*iii*) Nuclear-substituted aralkylamines, 21 – (*iv*) Benzyl derivatives of hydroxylamine, 25 – (*v*) Benzyl derivatives of hydrazine, 28 – (*vi*) Diazo compounds, triazenes and azides related to benzylamine, 29

2. Monohydric alcohols of the benzene series and their derivatives 30

   a. Methods of formation and preparation 30

      (*i*) From aralkyl halides, 30 – (*ii*) From aldehydes and ketones, 31 – (*iii*) From carboxylic acids and their derivatives, 32 – (*iv*) From olefins, 33 – (*v*) From cyclic ethers, 34 – (*vi*) Miscellaneous methods, 35

   b. Physical properties and reactions 35

      (*i*) Reaction with oxidising agents, 35 – (*ii*) Cleavage of the benzyl–oxygen bond, 37 – (*iii*) Wittig rearrangement, 37 – (*iv*) Rearrangement of benzyl methyl sulphide and benzyldimethylsulphonium bromide, 38 – (*v*) Oxidation of benzyl sulphides, 38 – (*vi*) Reaction of benzyl alcohols and benzyl ethers with thallium(III) trifluoroacetate, 38

   c. Phenyl-substituted alcohols 39

   d. Functional derivatives of the alcohols 44

      (*i*) Ethers, 44 – (*ii*) Benzyl esters, 45 – (*iii*) Inorganic esters of benzyl alcohol, 46 – (*iv*) Thio analogues of benzyl alcohols and their derivatives, 48

   e. Nuclear-substituted benzyl alcohols 49

Chapter 14. Aralkylamines and Aralkanols and their Oxidation Products: Monocarbaldehydes and Monoketones of the Benzene Series
by G. W. H. CHEESEMAN AND P. F. G. PRAILL

3. Monocarbaldehydes of the benzene series 55
a. Methods of formation and preparation 55

(*i*) From methylbenzenes, 55 – (*ii*) From primary alcohols, 56 – (*iii*) From benzyl halides, 57 – (*iv*) From benzylidene dihalides, 58 – (*v*) From benzylamines and benzylidene–amines, 58 – (*vi*) From arylamines, 59 – (*vii*) From carboxylic acids and their derivatives, 59 – (*viii*) From formylation reactions of the type ArH → ArCHO, 65 – (*ix*) From organo-metallic reagents, 66 – (*x*) From dihydro-1,3-oxazines, 67 – (*xi*) From olefins, 69 – (*xii*) From α,β-unsaturated amides and nitriles, 70 – (*xiii*) From 1,2-glycols, 70 – (*xiv*) From aldehydes and ketones, 71 – (*xv*) Miscellaneous methods, 72

b. General properties 73
c. Reactions of aromatic aldehydes 74

(*i*) Oxidation and reduction, 75 – (*ii*) The Cannizzaro reaction, 76 – (*iii*) Benzoin reaction, 76 – (*iv*) Condensation reactions, 76 – (*v*) Conversion of aromatic aldehydes to nitriles, 81 – (*vi*) Conversion of aromatic aldehydes to carboxamides, 81 – (*vii*) Decarbonylation, 82 – (*viii*) Aldehydes as acylating reagents, 82 – (*ix*) Halogenation, 82 – (*x*) The reactions of aromatic aldehydes and diazoalkanes, 82 – (*xi*) Conversion of aldehydes into epoxides, 83 – (*xii*) Paterno–Büchi reaction, 83 – (*xiii*) Reaction with bis(1,3-diphenylimidazolinylidene-2), 83

d. Individual arenecarbaldehydes and their derivatives 86

(*i*) Benzaldehyde and its homologues, 86 – (*ii*) Functional derivatives of benzaldehyde, 88 – (*iii*) Sulphur and selenium analogues, 90 – (*iv*) Reaction of benzaldehyde with ammonia and amines, 91 – (*v*) Halogenobenzaldehydes, 95 – (*vi*) Nitrobenzaldehydes, 95 – (*vii*) Hydroxyimino-, nitroso-, azoxy- and azobenzaldehydes, 99 – (*viii*) Aminobenzaldehydes, 101 – (*ix*) Benzaldehydesulphonic acids, 103

e. Aryl-substituted aliphatic aldehydes 103
4. Monoketones in the benzene series 107
a. Methods of formation and preparation 107

(*i*) From alkylbenzenes and secondary alcohols, 107 – (*ii*) From alkyl halides, 108 – (*iii*) From arylamines, 108 – (*iv*) From carboxylic acids and their derivatives, 109 – (*v*) From β-oxo-esters, 113 – (*vi*) From β-diketones, 113 – (*vii*) From dihydro-1,3-oxazines, 113 – (*viii*) From 1,2-glycols and epoxides, 114 – (*ix*) From olefins, 115 – (*x*) From acetylenes, 116 – (*xi*) From aldehydes and ketones, 116 – (*xii*) Miscellaneous reactions, 119

b. Alkyl aryl ketones 121

(*i*) General properties, 121 – (*ii*) Reactions of alkyl aryl ketones, 122 – (*iii*) Individual alkyl aryl ketones and their analogues, 130 – (*iv*) Nuclear substituted acetophenones, 134

Chapter 15. Phenolic Aralkylamines, Monohydric Alcohols, Monocarbaldehydes, Monoketones and Monocarboxylic Acids
by J. GRIMSHAW

1. Phenolic aralkylamines 141
a. Methods of preparation 141

b. Monohydroxyphenylalkylamines 142
c. Di- and tri-hydroxyphenylalkylamines 144
d. Hydroxybenzyl isothiocyanates and their parent glucosides 145
2. Hydroxyaryl alcohols 145
a. Methods of preparation 145
b. Properties and reactions 146
(*i*) Acidity, 146 – (*ii*) Removal of side-chain, 146 – (*iii*) Carbonium ion stabilisation, 146
c. Monohydroxyaryl alcohols 147
d. Di- and tri-hydroxyphenyl alcohols 149
3. Hydroxyaryl carbaldehydes 150
a. Methods of preparation 150
(*i*) The Reimer–Tiemann reaction, 150 – (*ii*) Gattermann reaction, 151 – (*iii*) Vilsmeier reaction, 151 – (*iv*) Alternative formylating agents, 152 – (*v*) Duff reaction, 153 – (*vi*) Use of organometallic reagents, 153 – (*vii*) By the diazonium reaction, 154 – (*viii*) Side-chain oxidations, 154 – (*ix*) Side-chain reduction, 155
b. Properties and reactions 155
(*i*) Chelation and intramolecular hydrogen bonding, 155 – (*ii*) Oxidation, 156 – (*iii*) Hydrogenation, 156 – (*iv*) Cyclisation reactions, 156
c. Monohydroxybenzaldehydes and their homologues 156
(*i*) Methods of preparation, 158 – (*ii*) Reactions, 158
d. Dihydroxybenzaldehydes and their homologues 161
(*i*) Derivatives of catechol, 161 – (*ii*) Derivatives of resorcinol, 163 – (*iii*) Derivatives of hydroquinone, 163
e. Tri- and tetra-hydroxybenzaldehydes 164
(*i*) Derivatives of pyrogallol, 164 – (*ii*) Derivatives of 1,3,4-trihydroxybenzene, 164 – (*iii*) Derivatives of phloroglucinol, 165 – (*iv*) Derivatives of tetrahydroxybenzene, 165
4. Hydroxyphenyl ketones 165
a. Methods of preparation 165
(*i*) Fries reaction, 166 – (*ii*) Friedel–Crafts reaction, 166 – (*iii*) Hoesch reaction, 166 – (*iv*) Diazo reaction, 167 – (*v*) Replacement of halogen, 167
b. Properties and reactions 167
c. Monohydroxyaryl ketones 168
5. Phenolic monocarboxylic acids: the hydroxybenzoic acids 171
a. Methods of preparation 171
(*i*) Kolbe–Schmitt reaction, 171 – (*ii*) Friedel–Crafts reaction, 173 – (*iii*) Reimer–Tiemann reaction, 173 – (*iv*) Direct introduction of hydroxyl, 173 – (*v*) From organometallic compounds, 173 – (*vi*) By side-chain oxidation, 174
b. Properties and reactions 174
(*i*) Acidity, 174 – (*ii*) Substitution, 174 – (*iii*) Macrocyclic ester formation, 175
c. 2-Hydroxybenzoic acids: salicylic acid and its derivatives 175
(*i*) Salicylic acid, 175 – (*ii*) Ethers and esters of salicylic acid, 177 – (*iii*) Salicyloyl chloride and amides, 178 – (*iv*) Derivatives of salicylic acid containing phosphorus, 179 – (*v*) Sulphur derivatives of salicylic acid, 180 – (*vi*) Nuclear-

substituted salicylic acids, 181 – (*vii*) Homologues of salicylic acid, 183 – (*viii*) Cyclic anhydrides of salicylic acid and its homologues, 183

d. Other monohydroxybenzoic acids 187
e. Dihydroxybenzoic acids 188

(*i*) Derivatives of catechol, 188 – (*ii*) Derivatives of resorcinol, 190 – (*iii*) Derivatives of hydroquinone, 192

f. Trihydroxybenzoic acids 193

(*i*) Derivatives of pyrogallol, 193 – (*ii*) Gallic acid and its derivatives, 193 – (*iii*) Derivatives of 1,2,4-trihydroxybenzene, 198 – (*iv*) Derivatives of phloroglucinol, 199

6. Phenolic monocarboxylic acids: mono-, di- and tri-hydroxyphenylacetic and homologous acids 199
a. Hydroxyphenylacetic acids 199

(*i*) Preparation, 199 – (*ii*) Individual compounds, 200

b. Hydroxyphenylpropionic acids 201

(*i*) Preparation. 201 – (*ii*) Individual acids, 201

c. Hydroxyphenylbutyric and higher acids 202

(*i*) Preparation, 202 – (*ii*) Individual acids, 202

Chapter 16. Depsides, Hydrolysable Tannins, Lignans, Lignin and Humic Acid
by J. GRIMSHAW

1. Depsides, depsidones and depsones 203
a. Occurrence and general structure 203
b. Isolation 205
c. Biosynthesis 205
d. Depsides 208

(*i*) Degradation, 208 – (*ii*) Spectral properties, 209 – (*iii*) Synthesis, 210 – (*iv*) Individual depsides, 212

e. Depsidones 216

(*i*) Degradation, 216 – (*ii*) Spectral properties, 216 – (*iii*) Synthesis, 217 – (*iv*) Individual depsidones, 218 – (*v*) Depsones, 224

2. Hydrolysable tannins 227
a. Introduction 227
b. Isolation and characterisation 227

(*i*) Isolation and identification, 227 – (*ii*) General properties of tannins, determination of molecular weight, 228

c. Galloylglucoses and gallotannin 228

(*i*) Depsides of gallic acid, 228 – (*ii*) Simple gallate esters of glucose, 230 – (*iii*) The gallotannins, 231 – (*iv*) Biosynthesis of gallic acid, 236

d. Ellagitannins 237

(*i*) Acid components of hydrolysed tannin extracts, 237 – (*ii*) The ellagitannins, 242 – (*iii*) Interrelationship of the ellagitannin acids, 251

3. Lignans 253
a. Introduction, biosynthesis 253

b. Occurrence and isolation 254
c. Individual lignans 255

(*i*) 1,4-Diarylbutanes, 255 – (*ii*) Tetrahydrofuroguaiacins, 259 – (*iii*) Pinoresinol and lariciresinol types, 260 – (*iv*) Phenyltetralins, 265 – (*v*) Lignans of podophyllum resin, 269

4. Lignin 271
a. Occurrence 271
b. Isolation 272
c. Structure 273

(*i*) Evidence from degradation products, 273 – (*ii*) Model for lignin biosynthesis, 275 – (*iii*) Constitutional model for lignin, 276

5. Humic acid 277

INDEX 281

Titles of other parts of Volume III

AROMATIC COMPOUNDS

Vol. III A: General Introduction. Mononuclear hydrocarbons and their halogeno derivatives, and derivatives with nuclear substituents attached through nonmetallic elements from Group VI of the Periodic Table

Vol. III B: Benzoquinones and related compounds: Derivatives of mononuclear benzenoid hydrocarbons with nuclear substituents attached through an element other than the non-metals in Groups VI and VII of the Periodic Table

Vol. III C: Nuclear-substituted benzenoid hydrocarbons with more than one nitrogen atom in a substituent group

Vol. III E: Monobenzenoid hydrocarbon derivatives with functional groups in separate side-chains; dihydric and polyhydric aralkanols and their oxidation products; monobenzenoid hydrocarbons with unsaturated side-chains and their derivatives

Vol. III F: Polybenzenoid hydrocarbons and their derivatives; hydrocarbon ring assemblies, polyphenyl-substituted aliphatic hydrocarbons and their derivatives

Vols. III G, H and I: Aromatic compounds with fused carbocyclic ring systems. Index

OFFICIAL PUBLICATIONS

B.P.	British (United Kingdom) Patent
F.P.	French Patent
G.P.	German Patent
Ger. Offen.	German Patent Application, open for inspection
Sw.P.	Swiss Patent
U.S.P.	United States Patent
U.S.S.R.P.	Russian Patent
B.I.O.S.	British Intelligence Objectives Sub-Committee Reports, H.M. Stationery Office, London.
C.I.O.S.	Combined Intelligence Objectives Sub-Committee Reports
F.I.A.T.	Field Information Agency, Technical Reports of U.S. Group Control Council for Germany
B.S.	British Standards Specification
A.S.T.M.	American Society for Testing and Materials
A.P.I.	American Petroleum Institute Projects
C.I.	Colour Index Number of Dyestuffs and Pigments

SCIENTIFIC JOURNALS AND PERIODICALS

With few obvious and self-explanatory modifications the abbreviations used in references to journals and periodicals comprising the extensive literature on organic chemistry, are those used in the World List of Scientific Periodicals.

LIST OF ABBREVIATED NAMES OF CHEMICAL FIRMS MENTIONED IN PATENT REFERENCES

A.G.F.A., Agfa A.G.	Aktiengesellschaft für Anilinfabrikation (Berlin)
B.A.S.F.	Badische Anilin- und Soda-Fabrik (Ludwigshafen)
Bayer	Farbenfabriken vorm. Friedrich Bayer und Co. (Leverkusen)
Cassella	Leopold Cassella und Co. (Frankfurt am Main)
C.F.M.	Compagnie française des Matières Colorantes (Paris)
CIBA	Gesellschaft für chemische Industrie (Basel)
Du Pont	E.I. Du Pont de Nemours and Co. (U.S.A.)
G.A.F.	General Anilin and Film Corporation (U.S.A.)
Geigy A.G.	J. R. Geigy S.A. (Basel)
Hoechst	Hoechst A.G. (see M.L.B.)
I.C.I.	Imperial Chemical Industries, Ltd. (London)
I.G.	(=Interessen Gemeinschaft Farbenindustrie) of the principal dyestuffs manufacturers in Germany
Kalle	Kalle und Co., A.G. (Biebrich am Rhein)
M.L.B.	Farbwerke vormals Meister, Lucius und Brüning (Hoechst)
Sandoz	Sandoz A.G. Chemische Fabrik (Basel)

LIST OF COMMON ABBREVIATIONS AND SYMBOLS USED

A	acid
Å	Ångström units
Ac	acetyl
a	axial
as, *asymm.*	asymmetrical
at.	atmosphere
B	base
Bu	butyl
b.p.	boiling point
C, mC and μC	curie, millicurie and microcurie
c, *C*	concentration
c.d.	circular dichroism
conc.	concentrated
crit.	critical
D	Debye unit, 1×10^{-18} e.s.u.
D	dissociation energy
D	dextro-rotatory; dextro configuration
DL	optically inactive (externally compensated)
d	density
dec. or decomp.	with decomposition
deriv.	derivative
E	energy; extinction; electromeric effect
E1, E2	uni- and bi-molecular elimination mechanisms
E1cB	unimolecular elimination in conjugate base
e.s.r.	electron spin resonance
Et	ethyl
e	nuclear charge; equatorial
f	oscillator strength
f.p.	freezing point
G	free energy
g.l.c.	gas liquid chromatography
g	spectroscopic splitting factor, 2.0023
H	applied magnetic field; heat content
h	Planck's constant
Hz	hertz
I	spin quantum number; intensity; inductive effect
i.r.	infrared
J	coupling constant in n.m.r. spectra
K	dissociation constant
k	Boltzmann constant; velocity constant
kcal.	kilocalories
L	laevorotatory; laevo configuration
M	molecular weight; molar; mesomeric effect
Me	methyl

m	mass; mole; molecule; *meta-*
ml	millilitre
m.p.	melting point
Ms	mesyl (methanesulphonyl)
[M]	molecular rotation
N	Avogadro number; normal
n.m.r.	nuclear magnetic resonance
n	normal; refractive index; principal quantum number
o	*ortho-*
o.r.d.	optical rotatory dispersion
P	polarisation; probability; orbital state
Pr	propyl
Ph	phenyl
p	*para-*; orbital
p.m.r.	proton magnetic resonance
R	clockwise configuration
S	counterclockwise config.; entropy; net spin of incompleted electronic shells; orbital state
S_N1, S_N2	uni- and bi-molecular nucleophilic substitution mechanisms
S_Ni	internal nucleophilic substitution mechanisms
s	symmetrical; orbital
sec	secondary
soln.	solution
symm.	symmetrical
T	absolute temperature
Tosyl	*p*-toluenesulphonyl
Trityl	triphenylmethyl
t	time
temp.	temperature (in degrees centigrade)
tert	tertiary
U	potential energy
u.v.	ultraviolet
v	velocity
α	optical rotation (in water unless otherwise stated)
[α]	specific optical rotation
α_A	atomic susceptibility
α_E	electronic susceptibility
ε	dielectric constant; extinction coefficient
μ	microns (10^{-4} cm); dipole moment; magnetic moment
μ_B	Bohr magneton
μg	microgram (10^{-6} g)
λ	wavelength
υ	frequency; wave number
χ, χ_d, χ_μ	magnetic, diamagnetic and paramagnetic susceptibilities
$\sim$	about

(+)	dextrorotatory
(−)	laevorotatory
⊖	negative charge
⊕	positive charge

Chapter 13

Aralkylamines and Aralkanols and their Oxidation Products: Aralkylamines and Aralkanols and their Derivatives

G. W. H. CHEESEMAN AND P. F. G. PRAILL*

The aralkylamines and aralkyl alcohols are formally derived from the aliphatic amines and alcohols described in Vol. I by substituting hydrogen atoms attached to carbon by one or more aryl groups. The simplest primary alcohols, $ArCH_2OH$, give rise by oxidation to the true aromatic carbaldehydes, ArCHO, and carboxylic acids, $ArCO_2H$, in which the aldehydic and acidic functions are attached directly to the aromatic nucleus. Higher primary alcohols such as β-phenylethyl alcohol, $PhCH_2 \cdot CH_2OH$, by oxidation give the arylated aliphatic carbaldehydes and acids, while secondary alcohols such as α-phenylethyl alcohol, $PhCHOH \cdot CH_3$, and 1-phenylpropan-2-ol, $PhCH_2 \cdot CHOH \cdot CH_3$, give ketones. Many of the halogen compounds corresponding with these alcohols, aldehydes and ketones are described in Vol. III A, pp. 271 *et seq.* This chapter is concerned with the chemistry of monoaryl derivatives of the aliphatic amines and their simple derivatives, monoarylated aliphatic alcohols and the structurally related carbaldehydes and ketones.

1. Aralkylamines and their derivatives

(a) Aralkylamines; methods of formation and preparation

An excellent critical account of amine synthesis has been written by *P. A. S. Smith*, "Open-Chain Nitrogen Compounds", Benjamin, New York, 1965, Vol. 1, Chapter 2, and a valuable comprehensive review of this topic

*Manuscript date, March 1973.

appears in *Houben–Weyl*, "Methoden der organischen Chemie", Thieme, Stuttgart, 1957, 4th Edn., Vol. 11/1. Amine synthesis has also been reviewed by *M. S. Gibson*, in "The Chemistry of the Amino Group", Ed. *S. Patai*, Interscience, London, 1968, Chapter 2.

(i) Amines by aralkylation

(*1*) Primary, secondary and tertiary aralkylamines have been prepared by the aralkylation of ammonia, primary and secondary amines. Reactions of this type lead to the expected mixture of products, the composition being dependent on conditions.

For example, reaction of benzyl chloride with excess of liquid ammonia gives 53% of benzylamine and 39% of dibenzylamine, whereas reaction of benzyl chloride with 18% ethanolic ammonia at 100° gives 9% of benzylamine, 35% of dibenzylamine and 48% of tribenzylamine (*J. von Braun*, Ber., 1937, **70**, 979). Treatment of benzyl chloride with acetamide gives *N*-benzylacetamide and this on acid hydrolysis yields benzylamine in 65% overall yield (*M. A. Phillips*, J. Soc. chem. Ind., 1947, **66**, 325):

$$PhCH_2Cl \rightarrow PhCH_2NHCOMe \rightarrow PhCH_2NH_2$$

The use of the more weakly nucleophilic acetamide rather than ammonia is recommended in cases where monoaralkylation is required.

Primary and secondary amines may be alkylated or phenylated under mild conditions using bromoalkanes or bromobenzene, respectively, in the presence of naphthalenelithium in tetrahydrofuran (*K. Suga et al.*, Chem. and Ind., 1969, 78).

(*2*) From N-*benzylphthalimides*. These are readily prepared and on reaction with hydrazines are converted into the corresponding benzylamines (*H. R. Ing* and *R. H. F. Manske*, J. chem. Soc., 1926, 2348; *J. W. Baker*, *W. S. Nathan* and *C. W. Shoppee*, *ibid.*, 1935, 1847).

A related method is to form a saccharin derivative of the primary halide which can be further transformed into a secondary amine as shown (*K. Abe* and *S. Yamamoto*, J. pharm. Soc. Japan, 1953, **73**, 1322):

$ArCH_2X$ —(Saccharin / K_2CO_3)→ [C=O, SO_2, NCH_2Ar] —((1) KOH; (2) ROTs; (3) HCl - H_2O)→ $ArCH_2NHR$

(*3*) From the *quaternary salts from aralkyl halides and hexamethylenetetramine* which are transformed into aralkylamines either by treatment with concentrated hydrochloric acid and ethanol or by successive treatment with sulphur dioxide and dilute hydrochloric acid (*G. Spielberger*, in *Houben–Weyl*, "Methoden der organischen Chemie", 4th Edn., Thieme, Stuttgart, Vol. 11/1, 1957, p. 106; *J. Graymore*, J. chem. Soc., 1947, 1116):

$$[ArCH_2-N_4(CH_2)_6]^{\oplus} X^{\ominus} \xrightarrow{HCl/EtOH} ArCH_2NH_3^{\oplus} Cl^{\ominus}$$

$$[ArCH_2-N_4(CH_2)_6]^{\oplus} X^{\ominus} \xrightarrow{SO_2} ArCH_2NHCH_2OSO_2H \xrightarrow{H_2O/HCl} ArCH_2NH_3^{\oplus} Cl^{\ominus}$$

(*4*) In the alkylation of primary amines to secondary amines a variety of protecting groups such as benzenesulphonyl, phenylcyano and benzylidene have been employed (*H. Decker* and *P. Becker*, Ann., 1913, **395,** 362), *e.g.:*

$$PhCH_2 \cdot CH_2NH_2 \xrightarrow{PhCHO} PhCH_2 \cdot CH_2N{=}CHPh \xrightarrow{MeI}$$

$$PhCH_2 \cdot CH_2 \overset{\oplus}{N}(Me){=}CHPh \xrightarrow{H_2O} PhCH_2 \cdot CH_2NHMe$$

(*5*) Tertiary amines are readily obtained by the aralkylation of secondary amines (*Smith* and *R. N. Leoppky*, J. Amer. chem. Soc., 1967, **89,** 1147), *e.g.:*

$$(PhCH_2)_2NH + [4]NO_2C_6H_4 \cdot CH_2Br \xrightarrow[CH_2OH \cdot CH_2OH,\ 150°/24\,h]{K_2CO_3} (PhCH_2)_2NCH_2 \cdot C_6H_4NO_2[4] \quad (80\%)$$

The following sequence is illustrative of a further practical synthesis of tertiary amines from secondary amines (*M. Sekiya* and *Y. Terao*, Chem. pharm. Bull. Japan, 1970, **18,** 947):

$$Me(PhCH_2)NH + CH_2O + \text{succinimide (HN)} \longrightarrow Me(PhCH_2)NCH_2N\text{(succinimido)} \xrightarrow{PhCH_2MgBr} Me(PhCH_2)NCH_2 \cdot CH_2Ph \quad (53\%)$$

(ii) Amines by reduction

Aralkylamines can be prepared by reduction of compounds containing a wide variety of nitrogen-containing functional groups and also by the reductive alkylation of ammonia and of primary and secondary amines with aldehydes and ketones.

(*1*) The reduction of *nitro compounds* is not so important a method for th preparation of aralkylamines as for the preparation of aromatic amines because of the more limited availability of starting materials. Examples of successful conversions of this type are:

$$PhCH(NO_2) \cdot CH_2 \cdot CH_3 \xrightarrow{Zn/alkali} PhCH(NH_2) \cdot CH_2 \cdot CH_3$$

(*M. Konowalov*, J. Russ. phys. Chem., 1893, **25,** 523; Chem. Zentr., 1894, I, 465);

$$PhCH{=}CHNO_2 \xrightarrow{LiAlH_4} PhCH_2 \cdot CH_2NH_2 \quad (60\%)$$

(*R. F. Nystrom* and *W. G. Brown*, J. Amer. chem. Soc., 1948, **70,** 3738).

(*2*) *Azides* can be reduced catalytically or with lithium tetrahydridoaluminate. Since azides are derivable from alcohols or olefins *via* the reaction of alkyl halides or alkyl toluenesulphonates with sodium azide, this is a synthesis of potentially wide scope (*A. Bertho* and *J. Maier*, Ann., 1932, **498,** 50):

$$PhCH_2N_3 + H_2 \xrightarrow{Pt} \underset{(93\%)}{PhCH_2NH_2} + N_2$$

$$PhCH_2 \cdot CH_2N_3 \xrightarrow{LiAlH_4} \underset{(89\%)}{PhCH_2 \cdot CH_2NH_2} + N_2$$

(*J. H. Boyer*, J. Amer. chem. Soc., 1951, **73,** 5865).

(*3*) *Nitriles* can be reduced catalytically to primary amines, the primary amines formed are likely to be contaminated with secondary amines due to the equilibration of primary amine with intermediate imine:

$$RC{\equiv}N \xrightarrow{H_2} RCH{=}NH \xrightarrow{H_2} RCH_2NH_2$$

$$RCH_2NH_2 + RCH{=}NH \rightleftharpoons RCH_2N{=}CHR + NH_3$$

$$RCH_2N{=}CHR \xrightarrow{H_2} RCH_2NHCH_2R$$

The formation of secondary amine can be suppressed either by carrying out the reduction in the presence of excess of ammonia or in the presence of acetic anhydride which reacts with the primary amine before it can equilibrate with the intermediate imine:

$$PhCH_2CN \xrightarrow[\text{Raney Ni, 120-130°}]{H_2\text{(140 atm.)}, NH_3} \underset{(83-87\%)}{PhCH_2 \cdot CH_2NH_2}$$

(*J. C. Robinson* and *H. R. Snyder*, Org. Synth., Coll. Vol. 3, 1955, p. 720);

$$[4]MeC_6H_4 \cdot CN \xrightarrow[(CH_3 \cdot CO)_2O]{H_2/Pt} \underset{(88\%)}{[4]MeC_6H_4 \cdot CH_2NHCOMe}$$

(*W. H. Carothers* and *G. A. Jones*, J. Amer. chem. Soc., 1925, **47,** 3051).

Amine hydrochlorides are obtained by hydrogenation of nitriles with a platinum catalyst in chloroform (*J. A. Sacrist* and *M. W. Logue*, J. org. Chem., 1972, **37,** 335):

$$ArCN \xrightarrow[CHCl_3]{PtO_2-H_2} ArCH_2\overset{\oplus}{N}H_3\overset{\ominus}{Cl}$$

Nitriles are also reduced to amines with sodium and alcohol, lithium tetrahydridoaluminate (*Nystrom* and *Brown*, *loc. cit.*), Raney nickel alloy (*B. Staskun* and *T. van Es*, J. chem. Soc., 1966, 531), or with diborane. The latter reagent is conveniently generated from sodium tetrahydridoborate and boron trifluoride etherate and is effective for the selective reduction of the nitrile group (*H. C. Brown* and *B. C. Subba Rao*, J. Amer. chem. Soc., 1960, **82,** 681):

$$[3]NO_2C_6H_4\cdot CN \xrightarrow{B_2H_6} [3]NO_2C_6H_4\cdot CH_2NH_2$$

(isolated as the hydrochloride in 79% yield)

Nitriles, nitro compounds and amides can be reduced to primary amines with sodium tetrahydridoborate–transition metal systems in hydroxylic as well as non-hydroxylic solvents (*T. Satoh et al.*, Tetrahedron Letters, 1969, 4555). An alternative way of converting nitriles into primary amines is by conversion into imidic ester hydrochloride followed by electrolytic reduction (*H. Wenker*, J. Amer. chem. Soc., 1935, **57,** 772):

$$ArCN \xrightarrow{ROH-HCl} ArC(OR){=}\overset{\oplus}{N}H_2\overset{\ominus}{Cl} \xrightarrow{Electrolysis} ArCH_2NH_2$$

(*4*) From *amides*. The catalytic hydrogenation of amides to amines requires more forcing conditions than for nitriles; copper chromite has been commonly used as catalyst. This transformation is conveniently carried out with lithium tetrahydridoaluminate or a solution of diborane in tetrahydrofuran solution (*H. C. Brown* and *P. Heim*, J. Amer. chem. Soc., 1964, **86,** 3566):

$$PhCONPr^i_2 \xrightarrow{B_2H_6} PhCH_2NPr^i_2 \quad (98\%)$$

The latter procedure is effective for the reduction of primary, secondary, and especially tertiary amides to the corresponding amines and gives yields approaching 100%. A two step procedure for the reduction of secondary or tertiary amides to amines is illustrated by the conversion of *N*-ethylbenzamide to *N*-ethylbenzylamine (*R. F. Borch*, Tetrahedron Letters, 1968, 61):

$$PhCONHEt \xrightarrow[CH_2Cl_2]{Et_3O^{\oplus}BF_4^{\ominus}} PhC(OEt){=}NEt \xrightarrow[EtOH]{NaBH_4} PhCH_2NHEt \quad (92\%)$$

On treatment of *N,N*-diethylbenzamide and *N,N*-dimethyl-*β*-phenylpropionamide with sodium tetrahydridoborate in boiling pyridine, the corresponding amines are formed in 55 and 51% yield, respectively (*Y. Kikugawa, S. Ikegami* and *S. Yamada*, Chem. Pharm. Bull. Japan, 1969, **17,** 98).

Aluminium hydride is also an effective reagent for the conversion of amides to amines (*N. M. Yoon* and *Brown*, J. Amer. chem. Soc., 1968, **90,** 2927).

(*5*) *Oximes* are convenient starting materials from which to prepare primary aralkylamines. The most widely used methods of reduction are treatment with excess of sodium and alcohol or catalytic hydrogenation. The same precautions to minimise secondary amine formation are required as in the case of the catalytic reduction of nitriles. Alternatively this reduction can be achieved with aluminium hydride, which is conveniently prepared from lithium tetrahydridoaluminate and sulphuric acid:

$$2\ LiAlH_4 + H_2SO_4 \longrightarrow 2\ AlH_3 + 2\ H_2 + Li_2SO_4$$

$$PhC(=NOH)Me \longrightarrow \underset{(82\%)}{PhCH(NH_2)Me} + \underset{(4.3\%)}{PhNHCH_2Me}$$

Raney alloy, and Raney nickel and sodium hypophosphite, have been used for the reduction of oximes (*Staskun* and *Van Es, loc. cit.*).

The reduction of the *O*-methyloximes of aldehydes and ketones (obtained by the reaction of aldehydes and ketones with *O*-methylhydroxylamine hydrochloride) with diborane in tetrahydrofuran at 25° is a convenient method for the preparation of primary and secondary amines (*H. Feuer* and *D. M. Braunstein*, J. org. Chem., 1969, **34**, 1817).

$$R^1R^2C{=}N{-}OMe \xrightarrow{B_2H_6/THF/25°} R^1R^2CH{-}NH_2$$

The reduction of the corresponding oximes with diborane is carried out at 105–110°. Sodium bis(2-methoxyethoxy)aluminium hydride also reduces aldoximes and ketoximes to primary amines in satisfactory yield (*M. Černý et al.*, Coll. Czech. chem. Comm., 1969, **34**, 1033).

(*6*) *Schiff bases* are reduced to secondary amines catalytically, with lithium tetrahydridoaluminate or sodium tetrahydridoborate (*Nystrom* and *Brown*, *loc. cit.*):

$$PhN{=}CHPh \xrightarrow{LiAlH_4} \underset{(93\%)}{PhNHCH_2Ph}$$

Reduction of a methanolic solution of *N*-benzylideneaniline with sodium tetrahydridoborate similarly gives phenylbenzylamine in 97% yield (*J. H. Billman* and *A. C. Diesing*, J. org. Chem., 1957, **22**, 1068).

(*7*) *Isocyanates and isothiocyanates* are reduced by lithium tetrahydridoaluminate to *N*-methylarylamines (*A. E. Finholt, C. D. Dean* and *C. L. Agre, ibid.*, 1953, **18**, 1338):

$$PhNCO \xrightarrow{LiAlH_4} \underset{(86\%)}{PhNHMe}$$

(*8*) *Reductive alkylation.* In this process a mixture of ammonia or a primary or secondary amine and aldehyde or ketone is hydrogenated over a Raney nickel or platinum catalyst (*W. S. Emerson*, Org. Reactions, 1948, **4**, 174). Alternative methods of reduction may be used; method (*6*) above corresponds to the case of reductive alkylation of a primary amine with an aldehyde and with isolation of the intermediate.

The conversion:

$$PhCOMe + NH_3 + H_2 \xrightarrow[H_2 \text{ pressure}]{\text{Raney Ni}} PhCH(NH_2)Me + H_2O$$
(44-52%)

(*Robinson* and *Snyder*, *loc. cit.*) can also be carried out using formic acid as the reducing agent (*A. W. Ingersoll*, Org. Synth., Coll. Vol. 2, 1943, p. 503):

$$PhCOMe + NH_3 + HCO_2H \longrightarrow PhCH(NH_2)Me + H_2O + CO_2$$
(60-66%)

The experimental procedure is to heat the ketone with ammonium formate; the formyl derivative of the amine is produced initially and subsequently hydrolysed. This general method, known as the Leuckart Reaction, has been reviewed by *M. L. Moore* in Org. Reactions, 1949, **5**, 301. The reaction of primary or secondary amines with aldehydes or ketones and sodium tetrahydridoborate in an acetate buffer at 0° yields secondary or tertiary amines (*K. A. Schellenberg*, J. org. Chem., 1963, **28**, 3255):

$$PhNH_2 + PhCHO \xrightarrow{NaBH_4} PhNHCH_2Ph$$
(83%)

Lithium cyanohydridoborate has been recommended as a reagent for carrying out this type of reaction. Its use obviates the necessity of high-pressure hydrogenation (*e.g.*, *R. F. Borch* and *H. D. Durst*, J. Amer. chem. Soc., 1969, **91**, 3996):

$$PhCHO \xrightarrow[MeOH - pH\ 5\ to\ 6]{EtNH_2-LiBH_3CN} PhCH_2NHEt$$
(80%)

Benzaldehyde has been converted into benzylamine by condensation with benzyl carbamate and subsequent low-pressure hydrogenation in the presence of a palladium catalyst (*A. E. Martell* and *R. M. Herbst*, J. org. Chem., 1941, **6**, 878):

$$PhCHO \xrightarrow[(2)\ H_2-Pd]{(1)\ NH_2CO_2CH_2Ph} PhCH_2NH_2$$

(9) *Hydrogenolysis of quaternary salts.* In general the cleavage of quaternary salts is best effected with lithium tetrahydridoaluminate (*A. C. Cope et al.*, J. Amer. chem. Soc., 1960, **82**, 4651):

$$R_4N^{\oplus} \xrightarrow{LiAlH_4} R_3N + RH$$

This possibility is useful when the purification of a tertiary amine is conveniently carried out by crystallisation of a quaternary salt. Reduction of the methiodide of 1,2,3,4-tetrahydro-2-methylisoquinoline with sodium and liquid ammonia gives *N,N*-dimethyl-2-*o*-tolylethylamine (*D. B. Clayson*, J. chem. Soc., 1949, 2016):

$$\text{(tetrahydroisoquinolinium, } N^{\oplus}(Me)_2\ I^{\ominus}) \longrightarrow \text{o-MeC}_6H_4CH_2{\cdot}CH_2N(Me)_2$$

(iii) Degradative methods

These methods are widely employed for the conversion of aralkyl carboxylic acids to aralkylamines. These include the Hofmann degradation of amides (*E. S. Wallis* and *J. F. Lane*, Org. Reactions, 1946, **3,** 267), the Curtius degradation of acyl azides (*Smith, ibid.*, p. 337) and the Schmidt degradation of carboxylic acids (*H. Wolff, ibid.*, p. 307):

$$RCONH_2 \xrightarrow[OH^{\ominus}]{X_2} RCONHX \longrightarrow RN{=}C{=}O \longrightarrow RNH_2 + CO_3^{2\ominus}$$

(X = halogen)

$$RCON_3 \xrightarrow{\Delta} RN{=}C{=}O \xrightarrow{H_2O} RNH_2 + CO_2$$

$$RCO_2H + HN_3 \xrightarrow{\text{conc. } H_2SO_4} RNH_2 + N_2 + CO_2$$

(iv) Miscellaneous methods

(*1*) From *organometallic compounds.* The addition of alkylmagnesium halides to azomethines has been used (*M. Busch*, Ber., 1904, **37,** 2691; *W. J. Hickinbottom*, J. chem. Soc., 1934, 322):

$$PhCH{=}NPh \xrightarrow[(2)\ H_2O]{(1)\ MeMgI} \underset{(79\%)}{PhMeCHNHPh}$$

Reaction of an ethereal solution of chloramine with the Grignard reagents from benzyl and β-phenylethyl chlorides gives benzylamine and β-phenylethylamine in 85 and 74% yield, respectively (*G. H. Coleman* and *C. R. Hauser*, J. Amer. chem. Soc., 1928, **50,** 1193):

$$RMgCl + ClNH_2 \longrightarrow RNH_2 + MgCl_2$$

The corresponding reaction has been achieved with Grignard reagents and β-chloroethylamines (*P. M. G. Bavin et al.*, J. med. Chem., 1966, **9,** 790), *e.g.:*

$$PhMgBr + ClCH_2{\cdot}CH_2NMe_2 \longrightarrow PhCH_2{\cdot}CH_2NMe_2$$

The adduct obtained by reaction of a Grignard reagent and a nitrile may be reduced with lithium tetrahydridoaluminate to give a primary amine (*A. Pohland* and *H. R. Sullivan*, J. Amer. chem. Soc., 1953, **75,** 5898):

$$PhMgBr + CH_3{\cdot}CH_2{\cdot}CN \longrightarrow PhC({=}NMgBr)CH_2{\cdot}CH_3 \longrightarrow PhCH(NH_2)CH_2{\cdot}CH_3 (80\%)$$

(*2*) From *organoboranes.* Organoboranes react with chloramine or hydroxylamine *O*-sulphonic acid to give primary amines. Since organoboranes are derived from olefins and diborane, this method represents the synthesis of amines from olefins (*M. W. Rathke et al.*, *ibid.*, 1966, **88,** 2870):

$$R_3B \xrightarrow{NH_2OSO_3H} RNH_2$$

(*3*) *Amidomethylation.* The scope of this reaction has been reviewed by *H. E. Zaugg* and *W. B. Martin* (Org. Reactions, 1965, **14,** 52).

For example, amidomethylation of *m*-xylene with acetonitrile and formaldehyde in 85% phosphoric acid yields 60–66% of *N,N'*-diacetyl-4,6-di-(aminomethyl)-1,3-xylene, alternatively *N*-(2,4-dimethylbenzy)formamide is prepared in 25% yield from *m*-xylene, formaldehyde and hydrogen cyanide (*C. L. Parris* and *R. M. Christenson*, J. org. Chem., 1960, **25,** 1888):

Me, Me, CH_2NHCHO ← Me, Me → Me, $AcNHCH_2$, Me, CH_2NHAc

A further modification is to utilise the hydroxymethyl derivatives of amides. This reaction is known as the Tscherniac–Einhorn reaction and commonly the *N*-hydroxymethylamide of chloroacetic acid is used (*A. Einhorn et al.*, Ann., 1905, **343,** 207; 1908, **361,** 113):

$$PhCO_2H + ClCH_2 \cdot CONHCH_2OH \xrightarrow{H_2SO_4} C_6H_4(CO_2H)(CH_2NHCO \cdot CH_2Cl) \xrightarrow{dil.HCl} C_6H_4(CO_2H)(CH_2NH_2)$$

(50%)

(*4*) Primary amines such as benzylamine and α- and β-phenylethylamine are converted into the corresponding secondary amines by boiling with Raney nickel in benzene (*K. Kindler*, *G. Malamed* and *D. Matties*, *ibid.*, 1961, **644,** 23):

$$PhCH_2NH_2 \longrightarrow PhCH{=}NH \longrightarrow (PhCH_2)_2NH$$

(85%)

(*5*) β-Substituted phenylethylamines are formed by the aluminium chloride-promoted reaction of unsaturated amines and benzene and its homologues (*A. W. Weston*, *A. Wayne Ruddy* and *C. M. Suter*, J. Amer. chem. Soc., 1943, **65,** 674) *e.g.:*

$$C_6H_6 + CH_2{=}CH \cdot CH_2NH_2 \longrightarrow CH_3 \cdot CHPh \cdot CH_2NH_2$$

(94%)

A related synthesis is the aluminium chloride-catalysed reaction of arenes with aziridines (*N. Milstein*, J. heterocycl. Chem., 1968, **5,** 339):

$$C_6H_5Cl + \underset{H_2C-CH_2}{\overset{H}{N}} \longrightarrow o\text{-}ClC_6H_4CH_2{\cdot}CH_2NH_2 + p\text{-}ClC_6H_4CH_2{\cdot}CH_2NH_2$$

(*6*) Amine radical ions generated from protonated *N*-chloroamines and reducing metal salts, simultaneously attack aromatic rings and benzylic positions (*F. Minisci, R. Galli* and *R. Bernardi*, Tetrahedron Letters, 1966, 699; *Minisci*, Synthesis, 1973, 1):

$$R_2\overset{\oplus}{N}H-Cl \quad + \quad Fe^{2\oplus} \quad \rightarrow \quad R_2\overset{\oplus}{N}H \quad + \quad Cl^{\ominus} \quad + \quad Fe^{3\oplus}$$

(*7*) *Ritter reaction.* This involves the acid-catalysed addition of a secondary or tertiary alcohol or olefin to a nitrile or hydrogen cyanide (*J. J. Ritter* and *J. Kalish*, Org. Synth., 1964, **44**, 44):

$$PhCH_2{\cdot}CMe_2OH \; + \; HCN \xrightarrow{H_2SO_4} \underset{(65-70\%)}{PhCH_2{\cdot}CMe_2NHCHO} \xrightarrow{NaOH} \underset{(75\%)}{PhCH_2{\cdot}CMe_2NH_2}$$

The scope of this reaction has been reviewed (*L. I. Krimen* and *D. J. Cota*, Org. Reactions, 1969, **17**, 213); it is of special value for the preparation of tertiary amines.

Alcohols have also been converted into amines by conversion into sulphamate esters, thermal rearrangement of these esters and final hydrolysis to the amine (*E. H. White* and *C. A. Elliger*, J. Amer. chem. Soc., 1965, **87**, 5261):

$$ROH \xrightarrow[(2)\ Me_2NSO_2Cl]{(1)\ NaH-MeOCH_2{\cdot}CH_2OMe} ROSO_2NMe_2 \xrightarrow{\Delta}$$

$$R-\overset{Me}{\underset{Me}{N^{\oplus}}}-SO_3^{\ominus} \xrightarrow[(2)\ NaOH]{(1)\ HCl-H_2O} RNHMe_2$$

N,N-Dimethylbenzylamines are obtained by heating benzylic alcohols with hexamethylphosphoric triamide (HMPT) (*R. S. Monson* and *D. N. Priest*, Chem. Comm., 1971, 1018).

(*8*) A further synthesis of *N*-substituted benzylamines involves the alkylation of benzonitrile with dialkoxycarbonium tetrafluoroborates. The resulting nitrilium salts after ethanolysis and sodium tetrahydridoborate reduction afford alkylbenzylamines (*R. F. Borch*, J. org. Chem., 1969, **34**, 627). Dialkoxycarbonium tetrafluoroborates are readily available from the reaction of the corresponding orthoesters and boron trifluoride:

$$PhC{\equiv}N \; + \; \overset{\oplus}{H}C(OR)_2\,BF_4^{\ominus} \longrightarrow PhC{\equiv}\overset{\oplus}{N}-R\ BF_4^{\ominus} \xrightarrow{EtOH}$$

$$PhC(OEt){=}N-R \xrightarrow{NaBH_4} PhCH_2NHR$$

(b) Physical properties of aralkylamines

Benzylamine resembles simple alkylamines in being freely soluble in water and in combining with atmospheric carbon dioxide. It has a p*K*a value of 9.34 at 25° in water and is therefore a weaker base than methylamine (p*K*a 10.62) and a much stronger base than aniline (p*K*a 4.58). A monograph by *D. D. Perrin* ("Dissociation Constants of Organic Bases in Aqueous Solution", Butterworths, London, 1965) tabulates the p*K*a values of many bases related to benzylamine. See also *J. W. Smith*, in "The Chemistry of the Amino Group", ed. S. Patai, Interscience, London, 1968, p. 179.

A characteristic feature of the infrared spectra of primary and secondary amines is N–H stretching absorption at 3500–3300 cm^{-1}. Primary amines have two bands in this region whereas secondary amines generally show only one band. In the p.m.r. spectrum of benzylamine in $CDCl_3$ the methylene resonance is at 6.44 τ. The methyl resonance in the p.m.r. spectrum of *N,N*-dimethylbenzylamine in deuterochloroform is at 7.78 τ (*J. C. N. Ma* and *E. W. Warnhoff*, Canad. J. Chem., 1965, **43,** 1849). The resonances of N–H protons are variable due to differences in the degree of hydrogen bonding. Benzylamine fragments in the mass spectrometer with successive loss of $H^{\oplus}$ and HCN (*A. P. Bruins, N. M. M. Nibbering* and *Th. J. de Boer*, Tetrahedron Letters, 1972, 1109).

O.r.d. and c.d. measurements on α- and β-phenylethylamines indicate that the absolute configuration of the asymmetric centre may be assigned from the o.r.d. curve. Compounds of the *S*-configuration exhibit a positive multiple Cotton effect between 270 and 250 nm. (−)-α-Phenylethylamine is configurationally related to L-(+)-alanine and is thus correctly described as *S*-(−)-α-phenylethylamine:

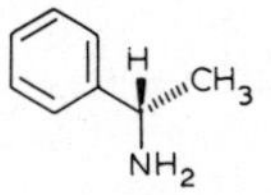

H CH_3 NH_2 CH_2 H CH_3 NH_2

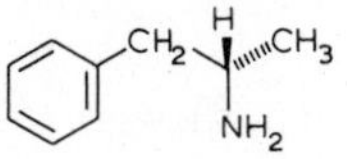

(+)-α-Methyl-β-phenylethylamine was also shown to have *S*-configuration (*J. Cymerman Craig, R. P. K. Chan* and *S. K. Roy*, Tetrahedron, 1967, **23,** 3573).

(c) Reactions of aralkylamines

Aralkylamines resemble alkylamines in many of their reactions and only some of the special features of their reactivity are mentioned in the following discussion.

(i) Acylation

Benzylamines may be converted into mono- or di-acetyl derivatives depending on the conditions, thus treatment with boiling acetic anhydride gives diacetyl derivatives (*R. P. Mariella* and *K. H. Brown*, J. org. Chem., 1971, **36,** 735). Benzyldialkylamines of the type $ArCH_2NR_2$ react with acetic anhydride to give the benzyl acetate and the *N*,*N*-dialkylacetamide, $R_2NCO{\cdot}CH_3$. This conversion may also be effected *via* an intermediate quaternary salt (*W. R. Brasen* and *Hauser*, Org. Synth., 1954, **34,** 58):

$$\text{o-}MeC_6H_4CH_2NMe_2 \xrightarrow{EtBr} \text{o-}MeC_6H_4CH_2\overset{\oplus}{N}Me_2Et\ Br^{\ominus} \xrightarrow[MeCO_2H]{MeCO_2K} \text{o-}MeC_6H_4CH_2OAc\ (88-91\%)$$

(ii) Nitrosation

Aqueous nitrous acid converts benzylamine into benzyl alcohol, however, diazotisation of benzylamine in anhydrous dimethyl sulphoxide gives benzaldehyde in 82% yield (*K. H. Scheit* and *W. Kampe*, Angew. Chem. internat. Edn., 1965, **4,** 787). Tertiary amines undergo complex reactions with nitrous acid resulting in oxidative dealkylation *(Smith* and *Leoppky, loc. cit.):*

$$R_2N-CHAr_2 \xrightarrow{HONO} R_2N-NO + Ar_2C{=}O$$

N-Nitrosodibenzylamines give good yields of hydrocarbon product when treated with sodium dithionite and base or lithium in liquid ammonia (*C. G. Overberger*, *J. C. Lombardino* and *R. C. Hiskey*, J. Amer. chem. Soc., 1958, **80,** 3009; see also *D. M. Lemal* and *T. W. Rave*, *ibid.*, 1965, **87,** 393):

$$(PhCH_2)_2N-NO \xrightarrow{Na_2S_2O_4/NaOH} PhCH_2{\cdot}CH_2Ph\ (77\%)$$

(iii) Reductive cleavage

The benzyl–nitrogen bond can be cleaved by electrolytic reduction, reduction with sodium amalgam (Emde reduction), reduction with sodium in liquid ammonia, or by hydrogenolysis (*W. H. Hartung* and *R. Simonoff*, Org. Reactions, 1953, **7,** 253). The relative ease of cleavage is

$$PhCH_2-\overset{|}{\underset{|}{N}}{}^{\oplus}- \ > \ PhCH_2-O- \ > \ PhCH_2-\overset{|}{N}-$$

$$C_6H_3(Me)_2CH_2\overset{\oplus}{N}Me_3\ I^{\ominus} \xrightarrow[H_2O]{Na/Hg} C_6H_3(Me)_3\ (85-90\%)$$

(*Brasen* and *Hauser*, *loc. cit.*, p. 56);

$$\text{1-benzyl-1,2,3-triazole (N-CH}_2\text{Ph)} \xrightarrow{\text{Na/liquid NH}_3} \text{1H-1,2,3-triazole (N-H)} \quad (77\%)$$

(*R. H. Wiley, K. F. Hussung* and *J. Moffat*, J. org. Chem., 1956, **21,** 190).

The use of benzyl as a protective group has been reviewed by *J. F. W. McOmie* in "Advances in Organic Chemistry, Methods and Results", Interscience, New York, 1963, **3,** 191.

Reaction of benzylamine with sodium hydride at 50–85° gives dibenzylamine and ammonia but this is not a very general reaction (*R. Baltzly* and *S. W. Blackman*, J. org. Chem., 1963, **28,** 1158).

(iv) Reactions with oxidising agents

A systematic study has been made of the mechanisms of oxidation of benzylamine and substituted benzylamines with permanganate; primary attack involves abstraction of benzylic hydrogen (*Min-Min Wei* and *Ross Stewart*, J. Amer. chem. Soc., 1966, **88,** 1974).

Oxidation of benzylamine with potassium permanganate in the presence of calcium sulphate gives benzaldehyde in 61% yield; substituted benzylamines are similarly oxidised and α-phenylethylamine is converted in 78% yield to acetophenone by the same procedure (*S. S. Rawalay* and *H. Shechter*, J. org. Chem., 1967, **32,** 3129. Oxidation of *N*-benzylanilines with manganese dioxide gives benzylideneanilines (*E. F. Pratt* and *T. P. McGovern*, *ibid.*, 1964, **29,** 1540):

$$PhCH_2NHPh \xrightarrow[\text{benzene}]{MnO_2} PhCH{=}NPh \quad (62\%)$$

Benzylamines are rapidly oxidised to the corresponding carbonyl compounds with potassium ferrate (K_2FeO_4) (*R. J. Audette, J. W. Quaill* and *P. J. Smith*, Tetrahedron Letters, 1971, 279; Chem. Comm., 1972, 38). Similar products are obtained from the oxidation of benzylic alcohols with this reagent.

Nickel peroxide oxidises benzylamines to nitriles in yields of 50–90%, presumably through the intermediate aldimines (*K. Nakagawa* and *T. Tsuji*, Chem. Pharm. Bull. Japan, 1963, **11,** 296). This conversion may also be effected with an active cobalt oxide (prepared from cobalt(II) sulphate, sodium hydroxide and ozonised oxygen) (*J. S. Belew, C. Garza* and *J. W. Mathieson*, Chem. Comm., 1970, 634). Dibenzylamine is oxidised by benzoyl peroxide to the benzoate of *N,N*-dibenzylhydroxylamine (*W. A. Waters*, "Mechanisms of Oxidation of Organic Compounds", Methuen, London, 1964, p. 135). The reaction of benzylamine and sulphur gives benzylammonium polysulphides, ammonia and *N*-benzylidenebenzylamine (*Y. Sasaki* and *F. P. Olsen*, Canad. J. Chem., 1971, **49,** 283). The *N*-oxides derived from *N,N*-dialkylbenzylamines rearrange on heating to hydroxylamine derivatives (*Cope et al.*, J. Amer. chem. Soc., 1949, **71,** 1628, 3423; *U. Schöllkopf, M. Patsch* and *H. Schäfer*, Tetrahedron Letters, 1964, 2515):

$$Me_2N(\rightarrow O)CH_2Ph \longrightarrow Me_2NOCH_2Ph \ (61\%)$$

$$Et_2N(\rightarrow O)CH_2Ph \longrightarrow Et_2NOCH_2Ph \ (34\%)$$

$$Et_2N(\rightarrow O)CH_2Ph \longrightarrow CH_2{=}CH_2 + EtN(OH)CH_2Ph \ (31\%)$$

(v) Nitration

In the series: PhMe, $PhCH_2\overset{\oplus}{N}Me_3$, $PhCH_2{\cdot}CH_2\overset{\oplus}{N}Me_3$, and $PhCH_2{\cdot}CH_2{\cdot}CH_2\overset{\oplus}{N}Me_3$ the % *meta* nitration is 3, 88, 19 and 5%, respectively (*J. Hine*, "Physical Organic Chemistry", McGraw-Hill, New York, 1962, 2nd Edn., p. 376).

(vi) Sommelet–Hauser rearrangement

Benzyltrimethylammonium salts rearrange on treatment with sodamide in liquid ammonia to 2-methylbenzyldimethylamines (*Brasen* and *Hauser*, *loc. cit.*, p. 61):

$$C_6H_5CH_2\overset{\oplus}{N}Me_3 \longrightarrow C_6H_5CH_2\overset{\oplus}{N}Me_2\overset{\ominus}{C}H_2 \longrightarrow \text{(6-methylene-5-(dimethylaminomethyl)cyclohexa-1,3-diene)} \longrightarrow 2\text{-}CH_3C_6H_4CH_2NMe_2 \ (90\text{–}95\%)$$

Protons are reversibly removed from the more acidic benzyl group and the species $Ph\overset{\ominus}{C}H\overset{\oplus}{N}Me_3$ has been intercepted by reaction with benzophenone. Sommelet–Hauser rearrangement only occurs when none of the alkyl groups of the quaternary ammonium ion carries the β-hydrogen required for an elimination reaction *e.g.*:

$$PhCH_2\overset{\oplus}{N}(CH_2{\cdot}CH_2{\cdot}CH_3)_3 \xrightarrow{NaNH_2/NH_3} PhCH_2N(CH_2{\cdot}CH_2{\cdot}CH_3)_2 + CH_2{=}CH{\cdot}CH_3$$

(vii) Stevens rearrangement

This involves base-catalysed intramolecular migration from a quaternary nitrogen atom; reaction is initiated by proton abstraction from a carbon activated, for example, by both a quaternary ammonium and a carbonyl group:

$$Me_2\overset{\oplus}{N}(CH_2Ph){-}CH_2{\cdot}COPh \longrightarrow Me_2N{-}CH(CH_2Ph){\cdot}COPh$$

Predominantly 1,2-rearrangement occurs on reaction of allylbenzyldimethylammonium salts with sodamide in liquid ammonia, with sodamide in benzene at 80°, a significant amount of 1,4-rearrangement also occurs:

$$\mathrm{Me-\overset{\overset{Me}{|}}{\underset{\underset{CH_2Ph}{|}}{N^{\oplus}}}-CH_2-CH{=}CH_2} \xrightarrow{\text{Liq. NH}_3} \mathrm{Me-\overset{\overset{Me}{|}}{N}-\underset{\underset{CH_2Ph}{|}}{CH}-CH{=}CH_2} \text{ (1,2-rearrangement)}$$

$$\xrightarrow{80^\circ} \mathrm{Me-\overset{\overset{Me}{|}}{N}-CH_2-CH{=}\underset{\underset{CH_2Ph}{|}}{CH}} \text{ (1,4-rearrangement)}$$

In addition to aralkyl groups, alkyl, allyl and phenacyl groups can migrate under these conditions.

Competition between the Stevens and Sommelet rearrangements is illustrated by the reaction of dibenzyldimethylammonium bromide (I) with phenyl-lithium in ether, which gives both the Stevens product, α,β-diphenylethyldimethylamine (II) and the Sommelet product, 2-methylbenzhydryldimethylamine (III):

$$\underset{\text{(I)}}{\mathrm{Me-\overset{\overset{Me}{|}}{\underset{\underset{CH_2Ph}{|}}{N^{\oplus}}}-CH_2Ph}} \longrightarrow \underset{\text{(II)}}{\mathrm{Me-\overset{\overset{Me}{|}}{N}-\underset{\underset{CH_2Ph}{|}}{CHPh}}} + \underset{\text{(III)}}{\mathrm{2\text{-}Me\text{-}C_6H_4\text{-}CH(Ph)NMe_2}}$$

When both the Stevens and Sommelet rearrangement involve the same ylid, increased temperature of reaction favours the Stevens rearrangement (*H. J. Shine*, in "Aromatic Rearrangements", Elsevier, Amsterdam, 1967, p. 324). Both the Stevens and Sommelet–Hauser rearrangements have been reviewed by *H. Zimmerman*, in "Molecular Rearrangements", ed. *P. de Mayo*, Interscience, New York, 1963, Part I, p. 378 and by *D. J. Cram*, in "Fundamentals of Carbanion Chemistry", Academic Press, New York, 1965, p. 223. The rearrangement of quaternary compounds to amines is also reviewed by *D. V. Banthorpe*, in "The Chemistry of the Amino Group", ed. *S. Patai*, Interscience, London, 1968, p. 612, and by *T. S. Stevens* in Prog. org. Chem., 1968, **7,** 48, and *S. H. Pine* in Org. Reactions, 1970, **18,** 403.

(d) Individual amines

The physical properties and derivatives of the phenylalkylamines are given in Table 1, the corresponding data for nuclear substituted aralkylamines are in Table 2.

(i) Aralkylamines

Benzylamine, $PhCH_2NH_2$, is conveniently prepared from the quaternary salt of benzyl chloride and hexamethylenetetramine (*B. Reichert* and *W. Dornis*, Arch. Pharm., 1944, **282,** 100; *A. Galat* and *G. Elion*, J. Amer. chem. Soc., 1939, **61,** 3585), also by treatment of phenacetyl chloride with sodium azide in benzene followed by treatment of the product with concentrated hydrochloric acid (*P. A. S. Smith*, Org. Reactions, 1946, **3,** 337). The preparation of optically active [α-^{2}H]benzylamine from [α-^{2}H]benzyl alcohol *via* the azide, has been described by *A. Streitwieser* and *J. R. Wolfe* (J. org. Chem., 1963, **28,** 3263). Benzylamine is prepared in 89% yield from benzaldehyde, ammonia and hydrogen at 90 atmospheres pressure at 40–70°, in the presence of a Raney nickel catalyst (*W. S. Emerson*, Org. Reactions, 1948, **4,** 199). Benzylamine readily forms salts with carboxylic acids, sulphonic acids, and nitrophenols which are suitable for characterisation purposes (*G. W. H. Cheeseman* and *R. C. Poller*, Analyst, 1961, **86,** 256). The benzoyl derivative is formed by Beckmann rearrangement of *N*-benzylbenzaldoxime (*E. Beckmann*, Ber., 1904, **37,** 4137):

$$PhCH{=}\underset{\downarrow \atop O}{N}CH_2Ph \xrightarrow{PhSO_2Cl} PhCONHCH_2Ph$$

Benzylamine reacts with α,β-epoxy acids to give exclusively α-benzylamino-β-hydroxy acids, the corresponding α-amino-β-hydroxy acids are obtained by hydrogenolysis with 30% palladium-on-charcoal (*Y. Liwschitz, Y. Rabinsohn* and *A. Haber*, J. chem. Soc., 1962, 3589). Benzylamine is used in reductive amination; the intermediate benzylamino compound formed undergoes hydrogenolysis to the amine (*J. Schmitt et al.*, Bull. Soc. chim. Fr., 1962, 1855). An improved procedure for the Hofmann isonitrile synthesis has been published. Thus benzylamine is converted to benzylisonitrile in 55% yield by reaction with dichlorocarbene (*W. P. Weber* and *G. W. Grokel*, Tetrahedron Letters, 1972, 1637).

α-**Phenylethylamine,** $PhCH(NH_2){\cdot}CH_3$, is prepared from acetophenone and ammonium formate (*A. W. Ingersoll*, Org. Synth., Coll. Vol. 2, 1943, p. 506) or by the catalytic reduction of acetophenone in liquid ammonia (*J. C. Robinson* and *H. R. Snyder*, Org. Synth., Coll. Vol. 3, 1955, p. 717).

α-Phenylethylamine has also been prepared from α-phenylethanol by application of the Ritter reaction. This involves reaction of the alcohol with potassium cyanide and concentrated sulphuric acid followed by acid-hydrolysis of the resulting formamide (*L. I. Krimen* and *D. J. Cota*, Org. Reactions, 1969, **17,** 213).

Detailed directions for the resolution of the ±-amine are given by *Ingersoll (loc. cit.)*, a newer procedure makes use of (+)-tartaric acid (*A. Ault*, Org. Synth., 1969, **49,** 93). The specific rotations for the (+)- and (−)-bases in methanol are $[\alpha]_D^{25}$

+ 39.2° and $[\alpha]_D^{15}$ −38°. A general method for the synthesis of α-aminoketones is illustrated by the conversion of α-phenylethylamine to ω-aminoacetophenone (*H. E. Baumgarten* and *J. M. Petersen*, Org. Synth., 1961, **41,** 82):

$$PhCH(NH_2)CH_3 \xrightarrow[(2)\ MeONa]{(1)\ ClOBu^t} PhC(OMe)\text{-}CH_2\text{-}NH\ (\text{aziridine}) \longrightarrow PhCO{\cdot}CH_2\overset{\oplus}{N}H_3\overset{\ominus}{Cl}$$

(55–72%)

β-**Phenylethylamine,** $PhCH_2{\cdot}CH_2NH_2$, has been reported to occur in many species of acacia (*E. P. White*, N.Z.J. Sci. Tech., 1944, **25B,** 139; *J. S. Fitzgerald*, Austr. J. Chem., 1964, **17,** 160). It is often encountered in the investigation of decaying protein as it is produced by the decarboxylation of phenylalanine. It is prepared by hydrogenation of benzyl cyanide in liquid ammonia (*Robinson* and *Snyder*, *loc. cit.*, p. 720), or by reduction of benzyl cyanide with lithium tetrahydridoaluminate and aluminium chloride in tetrahydrofuran (*R. F. Nystrom*, J. Amer. chem. Soc., 1955, **77,** 2544). β-Phenylethylamine has also been prepared by reduction of the *O*-benzoylhydroxamic acid derived from phenylacetic acid (*F. Winternitz* and *Ch. Wlotzka*, Bull. Soc. chim. Fr., 1960, 509):

$$PhCH_2{\cdot}CONHOCOPh \xrightarrow{LiAlH_4 - AlCl_3 - THF} PhCH_2{\cdot}CH_2NH_2$$

A general procedure for the de-amination of amines is illustrated by the conversion of β-phenylethylamine into the benzoate of β-phenylethyl alcohol (*E. White*, Org. Synth., 1967, **47,** 44):

$$PhCH_2{\cdot}CH_2NH_2 + PhCOCl \xrightarrow{C_5H_5N} PhCH_2{\cdot}CH_2NHCOPh \xrightarrow[AcONa]{N_2O_4}$$

$$PhCH_2{\cdot}CH_2N(NO)COPh \xrightarrow{\Delta} PhCH_2{\cdot}CH_2OCOPh$$

(56-59% based on *N*-β-phenylethylbenzamide)

β-Phenylethylamines are widely used as intermediates for isoquinoline synthesis. In the Bischler–Napieralski reaction cyclisation of an acyl or aroyl derivative leads to a 3,4-dihydroisoquinoline which is readily oxidised by mild reagents to an isoquinoline:

$$PhCH_2{\cdot}CH_2NHCOR \xrightarrow{2\ steps} \text{1-R-isoquinoline}$$

Alternatively, condensation of β-phenylethylamines with aldehydes yields 1,2,3,4-tetrahydroisoquinolines which are readily oxidised to isoquinolines:

$$PhCH_2CH_2NH_2 + RCHO \xrightarrow{\text{2 steps}} \text{1-R-isoquinoline}$$

β-Phenylethylamine and its homologues and their derivatives are of considerable importance in pharmacology. Structure–physiological activity relationships have been reviewed by *W. H. Hartung*, Ind. Eng. Chem., 1945, **37,** 125; see also *W. A. Sexton*, "Chemical Constitution and Biological Activity", Spon, London, 3rd Edn., 1963, p. 299 and *E. Werle*, in "Ullmann Encyklopädie der technischen Chemie", Urban and Schwarzenberg, München–Berlin, 3rd Edn., Vol. 4, 1953, p. 244.

1-**Amino**-1-**phenylpropane,** $PhCH(NH_2) \cdot CH_2 \cdot CH_3$, is obtained by hydrogenation of propiophenone in alcoholic ammonia (*C. Mignonac*, Compt. rend., 1921, **172,** 226) or by reduction of propiophenone oxime with sodium and alcohol, or by reduction of the adduct of phenylmagnesium bromide and propionitrile with lithium tetrahydridoaluminate (*A. Pohland* and *H. R. Sullivan*, J. Amer. chem. Soc., 1953, **75,** 5898). The ($\pm$)-base has been resolved by crystallisation of its tartrate (*P. Billon*, Ann. Chim. Fr., 1927, [x], **7,** 347). The specific rotations of the (+)- and (−)-bases are $[\alpha]_D^{17}$ $+20.15°$ and $[\alpha]_D^{17}$ $-19.85°$.

1-**Amino**-2-**phenylpropane,** $PhCH(CH_3) \cdot CH_2NH_2$, is prepared from α-methylstyrene by hydroboration, followed by reaction of the resulting organoborane with chloramine (*H. C. Brown et al.*, J. Amer. chem. Soc., 1964, **86,** 3565):

$$PhMeC{=}CH_2 \xrightarrow{BH_3} (PhMeCH \cdot CH_2)_3B \xrightarrow{ClNH_2} \underset{(52\%)}{PhMeCH \cdot CH_2NH_2}$$

2-**Amino**-1-**phenylpropane,** *amphetamine*, *benzedrine*, $PhCH_2 \cdot CH(NH_2) \cdot CH_3$, is prepared by reduction of benzyl methyl ketoxime (*D. H. Hey*, J. chem. Soc., 1930, 18) or by heating benzyl methyl ketone with formamide and followed by acid hydrolysis of the resulting formyl derivative of the base (*Q. Mingoia*, Ann. Chim. appl., 1940, **30,** 187). The ($\pm$)-base has been resolved by crystallisation of its tartrate (*W. Liethe*, Ber., 1932, **65,** 664) and the (+)-base has been obtained from (+)α-benzylethylcarbamide (*L. W. Jones* and *E. S. Wallis*, J. Amer. chem. Soc., 1926, **48,** 180). The dextrorotatory isomer is more biologically potent than the laevorotatory form. The free base is used in nasal sprays to shrink the mucous membrane and the sulphate is administered to stimulate the central nervous system in a variety of mental conditions, it is also used in reducing weight by depressing the appetite.

2-**Methylamino**-1-**phenylpropane,** *methedrine*, *pervitin*,

$$PhCH_2 \cdot \underset{}{\overset{\overset{\displaystyle Me}{|}}{C}}HNHMe$$

is prepared by reduction of a mixture of benzyl methyl ketone and methylamine (*K. Löffler*, Ber., 1910, **43,** 2031) or from the reductive fission of 4-methyl-5-phenylthiazole (*H. Erlenmeyer* and *M. Simon*, Helv., 1942, **25,** 528):

$$\begin{array}{c} PhC{=}C{\cdot}Me \\ | \quad\; | \\ S \quad N \\ \backslash\!\!\!\!\;\; C \!/\!\!/ \\ H \end{array} \xrightarrow{Na/EtOH} PhCH_2{\cdot}\overset{\displaystyle Me}{\overset{|}{C}}HNHMe + H_2S$$

The (±)-base has been resolved by crystallisation of its tartrate; the (+)-base, $[\alpha]_D^{20}$ + 17.9° in water, is also obtained by reduction of ephedrine, PhCHOH·CHNHMe·CH_3 (*A. Ogata*, C.A., 1920, **14**, 745; *H. Emde*, Helv., 1929, **12**, 373). It resembles amphetamine in its physiological activity.

1-**Amino**-3-**phenylpropane,** $PhCH_2{\cdot}CH_2{\cdot}CH_2NH_2$, is obtained by hydrogenation of hydrocinnamonitrile (*J. v. Braun, G. Blessing* and *F. Zobel*, Ber., 1923, **56,** 1990) together with $(PhCH_2{\cdot}CH_2{\cdot}CH_2)_2NH$; its *N*,*N*-dimethyl derivative is prepared by electrolytic reduction of hydrocinnamdimethylamide (*Kindler*, Ann., 1923, **431,** 220).

N-**Methylbenzylamine,** $PhCH_2NHMe$, is present together with other bases in the Chinese drug Ma Huang, and is prepared by the action of methylamine on benzyl chloride in aqueous ethanol at 60° (*E. L. Holmes* and *C. K. Ingold*, J. chem. Soc., 1925, **127,** 1813).

N-**Ethylbenzylamine,** $PhCH_2NHEt$, has been prepared by reductive amination of benzaldehyde with ethylamine in the presence of lithium cyanotrihydridoborate (*R. F. Borch* and *H. D. Durst*, J. Amer. chem. Soc., 1969, **91,** 3996), see also *Borch*, Tetrahedron Letters, 1968, 61).

N,*N*-**Dimethylbenzylamine,** $PhCH_2NMe_2$, is prepared by treatment of benzylamine with formic acid and formaldehyde (*M. L. Moore*, in Org. Reactions, 1949, **5,** 323; *S. H. Pine*, J. chem. Educ., 1968, **45,** 118), also by reduction of benzyltrimethylammonium iodide with lithium tetrahydridoaluminate (*Cope et al.*, J. Amer. chem. Soc., 1960, **82,** 4651). It reacts with nitrous acid to give benzaldehyde and *N*-nitrosodimethylamine and with acetic anhydride to give benzyl acetate and *N*,*N*-dimethylacetamide. The synthetic uses of various quaternary salts of the tertiary amine are summarised by *L. F. Fieser* and *M. Fieser*, in "Reagents in Organic Synthesis", Wiley, New York, 1967, p. 53. *Triton B*, benzyltrimethylammonium hydroxide, is a valuable catalyst for promoting base-catalysed condensation (*idem, ibid.*, p. 1252). The anodic methoxylation of *N*,*N*-dimethylbenzylamine gives mainly *N*-methoxymethyl-*N*-methylbenzylamine, $PhCH_2NMe{\cdot}CH_2OMe$, (*P. J. Smith* and *C. K. Mann*, J. org. Chem., 1968, **33,** 316). Oxidation with oxygen and platinum black gives *N*-benzyl-*N*-methylformamide (*G. T. Davis* and *D. H. Rosenblatt*, Tetrahedron Letters, 1968, 4085). The principal product obtained by reaction of benzyne with *N*,*N*-dimethylbenzylamine is *N*-methyl-*N*-(α-phenylethyl)aniline (*A. R. Lepley, R. H. Becker* and *A. G. Giumanini*, J. org. Chem., 1971, **36,** 1222):

$$PhCH_2NMe_2 \xrightarrow[\text{(from } n\text{-butyl-lithium and fluorobenzene)}]{\text{Benzyne}} PhNMe{\cdot}CHMePh$$

N-**Ethyl**-*N*-**methylbenzylamine,** $PhCH_2NMeEt$, is prepared from benzyl chloride and ethylmethylamine. It forms an *N*-oxide, the resolution of which has been described by *J. Meisenheimer*, Ann., 1922, **428,** 279.

N,*N*-**Diethylbenzylamine,** $PhCH_2NEt_2$, is prepared by reduction of *N*,*N*-diethylbenzamide with lithium tetrahydridoaluminate (*M. Mousseron et al.*, Bull. Soc. chim. Fr., 1952, [5], **19,** 1042) or by successive reaction of the amide with triethyloxonium tetrafluoroborate and sodium tetrahydridoborate (*Borch, loc. cit.*).

N-**Phenylbenzylamine,** *benzylaniline*, $PhCH_2NHPh$, is prepared by the reaction of benzyl chloride and aniline in the presence of sodium hydrogen carbonate (*F. G. Wilson* and *T. S. Wheeler*, Org. Synth., Coll. Vol. 1, 2nd Edn., 1947, p. 102). On pyrolysis it yields acridine (*H. Meyer* and *A. Hofmann*, Monatsh., 1916, **37,** 698).

Dibenzylamine, $(PhCH_2)_2NH$, is formed together with benzylamine on catalytic hydrogenation of hydrobenzamide (*F. Möller* and *R. Schröter*, in *Houben–Weyl*, Methoden der Organischen Chemie", 4th Edn., Vol. 11/1, 1957, p. 603):

$$PhCH(N{=}CHPh)_2 + 3H_2 \xrightarrow{Ni} (PhCH_2)_2NH + PhCH_2NH_2$$

It is also prepared from benzylamine *via* its anil (*M. A. Phillips*, J. Soc. chem. Ind., 1947, **66,** 325; see also *Emerson*, Org. Reactions, 1948, **4,** 199):

$$PhCH_2NH_2 \longrightarrow PhCH_2N{=}CHPh \xrightarrow{Zn/AcOH} (PhCH_2)_2NH \; (50\%)$$

Tribenzylamine, $(PhCH_2)_3N$, is prepared by the action of ethanolic ammonia on benzyl chloride at 100° (*J. v. Braun*, Ber., 1937, **70,** 979).

α-**Phenylethylaniline,** PhCHMeNHPh, and its homologues of the general formula PhCHRNHPh are prepared by the addition of alkylmagnesium halides to benzylideneaniline (*M. Busch*, Ber., 1904, **37,** 2691) or by the reduction of the anils of alkyl phenyl ketones.

N-**Methyl-β-phenylethylamine,** $PhCH_2 \cdot CH_2NHMe$, is obtained by electroreduction of the corresponding amide (*Kindler*, Ber., 1924, **57,** 774) or by reaction of β-phenylethyl chloride with methylamine (*G. Barger* and *A. J. Ewins*, J. chem. Soc., 1910, **97,** 2255).

N,N-**Dimethyl-β-phenylethylamine,** $PhCH_2 \cdot CH_2NMe_2$, is prepared by methylation of β-phenylethylamine with formaldehyde and formic acid (*R. N. Icke, B. B. Wisegarver* and *G. A. Alles*, Org. Synth., Coll. Vol. 3, 1955, p. 723); it has also been prepared from 1-phenylethanol *via* a sulphamate ester (*E. H. White* and *C. A. Elliger*, J. Amer. chem. Soc., 1965, **87,** 5261).

(ii) Carbonic acid derivatives of benzylamine and its homologues

Benzyl carbamate, $H_2NCO_2CH_2Ph$, m.p. 86°, is prepared by heating urea and benzyl alcohol or from benzyl chloroformate and ammonia (*J. Thiele* and *F. Dent*, Ann., 1898, **302,** 258).

Methyl N-*benzylcarbamate*, $PhCH_2NHCO_2Me$, m.p. 64°, is formed from phenylacetamide by reaction with bromine and sodium methoxide in methyl alcohol. The *ethyl ester*, $PhCH_2NHCO_2Et$, m.p. 46°, is similarly prepared (*R. A. Weerman* and *W. J. Jongkees*, Rec. Trav. chim., 1906, **25,** 243). It is also obtained by boiling the azide of phenylacetic acid with methanol (*E. Boetzelen*, J. prakt. Chem., 1901, [ii], **64,** 320) or by the reaction of benzylamine with chloroformic ester (*A. Hantzsch*, Ber., 1898, **31,** 180).

N-**Benzylurea,** $PhCH_2NHCONH_2$, m.p. 148°, is prepared in excellent yield by reaction of silicon tetra-isocyanate with benzylamine (*R. G. Neville* and *J. J. McGee*, Org. Synth., 1965, **45,** 72):

$$PhCH_2NH_2 + Si(NCO)_4 \longrightarrow (PhCH_2NHCONH)_4Si \xrightarrow[-SiO_2]{H_2O} PhCH_2NHCONH_2$$

or when urea is refluxed with benzylamine in aqueous solution (*T. L. Davis* and *K. C. Blanchard*, J. Amer. chem. Soc., 1923, **45,** 1819). The N-*nitroso deriv.*, $PhCH_2N(NO)CONH_2$, has m.p. 101° (*E. A. Werner*, J. chem. Soc., 1919, **115,** 1101). N,N′-*Dibenzylurea*, $(PhCH_2NH)_2CO$, m.p. 167°, is prepared by heating urea and benzylamine at 160–170° (*Davis* and *Blanchard*, *loc. cit.*). N-*Benzylthiourea*, $PhCH_2NHCSNH_2$, m.p. 161°, is prepared from benzyl isothiocyanate and ammonia or from benzylammonium thiocyanate ($PhCH_2\overset{\oplus}{N}H_3\overset{\ominus}{N}CS$) (*A. E. Dixon*, J. chem. Soc., 1891, **59,** 555).

S-Benzylthiourea hydrochloride, S-*benzylthiuronium chloride*, $PhCH_2SC(NH_2)=\overset{\oplus}{N}H_2Cl^{\ominus}$, m.p. 140–143° and 172–174°, is prepared from thiourea and benzyl chloride (*E. Chambers* and *G. W. Watts*, J. org. Chem., 1941, **6,** 376). It gives crystalline salts of low solubility with most carboxylic acids, which are often prepared for characterisation purposes. The reagent suffers from the disadvantage that it is readily decomposed in alkaline solution to phenylmethanethiol which can be recognised by its characteristic unpleasant smell:

$$PhCH_2SC(NH_2)_2^{\oplus}\ Cl^{\ominus} + RCO_2Na \longrightarrow PhCH_2SC(NH_2)_2^{\oplus}\ RCO_2^{\ominus} + NaCl$$

The melting points of the aliphatic acid salts are over a relatively narrow range and mixed melting points are frequently necessary for positive identification.

4-**Benzylsemicarbazide,** $PhCH_2NHCONH \cdot NH_2$, m.p. 111°, is prepared by the acid hydrolysis of the product of the action of benzylamine on acetone semicarbazone (*F. L. Wilson, I. V. Hopper* and *A. B. Crawford*, J. chem. Soc., 1922, **121,** 867).

Benzyl isothiocyanate, *benzyl mustard oil*, $PhCH_2NCS$, a yellow oil, b.p. 243°, is present as a glycoside in the essential oils of many cresses (*J. Gadamer*, Ber., 1899, **32,** 2336). It is prepared from benzylamine and thiocarbonyl chloride (*G. M. Dyson* and *H. J. George*, J. chem. Soc., 1924, **125,** 1705). *β-Phenylethyl isothiocyanate*, $PhCH_2 \cdot CH_2NCS$, is obtained by the hydrolysis of the glycoside present in turnips and in nasturtiums. (See *F. Challenger*, in "Aspects of the Organic Chemistry of Sulphur", Butterworth, London, 1959, Chapter 4.)

(iii) Nuclear-substituted aralkylamines

The physical constants of a selection of these compounds and derivatives are presented in Table 2. Benzylamines may be prepared by the direct electrophilic substitution of highly activated aromatic compounds such as *m*-xylene and anisole (see amidomethylation, p. 9).

Halogen-substituted benzylamines are prepared from the reaction of halogen-substituted benzyl chlorides with ammonia, *e.g.* preparation of 4-chlorobenzylamine from 4-chlorobenzyl chloride, or by reduction of the corresponding nitrile, *e.g.* preparation

TABLE 1

	M.p. (°C)	*B.p. (°C/mm)*	n_D°
$PhCH_2NH_2$		185, 90/12	1.5401^{20}
(±)-$PhMeCHNH_2$		187.5, 87/24	1.5238^{25}
$PhCH_2 \cdot CH_2NH_2$		197–198	1.5290^{25}
(±)-$PhEtCHNH_2$		204, 88/16	1.5173^{25}
(±)-$PhMeCH \cdot CH_2NH_2$		114–116/35	1.5240^{20}
(±)-$PhCH_2 \cdot CHMe(NH_2)$		205	
(±)-$PhCH_2 \cdot CHMe(NHMe)$		209–210, 93/15	
$PhCH_2 \cdot CH_2 \cdot CH_2NH_2$		221.5/755 112–114/18	
$PhCH_2NHMe$		180, 78/14	
$PhCH_2NHEt$		199 (194)	
$PhCH_2NMe_2$		181, 66–67/15	
$PhCH_2NEtMe$		85–87/10	
$PhCH_2NEt_2$		211–212, 94/15	
$PhCH_2NHPh$	37–38	190/16	
$(PhCH_2)_2NH$		150–151/6 173–175/13	
$(PhCH_2)_3N$	91	218–219/14	
PhMeCHNHPh		183/20	
$PhCH_2 \cdot CH_2NHMe$		205	
$PhCH_2 \cdot CH_2NMe_2$		205, 98/22	
$PhCH_2 \cdot CH_2 \cdot CH_2NMe_2$		225, 99/14	
(±)-$PhMeCH \cdot CH_2NH_2$		210, 104/21	1.5255^{20}
$PhMe_2CNH_2$		196–197/762	1.5181^{25}
$PhCH_2 \cdot CMe_2(NH_2)$		80–82/10	

References
1 See text.
2 *K. Kindler*, Ann., 1923, **431,** 221.
3 *A. W. Weston, A. W. Ruddy* and *C. M. Suter*, J. Amer. chem. Soc., 1943, **65,** 674.
4 *M. Brander*, Rec. Trav. chim., 1918, **37,** 68.
5 *J. J. Ritter* and *J. Kalish*, Org. Synth., 1964, **44,** 44.

PHENYLALKYLAMINES

$d_{\circ}^{\circ}$	*Derivative, m.p. (°C)*	*Reference*
0.9797$^{20}_{0}$	*B,HCl*, 255–256; *picrate*, 194; *acetyl*, 60; *benzoyl*, 106.	1
0.9395^{15}	*B,HCl*, 158; *picrate*, 187–190; p-*toluenesulphonyl*, 81–82.	1
0.9450$^{24}_{4}$	*B,HCl*, 217; *picrate*, 170; *acetyl*, 51; *benzoyl*, 116.	1
0.9347$^{25}_{0}$	*B,HCl*, 194; *benzoyl*, 115–116.	1
		1
	B,HCl, 145–147; *picrate*, 143.	1
	B,HCl, 134–135; B,H_2PtCl_6, 205; *picrate*, 128.	1
0.9760$^{25}_{4}$	*B,HCl*, 218; B_2,H_2PtCl_6, 233; *picrate*, 152–153.	1
0.9450$^{18}_{15}$	*B,HCl*, 178; B_2,H_2PtCl_6, 197; *picrate*, 117–118.	1
0.9350$^{17}_{15}$	*B,HCl*, 184; *picrate*, 122–123.	1
	B, HCl, 195(175); B_2,H_2PtCl_6,192; *picrate*, 93; *methiodide*, 178–179.	1
	Picrate, 113.	1
	B_2,H_2PtCl_6, 203.	1
	B,HCl, 214–216(197); B_2,H_2PtCl_6, 155; *formyl*, 49; *acetyl*, 58; *benzoyl*, 102.	1
	B,HCl, 256; p-*toluenesulphonyl*, 158–159.	1
	B,HCl, 227–228; *methiodide*, 184; *picrate*, 191.	1
	B,HCl, 184–185.	1
	B,HCl, 157; *picrate*, 143.	1
	B,HCl, 205; *picrate*, 136.	1
	B,HCl, 146; B_2, H_2PtCl_6, 152; *methiodide*, 179.	2
0.9433$^{20}_{4}$	*B, HCl*, 146–147; *picrate*, 182; *benzoyl*, 85.	3
0.9424$^{20}_{0}$	*B,HCl*, 235.5; *benzoyl*, 159.	4
		5

of 4-bromobenzylamine from 4-bromobenzonitrile. The nitrobenzylamines are conveniently prepared by reaction of the corresponding nitrobenzyl chlorides with ammonia or *via* the *N*-nitrobenzylphthalimides.

Nitration of benzylamine gives a mixture of 49% of 3-nitrobenzylamine and 43% of 4-nitrobenzylamine (*F. R. Goss, C. K. Ingold* and *I. S. Wilson,* J. chem. Soc., 1926, 2440). Nitration of β-phenylethylamine gives mainly the 4-nitro derivative (*Goss, W. Hanhart* and *Ingold, ibid.,* 1927, 250).

Aminobenzylamines are formed by the reduction of the corresponding nitrobenzylamines (*S. Gabriel,* Ber., 1887, **20,** 2229, 2870; *Gabriel* and *J. Colman, ibid.,* 1904, **37,** 3644; *H. Salkowski, ibid.,* 1889, **22,** 2142).

2-**Aminobenzylamine** condenses with aldehydes to yield 1,2,3,4-tetrahydroquinazolines (*M. Busch,* J. prakt. Chem., 1896, [ii], **53,** 414) and on heating with formic acid and sodium formate it gives 3,4-dihydroquinazoline:

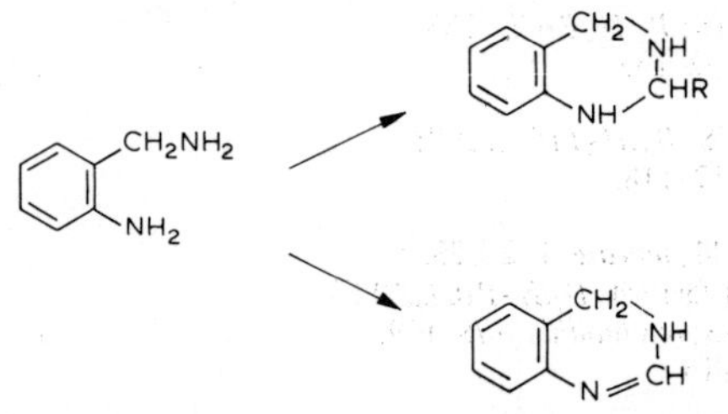

N-Alkyl and *N*-aryl derivatives of 2-aminobenzylamine undergo similar cyclisations. Thus 2-*aminobenzylaniline* (IV), m.p. 86–87°, is converted into the tetrahydroquinazoline V, on treatment with formaldehyde *(idem, loc. cit.).* Reaction of 2-aminobenzylaniline with carbonyl chloride gives the tetrahydroquinazolin-2-one VI, and with carbon disulphide in alcoholic alkali the tetrahydroquinazoline-2-thione VII, is formed (*idem,* Ber., 1892, **25,** 2853):

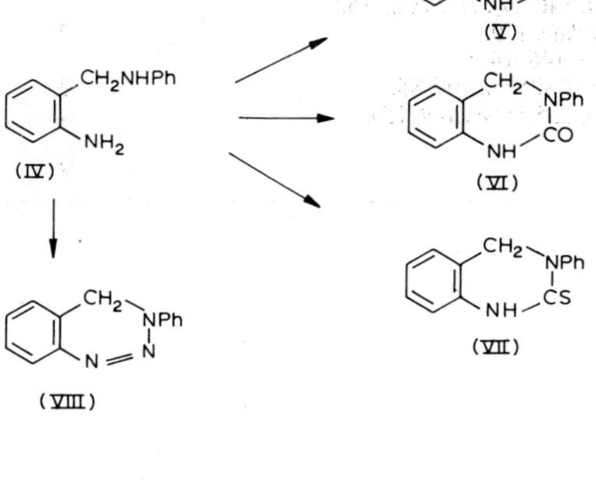

Reaction of 2-aminobenzylaniline with nitrous acid gives the triazine VIII (*idem, ibid.*, p. 449).

2-*Acetamidobenzylaniline* (IX), m.p. 126°, undergoes cyclisation on heating to give the dihydroquinazoline X (*C. Paal* and *F. Krecke*, Ber., 1891, **24,** 3051):

CH_2NHPh, NHCOMe (IX) → CH_2, NPh, CMe, N (X)

2-*Acetamidomethylaniline* (XI), m.p. 112–113°, is similarly cyclised by heating at 240° under nitrogen (*W. F. L. Armarego*, J. chem. Soc., 1961, 2697), to give the dihydroquinazoline XII:

$CH_2NHCOMe$, NH_2 (XI) → CH_2, NH, CMe, N (XII)

(*iv*) *Benzyl derivatives of hydroxylamine*

N-**Benzylhydroxylamine**, $PhCH_2NHOH$, m.p. 57°, is formed by the hydrolysis of *N*-benzylbenzaldoxime ($PhCH_2N(\rightarrow O){=}CHPh$) or *N*-benzylacetoxime (*Beckmann*, Ber., 1889, **22,** 438, 1532; *C. Neubauer*, Ann., 1897, **298,** 200), the hydrolysis of its *O*-benzyl ether (*R. Behrend* and *K. Leuchs, ibid.*, 1890, **257** 212), or by the reduction of benzaldoxime with diborane followed by acid hydrolysis (*H. Feuer, B. F. Vincent Jr.*, and *R. S. Bartlett*, J. org. Chem., 1965, **30,** 2877). It reduces Fehling's solution in the cold and on oxidation with dichromate it is converted into phenylnitrosomethane, $PhCH_2NO$. Reaction with nitrous acid in aqueous solution at 0° gives *benzylnitrosohydroxylamine*, m.p. 77–78° (*Behrend* and *E. König*, Ann., 1891, **263,** 217). Treatment of *N*-benzylhydroxylamine hydrochloride with sulphur dioxide gives benzylsulphamic acid, $PhCH_2NHSO_3H$ (*M. Schmidt*, J. prakt. Chem., 1891, [ii], **44,** 514). *N*-Benzylhydroxylamine reacts with benzaldehyde to give *N*-benzylbenzaldoxime (*Beckmann, loc. cit.*):

$$PhCHO \quad + \quad PhCN_2NHOH \quad \rightarrow \quad \underset{\substack{| \\ O}}{PhCH_2NCHP} \quad + \quad H_2O$$

O-**Benzylhydroxylamine,** $PhCH_2ONH_2$, b.p. 118–119°/30 mm, *hydrochloride* 230–235° decomp., is prepared by the acid hydrolysis of *O*-benzylacetoxime (*Behrend* and *Leuchs, loc. cit.*, p. 206) or from sodium benzyloxide and chloramine (*W. Theilacker* and *K. Ebke*, Angew. Chem., 1956, **68,** 303). On heating at 160°, it is transformed into benzaldoxime benzyl ether.

N,N-**Dibenzylhydroxylamine,** $(PhCH_2)_2NOH$, m.p. 123°, is prepared by the action of benzyl chloride on hydroxylamine (*Behrend* and *Leuchs, loc. cit.* p. 217; *B.A.S.F.*, F.P. 1,481,069/1967) or by the benzylation of *N*-benzylhydroxylamine. It reduces Fehl-

TABLE 2

Benzylamine	*M.p. (°C)*	*B.p. (°C/mm)*	n_D°
2-Methyl-	–20	206/745, 80–82/15	1.5436^{19}
3-Methyl-		205/750, 96/20	
4-Methyl-	12.6–13.2	204/739	1.5364^{20}
2,4-Dimethyl-		218–219	
3,5-Dimethyl-		221/758	1.5305^{20}
4-Isopropyl-		225–227/724, 110/12	1.5182^{17}
2,4,5-Trimethyl-	52 64.5	(two forms)	
2,4,6-Trimethyl-			
2-Chloro-		103–104/11	
3-Chloro-		110–112/17	
4-Chloro-		215/734 106–107/15	
2-Bromo-		118/9	
3-Bromo-		244–245 corr., 84/14	
4-Bromo-	20	249.5–251.5 (corr.), 126–127/15	
2-Iodo-			
3-Iodo-		132/8	
4-Iodo-	45		
2-Nitro-			
3-Nitro-			
4-Nitro-	40		
2-Amino-	60–62		
4-Amino-		268–270	

NUCLEAR SUBSTITUTED BENZYLAMINES

$d^{\circ}_{\circ}$	*Derivative, m.p. (°C)*	*Reference*
0.9768^{19}_{0}	*B,HCl*, 225; B_2,H_2PtCl_6, 220–223 decomp.; *acetyl*, 76; *benzoyl*, 88(115–116).	1,2
0.9654^{20}_{0}	*B,HCl*, 208; $B_2H_2PtCl_6$, 214; *benzoyl*, 150(70).	1,3
0.9520^{20}_{0}	*B,HCl*, 235; *picrate*, 204 decomp.; *acetyl*, 111–112; *benzoyl*, 137.	1,3a
	B,HCl, 212; B_2,H_2PtCl_6, 228–229 decomp.; *benzoyl*, 98.	4
0.9500^{20}_{0}	*B,HCl*, 245; B_2,H_2PtCl_6, 206; *benzoyl*, 78.	5
	B,HCl, 239–240; *benzoyl*, 93.	6
	B,HCl, 275; *acetyl*, 143.5.	
	B,HCl, 240–242; B_2, H_2PtCl_6, 208–209 (decomp.).	7
	B,HCl, 315; *acetyl*, 186–187.	8
	B,HCl, 215–216; *picrate*, 217 (decomp.); *benzoyl*, 116–117.	9
	B,HCl, 225; *picrate*, 203; *benzoyl*, 114.	10
	B,HCl, 261.5 (corr.); B_2,H_2PtCl_6, 244; *picrate*, 210; *benzoyl*, 143.	11
	B,HCl, 241–242(208).	12
	B,HCl, 218.5 (corr.); B_2,H_2PtCl_6, 250.5–251.5 (corr.) decomp.; *picrate*, 205, decomp.; *benzoyl*, 135.5.	12
	B,HCl, 260; *picrate*, 221; *benzoyl*, 143.	12
	benzoyl, 154.	13
	Acetyl, 114.5; *benzoyl*, 132.	14
	B,HCl, 240; *acetyl*, 132.	15
	B,HCl, 248 decomp.; *picrate*, 206–208; *acetyl*, 99; *benzoyl*, 110.	16
	B,HCl, 225 decomp.; *acetyl*, 91; p-*nitrobenzylidene*, 115.	16
	B,HCl, 256 decomp.; *picrate*, 194; *acetyl*, 133; *benzoyl*, 155–156; p-*nitrobenzylidene*, 150.	16
	B,2HCl, 215–218; αN-*diacetyl*, 112.5–113.5.	17
		18

References to Table 2
1 *J. B. Shoesmith* and *R. H. Slater*, J. chem. Soc., 1924, **125,** 2280.
2 *H. Strassmann*, Ber., 1888, **21,** 577.
3 *H. Rupe* and *F. Bernstein*, Helv., 1930, **13,** 462.
3a *H. C. Brown* and *B. C. Subba Rao*, J. Amer. chem. Soc., 1956, **78,** 2582.
4 *T. Curtius* and *E. Haager*, J. prakt. Chem., 1900,[*ii*], **62,** 113.
5 *M. Konowalow*, Ber., 1895, **28,** 1863.
6 *H. Goldschmidt* and *A. Gessner*, *ibid.*, 1887, **20,** 2414.
7 *R. Willstätter* and *H. Kubli*, *ibid.*, 1909, **42,** 4156.
8 *R. C. Fuson* and *J. J. Denton*, J. Amer. chem. Soc., 1941, **63,** 656.
9 *H. Franzen*, Ber., 1905, **38,** 1415.
10 *C. W. Shoppee*, J. chem. Soc., 1932, 701.
11 *J. Braun*, *M. Kühn* and *J. Weismantel*, Ann., 1926, **449,** 266.
12 *Rupe* and *Bernstein*, Helv., 1930, **13,** 466.
13 *C. F. Mabery* and *F. C. Robinson*, Amer. Chem. J., 1882, **4,** 103.
14 *Shoppee*, J. chem. Soc., 1932, 702.
15 *Idem*, *ibid.*, 1931, 1235.
16 *H. R. Ing* and *R. M. F. Manske*, *ibid.*, 1926, 2348.
17 *S. Gabriel* and *J. Colman*, Ber., 1904, **37,** 3643.
18 *H. Salkowski*, *ibid.*, 1889, **22,** 2142.

ing's solution and is oxidised by mercuric oxide to *N*-benzylbenzaldoxime (*A. Angeli et al.*, Atti Accad. Lincei, 1911, [v], **20,** 1554).

O,N-**Dibenzylhydroxylamine,** $PhCH_2NHOCH_2Ph$, b.p. 146°/3 mm, and O,N,N-*tribenzylhydroxylamine*, $(PhCH_2)_2NOCH_2Ph$, oil, are both formed by heating *O*-benzylhydroxylamine with benzyl chloride in alcohol (*Behrend* and *Leuchs*, *loc. cit.*, p. 208).

(*v*) *Benzyl derivatives of hydrazine*

Benzylhydrazine, $PhCH_2NH{\cdot}NH_2$, b.p. 135°/29 mm, is prepared by the reduction of benzaldehyde hydrazone with sodium amalgam or by the acid hydrolysis of benzylidenebenzylhydrazine (*T. Curtius*, Ber., 1900, **33,** 2460; J. prakt. Chem., 1900, [ii], **62,** 94). It is also formed by reduction of sodium benzyldiazotate with aluminium turnings and aqueous alkali (*J. Thiele*, Ann., 1910, **376,** 255). It yields a stable *nitrosamine*, $PhCH_2N(NO)NH_2$, m.p. 71° (*A. Wohl* and *C. Oesterlin*, Ber., 1900, **33,** 2736).

N-**Benzyl-***N′*-**phenylhydrazine,** $PhCH_2NH{\cdot}NHPh$, m.p. 35°, is prepared by reduction of benzaldehyde phenylhydrazone catalytically (*C. Paal* and *C. Amberger*, G.P. 346,949; Friedländer, 1926, **14,** 1468) or with sodium amalgam (*P. Jacobson*, Ann., 1922, **427,** 220). N-*Benzyl*-N-*phenylhydrazine*, $PhCH_2NPh{\cdot}NH_2$, b.p. 207–208°/10 mm, is obtained from benzyl chloride and phenylhydrazine (*G. Minunni*, Gazz., 1892, **22,** [ii], 219) or by the reduction of benzylphenylnitrosoamine with zinc and acetic acid (*O. Antrick*, Ann., 1885, **227,** 361). It is used with advantage for isolating and characterising sugars as hydrazones (*O. Ruff* and *G. Ollendorff*, Ber., 1900, **33,** 1798; *R. Ofner*, Monatsh., 1904, **25,** 592). It is oxidised by mercuric oxide to *dibenzyldiphenyltetrazene*, $(PhCH_2NPh{\cdot}N{:}N{\cdot}NPhCH_2Ph)$, m.p. 145° (*Minunni*, *loc. cit.*, p. 224).

N,N′-**Dibenzylhydrazine,** $PhCH_2NH{\cdot}NHCH_2Ph$, m.p. 47°, is obtained by the electrolytic reduction of benzalazine (*J. Thiele*, Ann., 1910, **376,** 261). N,N-*Dibenzylhydrazine*, m.p. 65°, is produced by the action of benzyl chloride on hydrazine (*Busch* and *B. Weiss*, Ber., 1900, **33,** 2702) or by the reduction of dibenzylnitrosamine (*Curtius* and *H. Franzen*, Ber., 1901, **34,** 558). It is oxidised to *tetrabenzyltetrazene*, $(PhCH_2)_2N\cdot N{:}N\cdot N(CH_2Ph)_2$, m.p. 97°, with mercuric oxide.

Tribenzylhydrazine (*hydrochloride*, m.p. 181°) is obtained by the reduction of benzylidenedibenzylhydrazine with sodium amalgam (*Franzen* and *F. Kraft*, J. prakt. Chem., 1911, [ii], **84,** 137) or from benzyl chloride and hydrazine hydrate in boiling alcohol (*J. Kenner* and *J. Wilson*, J. chem. Soc., 1927, 1112), a reaction which also produces some **tetrabenzylhydrazine,** m.p. 139°. The latter compound is also obtained by the action of benzyl bromide on tribenzylhydrazine (*H. Wieland* and *E. Schamberg*, Ber., 1920, **53,** 1334).

(vi) Diazo compounds, triazenes and azides related to benzylamine

Toluene-ω-diazotates *(benzyldiazotates)*, $PhCH_2N{:}NOH$. The potassium salt, $PhCH_2N{:}NOK{,}H_2O$, is formed by the action of very concentrated potassium hydroxide on ethyl *N*-benzyl-*N*-nitrosocarbamate, $PhCH_2N(NO)CO_2Et$. It is a white powder, very sensitive to moisture forming mainly phenyldiazomethane with some benzyl alcohol and nitrogen.

A *sodium salt*, $PhCH_2N{:}NONa$, of quite different properties, and presumably therefore derived from a stereoisomeric form, has been isolated. It forms colourless needles, soluble in cold water, but decomposed on warming with water. It is prepared by the action of ethyl nitrite on *N*-benzyl-*N*-nitrosohydrazine in alkaline solution, alternatively by the reduction of *benzylnitroamine*, $PhCH_2NH\cdot NO_2$, m.p. 39°, with aluminium and alkali (*Thiele*, Ann., 1910, **376,** 255). It is reduced to benzylhydrazine and forms benzyl alcohol on treatment with dilute sulphuric acid.

1-**Benzyl**-3-**methyltriazene,** $PhCH_2N{:}N\cdot NHMe$, a colourless mobile oil, decomposed by acids, is formed by the reaction of methylmagnesium iodide on benzyl azide (*O. Dimroth*, Ber., 1905, **38,** 684); *copper salt*, m.p. 114°; *silver salt*, isolated as colourless needles, m.p. 125°, which darken on exposure to light.

1-*Benzyl*-3-*phenyltriazene*, $PhCH_2N{:}N\cdot NHPh$, m.p. 75°, is prepared by the action of phenylmagnesium bromide on benzyl azide *(Dimroth, loc. cit.)*, or by treating benzylamine with benzenediazonium chloride (*H. Goldschmidt* and *J. Holm*, Ber., 1888, **21,** 1016). It is decomposed by ice-cold dilute hydrochloric acid to benzyl chloride, aniline and nitrogen.

3-*Benzyl*-1,3-*diphenyltriazene*, $PhCH_2NPh\cdot N{:}NPh$, yellow needles, m.p. 81°, is prepared from benzenediazonium chloride and benzylaniline. 3,3-*Dibenzyl*-1-*phenyltriazene*, pale yellow needles, m.p. 83°, is prepared similarly from dibenzylamine (*L. Vignon* and *A. Simonet*, Bull. Soc. chim. Fr., 1905, [iii], **33,** 657).

Benzyl azide, *ω-azidotoluene*, $PhCH_2N_3$, b.p. 83°/16 mm, is prepared by heating benzyl chloride with sodium azide in alcohol (*Curtius* and *G. Ehrhart*, Ber., 1922, **55,** 1565). Benzyl azide if heated with water gives the radical $PhCH_2N\cdot$; this rearranges

$$PhCH_2N_3 \xrightarrow{-N_2} PhCH_2N\cdot \begin{cases} \nearrow PhCH{=}NH \xrightarrow{H_2O} PhCHO + NH_3 \\ \searrow PhN{=}CH_2 \xrightarrow{H_2O} PhNH_2 + CH_2O \end{cases}$$

partly to benzylidene-imine and partly to methylene-aniline. Hydrolysis of these intermediates gives the products shown. On boiling in diphenyl ether, the principal products of decomposition are the imidazole XIII, the triazine XIV and *N*-benzylideneaniline, PhCH:NPh (*B. Coffin* and *R. F. Robbins*, J. Chem. Soc., 1964, 5901):

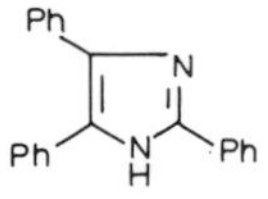

(XIII) 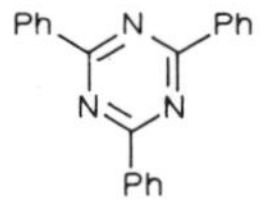 (XIV) 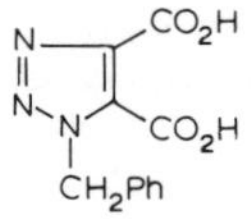 (XV)

At 170°, benzyl azide reacts with diethyl malonate to form nitrogen and diethyl benzylaminomalonate, $PhCH_2NHCH(CO_2Et)_2$ (*Curtius* and *Ehrhart*, *loc. cit.*). The triazole XV is readily formed from benzyl azide and acetylenedicarboxylic acid (*R. H. Wiley*, *K. F. Hussung* and *J. Moffat*, J. org. Chem., 1965, **21**, 190).

2. Monohydric alcohols of the benzene series and their derivatives

The simplest compound of this group is benzyl alcohol, $PhCH_2OH$. Primary, secondary, and tertiary alcohols are known; they resemble aliphatic alcohols in many of their general reactions.

(a) Methods of formation and preparation

(i) From aralkyl halides

(*1*) by refluxing with an aqueous alkali such as potassium carbonate, or by treatment with lead oxide and water (*G. S. Mironov*, *M. I. Farberov* and *L. S. Yukhtina*, Zhur. priklad. Khim., 1967, **40,** 2339; C.A., 1968, **68,** 58817s), *e.g.*:

(*2*) by heating with the salt of a carboxylic acid, *e.g.* potassium or silver acetate, to give an ester which by hydrolysis gives the alcohol (*W. W. Hartman* and *E. J. Rahrs*, Org. Synth., Coll. Vol. 3, 1955, p. 652);

(*3*) by oxidation with air of the derived Grignard reagent followed by hydrolysis of the resulting magnesyl halide (*L. Bert*, Bull. Soc. chim. Fr., 1925, [iv], **37,** 1577).

$$ArCH_2MgCl \xrightarrow{O_2} ArCH_2OMgCl \xrightarrow{H_2O} ArCH_2OH$$

(*4*) by reaction of the derived Grignard reagent with an appropriate aldehyde, ketone, ester, acid chloride or epoxide. With both aliphatic aldehydes and ketones, alcohols of the general formula $ArCH_2C(OH)RR'$ (R and R' = H or alkyl) are formed and with ethylene chlorohydrin, alcohols of the type $ArCH_2 \cdot CH_2OH$ result. Benzylmagnesium chloride reacts abnormally with formaldehyde to give 2-methylbenzyl alcohol (*M. Tiffeneau* and *R. Delange*, Compt. rend., 1903, **137,** 573).

(ii) From aldehydes and ketones

(*1*) by reduction with hydrogen and a metal catalyst (*K. Kindler, H.-G. Helling* and *E. Sussner*, Ann., 1957, **605,** 200). Certain soluble rhodium catalysts can function as homogeneous catalysts for the hydrogenation of ketones (*R. R. Schrock* and *J. A. Osborn*, Chem. Comm., 1970, 567):

$$\text{PhCOMe} \xrightarrow{H_2 - RhH_2[PhPMe_2]_2^{\oplus}\ ClO_4^{\ominus}} \text{PhMeCHOH}$$

(*2*) by reduction using an aluminium alkoxide and a suitable alcohol as solvent (Meerwein–Ponndorf–Verley method, *A. L. Wilds*, Org. Reactions, 1944, **2,** 178) or with lithium isopropoxide in isopropanol (*D. N. Kirk* and *A. Mudd*, J. chem. Soc., C, 1965, 805);

(*3*) by reduction with sodium and a moist solvent;

(*4*) by reduction with a complex metal hydride such as lithium tetrahydridoaluminate (*R. F. Nystrom* and *W. G. Brown*, J. Amer. chem. Soc., 1947, **67,** 1197, 2548), lithium tri-*tert*-butoxyhydridoaluminate (*H. C. Brown* and *R. F. McFarlin, ibid.*, 1958, **80,** 5372), lithium cyanotrihydridoborate (*R. F. Borch* and *H. D. Durst, ibid.*, 1969, **91,** 3996), triphenylstannane (*H. G. Kuivila* and *O. F. Beumel, ibid.*, 1961, **83,** 1246). Some representative aromatic aldehydes have been reduced to the corresponding alcohols in high yield by di-imide generated from potassium azodicarboxylate (*D. C. Curry, B. C. Uff* and *N. D. Ward*, J. chem. Soc., C, 1967, 1120; see also *E. E. van Tamelen, M. Davis* and *M. F. Deem*, Chem. Comm., 1965, 71), or calcium hydridotrimethoxyborate, $Ca[BH(OCH_3)_3]_2$ (*G. Hesse* and *H. Jäger*, Ber., 1959, **92,** 2022) or sodium dihydridobis(2-methoxyethoxy) aluminate, $NaAlH_2(OCH_2 \cdot CH_2OMe)_2$ (*M. Čapka et al.*, Coll. Czech. chem. Comm., 1969, **34,** 118) or pyridine–borane (*R. P. Barnes, J. H. Graham* and *D. M. Taylor*, J. org. Chem., 1958, **23,** 1561);

(*5*) by reaction with alkylmagnesium halides to give alcohols of the general formula $ArC(OH)RR'$ (R and R' = H or alkyl) (see for example, *L. A. Brooks*, J. Amer. chem. Soc., 1944, **66,** 1295).

(*6*) by reaction of aromatic aldehydes with aqueous alkali to give a mixture of the corresponding alcohol and acid (Cannizzaro reaction, p. 76).

(iii) From carboxylic acids and their derivatives

(*1*) Aromatic acids such as benzoic acid and its nuclear substituted derivatives and their esters, chlorides or anhydrides, are reduced smoothly by lithium tetrahydridoaluminate in ether to benzyl alcohol or nuclear-substituted benzyl alcohols (*Nystrom* and *Brown, loc. cit.; R. Adams et al.*, J. Amer. chem. Soc., 1949, **71**, 1624; *B. O. Field* and *J. Grundy*, J. chem. Soc., 1955, 1110; *J. S. Clovis* and *G. S. Hammond*, J. org. Chem., 1962, **27**, 2284). Sodium dihydridobis(2-methoxyethoxy)aluminate has similar reducing properties (*M. Černý et al.*, Coll. Czech. chem. Comm., 1969, **34**, 1025). Carboxylic acids and their esters are also reduced by a reagent prepared by the addition of aluminium chloride to a solution of sodium tetrahydridoborate in diglyme (*H. C. Brown* and *B. C. Subba Rao*, J. Amer. chem. Soc., 1956, **78**, 2582), *e.g.*:

$$p\text{-}NO_2C_6H_4CO_2H \xrightarrow[\text{diglyme}]{NaBH_4 - AlCl_3} p\text{-}NO_2C_6H_4CH_2OH \quad (82\%)$$

tert-Butylamine–borane, $(CH_3)_3CHN_2 \cdot BH_3$, reduces benzoyl chloride to benzyl alcohol in 92% yield (*H. Nöth* and *H. Beyer*, Ber., 1960, **93**, 1078). These methods can also be used for the reduction of aryl-substituted aliphatic carboxylic acids and their esters.

(*2*) Alternative procedures for the reduction of aromatic carboxylic acid esters include the use of sodium tetrahydridoborate and lithium bromide (*H. C. Brown, E. J. Mead* and *Subba Rao*, J. Amer. chem. Soc., 1955, **77**, 6209) and aluminium amalgam (*J. N. Ray, A. Mukherji* and *N. Gupta*, J. Indian chem. Soc., 1961, **38**, 705). The reduction of thioesters with Raney nickel to the corresponding primary alcohols is a fairly widely applicable procedure (*O. Jeger et al.*, Helv., 1946, **29**, 684).

(*3*) Aralkyl carboxylic acid esters can be reduced with sodium and an alcohol (*L. Bouveault* and *G. Blanc*, Compt. rend., 1903, **136**, 1676; **137**, 60; Bull. Soc. chim. Fr., 1904, [iii] **31**, 748) also with sodium tetrahydridoborate in methanol under reflux (*M. S. Brown* and *H. Rapoport*, J. org. Chem., 1963, **28**, 3261).

(4) Electrolytic reduction cf aromatic acids in alcoholic sulphuric acid furnishes the corresponding alcohols (*J. Tafel* and *G. Friederichs*, Ber., 1904, **37**, 3187; *C. Mettler*, *ibid.*, 1905, **38**, 1745; 1906, **39**, 2933; *C. Marie et al.*, Bull. Soc. chim. Fr., 1919, [iv], **25**, 512; *E. Bauer* and *E. Müller*, Z. Elektrochem., 1928, **34**, 98; *F. Fichter*, Helv., 1929, **12**, 821; *S. Swann, Jr.* and *G. D. Lucker*, Trans. electrochem. Soc., 1939, **75**, 411). If the esters

of the aromatic acids are used instead of the free acids, benzyl ethers are also formed. Electrolytic reduction of benzoates in methanol in the presence of tetramethylammonium chloride gives benzyl alcohol (*L. Horner* and *H. Neumann*, Ber., 1965, **98,** 3462).

(5) Reaction of aromatic esters with two moles of alkylmagnesium halide results in tertiary alcohols of the general type $ArC(OH)R_2$. An alternative procedure is to add a mixture of ester and alkyl halide to a suspension of lithium in ether (*P. J. Pearce, D. H. Richards* and *N. F. Scilly*, Chem. Comm., 1970, 1160).

(6) Rhenium(VI) oxide is reported to be an efficient catalyst for the reduction of carboxylic acids to alcohols (*H. Smith Broadbent* and *W. J. Bartley*, J. org. Chem., 1963, **28,** 2345), *e.g.:*

$$PhCO_2H \xrightarrow{ReO_3 - H_2 - 205\ atm} PhCH_2OH$$

(7) Reduction of tertiary amides with excess of lithium tetrahydridoaluminate gives the corresponding primary alcohols (*F. Weygand* and *R. Mitgau*, Ber., 1955, **88,** 301):

$$ArCONMePh \xrightarrow{LiAlH_4 - THF} ArCH_2OH$$

N-substituted benzamides have been reduced to benzyl alcohol by electrolysis in methanol in the presence of tetramethylammonium chloride (*Horner* and *Neumann, loc. cit.*).

(iv) From olefins

Alcohols may be derived from olefins by hydration. Anti-Markownikoff addition of water is achieved by the hydroboration–oxidation procedure, and Markownikoff addition by successive reaction with mercuric acetate and sodium tetrahydridoborate.

$$PhMeC{=}CH_2 \xrightarrow{NaBH_4 - BF_3} \xrightarrow{H_2O_2/OH^{\ominus}} \underset{(92\%)}{PhMeC \cdot CH_2OH}$$

(*G. Zweifel* and *H. C. Brown*, Org. Reactions, 1963, **13,** 38);

$$PhMeC{=}CH_2 \xrightarrow{Hg(OAc)_2} \xrightarrow{NaBH_4} \underset{(95\%)}{PhMe_2C(OH)}$$

(*Brown* and *P. Geoghegan*, J. Amer. chem. Soc., 1967, **89,** 1522).

Styrenes may be converted into 3-arylalkan-1-ols by reaction with formaldehyde to give a 1,3-dioxane, followed by hydrogenolysis of the product, illustrated by the following preparation of 3-phenylpropan-1-ol (*W. S. Emerson et al.*, *ibid.*, 1950, **72,** 5314).

$$PhCH{=}CH_2 \xrightarrow[H_2SO_4]{HCHO} \text{(4-phenyl-1,3-dioxane)} \xrightarrow[\text{copper chromite}]{H_2\,(1500\text{-}2600\ \text{p.s.i.})} Ph(CH_2)_3OH$$

Diethylaluminium chloride adds to olefinic double bonds with elimination of ethylene, oxidation of the product with alkaline peroxide then giving an alcohol (*A. Alberola*, Tetrahedron Letters, 1970, 3471), *e.g.*:

$$Ph(Me)C{=}CH_2 + Et_2AlCl \longrightarrow Ph(Me)CH{\cdot}CH_2Al(Cl)Et + C_2H_4$$

$$\xrightarrow{H_2O_2 - OH^{\ominus}} Ph(Me)CH{\cdot}CH_2OH \quad (85\%)$$

(85%)

(v) From cyclic ethers

(*1*) Reduction of epoxides catalytically or with lithium tetrahydridoaluminate gives alcohols; *e.g.*, reduction of styrene oxide with lithium tetrahydridoaluminate gives α-phenylethanol but treatment with lithium tetrahydridoaluminate and aluminium chloride results in alternative ring fission to give predominantly β-phenylethanol (*E. L. Eliel* and *D. W. Delmonte*, J. Amer. chem. Soc., 1956, **78,** 3226).

(*2*) Friedel–Crafts alkylation of aromatic compounds with 1,2-epoxyalkanes gives rise to β-arylethanols (*F. Johnson*, in "Friedel Crafts and Related Reactions", ed. *G. A. Olah*, Interscience, New York, 1965, Vol. IV, Chapter 47):

$$ArH + RCH{-}CH_2\ (\text{epoxide, O bridging}) \longrightarrow RCHAr{\cdot}CH_2OH$$

(*3*) Aluminium chloride-promoted reaction of trimethylene oxide and benzene gives 3-phenylpropan-1-ol (*S. Searles*, J. Amer. chem. Soc., 1954, **76,** 2313):

$$C_6H_6 + \text{(trimethylene oxide: } CH_2{-}CH_2,\ CH_2{-}O) \xrightarrow{AlCl_3} PhCH_2{\cdot}CH_2{\cdot}CH_2OH$$

$$PhCH{=}CH{\cdot}CH_2OH \xrightarrow{H_2\text{-}\,\text{catalyst}} PhCH_2{\cdot}CH_2{\cdot}CH_2OH$$

(vi) Miscellaneous methods

(*1*) Unsaturated alcohols are reduced to saturated alcohols by catalytic hydrogenation. Cinnamyl alcohol is reduced in this way to 3-phenylpropan-1-ol (*F. Straus* and *H. Grindel*, Ann., 1924, **439,** 276).

(*2*) Reaction of aralkylamines with nitrous acid gives the corresponding alcohols, for example, the conversion of benzylamine into benzyl alcohol.

(*3*) Toluene has been oxidised to benzyl alcohol with argentic picolinate in dimethyl sulphoxide (*J. B. Lee* and *T. G. Clarke*, Tetrahedron Letters, 1967, 415).

(b) Physical properties and reactions

The infrared spectra of alcohols show characteristic O–H and C–O stretching bands in the region of 3500–3100 cm^{-1} and 1250–1000 cm^{-1}, respectively. In the p.m.r. spectrum of benzyl alcohol in $CDCl_3$, the methylene resonance is at 5.60 τ.

The mass spectrum of benzyl alcohol shows a prominent molecular ion and a $(M-1)^{\oplus}$ ion due to random loss of hydrogen. The $(M-1)^{\oplus}$ ion loses carbon monoxide to give the benzonium ion $(C_6H_7)^{\oplus}$ (*m/e* 79) which in turn loses two hydrogens to provide the phenyl cation (*m/e* 77). Other prominent ions in the mass spectrum are the tropylium ion (*m/e* 91) and the benzoylium ion ($PhC\equiv\overset{\oplus}{O}$) (*m/e* 105). The mass spectra of benzyl alcohols are discussed by *H. Budzikiewicz*, *C. Djerassi* and *D. H. Williams*, in "Interpretation of Mass Spectra of Organic Compounds", Holden-Day, San Francisco, 1964, p. 169.

The reactions of benzyl alcohol and related primary, secondary and tertiary alcohols in general resemble those of aliphatic alcohols and only some selected features of the reactivity of benzyl alcohols and their derivatives will be discussed.

(i) Reaction with oxidising agents

Benzylic alcohols are extremely easily oxidised by a wide variety of reagents and are therefore convenient precursors of substituted benzaldehydes.

Oxidation of benzylic alcohols with dipyridine-chromium(VI) oxide in dichloromethane at room temperature appears to be a promising procedure (*J. C. Collings*, *W. W. Hess* and *F. J. Frank*, Tetrahedron Letters, 1968, 3363; see *L. F. Fieser* and *M. Fieser*, "Reagents in Organic Synthesis", Wiley, New York, 1967, p. 146, for details of the preparation of the reagent). Pyridine dichromate in pyridine has similar oxidising properties and is a less hazardous reagent (*W. M. Coates* and *J. R. Corrigan*, Chem. and Ind., 1969, 1594). A reproducible procedure for the activation of manganese dioxide by azeotropic removal of water by distilling with benzene gives a reagent suitable for the oxidation of benzylic alcohols (*I. M. Goldman*, J. org. Chem., 1969, **34,** 1979). Oxidation of alcohols with dimethyl sulphoxide and acetic anhydride has been reported (*J. D. Albright* and *L. Goldman*, J. Amer. chem. Soc., 1967, **89,** 2416). This procedure is especially useful for the oxidation of sterically hindered hydroxyl groups. 1-Chlorobenzotriazole is a readily prepared oxidant and also suitable for oxidation of benzylic alcohols (*C. W. Rees* and *R. C. Storr*, J. chem. Soc., C, 1969, 1474):

$$R^1R^2CHOH + \text{1-chlorobenzotriazole} \longrightarrow R^1R^2C{=}O + \text{benzotriazolium } Cl^{\ominus}$$

A further reagent which has been used for the oxidation of alcohols to aldehydes and ketones is silver(II) picolinate (*J. B. Lee et al.*, Canad. J. Chem., 1969, **47,** 1649). Iodosobenzene also appears to be a useful reagent for the selective oxidation of primary alcohols (*T. Takaya*, *H. Enyo* and *E. Imoto*, Bull. chem. Soc. Japan, 1968, **41,** 1032). Iodonium nitrate also oxidises benzyl alcohol to benzaldehyde in high yield (*U. E. Diner*, J. chem. Soc., C, 1970, 676). Diethyl azodicarboxylate is an effective and mild oxidising agent for both primary or secondary alcohols (*F. Yoneda*, *K. Suzuki* and *Y. Nitta*, J. org. Chem., 1967, **32,** 727). Ceric ammonium nitrate is probably the reagent of choice, the merits of this reagent compared with other oxidising agents have been discussed by *W. S. Trahanovsky*, *L. B. Young* and *G. L. Brown* (J. org. Chem., 1967, **32,** 3865; these authors give a comprehensive bibliography of reagents used for the oxidation of benzylic alcohols and related compounds):

$$ArCH_2OH + 2\ Ce(IV) \longrightarrow ArCHO + 2\ Ce(III) + 2\ H^{\oplus}$$

Oxidation of benzyl alcohol alone with nickel peroxide gives benzaldehyde but treatment of a mixture of benzyl alcohol and ammonia with this reagent gives benzamide. This is thought to be formed *via* oxidation of the intermediate aldehyde ammonia (*K. Nakagawa*, *H. Onoue* and *K. Minami*, Chem. Comm., 1966, 17):

$$PhCH_2OH \longrightarrow PhCHO \longrightarrow PhCH(OH)NH_2 \longrightarrow PhCONH_2 \quad (71\%)$$

Electrolytic oxidation of benzylic alcohols, ethers and esters ($ArCH_2X$) at a platinum electrode in acetonitrile gives ArCHO and XH as the major products (*E. A. Mayeda*, *L. L. Miller* and *J. F. Wolf*, J. Amer. chem. Soc., 1972, **94,** 6812). Benzyl alcohols are rapidly oxidised to the corresponding benzaldehydes on treatment with hot nitric acid in aqueous 'glyme'. Pure aldehydes are isolated in 86–96% yield (*A. McKillop* and *M. E. Ford*, Synthetic Comm., 1972, **2,** 307).

(ii) Cleavage of the benzyl–oxygen bond

The benzyl–oxygen bond in benzyl alcohols, ethers and esters is cleaved by a variety of reagents as illustrated by the following equations:

$$ArCH_2OH \xrightarrow{H_2/CuCrO_2} ArCH_3 + H_2O$$

$$RCO_2 \cdot CH_2Ph \xrightarrow{H_2/Pt} RCO_2H + PhCH_3$$

$$RCO_2 \cdot CH_2Ph \xrightarrow{NaI} RCO_2Na + PhCH_2I$$

$$RNHCO \cdot CH_2Ph \xrightarrow{HX} RNH_2 + CO_2 + PhCH_2X$$

The benzyl–oxygen bond in α-substituted benzyl alcohols and their ethers survives hydrogenation in the presence of a 5% rhodium-on-alumina catalyst and good yields of cyclohexane derivatives are formed; the reaction fails with benzyl alcohol itself (*J. H. Stocker*, J. org. Chem., 1962, **27,** 2288).

Benzyl ethers are cleaved by treatment with asymmetrical dichlorodimethyl ether in the presence of zinc chloride (*A. Reiche* and *H. Gross*, Ber., 1959, **92,** 83).

Benzylic alcohols are converted into halides under neutral conditions by use of the 1:1-complexes of *N*-bromo- and *N*-chloro-succinimide with dimethyl sulphide (*E. J. Corey, C. U. Kim* and *M. Takeda*, Tetrahedron Letters, 1972, 4339), *e.g.*:

$$Me_2\overset{\oplus}{S}\text{—N(succinimide)}\ Cl^{\ominus} + R^1R^2CHOH \xrightarrow{CH_2Cl_2} Me_2\overset{\oplus}{S}\text{—}OCHR^1R^2\ Cl^{\ominus} \longrightarrow R^1R^2CHCl + Me_2SO$$

(iii) Wittig rearrangement

Treatment of benzyl ethers with a strong base such as phenyl-lithium or sodamide results in a 1,2-shift and an alkoxide is formed:

$$PhCH_2OMe \xrightarrow{PhLi} Ph\overset{\ominus}{C}HOMe\overset{\oplus}{Li} \longrightarrow PhMeCH\overset{\ominus}{O}\overset{\oplus}{Li} \xrightarrow{H_2O} PhMeCHOH$$

This rearrangement has been reviewed by *H. E. Zimmerman* in "Molecular Rearrangements", Interscience Publishers, New York, 1963, Vol. 1, p. 372 and by *D. J. Cram*, in "Fundamentals of Carbanion Chemistry", Academic Press, New York, 1965, p. 230.

(iv) Rearrangement of benzyl methyl sulphide and benzyldimethylsulphonium bromide

These rearrangements can also be brought about with strong base (*C. R. Hauser, S. W. Kantor* and *W. R. Brasen*, J. Amer. chem. Soc., 1953, **75**, 2660):

$C_6H_5CH_2SMe \xrightarrow{KNH_2/ether}$ 2-Me-$C_6H_4CH_2SH$

$C_6H_5CH_2\overset{\oplus}{S}Me_2 \xrightarrow{NaNH_2/liq.\ NH_3}$ 2-Me-$C_6H_4CH_2SMe$

The latter reaction is analogous to the rearrangement of benzyltrimethylammonium salts (see Sommelet–Hauser rearrangement, p. 14).

(v) Oxidation of benzyl sulphides

Benzyl sulphides are commonly oxidised to sulphoxides and sulphones on treatment with hydrogen peroxide. However, oxidation with dimethyl sulphoxide and benzoyl chloride yields the corresponding aldehyde. The α-chlorosulphide is formed as an intermediate in this reaction (*R. Oda* and *Y. Hayashi*, Tetrahedron Letters, 1967, 3141), *e.g.*:

$$[4]ClC_6H_4 \cdot CH_2SCH_3 \rightarrow [4]ClC_6H_4 \cdot CHO \quad (77\%)$$

(vi) Reaction of benzyl alcohols and benzyl ethers with thallium(III) trifluoroacetate

Benzyl alcohol and benzyl methyl ether have been converted quantitatively into their 2-iodo derivatives by treatment with thallium(III) trifluoroacetate (TTFA) and reaction of the resulting arylthallium bistrifluoroacetate with aqueous potassium iodide:

$C_6H_5CH_2OH \xrightarrow{TTFA}$ 2-$Tl(OCO \cdot CF_3)_2$-$C_6H_4CH_2OH \xrightarrow{aq.\ KI}$ 2-I-$C_6H_4CH_2OH$

By appropriate choice of conditions it is possible with these reagents to convert 2-phenylethanol into predominantly its *ortho-*, *meta-*, or *para-*iodo derivative (*E. C. Taylor et al.*, J. Amer. chem. Soc., 1970, **92**, 2173).

Illustrative of a general synthesis of aromatic nitriles is the following conversion of benzyl methyl ether into 2-cyanobenzyl methyl ether (*idem, ibid.*, p. 3520):

$$\text{C}_6\text{H}_5\text{CH}_2\text{OMe} \xrightarrow{\text{TTFA}} o\text{-}(\text{CH}_2\text{OMe})\text{C}_6\text{H}_4\text{Tl(OCO}\cdot\text{CF}_3)_2 \xrightarrow[(2)\ h\nu]{(1)\ \text{aq. KCN}} o\text{-}(\text{CH}_2\text{OMe})\text{C}_6\text{H}_4\text{CN}\ (55\%)$$

(c) Phenyl-substituted alcohols

Physical constants and derivative m.p. data of individual alcohols are summarised in Table 3.

Benzyl alcohol, $PhCH_2OH$, is a colourless liquid with a faint aromatic odour, soluble in about 25 parts of water and miscible with alcohol and ether. It is present partly free and largely as the benzoate in tuber rose oil and acacia blossom oil; as acetate in jasmine oil. In Tolu and Peru balsams it occurs as benzoate or cinnamate and as the latter in liquid storax.

Benzyl alcohol is manufactured from benzyl chloride and sodium carbonate, and in this process dibenzyl ether is obtained as a by-product (*L. T. Rosenberg*, U.S.P. 2,221,882/1938; *Fabr. v. F. Bayer and Co*., G.P. 343,930/1919).

Benzyl alcohol can be readily esterified by treatment with an acid in the presence of a catalyst such as zinc chloride or boron trifluoride. A number of esters such as benzyl acetate and benzyl benzoate are prepared by reaction of benzyl chloride and a metal salt of the appropriate acid. The hydroxyl group may be replaced with halogen by treatment with concentrated aqueous hydrogen halide solutions or with phosphorus halides, *e.g.*:

$$PhCH_2OH \quad + \quad HI \xrightarrow{25^\circ} PhCH_2I \quad + \quad H_2O$$

The hydroxyl group may be replaced by an amino group by passing ammonia into an ethereal solution of benzyl alcohol in the presence of aluminium chloride. Monobenzylation of aromatic amines is achieved by heating the amine with benzyl alcohol in the presence of potassium hydroxide and distilling off the water as it is formed. The reaction is accelerated by the addition of benzaldehyde (*Y. Sprinzak*, J. Amer. chem. Soc., 1956, **78,** 3207). Benzyl alcohol forms acetals which are usually prepared by treatment with an excess of aldehyde in the presence of an acid catalyst.

The action of acid catalysts on benzyl alcohol results in a variety of products chief of which are dibenzyl ether, 2- and 4-benzylbenzyl alcohol, and polymeric benzylbenzyls $(PhCH_2)_x$. Benzyl alcohol reacts with benzene in the presence of Friedel–Crafts catalysts to give diphenylmethane, and with phenol under similar conditions, it yields a mixture of 2- and 4-benzylphenol.

Dehydrogenation of benzyl alcohol over copper oxide gives up to 95% of benzaldehyde (*R. Davies* and *H. H. Hodgson*, J. chem. Soc., 1943, 282). Benzyl alcohol can be reduced to toluene by hydrogenation over a palladium-on-charcoal catalyst (*R. Baltzly* and *J. S. Buck*, J. Amer. chem. Soc., 1943, **65,** 1948), or by reaction with Raney nickel alloy and alkali (*D. Papa*, *E. Schwenk* and *B. Whitman*, J. org. Chem.,

TABLE 3

PHENYL-SUBSTITUTED ALCOHOLS

	m.p. (°C)	*b.p.* (°C/mm)	n_D°	$d_{\circ}^{\circ}$	*Derivative, m.p.* (°C)
$PhCH_2OH$	–15.3	205.8	1.5408^{20}	1.0455_4^{20}	4-*Nitrobenzoyl*, 85; *phenylcarbamate*, 78
$PhCH_2 \cdot CH_2OH$	–25.8	219.8, 99–100/10	1.5323^{20}	1.0242_{15}^{15}	4-*Nitrobenzoyl*, 106–108; *phenylcarbamate*, 80
(±)-PhMeCHOH		204, 100/18	1.5260^{15}	1.008_4^{15}	*Phenylcarbamate*, 94 (62)
$PhCH_2 \cdot CH_2 \cdot CH_2OH$		236, 119/12	1.5357^{25}	1.007_{17}^{17}	*Phenylcarbamate*, 47; 4-*nitrobenzoyl*, 45–46
(±)-$PhCH_2 \cdot CH(OH)Me$		219–221, 116/7.5	1.5210^{20}	0.997^{20}	*Phenylcarbamate*, 92; 4-*nitrobenzoyl*, 61
(±)-PhEtCHOH		217–221, 105–108/10	1.5208^{18}	0.993_0^{20}	4-*Nitrobenzoyl*, 60
$PhMe_2C(OH)$	29–30	202, 94/13		1.007_4^{16}	α-*Naphthylcarbamate*, 100–101; 4-*nitrobenzoyl*, 65
(±)-$PhMeCH \cdot CH_2OH$		114/14	1.5221^{25}		
$Ph(CH_2)_3 \cdot CH_2OH$		140/17	1.4310^{16}	0.988_4^{14}	*Phenylcarbamate*, 53
(±)-$PhEtCH \cdot CH_2OH$		122–123/18			*Phenylcarbamate*, 58
$PhMe_2C \cdot CH_2OH$		131/30	1.5261^{20}		*Phenylcarbamate*, 59,5–60.5; α-*naphthylcarbamate*, 91.5–92.5
(±)-$PhPr^nCHOH$	16	116–118/20	1.5166^{22}	0.9739_4^{26}	*Benzoyl*, 111
(±)-$PhPr^iCHOH$		112–113/15	1.5193^{14}	0.9869_4^{14}	*Acid phthalate*, 133.7–134
(±)-$PhCH_2 \cdot CH_2 \cdot CHOH \cdot CH_3$		124–126/15	1.5159^{20}	0.9798^{20}	*Phenylcarbamate*, 116–117
$PhCH_2 \cdot CMe_2(OH)$	24	127–128/14	1.5194^{14}	0.9822_4^{14}	*Phenylcarbamate*, 96
(±)-$PhMePr^nC(OH)$		112–113/14		0.9723_4^{21}	
(±)-$PhCH_2 \cdot CMeEt(OH)$		113–115/14.5	1.5182^{20}	0.9754_0^{20}	*Acid phthalate*, 86
$PhCH_2CH_2 \cdot CMe_2(OH)$		121/13			*Phenylcarbamate*, 143–144
$PhEt_2C(OH)$		112/15	1.5172^{20}	0.9836_4^{19}	

For references see text.

1942, **7,** 587). Benzyl alcohol undergoes the following interesting reaction when the mixture is heated under reflux:

$$C_6H_5CH_2OH \xrightarrow[\text{2-dichlorobenzene}]{CH_2{=}C(OMe)NMe_2} o\text{-}Me\text{-}C_6H_4\text{-}CH_2{\cdot}CONMe_2\ (50\%) + o\text{-}(CH_2{\cdot}COMe)\text{-}C_6H_4\text{-}CH_2{\cdot}CONHMe_2\ (20\%)$$

The reagent exists in equilibrium with dimethylacetamide dimethylacetal (*A. E. Wick et al.*, Helv., 1964, **47,** 2425):

$$CH_3\text{–}C(OMe)_2NMe_2 \quad \rightleftharpoons \quad CH_2{=}C(OMe)NMe_2 \quad + \quad MeOH$$

β-**Phenylethyl alcohol,** 2-*phenylethan*-1-*ol*, *β-phenethyl alcohol*, $PhCH_2{\cdot}CH_2OH$, is a colourless liquid with a faint rose odour which becomes more pronounced on dilution with water. It is soluble in about 50 parts of water and is miscible with alcohol and ether. It occurs free or as an ester in various essential oils such as rose, carnation, orange blossom and geranium. *β*-Phenylethanol is used together with citronellol and geraniol as the basis of all rose-type perfumes.

At the present time practically all *β*-phenylethyl alcohol is produced from the aluminium chloride catalysed reaction of benzene and ethylene oxide (*L. Valik* and *I. Valik*, U.S.P. 1,944,959/1934):

$$C_6H_6 \quad + \quad \underset{\diagdown O \diagup}{CH_2\text{—}CH_2} \quad \longrightarrow \quad PhCH_2{\cdot}CH_2OH$$

It was previously obtained by Bouveault–Blanc reduction of ethyl phenylacetate or from the Grignard reaction between phenylmagnesium bromide and ethylene chlorohydrin. On heating with powdered potassium hydroxide, *β*-phenylethyl alcohol is dehydrated to styrene (*S. Sabetay*, Bull. Soc. chim. Fr., 1929, [iv], **45,** 72), whereas with diluted acids it is converted into bis-*β*-phenylethyl ether (*J. B. Senderens*, Compt. rend., 1926, **182,** 612; 1929, **188,** 1074). Dehydrogenation of *β*-phenylethanol over finely divided catalysts such as copper and silver gives phenylacetaldehyde. It can be hydrogenated to *β*-cyclohexylethanol (*H. Adkins* and *H. R. Billica*, J. Amer. chem. Soc., 1948, **70,** 695).

(±)α-**Phenylethyl alcohol,** 1-*phenylethan*-1-*ol*, *methylphenylmethanol*, PhCH(OH)Me, is prepared by reaction of methylmagnesium iodide with benzaldehyde (*A. M. Ward*, J. chem. Soc., 1927, 452) or from phenylmagnesium bromide and acetaldehyde (*J. B. Conant* and *A. H. Blatt*, J. Amer. chem. Soc., 1928, **50,** 544; *F. Ashworth* and *G. N. Burkhardt*, J. chem. Soc., 1928, 1798). It is conveniently prepared from acetophenone by hydrogenation (*J. v. Braun* and *G. Kochendörfer*, Ber., 1923, **56,** 2174; *F. Straus* and *H. Grindel*, Ann., 1924, **439,** 298) or by reduction with sodium and alcohol (*A. Klages* and *P. Allendorff*, Ber., 1898, **31,** 1003), with lithium isopropoxide in isopropanol (*D. N. Kirk* and *A. Mudd*, J. chem. Soc., C, 1969, 804), or with magnesium and methyl alcohol (*L. Zechmeister* and *P. Rom*, Ann., 1929, **468,** 124; see also *V. N. Ipatieff* and *V. Haensel*, J. Amer. chem. Soc., 1942, **64,** 520). It is also obtained by reduction of styrene oxide with lithium tetrahydridoaluminate (*R. F. Nystrom* and *W. G. Brown*, *ibid.*, 1948, **70**, 3738).

It is oxidised to acetophenone. For the action of hydrochloric acid in yielding styrene and the ether see *Ward* (*loc. cit.*, p. 455) and *Senderens* (Compt. rend., 1926, **182**, 613).

By crystallisation of the brucine salt of the hydrogen phthalate of racemic methylphenylmethanol it is possible to prepare the (−)-alcohol rapidly and in good yield (*A. J. H. Houssa* and *J. Kenyon*, J. chem. Soc., 1930, 2260).

3-**Phenylpropan**-1-**ol,** *hydrocinnamyl alcohol*, $PhCH_2 \cdot CH_2 \cdot CH_2OH$, is present as the cinnamyl ester in liquid storax, in Peru balsam and in other balsams (*W. v. Miller*, Ann., 1877, **188,** 202; *A. Hellstrom*, Arch. Pharm., 1905, **243,** 235). It is prepared in 87% yield by reduction of cinnamaldehyde with lithium tetrahydridoaluminate (*Nystrom* and *Brown*, J. Amer. chem. Soc., 1948, **70,** 3738), in 90% yield by the reduction of ethyl hydrocinnamate with sodium and phenol in the presence of a small amount of quinoline (*W. Enz*, Helv., 1961, **44,** 206) or in 79% yield by reaction of benzylmagnesium chloride with ethylene oxide (*R. C. Huston* and *A. G. Agett*, J. org. Chem., 1941, **6,** 123).

(±)-1-**Phenylpropan**-2-**ol,** *benzylmethylmethanol*, $PhCH_2 \cdot CH(OH)Me$, is prepared by the reduction of benzyl methyl ketone with sodium in boiling alcohol (*R. H. Pickard* and *Kenyon*, J. chem. Soc., 1914, **105,** 1124) or by the reaction of phenylmagnesium bromide with propylene oxide (*M. S. Newman*, J. Amer. chem. Soc., 1940, **62,** 2295). It has been resolved by crystallisation of the brucine salt of the hydrogen phthalate (*Kenyon, H. Phillips* and *V. P. Pitman*, J. chem. Soc., 1935, 1083.

(±)-1-**Phenylpropan**-1-**ol,** *ethylphenylmethanol*, $PhCH(OH) \cdot CH_2 \cdot CH_3$, has been prepared by the catalytic reduction of propiophenone (*Ipatieff* and *Haensel*, *loc. cit.*; see also *H. Kindler, W. G. Helling* and *E. Sussner*, Ann., 1957, **605,** 200) or by the reaction of benzaldehyde with ethylmagnesium iodide. The (+)-alcohol obtained from the brucine salt of the (+)-hydrogen succinate has $[\alpha]_D^{17}$ + 27.35° (*Pickard* and *Kenyon*, J. chem. Soc., 1911, **99,** 71).

2-**Phenylpropan**-2-**ol,** *dimethylphenylmethanol*, $PhC(OH)Me_2$, is prepared by the action of methylmagnesium iodide on acetophenone (*Klages*, Ber., 1902, **35,** 2637), or by hydration of α-methylstyrene ($MePhC{=}CH_2$) by successive reaction with mercuric acetate and sodium borohydride (*Brown* and *Geoghegan*, *loc. cit.*). It is very easily dehydrated.

(±)-2-**Phenylpropan**-1-**ol,** *hydratropic alcohol*, $PhCHMe \cdot CH_2OH$, is prepared by reduction of methylphenylacetamide (*Hauser et al.*, J. Amer. chem. Soc., 1947, **69,** 592) or by hydration of α-methylstyrene using the hydroboration–oxidation procedure (*Zweifel* and *Brown*, *loc. cit.*), or by the aluminium chloride promoted reaction of benzene with 1,2-epoxypropane (*N. Milstein*, J. heterocycl. Chem., 1968, **5,** 337). The stereospecific alkylation of benzene with (+)-1,2-epoxypropane has been achieved using this method (*T. Nakajima et al.*, Bull. chem. Soc., Japan, 1967, **40,** 2980). It is separated into optically active forms $[\alpha]_D^{20}$ ± 15.2° *via* the brucine and cinchonidine salts of its hydrogen 3-nitrophthalate (*J. B. Cohen, J. Marshall* and *H. E. Woodman*, J. chem. Soc., 1915, **107,** 900).

Aryl derivatives of higher aliphatic primary alcohols are prepared in general by reduction of the esters or amide of the corresponding carboxylic acids or by the Grignard reaction using formaldehyde or ethylene oxide.

A large number of alcohols of types $ArCH_2{\cdot}CH_2OH$ and $Ar_2CH{\cdot}CH_2OH$ have been described by *P. Ramart-Lucas* and *P. Amagat* (Ann. Chim. Fr., 1927, [x], **8,** 263). See also *P. A. Levene* and *R. E. Marker* (J. biol. Chem., 1931, **93,** 762; 1935, **110,** 332).

4-**Phenylbutan**-1-**ol,** $Ph(CH_2)_3{\cdot}CH_2OH$, is prepared from 3-phenylpropylmagnesium bromide and formaldehyde (*P. W. Clutterbuck* and *Cohen*, J. chem. Soc., 1923, **123,** 2510).

(±)-2-**Phenylbutan**-1-**ol,** $PhEtCH{\cdot}CH_2OH$, is prepared in 75% yield by the reduction of phenylethylacetamide with sodium and ethanol *(Hauser et al., loc. cit.).*

2-**Methyl**-2-**phenylpropan**-1-**ol,** $PhMe_2C{\cdot}CH_2OH$, is prepared in 72% yield by oxidation of the Grignard reagent from the corresponding chloride (*F. C. Whitmore, C. S. Weisgerber* and *A. C. Shabica*, J. Amer. chem. Soc., 1943, **65,** 1471).

Secondary alcohols are prepared in general by reduction of the appropriate ketones or by the Grignard reaction with aldehydes. A series of alkylphenylmethanols has been prepared by reaction of benzaldehyde with alkylmagnesium halides (*Pickard* and *Kenyon*, J. chem. Soc., 1911, **99,** 45).

(±)-1-**Phenylbutan**-1-**ol,** *phenyl*-n-*propylmethanol*, $PhPr^nCHOH$, is prepared in 76% yield from phenylmagnesium bromide and butyraldehyde. The (−)-*alcohol*, $[\alpha]_D^{27}$ − 45.9° (*c*, 6.1, benzene), is obtained by crystallisation of the strychnine salt of the hydrogen phthalate (*K. Mislow* and *C. L. Hamermesh*, J. Amer. chem. Soc., 1955, **77,** 1593).

(±)-2-**Methyl**-1-**phenylpropan**-1-**ol,** *phenylisopropylmethanol*, $PhPr^iCHOH$, is prepared from isopropylmagnesium iodide and benzaldehyde (*V. Grignard*, Ann. Chim. Fr., 1901, [vii], **24,** 467 and resolved by crystallisation of the strychnine salt of the acid phthalate. (+)-*Alcohol*, $[\alpha]_D^{23}$ 48.3° (*c*, 6.7, ether) (*Cram* and *J. E. McCarty*, J. Amer. chem. Soc., 1957, **79,** 2872).

(±)-1-**Phenylbutan**-3-**ol,** *methyl-β-phenylethylmethanol*, $PhCH_2{\cdot}CH_2{\cdot}CHOH{\cdot}CH_3$, is prepared in 85% yield from β-phenylethylmagnesium bromide and acetaldehyde (*A. Brewin* and *E. E. Turner*, J. chem. Soc., 1930, 503).

2-**Methyl**-1-**phenylpropan**-2-**ol,** *benzyldimethylmethanol*, $PhCH_2{\cdot}CMe_2(OH)$, is prepared from phenylacetic acid ester and methylmagnesium iodide (*Klages*, Ber., 1904, **37,** 1723).

(±)-2-**Phenylpentan**-2-**ol,** *methylphenyl*-n-*propylmethanol*, $PhCMeEt(OH)$, is prepared from *n*-propylmagnesium iodide and acetophenone (*idem, ibid.*, 1902, **35,** 2643).

(±)-2-**Methyl**-1-**phenylbutan**-2-**ol,** *benzylethylmethylmethanol*, $PhCH_2{\cdot}CMeEt(OH)$, is prepared from benzylmagnesium chloride and ethyl methyl ketone (*M. T. Bogert* and *D. Davidson*, J. Amer. chem. Soc., 1934, **56,** 187).

2-**Methyl**-4-**phenylbutan**-2-**ol,** *dimethyl-β-phenylethylmethanol*, $PhCH_2{\cdot}CH_2{\cdot}CMe_2(OH)$, is prepared from β-phenylethylmagnesium bromide and acetone (*Klages*, Ber., 1904, **37,** 2314).

3-**Phenylpentan**-3-**ol,** *diethylphenylmethanol*, $PhCEt_2(OH)$, is prepared in 93% yield from ethylmagnesium iodide and benzoyl chloride (*H. Gilman, R. E. Fothergill* and *H. H. Parker*, Rec. Trav. chim., 1929, **48,** 750).

(d) Functional derivatives of the alcohols

The halides have already been described (Vol. III A, p. 271 *et seq.*).

(i) Ethers

Ethers are exemplified by those of benzyl alcohol which are best prepared by reaction of benzyl chloride with the appropriate alkoxides, generally in the alcohol as solvent. Alternatively they are formed by alkylation of benzyl alcohol.

Benzyl ethers have also been prepared from aldehydes and ketones by reduction with silanes in alcoholic acidic media (*M. P. Doyle, D. J. DeBruyn* and *D. A. Kooistra*, J. Amer. chem. Soc., 1972, **94,** 3659), *e.g.*:

$$PhCHO + Et_3SiH + MeOH \xrightarrow{H_2SO_4} \underset{(87\%)}{PhCH_2OMe} + Et_3SiOH$$

Benzyl ethers undergo oxidative cleavage to yield benzaldehyde, for example, on treatment with trichloroisocyanuric acid (*E. C. Juenge* and *D. A. Beal*, Tetrahedron Letters, 1968, 5819):

$$PhCH_2OEt \xrightarrow{\text{trichloroisocyanuric acid}} \underset{(53\%)}{PhCHO}$$

Benzylic ethers have also been cleaved to benzaldehydes by treatment with triphenylmethyl tetrafluoroborate in dichloromethane and quenching with aqueous sodium hydrogen carbonate (*D. H. R. Barton et al.*, Chem. Comm., 1971, 1109).

Benzyl methyl ether, $PhCH_2OMe$, b.p. 171°, 59–60°/12 mm, n_D^{20} 1.5008, d_4^{20} 0.9634, is obtained in 88% yield by reduction of benzaldehyde dimethyl acetal with a 1:4 mixture of lithium tetrahydridoaluminate and aluminium chloride (*E. L. Eliel, V. G. Badding* and *M. N. Rerick*, J. Amer. chem. Soc., 1962, **84,** 2371). Addition of an ethereal solution of benzyl methyl ether to a suspension of lithium in tetrahydrofuran gives benzyl-lithium, which is a useful metalating agent (*Gilman* and *G. L. Schwebke*, J. org. Chem., 1962, **27,** 4259). Benzyl methyl ether is isomerised to phenylmethylmethanol on treatment with phenyl-lithium in ethereal solution (*G. Wittig* and *L. Löhmann*, Ann., 1942, **550,** 260). Reaction of benzyl methyl ether with *asymm.* dichlorodimethyl ether (Cl_2CHOMe) in the presence of zinc chloride gives benzyl chloride (*A. Rieche* and *H. Gross*, Ber., 1959, **92,** 83).

Benzyl chloromethyl ether, $PhCH_2OCH_2Cl$, b.p. 105–109°/11 mm, is prepared from benzyl alcohol, formaldehyde and hydrogen chloride (*A. J. Hill* and *DeW. T. Keach*, J. Amer. chem. Soc., 1926, **48,** 257) and has been used for the hydroxymethylation of ketones *via* a process of sodium hydride-promoted alkylation followed by hydrogenolysis (*C. L. Graham* and *F. L. McQuillin*, J. chem. Soc., 1964, 4521).

α-Chlorobenzyl methyl ether, PhCHClOMe, b.p. 71–72°/0.1 mm, is prepared in 80% yield by treatment of benzaldehyde dimethyl acetal with acetyl chloride containing a little thionyl chloride (*F. Straus* and *H. Heinze*, Ann., 1932, **493,** 191).

Benzyl ethyl ether, b.p. 185°, n_D^{20} 1.4958, d_4^{20} 0.9478, is converted by treatment with benzene and boron trifluoride into a mixture of diphenylmethane, 1,4-dibenzylbenzene and polybenzylbenzenes (*M. J. O'Connor* and *F. J. Sowa*, J. Amer. chem. Soc., 1938, **60,** 125). On reaction with phenyl-lithium in ethereal solution it is transformed into benzyl alcohol (*Wittig* and *Löhmann*, *loc. cit.*).

α-Phenylethyl ethyl ether, PhMeCHOEt, b.p. 72–74°/15 mm, n_D^{25} 1.4834, is prepared by reduction of acetophenone diethyl acetal with lithium tetrahydridoaluminate and aluminium chloride (*Eliel*, *Badding* and *Rerick*, *loc. cit.*).

Benzyl phenyl ether, $PhCH_2OPh$, m.p. 286°, 124–125°/4 mm, is prepared by the action of benzyl chloride on phenol in aqueous alkali (*W. F. Short* and *M. L. Stewart*, J. chem. Soc., 1929, 554) or in acetone containing suspended potassium carbonate (*S. G. Powell* and *R. Adams*, J. Amer. chem. Soc., 1920, **42,** 656), or in aqueous dioxane (*H. Burton* and *P. F. G. Praill*, J. chem. Soc., 1951, 522). On heating with hydrogen chloride at 100°, or zinc chloride at 160° or with copper it rearranges to give 2- and 4-benzylphenol, 2,4-dibenzylphenol and other products (*Short* and *Stewart*, *loc. cit.*; *J. van Alphen*, Rec. Trav. chim., 1927, **46,** 804; *O. Behaghel* and *H. Freienschner*, Ber., 1934, **67,** 1368; *W. J. Hickinbottom*, Nature, 1938, **142,** 830; 1939, **143,** 520).

Dibenzyl ether, $(PhCH_2)_2O$, b.p. 170°/16 mm, n_D^{20} 1.5618, d_4^{20} 1.0456, is formed by heating benzyl alcohol with 30% sulphuric acid at 200–230° or by the action of more concentrated sulphuric acid at lower temperatures (*J. Meisenheimer*, Ber., 1908, **41,** 1421; see also *C. W. Lowe*, J. chem. Soc., 1887, **51,** 700). It is also obtained as a by-product in the manufacture of benzyl alcohol by the action of alkali on benzyl chloride (see above). It is isomerised to benzylphenylmethanol by the action of phenyllithium (*Wittig* and *Löhmann*, *loc. cit.*), or with potassium amide in liquid ammonia (*C. R. Hauser* and *S. W. Kantor*, J. Amer. chem. Soc., 1951, **73,** 1437).

(ii) Benzyl esters

Benzyl acetate, $PhCH_2OAc$, b.p. 216°, 93–94°/10 mm, n_D^{20} 1.5029, d_4^{18} 1.5063, is the main constituent in jasmine oil, and is prepared by the action of benzyl chloride on anhydrous sodium acetate. It is also obtained by oxidation of toluene in acetic acid solution with lead tetra-acetate (*G. W. K. Cavill* and *D. H. Solomon*, J. chem. Soc., 1954, 3943) or with a palladium–stannous acetate catalyst and air (*D. R. Bryant*, *J. E. McKeon* and *B. C. Ream*, J. org. Chem., 1968, **33,** 4123; 1969, **34,** 1106) or by treating a mixture of benzaldehyde and acetaldehyde with aluminium isopropoxide (*I. Lin* and *A. R. Day*, J. Amer. chem. Soc., 1952, **74,** 5133).

Benzyl benzoate, $PhCH_2OCOPh$, m.p. 21°, b.p. 156°/4.5 mm, n_D^{20} 1.5681, d_4^{18} 1.114, occurs in Peru and Tolu balsams, and is most conveniently prepared by the action of sodium benzyl oxide on benzaldehyde (*O. Kamm* and *W. F. Kamm*, Org. Synth., Coll. Vol. 1, 2nd Edn., 1947, p. 104). It has been prepared in 94% yield by the reaction of sodium benzoate with benzyl chloride in the presence of triethylamine (*F. C. Whit-*

more, et al., Ind. Eng. Chem., 1947, **39,** 1300). Benzyl benzoate is used for external application in the treatment of scabies; it is also used in certain preparations such as dimercaprol injection.

Benzyl salicylate, $PhCH_2OCO \cdot C_6H_4OH[2]$, b.p. 186°/10 mm, n_D^{20} 1.582, d_4^{15} 1.179, is prepared by heating sodium salicylate with a slight excess of benzyl chloride and a small amount of diethylamine at 130–140° (*E. H. Volwiler* and *E. B. Vliet*, J. Amer. chem. Soc., 1921, **43,** 1672).

Dibenzyl succinate, $(PhCH_2OCO \cdot CH_2)_2$, m.p. 49–50°, b.p. 238°/14 mm, is prepared by boiling excess of an aqueous solution of sodium succinate with benzyl chloride (*M. Gomberg* and *C. C. Buchler*, *ibid.*, 1920, **42,** 2066).

Benzyl chloroformate, $PhCH_2OCOCl$, b.p. 103°/20 mm, is prepared by the action of equimolecular quantities of phosgene and benzyl alcohol in toluene (*H. E. Carter*, *R. L. Frank* and *H. W. Johnston*, Org. Synth., Coll. Vol. 3, 1955, p. 167). It is used for acylating amino groups in peptide syntheses, since the benzyloxycarbonyl group can be smoothly removed by hydrogenation as toluene and carbon dioxide.

Benzyl trichloroacetate, $PhCH_2OCO \cdot CCl_3$, b.p. 148–149°/15 mm, n_D^{14} 1.5341, is prepared from benzyl alcohol and trichloroacetyl chloride (*H. Hibbert* and *M. E. Greig*, Canad. J. Res., 1931, **4,** 254). On heating at 220° under atmospheric pressure it decomposes to benzaldehyde and chloral which represents an apparent reversal of the Cannizzaro Reaction.

Dibenzyl carbonate, $(PhCH_2O)_2CO$, m.p. 29°, b.p. 200–201°/13.5 mm, is prepared by treatment of potassium benzyl oxide with 0.5 mole of phosgene in toluene (*S. T. Bowden*, J. chem. Soc., 1939, 310).

Benzyl tosylate, $PhCH_2OSO_2C_6H_4Me$, m.p. 58.5–59°, and a number of substituted benzyl tosylates have been prepared by the addition of *p*-toluenesulphonyl chloride to a preformed suspension of sodium benzyl oxide in ether (*J. K. Kochi* and *G. S. Hammond*, J. Amer. chem. Soc., 1953, **75,** 344).

(iii) Inorganic esters of benzyl alcohol

Dibenzyl sulphite, $(PhCH_2O)_2SO$, is formed by reaction of benzyl alcohol with thionyl chloride in pyridine (*M. M. Richter*, Ber., 1916, **49,** 2342). It decomposes on heating above 130° at 13 mm to dibenzyl ether and sulphur dioxide.

Benzyl hydrogen sulphate, $PhCH_2OSO_3H$, is isolated as its crystalline barium or potassium salts (*M. Delépine*, Bull. Soc. chim. Fr., 1899, [iii], **21,** 1059; *A. Verley*, *ibid.*, 1901, [iii], **25,** 49).

Benzylthiosulphuric acid, $PhCH_2S \cdot SO_3H$. The sodium salt is prepared from benzyl chloride and sodium thiosulphate in aqueous alcohol. The free *acid*, m.p. 74–75°, is obtained by treating the barium salt with sulphuric acid (*A. Purgotti*, Gazz. 1890, **20,** 25). The reactions of the mono-esters of thiosulphuric acid have been examined in detail by *T. S. Price* and *D. F. Twiss* (J. chem. Soc., 1907, **91,** 2021; 1908, **93,** 1399; 1914, **105,** 1140) and by *H. B. Footner* and *S. Smiles* (*ibid.*, 1925, **127,** 2889), and their structure as *S*-esters established.

Benzyl hyponitrite, $PhCH_2ON{:}NOCH_2Ph$, m.p. 48–49°, is prepared by the action of benzyl iodide on silver hyponitrite (*J. R. Partington* and *C. C. Shah*, *ibid.*, 1932, 2589). Benzyl hyponitrite is an effective catalyst for promoting the polymerisation of methyl methacrylate; it decomposes on heating to yield benzyloxy radicals and nitrogen:

$$PhCH_2ON{=}NOCH_2Ph \longrightarrow 2\,PhCH_2O\cdot + N_2$$

Benzyl nitrite, $PhCH_2ONO$, b.p. 71°/18 mm, $n_D^{24.7}$ 1.4989, $d_4^{24.7}$ 1.075. is prepared by the addition of an aqueous solution of aluminium sulphate to a mixture of benzyl alcohol and aqueous sodium nitrite. It decomposes on storage to give benzaldehyde and benzoic acid (*A. Chrétien* and *Y. Longi*, Compt. rend., 1945, **220,** 746).

Benzyl nitrate, $PhCH_2ONO_2$, b.p. 72.5–73.5°/4–5 mm, n_D^{25} 1.5180, is prepared from benzyl chloride and powdered silver nitrate (*G. R. Lucas* and *L. P. Hammet*, J. Amer. chem. Soc., 1942, **64,** 1929). It reacts with hydroxide ion to give nitrite ion and benzaldehyde.

Benzyl dihydrogen phosphite, $PhCH_2OP(OH)_2$, is prepared as a yellow oil by partial hydrolysis of dibenzyl hydrogen phosphite; it forms a crystalline *ammonium salt*, m.p. 154° (*J. Baddiley et al.*, J. chem. Soc., 1949, 820).

Dibenzyl hydrogen phosphite, $(PhCH_2O)_2POH$, m.p. 17°, b.p. 165°/0.1 mm, n_D^{20} 1.5545, is prepared by reaction of a mixture of benzyl alcohol and dimethylaniline with a solution of phosphorus trichloride in benzene followed by treatment with water (*F. R. Atherton* and *A. R. Todd*, *ibid.*, 1948, 674).

Tribenzyl phosphite, $(PhCH_2O)_3P$, is prepared by reaction of a solution of 3 moles of benzyl alcohol and 3 moles of pyridine in ether with a solution of 1 mole of phosphorus trichloride in ether *(Baddiley et al., loc. cit.)*.

Benzyl dihydrogen phosphate, $PhCH_2OPO(OH)_2$, is prepared by boiling a solution of benzyl alcohol in ether with polyphosphoric acid and is isolated as its crystalline barium salt (*E. Cherbuliez* and *H. Weniger*, Helv., 1946, **29,** 2006).

Dibenzyl hydrogen phosphate, $(PhCH_2O)_2P(O)OH$, m.p. 79–80°, is prepared by reaction of tribenzyl phosphate with 4-methylmorpholine and treatment of the resulting quaternary salt with dilute sulphuric acid. It forms a *cyclohexylamine salt*, m.p. 173° (*Baddiley et al.*, J. chem. Soc., 1949, 818).

Tribenzyl phosphate, $(PhCH_2O)_3PO$, m.p. 64°, results from the interaction of benzyl chloride and silver phosphate in ether (*W. Lössen* and *A. Köhler*, Ann., 1891, **262,** 211).

Tribenzyl pyrophosphate, $(PhCH_2O)_2PO — O — PO(OH)OCH_2Ph$, is obtained as its silver salt by boiling tetrabenzyl pyrophosphate with 4-methylmorpholine in benzene and treatment of the reaction product with silver nitrate *(Baddiley et al., loc. cit.)*.

Tetrabenzyl pyrophosphate, $(PhCH_2O)_2PO—O—PO(OCH_2Ph)_2$, m.p. 60–61,° is prepared by treatment of dibenzyl hydrogen phosphite with carbon tetrachloride in aqueous potassium hydroxide (*Atherton* and *Todd*, J. chem. Soc., 1947, 677). Dibenzylchlorophosphonate $(PhCH_2O)_2P(O)Cl$, is the intermediate in this reaction.

Diethyl benzylphosphonate, $PhCH_2PO(OEt)_2$, prepared from benzyl bromide and triethyl phosphite, reacts readily with aromatic and heteroaromatic aldehydes to give *trans*-stilbenes (*E. J. Seus* and *C. V. Wilson*, J. org. Chem., 1961, **26,** 5243).

$$RCHO + PhCH_2PO(OEt)_2 \xrightarrow[DMF]{NaOMe} RCH{=}CHPh + (EtO)_2P(=O)OH$$

Tribenzyl arsenite, $(PhCH_2O)_3As$, b.p. 290°/35 mm, n_D^{21} 1.5800, is prepared by boiling benzyl alcohol with arsenious oxide in benzene (*P. Pascal* and *A. Dupire*, Compt. rend., 1932, **195,** 14).

Tetrabenzyl orthosilicate, $(PhCH_2O)_4Si$, m.p. 33°, b.p. 260–262°/1 mm, is prepared from benzyl alcohol and silicon tetrachloride (*B. Helferich* and *J. Hausen*, Ber., 1924, **57,** 796).

Tribenzyl borate, $(PhCH_2O)_3B$, b.p. 206°/4 mm, is prepared by heating benzyl alcohol and boric oxide at 180° (*R. L. Shriner* and *A. Berger*, J. org. Chem., 1941, **6,** 315).

(iv) Thio analogues of benzyl alcohols and their derivatives

α-Toluenethiol*, *phenylmethanethiol*, $PhCH_2SH$, b.p. 194°, n_D^{25} 1.5729, d_4^{25} 1.5729, is a liquid with a leek-like odour. It is prepared by the reaction of benzyl halides with potassium hydrogen sulphide (*H. Scheibler* and *J. Voss*, Ber., 1920, **53,** 382; *W. S. Hoffman* and *E. E. Reid*, J. Amer. chem. Soc., 1923, **45,** 1833), or by reaction of benzyl halides with thiourea followed by treatment with strong aqueous potassium hydroxide (*H. J. Backer* and *N. D. Dijkstra*, Rec. Trav. chim., 1948, **67,** 889). It readily yields dibenzyl disulphide on mild oxidation; more vigorous oxidation gives benzaldehyde and benzoic acid. On treatment with excess of peracetic acid in acetonitrile it is converted into α-toluenesulphonic acid (*C. J. Cavallito* and *D. M. Fruehauf*, J. Amer. chem. Soc., 1949, **71,** 2248). It is reduced to toluene on warming with Raney nickel in aqueous ethanolic sodium hydroxide (*D. Papa, E. Schwenk* and *H. F. Ginsberg*, J. org. Chem., 1949, **14,** 728). It reacts with benzaldehyde in the presence of hydrogen chloride to form the dibenzyl dithioacetal, $PhCH(SCH_2Ph)_2$ (*E. Fromm* and *E. Junius*, Ber., 1895, **28,** 1111). Its reactions with other saturated and unsaturated ketones and aldehydes and with oxo acids are described by *T. Posner* (Ber., 1901, **34,** 2643; 1902, **35,** 799, 2344; 1903, **36,** 298; 1904, **37,** 502). It forms a *mercuric salt*, $Hg(SCH_2Ph)_2$, needles, m.p. 121°, a mercurichloride, $ClHgSCH_2Ph$, and a lead salt, $Pb(SCH_2 \cdot Ph)_2$. Phenylmethanethiol is converted in 55% yield to benzyl thiosulphate, $PhCH_2S \cdot SO_3H$, by reaction with dicyclohexylcarbodiimide and sulphuric acid (*R. O. Mumma, K. Fujitani* and *C. P. Hoiberg*, J. Chem. and Engineering Data, 1970, **15,** 358).

Benzyl methyl sulphide, $PhCH_2SMe$, b.p. 198°, 93–94°/14 mm, is prepared by reaction of phenylmethanethiol with dimethyl sulphate in aqueous ethanolic sodium hydroxide (*T. Thomson* and *T. S. Stevens*, J. chem. Soc., 1932, 69). Oxidation of benzyl methyl sulphide with hydrogen peroxide in acetone gives *benzyl methyl sulphoxide*, m.p. 54°, whereas with hydrogen peroxide in 50% acetic acid the corresponding sulphone is formed (*M. Gazdar* and *Smiles*, *ibid.*, 1908, **93,** 1833; *S. Hünig* and *O. Boes*, Ann., 1953, **579,** 23). *Benzyl methyl sulphone*, m.p. 125°, is also prepared by the reaction of sodium benzylsulphinate with methyl iodide (*Fromm* and *S. de Palma*, Ber., 1906, **39,** 3315; *C. K. Ingold et al.*, J. chem. Soc., 1927, 818). *Benzyl ethyl sulphide*, b.p. 222°; *sulphone*, m.p. 84°.

Benzyldichloromethyl sulphide and **benzyltrichloromethyl sulphide** have been prepared from *S*-benzyl thioformate ($PhCH_2SCHO$) and phosphorus pentachloride. The dichloromethyl sulphide is obtained at 35–45° and the trichloromethyl sulphide at 100–

*The older name "benzylmercaptan" is not acceptable in I.U.P.A.C. rules.

110° (*D. H. Holsboer* and *A. P. M. van der Veek*, Rec. Trav. chim., 1972, **91**, 351).

Benzyl phenyl sulphide, $PhCH_2SPh$, m.p. 40–41°, b.p. 197°/25 mm, is formed by reaction of benzyl chloride with thiophenol in ethanolic sodium ethoxide (*Shriner, H. C. Struck* and *W. J. Jorison*, J. Amer. chem. Soc., 1930, **52**, 2060). Reaction of methyl benzenesulphenate (PhSOMe) with phenylmethanethiol gives benzyl phenyl sulphide in 87% yield; this method is general for preparing unsymmetrical disulphides (*D. A. Armitage, M. J. Clark* and *C. C. Tso*, J. chem. Soc., Perkin Trans. I, 1972, 680). The *sulphoxide*, m.p. 122–123°, is obtained by treatment of the sulphide with aqueous hydrogen peroxide in acetone and the *sulphone*, m.p. 146°, either by oxidation of the sulphoxide or by reaction of benzyl chloride with sodium benzenesulphinate (*Fromm*, Ber., 1908, **41**, 3403).

Benzyl 4-tolyl sulphide, m.p. 44° (*H. Gilman* and *W. B. King*, J. Amer. chem. Soc., 1925, **47**, 1140); *sulphoxide*, m.p. 137°; *sulphone*, m.p. 145°.

Dibenzyl sulphide, $(PhCH_2)_2S$, m.p. 50°, is prepared by the reaction of benzyl chloride with sodium sulphide in aqueous ethanol (*Shriner, Struck* and *Jorison, loc. cit.*). It is oxidised to the *sulphoxide*, m.p. 133° and then to the *sulphone*, m.p. 151° (*A. E. Wood* and *E. G. Travis*, J. Amer. chem. Soc., 1928, **50**, 1227; *L. N. Lewin*, J. prakt. Chem., 1928, [ii], **119**, 213). Dibenzyl sulphoxide is cleaved by aqueous acid giving benzaldehyde and α-toluenethiol:

$$PhCH_2SOCH_2Ph \xrightarrow{H_3O^{\oplus}} PhCHO + PhCH_2SH$$

This is an example of a reaction generally referred to as a Pummerer rearrangement (see *I. Durst*, Adv. org. Chem., **6**, 357). Treatment of dibenzyl sulphide, dibenzyl disulphide and dibenzyl sulphoxide with potassium *tert*-butoxide in dimethylformamide yields stilbene as the olefinic decomposition product (*T. J. Wallace et al.*, J. chem. Soc., 1965, 1271). Dibenzyl sulphide forms additive compounds with chlorine, bromine and iodine, the *iodide*, $(Ph{\cdot}CH_2)_2SI_2$, forming red crystals, m.p. 65°. It forms sulphonium salts with alkyl halides.

Dibenzyl disulphide, m.p. 74°, may be obtained by reaction of benzyl chloride with sodium disulphide, or by oxidation of phenylmethanethiol with iodine in aqueous alcoholic sodium hydroxide (*G. Bulmer* and *F. G. Mann, ibid.*, 1945, 666). Dibenzyl disulphide has also been prepared by reaction of *N*-benzylthiophthalimide with phenylmethanethiol; the preparation of disulphides from sulphenimides and thiols is general (*K. S. Boustany* and *A. B. Sullivan*, Tetrahedron Letters, 1970, 3547; *D. N. Harpp et al., ibid.*, 1970, 3551).

(e) Nuclear-substituted benzyl alcohols

Physical constants and derivative m.p. data for some alkyl-substituted benzyl alcohols are given in Table 4.

The halogen-substituted benzyl alcohols are obtained by the methods given for benzyl alcohols. Those most generally used are the alkaline hydrolysis of the substituted benzyl halides (*S. C. J. Olivier*, Rec. Trav. chim.,

1926, **45,** 301), the hydrogenation of the corresponding aldehydes and the Cannizzaro reaction (*W. H. Carothers* and *R. Adams,* J. Amer. chem. Soc., 1924, **46,** 1682; *J. B. Shoesmith* and *R. H. Slater*, J. chem. Soc., 1926, 219). Reduction of 4-chlorobenzoic acid with sodium tetrahydridoborate and boron trifluoride etherate in "diglyme" gives 4-chlorobenzyl alcohol in excellent yield. An excellent yield of the alcohol is also obtained by reduction of ethyl 4-chlorobenzoate with sodium tetrahydridoborate and lithium bromide in "diglyme" (*H. C. Brown, E. J. Mead* and *B. C. Subba Rao,* J. Amer. chem. Soc., 1955, **77,** 681, 6209), or with sodium tetrahydridoaluminate and aluminium trichloride in "diglyme" (*Brown* and *Rao, ibid.,* 1956, **78,** 2582).

TABLE 4

ALKYL-SUBSTITUTED BENZYL ALCOHOLS

-benzyl alcohol	*m.p. (°C)*	*b.p. (°C/mm)*	*Derivative, m.p. (°C)*	*Ref.*
2-Methyl- (2-tolylmethanol, α-hydroxy-*o*-xylene)	36	223	4-*Nitrobenzoyl*, 101	1,2
3-Methyl-		218	4-*Nitrobenzoyl*, 89	2
4-Methyl-	61	217	3,5-*Dinitrobenzoyl*, 118	2,3
2,3-Dimethyl-	66	125/12		1
2,4-Dimethyl-	22	151/44	*Phenylcarbamate*, 79	4
2,5-Dimethyl-		143/37	*Phenylcarbamate*, 86	4,5
3,4-Dimethyl-	63			4,5a
3,5-Dimethyl-		218–221/745		6
4-Isopropyl-	28	246	*Phenylcarbamate*, 62.5	7
2,3,4-Trimethyl-	50			8
2,3,6-Trimethyl-	85			9
2,4,5-Trimethyl-	168			10
2,4,6-Trimethyl-	89			11
3,4,5-Trimethyl-	78			10

References

1 *W. R. Brasin* and *C. R. Hauser*, Org. Synth., Coll. Vol. 4, 1963, p. 582.
2 *J. B. Shoesmith* and *R. H. Slater*, J. chem. Soc., 1924, **125,** 2280.
3 *D. Davidson* and *M. Weiss*, Org. Synth., Coll. Vol. 2, 1943, p. 590.
4 *M. Sommelet*. Compt. rend., 1913, **157,** 1445.
5 *W. H. C. Rueggeberg et al.*, J. Amer. chem. Soc., 1945, **67,** 2154.
5a *F. Weygand* and *R. Mitgau*, Ber., 1955, **88,** 301.
6 *H. D. Law*, J. chem. Soc., 1907, **91,** 758.
7 *L. Palfray et al.*, Compt. rend., 1936, **203,** 1523.
8 *T. Reichstein et al.*, He.v., 1936, **19,** 416.
9 *L. I. Smith* and *C. L. Agre*, J. Amer. chem. Soc., 1938, **60,** 653.
10 *H. Krömer*, Ber., 1891, **24,** 2411.
11 *W. T. Nauta* and *J. W. Dienske*, Rec. Trav. chim., 1936, **55,** 1000.

4-Nitrobenzyl alcohol is also prepared in excellent yield from 4-nitrobenzoic acid by reduction with sodium tetrahydridoborate and boron trifluoride in "diglyme". In general the Cannizzaro reaction is not suitable for the preparation of nitrobenzyl alcohols; the reduction of nitrobenzyl aldehydes to the corresponding benzyl alcohols is best achieved with alcoholic solutions of aluminium alkoxides (*H. Meerwein* and *R. Schmidt*, Ann., 1925, **444,** 233; *H. J. Backer* and *N. Dost*, Rec. Trav. chim., 1949, **68,** 1143). 3-Nitrobenzaldehyde has been converted to 3-nitrobenzyl alcohol in 82% yield by reduction with sodium tetrahydridoborate in methanol (*S. W. Chaikin* and *W. G. Brown*, J. Amer. chem. Soc., 1949, **71,** 122). The alkaline hydrolysis of 4-nitro- and 4-iodo-benzyl acetates has been used successfully for the preparation of the corresponding alcohols (*W. W. Hartman* and *E. J. Rahrs*, Org. Synth., Coll. Vol. 3, 1955, p. 652). The controlled oxidation of nitrotoluenes to nitrobenzyl alcohols using either manganese dioxide and sulphuric acid (G.P. 212,949/1908) or electrolytic methods (*K. Elbs*, Z. Elektrochem., 1896, **2,** 522) has also been achieved.

The following are the constants of some mono-substituted benzyl alcohols: **chlorobenzyl alcohol,** 2-, m.p. 69°; 3-, b.p. 242°; 4-, m.p. 75°; *bromobenzyl alcohol*, 2-, m.p. 80°; 3-, b.p. 250°; 4-, m.p. 76°; *iodobenzyl alcohol*, 2-, m.p. 90°; 3-, b.p. 155°/10 mm; 4-, m.p. 72°; **nitrobenzyl alcohol,** 2-, m.p. 74°; 3-, m.p. 15° and 30.5°; 4-, m.p. 93°. 2-, 3- and 4-**Nitrobenzyl nitrates** have been prepared from the corresponding benzyl halides and silver nitrate (*J. W. Baker* and *T. G. Heggs*, J. chem. Soc., 1955, 616); they have m.p.'s of 28°, 43° and 68°, respectively.

2-Nitrobenzyl alcohol is reduced by zinc dust and aqueous ammonium chloride to 2-**hydroxylaminobenzyl alcohol,** $HOCH_2{\cdot}C_6H_4NHOH$, m.p. 104°, which can be oxidised to 2-*azoxybenzyl alcohol*, $HOCH_2{\cdot}C_6H_4N(O){:}NC_6H_4{\cdot}CH_2OH$, m.p. 123°, and 2-*nitrosobenzyl alcohol*, $ONC_6H_4{\cdot}CH_2OH$, m.p. 101°. The latter compound loses the elements of water on boiling to give anthranil (p. 99) (*E. Bamberger*, Ber., 1903, **36,** 836). 2-*Hydrazobenzyl alcohol, 2,2'-dihydroxymethylhydrazobenzene*, $(HOCH_2{\cdot}C_6H_4)_2N_2H_2$, m.p. 104°, is formed by reduction of 2-nitrobenzyl alcohol by zinc dust and aqueous alkali.

3-Nitrobenzyl alcohol, is reduced to the azoxy compound by sodium arsenite and to 3,3'-*dihydroxymethylazobenzene*, m.p. 117°, by zinc dust and aqueous alkali. 4-Nitrobenzyl alcohol similarly gives rise to the azo and azoxy compounds by suitable methods of reduction (*W. M. Cumming et al.*, J. roy. Tech. Coll. Glasgow, 1932, **2,** 596).

Reduction of the nitrobenzyl alcohols under more vigorous conditions gives the corresponding aminobenzyl alcohols. These are also formed by the electro-reduction of the nitro- and amino-benzoic acids (*C. Mettler*, Ber., 1905, **38,** 1751).

2-**Aminobenzyl alcohol,** $NH_2C_6H_4{\cdot}CH_2OH$, m.p. 82°, b.p. 160°/10 mm, is prepared by electrolytic reduction of anthranilic acid (*G. H. Coleman* and *H. L. Johnson*, Org. Synth., Coll. Vol. 3, 1955, p. 60; see also *B. Beilenson* and *F. M. Hamer*, J. chem.

Soc., 1942, 98). It is also formed by the reduction of anthranilic esters with sodium amalgam (*S. Langguth*, Ber., 1905, **38,** 2062), the reduction of anthranilic acid with lithium tetrahydridoaluminate or the reduction of 2-nitrobenzyl alcohol with sodium dithionite (*A. Reissert*, *ibid.*, 1928, **61,** 2555). Many derivatives of 2-aminobenzyl alcohol are readily cyclised to heterocyclic compounds.

2-*Aminobenzyl acetate* (I), is obtained by reduction of 2-nitrobenzyl acetate as an oil which isomerises on standing and more rapidly on heating, to 2-*acetamidobenzyl alcohol* (II), m.p. 116°. The latter compound is formed in quantitative yield by reduction of the benzoxazinone (III) with hydrogen and a palladium-on-carbon catalyst (*G. N. Walker*, J. Amer. chem. Soc., 1955, **77,** 6698). Brief warming of (II) with dilute acid gives a mixture of 2-aminobenzyl acetate and the benzoxazine (IV) (*K. von Auwers*, Ber., 1904, **37,** 2251):

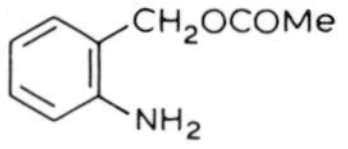

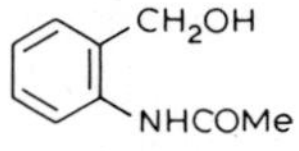

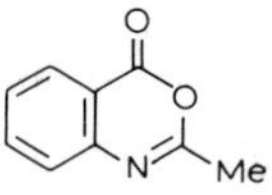

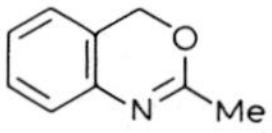

Other examples of ring formation are the production of the benzoxazinethiol (V) by boiling 2-aminobenzyl alcohol with carbon disulphide in alcohol and of the thiazinethiol (VI) by heating it with carbon disulphide and alkali; and the cyclisation of carbamide derivatives of 2-aminobenzyl alcohol (*C. Paal*, Ber., 1894, **27,** 1866, 2413, 2537):

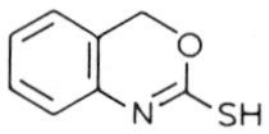

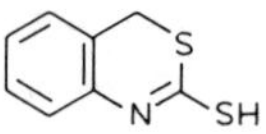

2-Benzeneazobenzyl alcohol (VII) forms 2-phenylindazole (VIII) under the influence of concentrated sulphuric acid.

3-*Aminobenzyl alcohol*, m.p. 92°, is formed by the electrolytic reduction of 3-nitrobenzyl alcohol (*Mettler*, *loc. cit.*). The N-*benzoyl* deriv. has m.p. 115°.

4-*Aminobenzyl alcohol*, m.p. 64°, is formed by the reduction of 4-nitrobenzyl alcohol. It readily passes into an anhydro form $(-NHC_6H_4CH_2-)_x$ on treatment with acids (*W. Löb*, Ber., 1898, **31,** 2037; *J. Meyer et al.*, *ibid.*, 1900, **33,** 250; 1902, **35,** 739);

similar polymeric forms result from the reaction between aniline and formaldehyde in acid solution. The N-*benzoyl* deriv. has m.p. 150–151°.

β-4-**Nitrophenylethyl alcohol,** $NO_2C_6H_4 \cdot CH_2 \cdot CH_2OH$, has m.p. 64° and b.p. 177°/16 mm; it is formed by nitration of β-phenylethyl acetate and hydrolysis of the product (*E. Ferber, ibid.*, 1929, **62,** 185; *S. Sabetay et al.*, Bull. Soc. chim. Fr., 1929, [iv], **45,** 846). On reduction it gives β-4-*aminophenylethyl alcohol*, m.p. 108°.

β-2,4,6-**Trinitrophenylethyl alcohol,** has m.p. 112°, and is formed by the reaction of 2,4,6-trinitrotoluene with formaldehyde in weakly alkaline solution (*V. Vender*, Gazz. 1915, **45,** [ii], 97).

Chapter 14

Aralkylamines and Aralkanols and their Oxidation Products: Monocarbaldehydes and Monoketones of the Benzene Series

G. W. H. CHEESEMAN and P. F. G. PRAILL*

3. Monocarbaldehydes of the benzene series

The aldehydes to be discussed in this section are of two types, those like benzaldehyde, PhCHO, in which the aldehyde grouping is directly linked to an aromatic nucleus and those like phenylacetaldehyde, $PhCH_2 \cdot CHO$, in which the aldehyde function is in the side chain.

(a) Methods of formation and preparation

A comprehensive account of the synthesis of aldehydes is to be found in *Houben–Weyl*, "Methoden der organischen Chemie", Thieme, Stuttgart, 1954, 4th Edn., Vol. 7/1. More recent reviews are to be found in "The Chemistry of the Carbonyl Group", ed. *S. Patai*, Interscience, London, 1966, Chapters 2–6. Aldehyde synthesis has been reviewed also by *A. Carnduff*, in Quart. Reviews, 1966, **20**, 169.

(i) From methylbenzenes

(*1*) The oxidation of nuclear methyl groups in the liquid phase may be accomplished by using chromium trioxide and acetic anhydride containing sulphuric acid. A diacetate is formed which on acid hydrolysis yields the aldehyde (*S. V. Lieberman* and *R. Connor*, Org. Synth., Coll. Vol. 2, 1943, p. 441):

*Manuscript date, March 1973.

$$[4]BrC_6H_4\cdot CH_3 \xrightarrow[Ac_2O/H_2SO_4]{CrO_3} [4]BrC_6H_4\cdot CH(OAc)_2 \xrightarrow{H_2O/H_2SO_4} [4]BrC_6H_4\cdot CHO$$

(83–96%)

Chromyl chloride in carbon disulphide is an effective oxidant for the methyl groups in toluene and its homologues, the overall reaction may be represented:

$$PhCH_3 \longrightarrow (PhCH_3\cdot 2\,CrO_2Cl_2) \xrightarrow{H_2O} PhCHO$$

This reaction is known as the Étard reaction (*O. Bayer*, in *Houben–Weyl*, Vol. 7/1, p. 144; *O. H. Wheeler*, Canad. J. Chem., 1958, **36,** 667), and has been used to oxidise 2-, 3-, and 4-fluorotoluenes to the corresponding 2-, 3-, and 4-fluorobenzaldehydes in yields of 28, 30 and 56%, respectively (*G. Schiemann*, Z. phys. Chem., Leipzig, 1931, **156A,** 417).

The conversion of $-CH_3$ to $-CHO$ may also be achieved by the use of manganese dioxide or lead peroxide in dilute sulphuric acid at 40° (*R. Freund*, Chem. Ztg., 1927, **51,** 803) or alkylarenesulphonic acids may be oxidised in concentrated acid to give aldehyde sulphonic acids. Other reagents which have been used more recently for this purpose have been sodium persulphate in the presence of silver nitrate (*R. G. R. Bacon* and *J. R. Doggart*, J. chem. Soc., 1960, 1332), ceric ammonium nitrate (*L. A. Dust* and *E. W. Gill*, J. chem. Soc., C, 1970, 1630) and argentic picolinate in dimethyl sulphoxide (*J. B. Lee* and *T. G. Clarke*, Tetrahedron Letters, 1967, 415). Electrolytic oxidation has also been used (*H. D. Law* and *F. M. Perkin*, J. chem. Soc., 1907, **91,** 261).

(*2*) Aromatic hydrocarbons are oxidised to aldehydes and ketones in the gas phase with molecular oxygen.

Toluene is oxidised to benzaldehyde in the presence of molybdenum trioxide as catalyst at temperatures of 450–530°. Alternative catalysts are the oxides of tungsten, zirconium and tantalum; molybdate and vanadate catalysts have also been used. In the heterogeneous catalytic oxidation of *o*-xylene, a high xylene to oxygen ratio and catalysts such as zirconium, molybdenum and tungsten oxides, favours the production of *o*-tolualdehyde in preference to phthalic anhydride.

Carbonyl-forming reactions of this type have been reviewed by *C. F. Cullis* and *A. Fish* in "The Chemistry of the Carbonyl Group", ed. *S. Patai*, Interscience, London, 1966, p. 79.

(ii) From primary alcohols

(*1*) The catalytic dehydrogenation of primary alcohols to aldehydes over a heated catalyst, such as silver, copper and vanadium pentoxide has been reported by *C. Moureu* and *G. Mignonac* (Compt. rend., 1920, **171,** 652) and *R. R. Davies* and *H. H. Hodgson* (J. chem. Soc., 1942, 282).

(*2*) Benzyl alcohols are very easily oxidised to the corresponding benzaldehydes by a wide variety of reagents (see p. 35).

For example, dinitrogen tetroxide oxidation of benzyl alcohols in chloroform solution gives excellent yields of the corresponding benzaldehydes (*B. O. Field* and *J. Grundy*, *ibid.*, 1955, 1110):

$$ArCH_2OH + N_2O_4 \rightarrow ArCHO + N_2O_3 + H_2O$$

Further references to the oxidation of primary alcohols to aldehydes are to be found in "Oxidation in Organic Chemistry", ed. *K. B. Wiberg*, Academic Press, New York, 1965 and in "Mechanisms of Oxidation of Organic Compounds", by *W. A. Waters*, Methuen, London, 1964.

(iii) From benzyl halides

(*1*) *Sommelet reaction.* Benzyl halides can be converted into aldehydes by formation of a quaternary salt with hexamethylenetetramine and subsequent hydrolysis of the salt:

$$RCH_2X + C_6H_{12}N_4 \longrightarrow [RCH_2C_6H_{12}N_4]^{\oplus}X^{\ominus} \xrightarrow{H_2O} RCH_2NH_2 \xrightarrow[pH\ 3-6.5]{C_6H_{12}N_4} RCHO$$

Frequently the reaction can be performed in one operation without the isolation of intermediates (*S. J. Angyal*, Org. Reactions, 1954, **8,** 197). A large excess of hexamethylenetetramine is used in order to prevent the formation of *N*-methylbenzylamine.

(*2*) *Kröhnke aldehyde synthesis.* The benzyl halide is initially converted into a pyridinium salt, this on treatment with 4-nitrosodimethylaniline gives a nitrone which on acid hydrolysis yields the aldehyde (*F. Kröhnke*, Angew. Chem., intern. Edn., 1963, **2,** 380). This method is particularly useful for the preparation of di-*ortho*-substituted benzaldehydes (*K. Clarke*, J. chem. Soc., 1957, 3808):

NO2 / CH2Br / Br $\xrightarrow{C_5H_5N}$ NO2 / $CH_2-\overset{\oplus}{N}C_5H_5$ $Br^{\ominus}$ / Br $\xrightarrow[NaOH]{[4]ONC_6H_4NMe_2}$

NO2 / $CH{=}\overset{O}{\overset{\uparrow}{N}}-C_6H_4-NMe_2$ / Br $\xrightarrow{H_2O/H^{\oplus}}$ NO2 / CHO / Br (79%)

The preparation of aromatic aldehydes by this method is described by *A. Kalir* (Org. Synth., 1966, **46,** 81).

(*3*) *Oxidation with nitroalkanes.* A method due to *H. B. Hass* and *M. L. Bender* (J. Amer. chem. Soc., 1949, **71,** 1767; Org. Synth., 1950, **30,** 99) involves treatment of the benzyl halide with the sodium salt of 2- nitropropane:

$$ArCH_2X + [Me_2CNO_2]^{\ominus}Na^{\oplus} \rightarrow ArCHO + Me_2C{=}NOH + NaX$$

In the case of nine substituted benzyl halides the yields of the corresponding aldehydes varied from 68–77%, the method failed with 4-nitrobenzyl chloride.

(*4*) *By oxidation with nitrates, etc.* Benzyl halides can be converted into aldehydes by heating under reflux with the nitrates of copper, lead or calcium in dilute nitric acid (*O. Fischer* and *P. Grieff*, Ber., 1880, **13,** 669; *K. Schulze*, *ibid.*, 1884, **17,** 1530), or by heating with aqueous dichromate and sodium carbonate (*T. Posner* and *G. Schreiber*, *ibid.*, 1924, **57,** 1131; *G. Blanc*, G.P. 347,583/1918). Potassium *tert*-butyl hydroperoxide has been used as an alternative oxidising agent (*K. Kulka*, Amer. Perfumer, 1957, **70,** 37).

An alternative procedure involves conversion of the benzyl halide into the corresponding tosylate followed by treatment of the tosylate with a mixture of sodium bicarbonate and dimethyl sulphoxide (*N. Kornblum*, *W. J. Jones* and *G. J. Anderson*, J. Amer. chem. Soc., 1959, **81,** 4113). The oxidation of the benzylic halides themselves has also been carried out under these conditions and is complete at 100° in 5 minutes (see also *W. W. Epstein* and *F. W. Sweat*, Chem. Reviews, 1967, **67,** 247):

$$ArCH_2X \longrightarrow ArCH_2OTs \xrightarrow[100°;\ <5\ min]{NaHCO_3/DMSO} ArCHO$$

(*5*) *By carbonylation.* Primary bromides have been transformed into aldehydes containing an additional carbon by reaction with sodium tetracarbonylferrate(II), prepared *in situ* by reduction of iron pentacarbonyl with sodium amalgam (*M. P. Cooke*, J. Amer. chem. Soc., 1970, **92,** 6080), in the presence of triphenylphosphine at 25°. Benzylic bromides, however, give poor yields of aldehyde:

$$PhCH_2{\cdot}CH_2Br \xrightarrow[(2)\ AcOH]{(1)\ Na_2Fe(CO)_4 - Ph_3P - THF} PhCH_2{\cdot}CH_2{\cdot}CHO \quad (86\%)$$

(iv) From benzylidene dihalides

Benzylidene dihalides are conveniently prepared by light-induced halogenation of nuclear methyl groups, they give aldehydes on hydrolysis (*G. H. Coleman* and *G. E. Honeywell*, Org. Synth., Coll. Vol. **2,** 1943, p. 89), *e.g.*:

$$[4]BrC_6H_4{\cdot}CH_3 \xrightarrow{Br_2/h\nu} [4]BrC_6H_4{\cdot}CHBr_2 \xrightarrow{H_2O/CaCO_3} [4]BrC_6H_4{\cdot}CHO$$

(60-69% overall)

(v) From benzylamines and benzylidene-amines

Benzylamine has been oxidised to benzaldehyde and β-phenylethylamine to phenylacetaldehyde by treatment with *tert*-butyl hypochlorite (*W. E. Bachmann*, *M. P. Cava* and *A. S. Dreiding*, J. Amer. chem. Soc., 1954, **76,** 5554). Benzylamine has also been oxidised to benzaldehyde with manganese dioxide (*R. J. Highet* and *W. C. Wildman*, *ibid.*, 1955, **77,** 4399). Diazotisation of benzylamines in anhydrous dimethyl sulphoxide gives 60–80% yields of benzaldehydes (*K. H. Scheit* and *W. Kampe*, Angew. Chem., internat. Edn., 1965, **4,** 787):

$$ArCH_2NH_2 \rightarrow ArCH_2N_2^{\oplus} \rightarrow ArCH_2-O-\overset{\oplus}{S}(CH_3)_2 \rightarrow ArCHO$$

Benzylidene-amines, obtained by oxidation of certain benzylanilines or by condensing aromatic nitroso compounds with toluenes having reactive methyl groups, give aldehydes on hydrolysis (*E. T. Pratt* and *T. P. McGovern*, J. org. Chem., 1964, **29,** 1540):

$$ArCH_2NHPh \xrightarrow{MnO_2} ArCH{=}NPh \xrightarrow{H_2O/H^{\oplus}} ArCHO$$

(*O. Bayer*, in *Houben–Weyl*, Vol. 7/1, p. 152):

$$ArCH_3 \xrightarrow{[4]ONC_6H_4NMe_2} ArCH{=}NC_6H_4NMe_2 \xrightarrow{H_2O/H^{\oplus}} ArCHO$$

(vi) From arylamines

Aromatic amines can be converted into benzaldehydes by reaction of the corresponding diazonium salts with formaldoxime *e.g.:*

$$[4]ClC_6H_4NH_2 \longrightarrow [4]ClC_6H_4N_2^{\oplus} \xrightarrow[CuSO_4,\,Na_2SO_3]{CH_2{=}N-OH} [4]ClC_6H_4\cdot CH{=}N-OH$$

$$\xrightarrow{H_2O/H^{\oplus}} [4]\,ClC_6H_4\cdot CHO \quad (60\%)$$

Reaction with acetoxime leads to the formation of methyl ketones (*W. F. Beech*, J. chem. Soc., 1954, 1297; see also *S. D. Jolad* and *S. Rajogopal*, Org. Synth., 1966, **46,** 13).

(vii) From carboxylic acids and their derivatives

(*1*) *Aldehydes from carboxylic acids.* A limited number of aromatic carboxylic acids have been reduced to aldehydes with formic acid under pressure in the presence of titanium dioxide as catalyst (*Davies* and *Hodgson*, J. chem. Soc., 1943, 84). Other reagents which have been used are hydridodiisobutyl aluminium (*L. I. Zakharkin* and *I. M. Khorlina*, J. gen. Chem. U.S.S.R., 1964, **34,** 1021; C.A., 1964, **60,** 15724f) and lithium and methylamine in the presence of ammonium nitrate (*A. O. Bedenbaugh et al.*, J. Amer. chem. Soc., 1970, **92,** 5774). Electrolytic reduction of benzoic acid and salicylic acid to the corresponding aldehydes has also been achieved, but this method is not generally applicable (*C. Mettler*, Ber., 1908, **41,** 4148; *K. Tesh* and *A. Lowy*, Trans. Amer. electrochem. Soc., 1924, **45,** 47).

A newer route from carboxylic acids to aldehydes, involves conversion of acid to an oxazoline, methylation with methyl iodide and reduction to an oxazolidine followed by a terminal step of acid hydrolysis to give the aldehyde (*I. C. Nordin*, J. heterocycl. Chem., 1966, **3,** 531):

$$RCO_2H \xrightarrow{Me_2C(NH_2)\cdot CH_2OH} \text{(2-R-4,4-dimethyloxazoline)} \xrightarrow[(2)\ NaBH_4]{(1)\ MeI} \text{(2-R-3,4,4-trimethyloxazolidine)} \xrightarrow{H_2O\ -\ H^{\oplus}} RCHO$$

(*2*) *Acid chloride-based procedures* for the preparation of aldehydes include (*a*) *The Rosenmund Reduction* in which the acid chloride is reduced with hydrogen and a catalyst, commonly palladium on barium sulphate, deactivated with *e.g.* thiourea (*E. Mosettig* and *P. Mozingo*, Org. Reactions, 1948, **4,** 362; for experimental details see also *R. P. Barnes*, Org. Synth., Coll. Vol. 3, 1955, p. 561):

$$RCOCl \quad + \quad H_2 \quad \rightarrow \quad RCHO \quad + \quad HCl$$

A modification of the experimental procedure for carrying out the Rosenmund reaction, using hydrogen under 3–4 atmospheres pressure has been described by *A. I. Rachlin, H. Gurien* and *D. P. Wagner* (Org. Synth., 1972, **51,** 8). Lithium tri-*tert*-butoxyhydridoaluminate is a selective reagent for the reduction of acid chlorides. It is readily prepared in ether from *tert*-butanol and lithium tetrahydridoaluminate (*H. C. Brown* and *B. C. Subba Rao*, J. Amer. chem. Soc., 1958, **80,** 5377):

$$RCOCl \quad + \quad LiAl(OBu^t)_3H \quad \rightarrow \quad RCHO \quad + \quad Al(OBu^t)_3 \quad + \quad LiCl$$

Hydrido(tri-*n*-butylphosphine)copper(I), [$CuHPBu_3$], reduces benzoyl chloride to benzaldehyde in 50% yield; the reagent is prepared by reduction of copper(I) bromide with hydridodiisobutylaluminium at −50° and complexing the copper hydride produced with tributylphosphine (*G. M. Whitesides et al., ibid.*, 1969, **91,** 6542). Tri-*n*-butylhydridostannane reduces acyl halides to aldehydes and esters; the factors governing product distribution have been studied (*H. G. Kuivila* and *E. J. Walsh Jr., ibid.*, 1966, **88,** 571):

$$RCOCl \rightarrow RCHO + RCH_2OCOR$$

Aroyl chlorides have been reduced with triethylsilane in the presence of a palladium on charcoal catalyst. Yields of aldehyde were in the range of 40–70% (*J. D. Citron*, J. org. Chem., 1969, **34,** 1977):

$$RCOCl + Et_3SiH \xrightarrow{Pd/C} RCHO + Et_3SiCl$$

Irradiation of benzoyl bromide in ether gives benzaldehyde in 80% yield (*U. Schmidt*, Angew. Chem., internat. Edn., 1965, **4,** 146):

$$PhCOBr \xrightarrow{h\nu - Et_2O} PhCHO$$

(*b*) *The Grundmann procedure.* This involves the following reaction sequence (*Mosettig*, Org. Reactions, 1954, **8,** 225):

$$RCOCl \xrightarrow{CH_2N_2} RCO{\cdot}CHN_2 \xrightarrow{AcOH} RCO{\cdot}CH_2OAc \xrightarrow[(2)\ Hydrolysis]{(1)\ Al(OPr^i)_3} RCHOH{\cdot}CH_2OH \xrightarrow{Pb(OAc)_4} RCHO$$

It is valuable for the preparation of phenylacetaldehydes.

(*c*) Alternatively acid chlorides are converted into aldehydes through the intermediacy either of Reissert compounds, thiol esters (*idem, ibid.*, p. 220), or acyl phosphonates (*I. Shahak* and *E. D. Bergmann*, Israel J. Chem., 1966, **4,** 225). The latter compounds are readily prepared from the acid chloride and triethyl phosphite. The acyl phosphonate is then reduced with sodium tetrahydridoborate and alkaline hydrolysis of the product in the presence of hydroxylamine gives the aldoxime:

$$RCOCl + P(OEt)_3 \longrightarrow RCOP(O)(OEt)_2 \longrightarrow RCH(OH)P(O)(OEt)_2 \xrightarrow{OH^{\ominus}/NH_2OH} RCH{=}NOH$$

Aldehydes have been prepared also from acid chlorides by reduction of ester mesylates with sodium tetrahydridoborate (*M. R. Johnson* and *B. Rickborn*, Org. Synth., 1972, **51,** 11):

$$RCOCl + \underset{\textit{erythro}\text{-2,3-butanediol monomesylate}}{CH_3{-}CH(OH){-}CH(OSO_2CH_3){-}CH_3} \longrightarrow CH_3{-}CH(OCOR){-}CH(OSO_2CH_3){-}CH_3$$

$$\xrightarrow{NaBH_4\text{-pyridine}} \text{2-R-2-H-4,5-dimethyl-1,3-dioxolane} \xrightarrow{H_3O^{\oplus}} RCHO$$

An alternative procedure involves the conversion of the aroyl chloride into an α-oxophosphonate ester followed by reduction of the ester at pH 6–7 with sodium tetrahydridoborate (*L. Horner* and *H. Roder*, Ber., 1970, **103,** 2984):

$$\mathrm{RCOCl} + \mathrm{P(OR^1)_3} \longrightarrow \mathrm{RCOP(=O)(OR^1)_2} + \mathrm{R^1Cl}$$

$$\mathrm{RCOP(=O)(OR^1)_2} \xrightarrow{\mathrm{NaBH_4}} \mathrm{RCH(OH)P(=O)(OR^1)_2} \xrightarrow{\mathrm{OH^{\ominus}}} \mathrm{RCHO}$$

Acid chlorides can be converted into the corresponding homologous aldehyde by formation of *S*-esters, obtained by successive reaction with diazomethane and ethanethiol. Reduction of the *S*-ester with Raney nickel in the presence of 1,2-dianilinoethane gives an intermediate tetrahydroimidazole which is readily hydrolysed to the aldehyde (*F. Weygand* and *H. J. Bestmann*, *ibid.*, 1959, **92,** 528; *Bestmann* and *H. Schulz*, *ibid.*, 1959, **92,** 530) (*cf.* reduction of nitriles on p. 64):

$$\mathrm{RCOCl} \xrightarrow[(2)\ \mathrm{EtSH},\ h\nu]{(1)\ \mathrm{CH_2N_2}} \mathrm{RCH_2{\cdot}COSEt} \xrightarrow[\text{1,2-dianilinoethane}]{\text{Raney Ni}}$$

$$\mathrm{RCH_2{\cdot}CH}\begin{matrix}\mathrm{N(Ph)}\\ \mathrm{N(Ph)}\end{matrix}\!\!\big] \xrightarrow{\mathrm{H_3O^{\oplus}}} \mathrm{RCH_2{\cdot}CHO}$$

(*3*) *Reduction of esters.* (*a*) Reduction of *acylmalonic esters* with sodium tetrahydridoborate is a general method for aldehyde synthesis. The required starting materials are conveniently obtained from the acid as shown (*H. Muxfeldt*, *W. Rogalski* and *G. Klauenberg*, *ibid.*, 1965, **98,** 3040):

$$\mathrm{RCO_2H} \xrightarrow[(2)\ \mathrm{ClCO_2Et}]{(1)\ \mathrm{Et_3N}} \mathrm{RCO{\cdot}OCO_2Et} \xrightarrow[\text{malonic ester}]{\text{magnesium}}$$

$$\mathrm{RCO{\cdot}CH(CO_2Et)_2} \xrightarrow{\mathrm{NaBH_4}} \mathrm{RCHO}$$

(*b*) *Phenyl esters* are reduced with lithium tri-*tert*-butoxyhydridoaluminate to the corresponding aldehyde ((*P. M. Weissman* and *H. C. Brown*, J. org. Chem., 1966, **31,** 283):

$$\mathrm{RCO_2Ph} \rightarrow \mathrm{RCHO}$$

(*c*) A more generally applicable procedure for the reduction of esters involves the use of sodium hydridobis(2-methoxyethoxy)aluminate at temperatures in the range −50 to −70°. At these temperatures the resulting aldehyde is practically inert to further reduction (*M. Fieser* and *L. F. Fieser*, "Reagents in Organic Synthesis", Vol. 3, Wiley–Interscience, 1972, p. 261):

$$2\,RCO_2R^1 \quad + \quad NaAlH_2(OCH_2 \cdot CH_2OMe)_2 \quad \rightarrow$$

$$\rightarrow \quad 2\,RCHO \quad + \quad NaA(OR^1)_2(OCH_2 \cdot CH_2OMe)_2$$

(*4*) The *reduction of tertiary amides*, (*a*) with lithium di- and tri-ethoxyhydridoaluminates constitutes a general method of aldehyde synthesis, *e.g.*:

$$RCONMe_2 \xrightarrow{LiAl(OEt)_2H_2} RCHO + Me_2NH$$

The required hydridoaluminates ar conveniently synthesised *in situ* from lithium tetrahydridoaluminate and ethanol (*Brown* and *A. Tsukamoto*, J. Amer. chem. Soc., 1964, **86,** 1089).

(*b*) Reduction of tertiary amides to aldehydes has been achieved with bis(3-methyl-2-butyl)borane (disiamylborane) (Sia_2BH) (*Brown et al.*, *ibid.*, 1970, **92,** 7161):

$$R{-}C(=O){-}NMe_2 \xrightarrow{Sia_2BH} R{-}CH(OBSia_2){-}NMe_2 \xrightarrow{H_2O} RCHO$$

Sodium tetrahydridoaluminate has also been recommended for the selective reduction of *N,N*-disubstituted amides to the corresponding aldehydes (*Zakharin, D. N. Maslin* and *V. V. Gavrilenko*, Tetrahedron, 1969, **25,** 5555).

(*c*) Alternative procedures involve reduction of acylpyrazoles or acylimidazoles with lithium tetrahydridoaluminate (*W. Reid, G. Deuschel* and *A. Kotelko*, Ann., 1961, **642,** 121):

$$RCON(\text{3,5-dimethylpyrazol-1-yl}) \xrightarrow{LiAlH_4} RCHO$$

(*H. A. Staab*, Angew. Chem. internat. Edn., 1962, **1,** 352):

$$RCON(\text{imidazol-1-yl}) \xrightarrow{LiAlH_4} RCHO$$

(*5*) *Desulphurisation of thioamides* with partially de-activated Raney nickel gives aldehydes (*G. R. Pettit* and *E. E. van Tamelen*, Org. Reactions, 1962, **12,** 356):

$$ArCSNH_2 \xrightarrow{Ni-EtOH} ArCHO$$

(*6*) The *Sonn–Müller aldehyde synthesis* is illustrated by the following sequence:

$$ArCO_2H \longrightarrow ArCONHPh \xrightarrow{PCl_5} ArCCl{=}NPh$$

$$\xrightarrow{SnCl_2} ArCH{=}NPh \xrightarrow{H_2O/H^{\oplus}} ArCHO$$

The scope of the reaction is reviewed by *Mosettig*, Org. Reactions, 1954, **8,** 240 and the experimental method illustrated by the preparation of *o*-tolualdehyde (*J. W. Williams, C. H. Witten* and *J. A. Krynitsky*, Org. Synth., Coll. Vol. **3,** 1955, p. 818).

(7) In the *McFadyen–Stevens aldehyde synthesis* aryl sulphonyl derivatives of the hydrazides of aromatic acids are decomposed by treatment with sodium carbonate:

$$ArCONH{\cdot}NHSO_2Ph \xrightarrow[\text{ethylene glycol}]{Na_2CO_3} ArCHO + N_2 + PhSO_2Na$$

The addition of powdered glass appears to facilitate the evolution of nitrogen (*M. S. Newman* and *E. G. Caflisch*, J. Amer. chem. Soc., 1958, **80,** 862; *Mosettig*, Org. Reactions, 1954, **8,** 232).

(8) *Reduction of nitriles* with anhydrous stannous chloride in ether saturated with hydrogen chloride gives an aldimine hydrochloride and this on hydrolysis yields the aldehyde:

$$RCN + HCl \longrightarrow RCCl{:}NH \xrightarrow{SnCl_2} RCH{:}\overset{\oplus}{N}H_2\overset{\ominus}{Cl} \longrightarrow RCHO$$

An alternative procedure involves the use of lithium triethoxyhydridoaluminate, prepared *in situ* by reaction of 1 mole of lithium tetrahydridoaluminate in ether with 3 moles of ethanol or 1.5 moles of ethyl acetate. With this reagent both aromatic and aliphatic nitriles of widely varying structural types have been transformed to aldehydes, generally in yields of 70–90% (*Brown* and *G. P. Garg*, J. Amer. chem. Soc., 1964, **86,** 1085). A useful procedure for the reduction of hindered nitriles involves stirring the nitrile in formic acid solution with moist Raney nickel at 75–80°; yields in the range of 60–75% are obtained (*B. Staskun* and *O. G. Backeberg*, J. chem. Soc., 1964, 5880; see also *T. van Es* and *Staskun*, *ibid.*, 1965, 5775; Org. Synth., 1972, **51,** 20). Aldehyde semicarbazones are obtained by the catalytic reduction of nitriles in the presence of semicarbazide, and by the catalytic reduction of nitriles in the presence of dianilinoethane similarly yields tetrahydroimidazoles. The latter derivatives are very easily hydrolysed with acid to give the free aldehyde: (*H. Plieninger* and *G. Werst*, Ber., 1955, **88,** 1956):

$$RCN + PhNHCH_2{\cdot}CH_2NHPh \xrightarrow{H_2/\text{Raney Ni}} RCH\begin{matrix}NH\\NH\end{matrix}\!\!\Big] \longrightarrow RCHO$$

(viii) From formylation reactions of the type ArH → ArCHO

(*1*) The *Gattermann–Koch synthesis* of aldehydes involves the reaction of an aromatic hydrocarbon with carbon monoxide and hydrogen chloride in the presence of aluminium chloride and cuprous chloride as an activator (*N. N. Crounse*, Org. Reactions, 1949, **5**, 290). The required mixture of carbon monoxide and hydrogen chloride may be conveniently generated from formic acid and chlorosulphonic acid and if the reaction is carried out under pressure no activator is required (*G. H. Coleman* and *D. Craig*, Org. Synth., Coll. Vol. 2, 1943, p. 583):

$$MeC_6H_5 + CO + HCl \xrightarrow{AlCl_3/CuCl} [4]MeC_6H_4{\cdot}CHO + HCl \quad (46-51\%)$$

A development of the Gattermann–Koch procedure is the use of formyl fluoride as a formylating reagent. This is generated prior to use either from formic acid and potassium hydrogen fluoride, or from acetic–formic anhydride and anhydrous hydrogen fluoride. Boron trifluoride is the preferred catalyst for the formylation of aromatic hydrocarbons (*G. A. Olah* and *S. J. Kuhn*, J. Amer. chem. Soc., 1960, **82**, 2380):

$$\text{mesitylene} + HCOF \xrightarrow[CS_2]{BF_3} \text{2,4,6-}Me_3C_6H_2{\cdot}CHO \quad (70\%)$$

(*2*) In the *Gattermann aldehyde synthesis* hydrogen cyanide is used in place of carbon monoxide. The product is an aldimine which on hydrolysis yields the aldehyde (*W. E. Truce*, Org. Reactions, 1957, **9**, 37):

$$ArH + HCN + HCl \xrightarrow[(2)\ H_2O]{(1)\ AlCl_3 \text{ or } ZnCl_2} ArCHO + NH_4Cl$$

(*3*) A more recent procedure, involves the *reaction of an aromatic hydrocarbon with an alkyl dichloromethyl ether* in the presence of a Friedel–Crafts catalyst:

$$ArH + Cl_2CHOR \rightarrow ArCH(Cl)OR \rightarrow ArCHO$$

The required dichloromethyl ethers are prepared by reaction of a formate ester and phosphorus pentachloride (*A. Rieche*, *H. Gross* and *E. Höft*, Ber., 1960, **93**, 88).

(*4*) In the *Vilsmeier–Haack aldehyde synthesis* an aromatic substrate is reacted with *N*-methylformanilide or dimethylformamide and phosphoryl chloride and the intermediate hydrolysed to give the aldehyde (*E. Campaigne* and *W. L. Archer*, Org. Synth., Coll. Vol. 4, 1963, p. 331):

$$Me_2NPh \xrightarrow[(2)\ H_2O]{(1)\ Me_2NCHO/POCl_3} [4]Me_2NC_6H_4{\cdot}CHO \quad (80-84\%)$$

(5) *Reaction of* N-*methyl*-N-*phenylcarbamoyl chloride*, PhNMeCOCl, with aromatic compounds in the presence of aluminium chloride gives amides which on reduction with lithium tetrahydridoaluminate afford aldehydes (*F. Weygand* and *R. Mitgau*, Ber., 1955, **88**, 301):

Me, Me → Me, Me, CONMePh → Me, Me, CHO

The Gattermann–Koch, Gattermann and Vilsmeier aldehyde syntheses have been reviewed by *Olah* and *Kuhn* in "Friedel–Crafts and Related Reactions", Interscience, New York, 1964, Vol. III, Part II, p. 1153.

(ix) From organo-metallic reagents

A general study has been made of the preparation of aldehydes from Grignard reagents. The use of ethoxymethyleneaniline was found to give higher yields than ethyl orthoformate (*L. E. Smith* and *J. Nichols*, J. org. Chem., 1941, **6**, 489):

$$RMgX + PhN{=}CHOEt \longrightarrow RCH{=}NPh + EtOMgX \xrightarrow{H_2O/H^{\oplus}} RCHO$$

An alternative approach is to react the Grignard reagent with the readily prepared methiodide I followed by acid hydrolysis of the product (*H. M. Fales*, J. Amer. chem. Soc., 1955, **77**, 5118):

$C_6H_4Me[4]$, N, ⊕N, Me, $I^{\ominus}$ (I) $\xrightarrow{RMgX}$ $C_6H_4Me[4]$, N, R, H, N, Me $\xrightarrow{H_2O/H^{\oplus}}$ RCHO

Another method for formylating Grignard reagents is by use of 2-oxazolines in the presence of hexamethylphosphoramide as illustrated below (*A. I. Meyers* and *E. W. Collington*, *ibid.*, 1970, **92**, 6672):

$PhCH_2{\cdot}CH_2MgCl{\cdot}2\,HMPA$ + Me, Me, O, ⊕N, Me, $I^{\ominus}$ ⟶ Me, Me, O, H, N, $CH_2{\cdot}CH_2Ph$, Me

$$\xrightarrow{H_2O/H^{\oplus}} PhCH_2{\cdot}CH_2{\cdot}CHO \quad (87\%)$$

Reaction of chloral with arylmagnesium bromides gives aryltrichloromethylmethanols and these on hydrolysis with aqueous carbonate give a mixture of aromatic aldehyde and arylglycollic acid, from which further amounts of aldehyde can be obtained (*M. Savariau*, Compt. rend., 1908, **146**, 297):

$$RMgX + CCl_3{\cdot}CHO \longrightarrow RCHOH{\cdot}CCl_3 \begin{cases} \nearrow RCHO + CHCl_3 \\ \searrow RCHOH{\cdot}CO_2H \end{cases}$$

Phenylmagnesium bromide has been converted into phenylacetaldehyde by the intermediate formation of allyl benzene (*E. B. Hersberg*, Helv., 1934, **17**, 351).

$$PhMgBr \xrightarrow{CH_2Br{\cdot}CH{=}CH_2} PhCH_2{\cdot}CH{=}CH_2 \longrightarrow PhCH_2{\cdot}CHO$$

The reaction of iron pentacarbonyl with aryl-lithiums at -50 to $-60°$ yields the corresponding aromatic aldehyde in good yield (*M. Ryang, I. Rhee* and *S. Tsutsumi*, Bull. chem. Soc. Japan, 1964, **37**, 341).

The addition of organolithium compounds to isocyanides and treatment of the resultant 1-lithioaldimine with deuterium oxide provides a convenient route to C-1 deuterated aldehydes:

$$R{-}N{=}C \xrightarrow{R^1Li} R{-}N{=}C{<}^{Li}_{R^1} \xrightarrow[(2)\ H_3O^{\oplus}]{(1)\ D_2O} R^1CDO$$

The isocyanide of choice is 1,1,3,3-tetramethylbutyl isocyanide (*H. M. Walborsky* and *G. E. Niznik*, J. Amer. chem. Soc., 1969, **91**, 7778). Full experimental details of this method are given by *Niznik, W. H. Morrison* and *Walborsky* (Org. Synth., 1972, **51**, 31).

(x) From dihydro-1,3-oxazines

(*1*) Substituted acetaldehydes may be prepared from 2,4,4,6-tetramethyl-4*H*,5*H*,6*H*-1,3-oxazine (*Meyers et al.*, J. Amer. chem. Soc., 1969, **91**, 763; see *idem*, Org. Preparations and Procedures, 1969, **1**, 193 for experimental details) as illustrated by the following sequence:

Me, Me, Me, Me (2,4,4,6-tetramethyl-dihydro-1,3-oxazine) $\xrightarrow[(2)\ R^1X]{(1)\ C_4H_9Li}$ 2-CH_2R^1-4,4,6-trimethyl-dihydro-1,3-oxazine $\xrightarrow{NaBH_4}$ 2-CH_2R^1-4,4,6-trimethyl-tetrahydro-1,3-oxazine (N–H) $\xrightarrow[acid]{aqueous\ oxalic}$ $OCH{\cdot}CH_2R^1$

The oxazine is commercially available and can be readily prepared by the method of *E. J. Tillmanns* and *J. J. Ritter*, J. org. Chem., 1957, **22**, 839). Other oxazines, for example the corresponding 2-vinyloxazine, have also been utilised for aldehyde synthesis (*A. C. Kovelsky* and *Meyers*, Org. Preparations and Procedures, 1969, **1**, 213):

$$\text{2-vinyl-4,4,6-trimethyl-oxazine} \xrightarrow[\text{MeI}]{\text{2 PhMgBr}} \text{2-(CHMe}\cdot\text{CH}_2\text{Ph)-oxazine} \xrightarrow[(2)\ \text{H}_2\text{O}-\text{H}^{\oplus}]{(1)\ \text{NaBH}_4} \text{OHC}\cdot\text{CHMe}\cdot\text{CH}_2\text{Ph}$$

(66% overall yield)

2-(1-Phenylvinyl)-4,4,6-trimethyl-4*H*,5*H*,6*H*-1,3-oxazine and the 2-benzyl derivative are convenient starting materials for the synthesis of α-phenylaldehydes (*idem*, Tetrahedron Letters, 1969, 4809; *I. R. Politzer* and *Meyers*, Org. Synth., 1972, **51**, 24).

A further development of this approach, which has the advantage of avoiding low-temperature steps, involves the intermediate preparation of a ketene-*N*,*O*-acetal (*Meyers* and *N. Nazarenko*, J. Amer. chem. Soc., 1972, **94**, 3243):

$$\xrightarrow{\text{MeI}} \quad \xrightarrow{\text{NaH}} \quad \xrightarrow{\text{RX}} \quad \xrightarrow[(2)\ \text{H}_2\text{O}-\text{H}^{\oplus}]{(1)\ \text{NaBH}_4} \text{RCH}_2\cdot\text{CHO}$$

(*2*) 2,4-Dimethylthiazole has also been utilised for aldehyde synthesis, the major advantage of this route is the neutral method of aldehyde release from the thiazolidine intermediate (*L. J. Altman* and *S. Richheimer*, Tetrahedron Letters, 1971, 4709; *cf. Meyers, R. Munavu* and *J. Durandetta*, *ibid.*, 1972, 3929):

$$\xrightarrow[(2)\ \text{PhCH}_2\text{Cl}]{(1)\ \text{BuLi}} \quad \xrightarrow{\text{Me}_3\overset{\oplus}{\text{O}}\text{BF}_4^{\ominus}} \quad \xrightarrow{\text{NaBH}_4} \quad \xrightarrow{\text{HgCl}_2} \text{PhCH}_2\cdot\text{CH}_2\cdot\text{CHO}$$

(*3*) A similar approach to aldehyde and ketone synthesis involves the use of 1,3-dithianes. The carbanions generated with *n*-butyl-lithium from 1,3-dithianes undergo ready alkylation; a final step of hydrolysis is required (*E. J. Corey* and *D. Seebach*, Org. Synth., 1970, **50**, 72; Angew. Chem., internat. Edn., 1965, **4**, 1075 and refs. cited therein); *W. Huurdeman* and *H. Wynberg* (Synthetic Comm., 1972, **2**, 7) recommend chloramine-T as a reagent for the conversion of the substituted 1,3-dithianes to carbonyl compounds:

$$\xrightarrow[(2)\ \text{R'X}]{(1)\ \text{BuLi}} \quad \longrightarrow \text{RR'C=O}$$

(*4*) Aldehydes have also been prepared from trithiane and primary alkyl halides (*Seebach* and *A. K. Beck*, Org. Synth., 1972, **51**, 39) *e.g.*:

$$\text{trithiane (CH}_2\text{)} \xrightarrow{\text{BuLi}} \text{trithianyl-Li} \xrightarrow{PhCH_2Br} \text{2-benzyl-trithiane}$$

$$\xrightarrow{MeOH-HgO-HgCl_2} PhCH_2{\cdot}CH(OMe)_2 \xrightarrow{H_2O-H^{\oplus}} PhCH_2{\cdot}CHO$$

(*5*) A further extension of this approach is to generate a carbanion by treatment of methyl methylthiomethyl sulphoxide with sodium hydride and react this species with benzyl halide to give an aldehyde dimethylthioacetal-*S*-oxide; the latter can be hydrolysed with a catalytic amount of acid to aldehyde and dimethyl disulphide (*K. Ogura* and *G. Tsuchihashi*, Tetrahedron Letters, 1971, 3151):

$$MeSOCH_2SMe \xrightarrow[(2)\ ArCH_2Br]{(1)\ NaH-THF} ArCH_2{\cdot}CH\begin{matrix} SMe \\ SOMe \end{matrix} \xrightarrow{H_2O-H^{\oplus}}$$

$$ArCH_2CHO + MeS{\cdot}SMe$$

(xi) From olefins

(*1*) The synthesis of aldehydes from olefins *via* ozonolysis and reduction of the ozonide, and by oxidation of 1,2-glycols with lead tetra-acetate or periodic acid has been reviewed by *O. Bayer* in *Houben–Weyl*, Vol. 7/1, p. 333 and 335. Sodium metaperiodate, in the presence of osmium tetroxide, is a useful reagent for the oxidation of olefins to aldehydes (*R. Pappo et al.*, J. org. Chem., 1956, **21**, 478).

(*2*) Bistriphenylsilyl chromate [$Ph_3SiOCr(=O)_2OSiPh_3$] oxidatively cleaves olefins giving the corresponding aldehydes and ketones

$$PhCH=CH_2 \rightarrow PhCHO + H_2CO$$

The reagent is easily prepared from triphenylsilanol and chromium trioxide (*L. M. Baker* and *W. L. Carrick*, *ibid.*, 1970, **35**, 774).

(*3*) Aldehydes may be prepared from olefins by a two step procedure involving first hydroboration using 9-borabicyclo[3.3.1]nonane. Treatment of the intermediate borane with carbon monoxide in the presence of lithium hydridotrimethoxyaluminate and oxidation of the product with buffered hydrogen peroxide gives the aldehyde (*H. C. Brown*, *E. F. Knights* and *R. A. Coleman*, J. Amer. chem. Soc., 1969, **91**, 2144):

$$PhCH=CH_2 + HB(C_8H_{14}) \longrightarrow PhCH_2{\cdot}CH_2{-}B(C_8H_{14}) \longrightarrow \underset{(84\,\%)}{PhCH_2{\cdot}CH_2{\cdot}CHO}$$

9-Borabicyclo[3.3.1]nonane (9-BBN) is readily prepared from 1,5-cyclooctadiene.

(*4*) 3-Arylaldehydes and ketones are generally available from the arylation of allylic alcohols with organopalladium compounds as illustrated by the following synthesis of 2-methyl-3-phenylpropionaldehyde (*R. F. Heck*, Org. Synth., 1972, **51,** 17):

$$PhHgOAc + Pd(OAc)_2 \longrightarrow [PhPdOAc] + Hg(OAc)_2$$

$$[PhPdOAc] + CH_2{=}C(CH_3){-}CH_2OH \longrightarrow PhCH_2{-}C(CH_3)(PdOAc){-}CH_2OH \longrightarrow$$

$$PhCH_2{-}CH(CH_3){-}CHO + Pd + AcOH$$

(*5*) Aldehydes are prepared by the oxidative aromatisation of the substituted cyclohexenes derived from Diels–Alder addition of 1,3-dienes and α,β-unsaturated carbonyl compounds (*J. C. Leffingwell* and *H. J. Bluhm*, Chem. Comm., 1969, 1151) *e.g.*[7

(diene, Me) + (CH₂=CH·CHO) → (cyclohexene, CHO, Me) (mixed isomers) —MnO_2→ (benzene ring, CHO, Me)

(xii) From α,β-unsaturated amides and nitriles

(*1*) The amides of some α,β-unsaturated acids are converted into aldehydes by the following route (*R. A. Weerman*, Ann., 1913, **401,** 1):

$$ArCH{=}CH{\cdot}CONH_2 \xrightarrow{NaOCl/MeOH} ArCH{=}CHNHCO_2Me \xrightarrow{6N\,H_2SO_4} ArCH_2{\cdot}CHO$$

A related synthesis is illustrated by the conversion of benzylmalonic ester to phenylacetaldehyde (*T. Curtius*, J. prakt. Chem., 1916, [ii], **94,** 323):

$$PhCH_2{\cdot}CH(CO_2Et)_2 \xrightarrow{NH_2{\cdot}NH_2} PhCH_2{\cdot}CH(CONH{\cdot}NH_2)_2 \xrightarrow[(2)\ EtOH]{(1)\ NaNO_2}$$

$$PhCH_2{\cdot}CH(NHCO_2Et)_2 \xrightarrow{2\%\ H_2SO_4} PhCH_2{\cdot}CHO$$

(*2*) α,β-Unsaturated nitriles are transformed into saturated aldehydes by reaction with octacarbonyldicobalt in acidic methanol (*H. Wakamatsu* and *K. Sakamaki*, Chem. Comm., 1967, 1140) *e.g.*:

$$PhCH{=}CH{\cdot}CN + Co_2(CO)_8 + 5\ HCl + H_2O \rightarrow$$

$$PhCH_2{\cdot}CH_2{\cdot}CHO + 2\ CoCl_2 + NH_4Cl + 8\ CO$$
(50%)

(xiii) From 1,2-glycols

Aldehydes may be prepared by rearrangement of suitable 1,2-glycols, or of the corresponding epoxides or halogenohydrins. These rearrangements are illustrated by the following examples:

$$\mathrm{Me_2C(OH){-}CH(OH)Ph} \xrightarrow{20\%\ H_2SO_4} \mathrm{Me_2C(Ph){-}CHO} + \mathrm{H_2O}$$

(*J. Levy* and *R. Pernot*, Bull. Soc. chim. Fr., 1931, [4], **49,** 1721):

$$\mathrm{PhCH_2(Me)C{-}CH_2\ (epoxide)} \xrightarrow{HCO_2H} \mathrm{PhCH_2{\cdot}CHMe{\cdot}CHO}$$

(*P. Ramart-Lucas* and *L. Labaune*, Ann. Chim. Fr., 1931, [x], **16,** 276):

$$\mathrm{PhCHI{\cdot}CH_2OH} \xrightarrow{AgNO_3} \mathrm{PhCH_2{\cdot}CHO}$$

Cleavage of 1,2-diols to aldehydes has been achieved in high yields using mercuric oxide and iodine, *e.g.* 1,2-dihydroxy-1,2-diphenylethane is converted in 98% yield to benzaldehyde, and 4,5-dihydroxy-1,8-diphenyloctane is converted in 99% yield to 4-phenylbutanal (*A. Goosen* and *H. A. H. Laue*, J. chem. Soc., C, 1969, 383). Silver-catalysed persulphate also is an effective reagent for cleaving 1,2-diols (*F. P. Greenspan* and *H. M. Woodburn*, J. Amer. chem. Soc., 1954, **76,** 6345).

(xiv) From aldehydes and ketones

(*1*) In the *Darzen's glycidic ester condensation*, α,β-epoxyesters (glycidic esters) are formed by the base-catalysed condensation of an aldehyde or a ketone with an α-halogeno ester. The glycidic ester on hydrolysis and decarboxylation yields an aldehyde or ketone of higher carbon content than the initial aldehyde or ketone (*C. F. H. Allen* and *J. Van Allan*, Org. Synth., Coll. Vol. 3, 1955, pp. 727 and 733):

$$\mathrm{PhCOMe} + \mathrm{ClCH_2{\cdot}CO_2Et} \xrightarrow{NaNH_2} \mathrm{Ph(Me)C{-}CH{\cdot}CO_2Et\ (epoxide)}\ (62\text{–}64\%)$$

$$\xrightarrow[(2)\ HCl]{(1)\ EtONa} \mathrm{PhCHMe{\cdot}CHO}\ (65\text{–}70\%)$$

The scope of this reaction has been reviewed by *M. S. Newman* and *B. J. Magerlein* (Org. Reactions, 1949, **5,** 413).

(*2*) Aromatic ketones of the type $\mathrm{PhCO{\cdot}CH_2R}$ are transformed into benzaldehyde and an aldehyde RCHO by acid-catalysed reaction with certain alkyl azides (*J. H. Boyer* and *L. R. Morgan*, J. Amer. chem. Soc., 1959, **81,** 3309):

$$\mathrm{PhCO{\cdot}CH_2{\cdot}CH_3} \xrightarrow{n\text{-}BuN_3\text{–}H_2SO_4\text{–}toluene} \mathrm{PhCHO}\ (62\%) + \mathrm{MeCHO}\ (50\%)$$

(*3*) The Wittig reaction may be utilised for aldehyde synthesis as follows (*S. G. Levine*, *ibid.*, 1958, **80,** 6150; *G. Wittig* and *E. Knauss*, Angew. Chem.,

1959, **71,** 127; see also *Wittig, W. Boll* and *K.-H. Krück*, Ber., 1962, **95,** 2514):

$$R_2CO + MeOCH{=}PPh_3 \longrightarrow R_2C{=}CHOMe \xrightarrow{H_2O/H^{\oplus}} R_2CH{\cdot}CHO$$

(*4*) Treatment of an aldehyde $R_2CH{\cdot}CHO$ with bromine and cyanide gives an epoxynitrile which on further treatment with trimethylaluminium gives an aldehyde $R_2CMe{\cdot}CHO$ (*J. Cantacuzène* and *J.-M. Normant*, Tetrahedron Letters, 1970, 2947):

$$PhCH(Me){\cdot}CHO \xrightarrow[(2)\ CN^{\ominus}]{(1)\ Br_2} PhC(Me)\overset{O}{—}CH{\cdot}CN \xrightarrow[(2)\ \Delta]{(1)\ AlMe_3} Ph{-}C(Me)_2{-}CHO$$

(xv) Miscellaneous methods

(*1*) B-*Alkyl-9-borabicyclo*[3.3.1]*nonanes* react readily with carbon monoxide in the presence of lithium tri-*tert*-butoxyhydridoaluminate without reduction of functional substituents (*H. C. Brown* and *R. A. Coleman*, J. Amer. chem. Soc., 1969, **91,** 4606) *e.g.*:

$$m\text{-}O_2NC_6H_4{-}CH_2{\cdot}CH_2{-}B\langle + CO + LiAlH(OBu^t)_3 \longrightarrow$$

$$m\text{-}O_2NC_6H_4{-}CH_2{\cdot}CH_2{-}CH(OAl(OBu^t)_3){-}B\langle \xrightarrow{[O]} m\text{-}O_2NC_6H_4{-}CH_2{\cdot}CH_2{\cdot}CHO$$

(*2*) Aldehydes are formed by the acid decomposition of *nitronic esters* or the corresponding salts of *primary nitroparaffins*. The nature of the products is strongly dependent on the acid concentration (*N. Kornblum* and *R. A. Brown, ibid.*, 1965, **87,** 1742):

$$p\text{-}O_2N{-}C_6H_4{-}CH{=}NO_2Et \xrightarrow{4\,N\ H_2SO_4} p\text{-}O_2N{-}C_6H_4{-}CHO \quad (80\%)$$

$$p\text{-}O_2N{-}C_6H_4{-}CH{=}NO_2Et \xrightarrow{31\,N\ H_2SO_4} p\text{-}O_2N{-}C_6H_4{-}CONHOH \quad (98\%)$$

Alternatively the salts of primary nitroparaffins can be converted into aldehydes by oxidation with potassium permanganate in the presence of magnesium sulphate (*H. Shecter* and *F. T. Williams Jr.*, J. org. Chem., 1962, **27,** 3699).

(*3*) *Benzyl ethers* are oxidised in high yield to benzaldehyde with aqueous bromine at 25° (*N. C. Deno* and *N. H. Potter*, J. Amer. chem. Soc., 1967, **89,** 3550; see also *D. G. Markees*, J. org. Chem., 1958, **23,** 1490) *e.g.*:

$$PhCH_2OMe \xrightarrow{Br_2-MeCO_2H-H_2O} PhCHO \quad (83\%)$$

(4) *Electrolysis of α-ethoxyphenylacetic acid* in methanol gives an acetal in 71% yield which is readily hydrolysed to benzaldehyde (*B. Wladislaw* and *A. M. J. Ayres, ibid.*, 1962, **27**, 281):

$$PhCH(OEt)\cdot CO_2H \xrightarrow{MeOH} PhCH(OMe)(OEt)\ (71\%) \xrightarrow{2\%\ HCl} PhCHO$$

(b) General properties

Benzaldehyde and the homologous aromatic aldehydes are liquids with an aromatic odour.

4-Amino-3-hydrazino-5-mercapto-1,2,4-triazole is a sensitive and specific reagent for the detection of aldehydes. The purple solid produced on reaction with benzaldehyde has been characterised as 6-mercapto-3-phenyl-*s*-triazolo[4,3-*b*]-*s*-tetrazine (*R. G. Dickinson* and *N. W. Jacobsen*, Chem. Comm., 1970, 1719):

$$\text{4-amino-3-hydrazino-5-mercapto-1,2,4-triazole} + PhCHO \longrightarrow \text{6-mercapto-3-phenyl-}s\text{-triazolo[4,3-}b\text{]-}s\text{-tetrazine}$$

Aromatic aldehydes have weakly basic properties; benzaldehyde is half protonated in 82% sulphuric acid. The carbonyl-stretching absorption in aromatic aldehydes is in the region 1695–1715 cm^{-1}. The chemical shift of the aldehydic proton is in the range 9.7–10.5δ (-0.5–0.3 τ). The *ortho* protons are deshielded by the carbonyl group. The ultraviolet spectrum of benzaldehyde in alcohol shows a maximum at 244 mμ ($\log_{10}\varepsilon$ 4.2) and the corresponding band in the spectrum of 2,4,6-trimethylbenzaldehyde has λ_{max} 265 mμ ($\log_{10}\varepsilon$ 4.1). This indicates that the *ortho* substituents do not lead to large deviations from coplanarity (*N. J. Leonard* and *E. R. Blout*, J. Amer. chem. Soc., 1950, **72**, 484).

The mass spectrum of benzaldehyde shows a prominent molecular ion, loss of aldehydic hydrogen gives a $(M-1)^{\oplus}$ ion which then expels carbon monoxide to give the phenyl cation (*m/e* 77). Further decomposition by elimination of acetylene gives a $C_4H_3^{\oplus}$ ion (*m/e* 51), The mass spectra of aromatic aldehydes are discussed by *H. Budzikiewicz, C. Djerassi* and *D. H. Williams* in "Interpretation of Mass Spectra of Organic Compounds", Holden-Day, San Francisco, 1964, p. 191.

(c) Reactions of aromatic aldehydes

Aromatic aldehydes undergo most of the general addition reactions common to aliphatic aldehydes (see *C. C. C.*, Vol. I C, p. 10 *et seq.*). For example, they form normal adducts with hydrogen cyanide, sodium bisulphite, Grignard reagents and related organometallic derivatives. Their carbonyl reactivity judged by measurements on the equilibria

$$ArCHO + HCN \rightleftharpoons ArCHOH{\cdot}CN$$

is lower than that of aliphatic aldehydes. Electron-withdrawing substituents in the *para*-position such as chlorine and nitro enhance the carbonyl reactivity, whereas electron- releasing substituents such as methoxyl and dimethylamino have the opposite effect.

Aromatic aldehydes are readily characterised by oxime, hydrazone, and semicarbazone formation. With *N*-substituted hydroxylamines they yield nitrones:

$$ArCHO + RNHOH \rightarrow ArCH{=}\overset{\overset{O^{\ominus}}{|}}{\underset{\oplus}{N}}{-}R + H_2O$$

Aldoximes are conveniently converted into the parent aldehyde by treatment with thallium nitrate in methanol (*A. McKillop et al.*, J. Amer. chem. Soc., 1971, **93,** 4918). Nitriles are produced in good yields when aldoximes are treated with 2,4,6-trichloro-*s*-triazine in the presence of pyridine at room temperature (*J. B. Chakrabarti* and *T. M. Hotten*, Chem. Comm., 1972, 1226).

Aromatic aldehydes react with ammonia differently from aliphatic aldehydes, thus benzaldehyde reacts with ammonia on a 3:2-molar basis and with elimination of water to give hydrobenzamide (*D. H. Hunter* and *S. K. Sim*, Canad. J. Chem., 1972, **50,** 669, 678):

$$3\ PhCHO + 2\ NH_3 \rightarrow PhCH{=}NCHPhN{=}CHPh + 3\ H_2O$$

The condensation products of aromatic aldehydes with both aromatic and aliphatic primary amines are known as Schiff bases:

$$ArCHO + NH_2R \rightarrow ArCH{=}NR + H_2O$$

They are more stable than the corresponding products derived from aliphatic aldehydes. The reaction of benzaldehyde and aqueous dimethylamine gives benzylidene bis-dimethylamine (*S. V. Lieberman*, J. Amer. chem. Soc., 1955, **77,** 1114). Aromatic aldazines are conveniently prepared by heating the aldehyde with hydrazine hydrate in polyphosphoric acid at 100°, this method is also applicable to the preparation of ketazines (*D. B. Mobbs* and *H. Suschitzky*, J. chem. Soc., C, 1971, 175). Aromatic aldehydes unlike aliphatic aldehydes do not polymerise, though under the influence of alkali cyanide they form dimeric products (see Benzoin reaction, p. 76). Some special points regarding the reactivity of aromatic aldehydes are discussed below.

(i) Oxidation and reduction

(*1*) Aromatic aldehydes are easily oxidised to the corresponding carboxylic acids; molecular oxygen, silver oxide, permanganate, dilute nitric acid, and organic peracids are some of the oxidising agents which have been used. They are not, however, as easily oxidised as aliphatic aldehydes, for example, benzaldehyde does not reduce Fehling's solution.

(*2*) Reduction of aromatic aldehydes can occur to yield alcohols, hydrocarbons or glycols according to the experimental conditions. Catalytic hydrogenation in the liquid phase with nickel or platinum catalysts gives the primary alcohol as the first product though under more vigorous conditions it is possible to obtain a hydrocarbon by complete reduction of formyl to methyl. Reduction of aromatic aldehydes to toluenes has also been achieved by use of lithium, liquid ammonia and tetrahydrofuran (*S. S. Hall, A. P. Bartels* and *A. M. Engman*, J. org. Chem., 1972, 37, 760). The reduction of aldehyde to primary alcohol may also be achieved with lithium tetrahydridoaluminate or sodium tetrahydridoborate. Aluminium alkoxides are also useful reagents for the selective reduction of aromatic aldehydes (*A. L. Wilds*, in Org. Reactions, 1944, **2,** 178):

$$[2]NO_2C_6H_4{\cdot}CHO \xrightarrow[HOPr^i]{Al(OPr^i)_3} \underset{(92\%)}{[2]NO_2C_6H_4{\cdot}CH_2OH}$$

When aromatic aldehydes react with metallic sodium in ether solution, pinacols are formed:

$$2\ PhCHO + 2\ Na \rightarrow PhCHONa{\cdot}CHPhONa \rightarrow PhCHOH{\cdot}CHPhOH$$

A novel method for pinacol formation involves treating benzaldehyde with magnesium turnings and trimethylchlorosilane in hexamethylphosphoramide (HMPA) (*T. H. Chan* and *E. Vinokur*, Tetrahedron Letters, 1972, 75).

The Wolff–Kishner method can be used for the reduction of formyl to methyl (*D. Todd*, in Org. Reactions, 1948, **4,** 379):

$$ArCHO \longrightarrow ArCH{=}N{\cdot}NH_2 \xrightarrow[or\ KOH]{NaOEt} ArCH_3 + N_2$$

Phenylcarbene (or a carbenoid species) is generated from benzaldehyde by reaction with zinc in dry ether in the presence of boron trifluoride. It can be trapped by reaction with simple olefins to give phenylcyclopropanes:

$$\text{PhCHO} \xrightarrow{\text{Zn-ether-BF}_3} \text{PhCH:} \xrightarrow{\text{RCH=CHR}} \text{PhCH}\begin{matrix}\text{CHR}\\ |\\ \text{CHR}\end{matrix}$$

(ii) The Cannizzaro reaction

In the presence of aqueous or alcoholic alkali, aromatic aldehydes undergo an intermolecular oxidation–reduction reaction to yield the salt of the corresponding acid and the corresponding primary alcohol *e.g.*

$$2\ \text{PhCHO} \quad + \quad \text{OH}^{\ominus} \quad \rightarrow \quad \text{PhCO}_2{}^{\ominus} \quad + \quad \text{PhCH}_2\text{OH}$$

This reaction is observed with nearly all aromatic aldehydes. Di-*ortho*-substituted benzaldehydes fail to react normally however and when there are halogen or nitro substituents in both *ortho* positions, the formyl group is removed as formate (*T. A. Geissman*, in *ibid.*, 1944, **2,** 94). 2,4-Dinitrobenzaldehyde undergoes an anomalous reaction when treated with dilute sodium hydroxide when it is converted into 2-nitro-4-nitrosophenol and sodium formate (*E. J. Forbes* and *M. J. Gregory*, J. chem. Soc., B, 1968, 207). Because formaldehyde is more easily oxidised than aromatic aldehydes, the latter can be reduced to the corresponding alcohol by heating a mixture of the aromatic aldehyde and formaldehyde with concentrated aqueous sodium hydroxide solution, *e.g.:*

$$\text{PhCHO} \quad + \quad \text{HCHO} \quad + \quad \text{OH}^{\ominus} \quad \rightarrow \quad \text{PhCH}_2\text{OH} \quad + \quad \text{HCO}_2{}^{\ominus}$$

Benzaldehyde is converted into benzyl benzoate by the catalytic action of an alkoxide *(Tischenko reaction)* (*O. Kamm* and *W. F. Kamm*, Org. Synth., Coll. Vol. 1, 2nd Edn., 1947, p. 104):

$$\text{PhCHO} \xrightarrow{\text{PhCH}_2\text{ONa}} \underset{(92\%)}{\text{PhCH}_2\text{OCOPh}} \qquad (92\%)$$

(iii) Benzoin reaction

Benzaldehyde and its homologues are converted into acyloins by the action of a small amount of potassium cyanide in boiling alcohol. For example, benzaldehyde is transformed in 92% yield to benzoin, PhCHOH·COPh, in this manner (*R. Adams* and *C. S. Marvel*, *ibid.*, 1947, p. 94). The scope of this reaction is reviewed by *W. S. Ide* and *J. S. Buck*, in Org. Reactions, 1948, **4,** 272 *et seq.*).

(iv) Condensation reactions

Aromatic aldehydes condense with a large variety of compounds containing an active methylene group, generally to form an unsaturated compound:

$$ArCHO + CH_2R'R'' \rightarrow ArCHOH \cdot CHR'R'' \rightarrow ArCH{=}CR'R'' + H_2O$$

although reaction may sometimes occur with two molecules of the methylene compound:

$$ArCHO + 2\ CH_2R'R'' \rightarrow ArCH(CHR'R'')_2 + H_2O$$

Condensation with benzaldehyde often serves as a diagnostic test for an active methylene group but 4-nitrobenzaldehyde and 4-chlorobenzaldehyde are superior for this purpose since they react more readily. The optimum conditions for the condensation of benzaldehyde with 4-nitrotoluene are reaction in dimethylformamide in the presence of sodium hydroxide at room temperature (*W. J. Farrissey, F. P. Recchia* and *A. A. R. Sayigh*, J. org. Chem., 1969, **34**, 2785). The following transformations are illustrative of the wide range of reactions in this category:

$$PhCHO + CH_3{\cdot}CO_2Et \xrightarrow{NaOEt} \underset{(68-74\%)}{PhCH{=}CH{\cdot}CO_2Et} + H_2O$$

(*C. S. Marvel* and *W. B. King*, Org. Synth., Coll. Vol. 1, 2nd Edn., 1947, p. 252);

$$PhCHO + CH_2N_3CO_2Et \xrightarrow{NaOEt} \underset{(43\%)}{PhCH{=}C(N_3)CO_2Et} + H_2O$$

(*H. Hemetsberger, D. Knittel* and *H. Weidmann*, Monatsh., 1969, **100**, 1599);

$$PhCHO + PhCH_2{\cdot}CN \xrightarrow{NaOEt} \underset{(83-91\%)}{PhCH{=}C(CN){\cdot}Ph} + H_2O$$

(*S. Wawzonek* and *E. M. Smolin*, Org. Synth., Coll. Vol. 3, 1955, p. 715);

$$PhCHO + CH_3{\cdot}COPh \xrightarrow{NaOH} \underset{(85\%)}{PhCH{=}CH{\cdot}COPh} + H_2O$$

(*E. P. Kohler* and *H. M. Chadwell*, *ibid*., Coll. Vol. 1, 2nd Edn., 1947, p. 78);

$$2\ PhCHO + CH_3{\cdot}CO{\cdot}CH_3 \xrightarrow{NaOH} \underset{(90-94\%)}{PhCH{=}CH{\cdot}CO{\cdot}CH{=}CHPh} + 2\ H_2O$$

(*C. R. Conrad* and *M. A. Dolliver*, *ibid*., Coll. Vol. 2, 1943, p. 167);

$$PhCHO + CH_3NO_2 \xrightarrow[(2)\ HCl]{(1)\ NaOH} \underset{(80-83\%)}{PhCH{=}CHNO_2} + H_2O$$

(*D. E. Worrall*, *ibid*., Coll. Vol. 1, 2nd Edn., 1947, p. 413);

$$PhCHO + CH_2(CO_2Et)_2 \xrightarrow[\text{benzoate}]{\text{Piperidine}} PhCH{=}C(CO_2Et)_2 + H_2O \quad (89\text{–}91\%)$$

(*C. F. H. Allen* and *F. W. Spangler*, *ibid.*, Coll. Vol. 3, 1955, p. 377);

$$PhCHO + (CH_2 \cdot CO_2Et)_2 \xrightarrow[\text{refluxing ether}]{\text{NaOEt}} PhCH{=}C(CO_2H) \cdot CH_2 \cdot CO_2H + H_2O \quad (35\%)$$

(*H. Stobbe*, Ber., 1908, **41**, 4350);

$$PhCHO + Ac_2O \xrightarrow{\text{AcOK}} PhCH{=}CH \cdot CO_2H + AcOH \quad (55\text{–}60\%)$$

(*J. R. Johnson*, Org. Reactions, 1942, **1**, 248);

$$PhCHO + CH_2(NHAc) \cdot CO_2H \xrightarrow[Ac_2O]{\text{AcONa}} PhCH{=}C{-}C{=}O \text{ (oxazolone ring: N=C(CH}_3\text{)–O)} + 2\,H_2O \quad (74\text{–}77\%)$$

(*R. M. Herbst* and *D. Shemin*, Org. Synth., Coll. Vol. 2, 1943, p. 1).

Benzaldehyde and chloroform react together in the presence of powdered potassium hydroxide or in presence of potassium *tert*-butoxide in butanol to give phenyl-(trichloromethyl)methanol. This is a general reaction of aromatic aldehydes:

$$PhCHO + CHCl_3 \xrightarrow{\text{KOH}} PhCH(OH) \cdot CCl_3$$

A development of this reaction has led to a one-step synthesis of α-methoxyphenylacetic acids. This involves adding methanolic potassium hydroxide to a mixture of the aromatic aldehyde and either chloroform or bromoform:

$$R{-}C_6H_4{-}CHO + CHX_3 + CH_3OH \xrightarrow{\text{KOH}} R{-}C_6H_4{-}CH(OCH_3){-}COOH$$

The methoxy acid can be prepared alternatively by the action of methanolic potassium hydroxide on the intermediate aryltrichloromethylmethanol (*W. Reeve*, Synthesis. 1971, 131).

Phenylacetic acid derivatives can be obtained from aromatic aldehydes using methyl methylthiomethyl sulphoxide (*K. Ogura* and *G. Tsuchihashi*, Tetrahedron Letters, 1972, 1383):

$$ArCHO + MeSCH_2SOMe \xrightarrow{\text{Triton B}} (Ar)(H)C{=}C(SMe)(SOMe) \xrightarrow{H^{\oplus} - ROH} ArCH_2 \cdot CO_2R$$

Aromatic aldehydes are converted into ketene dithioacetals by the following route; these are useful synthetic derivatives which can be converted into acids, aldehydes and acetylenes (*D. Seebach et al.*, Angew. Chem. internat. Edn., 1972, **11,** 443):

$$\text{1,3-dithiane} \xrightarrow[(2)\ Me_3SiCl]{(1)\ BuLi} \text{2-H-2-}SiMe_3\text{-1,3-dithiane} \xrightarrow[(2)\ ArCHO]{(1)\ BuLi} Ar(H)C{=}\text{(1,3-dithian-2-ylidene)}$$

Related condensation reactions of great synthetic utility are the *Wittig reaction* (*A. Maercker*, Org. Reactions, 1965, **14,** 396) *e.g.:*

$$PhCHO + Ph_3P{=}C_5H_8 \longrightarrow PhCH{=}C_5H_8\ (65\%) + Ph_3PO$$

and the *Reformatzky reaction* (*C. R. Hauser* and *D. S. Breslow*, Org. Synth., Coll. Vol. 3, 1955, 408), *e.g.:*

$$PhCHO + BrCH_2{\cdot}CO_2Et \xrightarrow{Zn} PhCHOH{\cdot}CH_2{\cdot}CO_2Et \quad (61\text{–}64\%)$$

An alternative procedure for forming β-hydroxy esters is to use an ethoxyvinyl ester (*H. H. Wasserman* and *S. H. Wentland*, Chem. Comm., 1969, 1216):

$$ArCHO + H_2C{=}C(OEt){-}OCOR \longrightarrow ArCH(OCOR){\cdot}CH_2{\cdot}CO_2Et \xrightarrow{\text{dilute KOH}} ArCHOH{\cdot}CH_2{\cdot}CO_2Et$$

A further alternative to the Reformatzky reaction is to condense thermally a silylated ketene acetal with an aromatic aldehyde followed by acid hydrolysis of the product (*P. L. Greger*, Tetrahedron Letters, 1972, 79):

$$ArCHO + (H)(R)C{=}C(OSiMe_3)(OR') \xrightarrow{\Delta} ArCHOH{\cdot}CR{=}C(OSiMe_3)(OR')$$

$$\xrightarrow{H^{\oplus} - H_2O} ArCHOH{\cdot}CHR{\cdot}CO_2R'$$

A reaction analogous to the Wittig reaction, for the conversion of aldehydes to methylenic olefins is illustrated in the following equation (*G. Cainelli et al.*, *ibid.*, 1967, 5163):

$$PhCHO + CH_2(MgI)_2 \rightarrow PhCH{=}CH_2 + (MgI)_2O$$

For a general discussion of the stereochemical aspects of the above olefin-forming reactions see *H. O. House*, in "Modern Synthetic Reactions", Benjamin, New York, 1965.

Aromatic aldehydes may be converted into dienes by treatment with triphenylcinnamylphosphonium chloride in the presence of lithium ethoxide (*R. N. McDonald* and *T. W. Campbell*, Org. Synth., 1960, **40,** 36):

$$PhCHO + PhCH{=}CHCH_2\overset{\oplus}{P}Ph_3\overset{\ominus}{Cl} \xrightarrow{LiOEt} PhCH{=}CH{-}CH{=}CHPh$$

trans, *trans*-1,4-diphenylbuta-1,3-diene (60-67%)

The condensation of aldehydes with alkoxycarbonyl-alkylidenephosphoranes is a route to the esters of α,β-unsaturated carboxylic acids (*H. J. Bestmann* and *H. Schulz*, Ber., 1962, **95,** 2921) *e.g.:*

$$PhCH_2{\cdot}C(=PPh_3){\cdot}CO_2Et + PhCHO \longrightarrow PhCH_2{\cdot}C(=CHPh){\cdot}CO_2Et \xrightarrow{H_2O/H^{\oplus}} PhCH_2{\cdot}C(=CHPh){\cdot}CO_2H$$

α-benzylcinnamic acid (42%)

Aldehydes react with diethyl β-cyclohexylaminovinylphosphonates in tetrahydrofuran and in the presence of sodium hydride to give α,β-unsaturated aldimines which can be hydrolysed using aqueous oxalic acid to afford α,β-unsaturated aldehydes. The reaction proceeds stereospecifically to give *trans*-formyl olefins (*W. Nagata* and *Y. Hayase*, J. chem. Soc., C, 1969, 460):

$$PhCHO + (EtO)_2P(=O){-}CH{=}CH{-}NH{-}C_6H_{11} \xrightarrow{NaH/THF}$$

$$PhCH{=}CH{-}CH{=}N{-}C_6H_{11} \longrightarrow (Ph)(H)C{=}C(H)(CHO)$$

trans-cinnamaldehyde (77%)

This conversion can also be achieved by the reaction of benzaldehyde with a metallated Schiff base (*G. Wittig* and *H. Reiff*, Angew. Chem. internat. Edn., 1968, **7,** 7):

$$PhCHO + LiCH_2CH{=}NC_6H_{11} \longrightarrow PhCH(OLi)CH_2CH{=}NC_6H_{11}$$

$$\xrightarrow{H_2O/H^{\oplus}} PhCH{=}CH{\cdot}CHO \quad (72\%)$$

The reaction of phosphonate carbanions containing electron-withdrawing groups with aldehydes and ketones constitutes a useful olefin synthesis. These reagents are in general more reactive than the analogous triarylphosphoranes (*W. S. Wadsworth* and *W. D. Emmons*, J. Amer. chem. Soc., 1961, **83,** 1733):

$$PhCHO + (EtO)_2P(=O)CH_2{\cdot}COPh \xrightarrow{NaH} PhCH{=}CH{\cdot}COPh \quad (61\%)$$

Acetonylidenetriphenylphosphorane reacts with aldehydes to furnish α,β-unsaturated methyl ketones (*C. D. Snyder* and *H. Rapoport*, *ibid.*, 1969, **91,** 731):

$$ArCHO + CH_3{\cdot}CO{\cdot}CH{=}PPh_3 \rightarrow ArCH{:}CH{\cdot}CO{\cdot}CH_3$$

A technique for aldehyde to diene conversion involves the generation of an anion from the diazaphospholidine oxide (II) (*E. J. Corey* and *D. E. Crane*, J. org. Chem., 1969, **34,** 3053):

$$\underset{(II)}{CH_2{=}CH{\cdot}CH_2{-}\overset{\ominus}{O}{-}\overset{\oplus}{P}(NH)_2C_2H_4} \xrightarrow[(3)\ H_2O\quad(4)\ \Delta]{(1)\ n\text{-}C_4H_9Li\quad(2)\ PhCHO} CH_2{=}CH{\cdot}CH{=}CHPh$$

A related condensation of aromatic aldehydes with the anion derived from (III) gives an allyl vinyl sulphide which is transformed as shown to an allyl(phenyl)acetaldehyde (*Corey* and *J. I. Schulman*, J. Amer. chem. Soc., 1970, **92,** 5522):

$$\underset{(III)}{CH_2{=}CH{-}CH_2{-}SCH_2{-}\overset{\oplus}{P}(\overset{\ominus}{O})(OEt)_2} \xrightarrow[(2)\ PhCHO\quad(3)\ H_2O]{(1)\ n\text{-}C_4H_9Li} CH_2{=}CH{\cdot}CHSCH{=}CHPh$$

$$\xrightarrow{\Delta\ -\ HgO} PhCH(CHO)CH_2{\cdot}CH{=}CH_2$$

(v) Conversion of aromatic aldehydes to nitriles

This has been achieved by treatment with ammonium dihydrogen phosphate and nitropropane in acetic acid (*H. M. Blatter*, *H. Lukaszewski* and *G. de Stevens*, *ibid.*, 1961, **83,** 2203); hydroxylamine hydrochloride and sodium formate in formic acid (*J. van Es*, J. chem. Soc., 1965, 1564); and ammonia and lead tetra-acetate in benzene (*K. M. Parameswarm* and *O. M. Friedman*, Chem. and Ind., 1965, 988).

Nitriles may also be synthesised from aldehydes using a procedure involving formation of an imine-cobalt complex which is subsequently oxidised with bromine. This reaction is also applicable to α,β-unsaturated and aliphatic aldehydes (*I. Rhee*, *M. Ryang* and *S. Tsutsumi*, Tetrahedron Letters, 1970, 3419):

$$RCHO + Co(NH_3)_6[Co(CO)_4]_2 \longrightarrow \left[\begin{matrix}R{-}CH{=}NH\\ \downarrow\\ Co(NH_3)_5\end{matrix}\right][Co(CO)_4]_2 \xrightarrow{Br_2} RCN$$

(vi) Conversion of aromatic aldehydes to carboxamides

Oxidation of aromatic aldehydes with manganese dioxide in the presence of amines and sodium cyanide leads to the corresponding carboxamides (*N. W. Gilman*, Chem. Comm., 1971, 733):

$$\text{ArCHO} + \text{HNR}_2 \xrightarrow[\text{NaCN}]{\text{MnO}_2} \text{ArCONR}_2$$

(vii) Decarbonylation

The decarbonylation of aromatic aldehydes by heating with 5% palladium-on-carbon appears to be a fairly general reaction (*J. O. Hawthorne* and *M. H. Wilt*, J. org. Chem., 1960, **25,** 2215).

(viii) Aldehydes as acylating reagents

The homolytic acylation of heterocyclic aromatic bases with aldehydes (aromatic and aliphatic) is a general reaction. Acylation of a number of heterocyclic systems has been achieved, it occurs at positions of high nucleophilic reactivity (*T. Caronna et al.*, J. chem. Soc., C, 1971, 1747 and papers cited therein):

$$\text{(benzothiazole)} + \text{PhCHO} \xrightarrow[\text{Bu}^t\text{OOH}]{\text{FeSO}_4} \text{2-COPh-benzothiazole} \quad (69\%)$$

(ix) Halogenation

Aromatic aldehydes undergo electrophilic substitution predominantly in the *meta* position. For example, iodination of benzaldehyde with iodine and silver sulphonate in aqueous sulphuric acid gives 3-iodobenzaldehyde (*A. N. Novikov*, J. Gen. Chem., U.S.S.R., 1954, **24,** 655, 665). Chlorination of benzaldehyde with chlorine and excess of aluminium chloride gives 43% of 3-chlorobenzaldehyde (*D. E. Pearson, H. W. Pope* and *W. W. Hargrove*, Org. Synth., 1960, **40,** 10). However, in the absence of catalyst substitution of the aldehydic hydrogen atom occurs (*Clarke* and *E. R. Taylor*, *ibid.*, Coll. Vol. 1, 2nd Edn., 1947, p. 155) 155) *e.g.*:

$$[2]\text{ClC}_6\text{H}_4\cdot\text{CHO} + \text{Cl}_2 \xrightarrow{140\text{–}160^\circ} [2]\text{ClC}_6\text{H}_4\cdot\text{COCl} + \text{HCl} \quad (70\text{–}72\%)$$

(x) Reactions of aromatic aldehydes and diazoalkanes

These have been reviewed by *C. D. Gutsche*, in Org. Reactions, 1954, **8,** 364. The products depend on the solvent and on the substituents in the ring. In the absence of methanol, reaction of benzaldehyde and diazomethane gives 97% of acetophenone; when electron-withdrawing groups are present, particularly in the *ortho* or *para* positions the product is predominantly the oxirane (styrene oxide) *e.g.*:

$$\text{2-}O_2N\text{-}C_6H_4\text{-}CHO \xrightarrow{CH_2N_2} \text{2-}O_2N\text{-}C_6H_4\text{-}\overset{O}{CH\text{—}CH_2}\ (65\%) + \text{2-}O_2N\text{-}C_6H_4\text{-}CO{\cdot}CH_3\ (16\%)$$

The reaction of 4-nitrobenzaldehyde and diazomethane in ether is, however, reported to give 25% of 4-nitrostyrene oxide and 60% of 4-nitroacetophenone (*G. Biglino* and *G. M. Nano*, Ann. Chim. Rome, 1967, **57,** 1533). The reaction of aldehydes and higher diazoalkanes furnishes ketones in excellent yield (*D. W. Adamson* and *J. Kenner*, J. chem. Soc., 1939, 181); *e.g.* benzaldehyde reacts with diazoethane to give 94% of propiophenone.

Dimethylsulphonium methylide, $[(CH_3)_2S{=}CH_2]$, is a very selective methylene transfer reagent and with benzaldehyde gives 75% of the oxirane (*E. J. Corey* and *M. Chaykovsky*, J. Amer. chem. Soc., 1962, **84,** 3782).

(xi) Conversion of aldehydes into epoxides

Reaction of aromatic aldehydes with hexamethylphosphorous triamide affords a one step method for preparing symmetrical and unsymmetrical epoxides (*V. Mark*, Org. Synth., 1966, **46,** 142) *e.g.*

$$2\ \text{3-}Cl\text{-}C_6H_4\text{-}CHO + [Me_2N]_3P \longrightarrow \text{3-}Cl\text{-}C_6H_4\text{-}\overset{O}{CH\text{—}CH}\text{-}C_6H_4\text{-}Cl\text{-3} + [Me_2N]_3PO$$

55-60% *trans*, 35-40% *cis*

(xii) Paterno–Büchi reaction

This involves oxetane formation as the result of light-induced cycloaddition of carbonyl compounds to olefins. Illustrative of this process is the photoaddition of benzaldehyde with 2-methylbut-2-ene (*G. Büchi, C. G. Inman* and *E. S. Lipinsky*, J. Amer. chem. Soc., 1954, **76,** 4327; *N. C. Yang, R. L. Loeschen* and *D. Mitchell*, *ibid.*, 1967, **89,** 5465):

$$PhCHO + CH_3{\cdot}CH{=}C(CH_3)_2 \longrightarrow \text{oxetane: O—C(H)(CH}_3\text{); Ph—C(H)—C(CH}_3)_2$$

(xiii) Reaction with bis(1,3-diphenylimidazolinylidene-2)

The leads to the formation of a dianilinoethane and since derivatives of this type are readily hydrolysed it makes possible the transformation of an aldehyde to the homologous α-ketoaldehyde (*H. W. Wanzlick*, Angew. Chem., internat. Edn., 1962, **1,** 75; Org. Synth., 1967, **47,** 14) *e.g.*:

TABLE 5

BENZALDEHYDE AND ITS HOMOLOGUES

Benzaldehyde	*m.p. (°C)*	*b.p. (°C/mm)*	n_D°	$d_\circ^\circ$	*Derivative, m.p. (°C)*	*Reference*
Parent	−55	179/751, 62/10	$1.5463^{17.6}$	1.0504_4^{15}	*Phenylhydrazone*, two isomers, m.p. 157–158 and 154–155; *2,4-dinitrophenylhydrazone*, 237; *semicarbazone*, 222.	1
2-Methyl-		197, 94/10	1.549^{19}	1.0386_4^{19}	*Phenylhydrazone*, 105–106; *2,4-dinitrophenylhydrazone*, 193–194; *semicarbazone*, 212 (196).	1,1a,1b
3-Methyl-		200, 93–94/17	$1.541^{21.4}$	$1.0189_4^{21.4}$	*Phenylhydrazone*, 90; *semicarbazone*, 233–234 (206)	1,1b
4-Methyl-		204–205, 106/10	$1.547^{16.7}$	$1.0194_4^{16.7}$	*Phenylhydrazone*, 121 (108) *2,4-dinitrophenylhydrazone*, 232.5–234.5; *semicarbazone*, 234 (215)	1,1b,1c,1d
2,3-Dimethyl-					*Oxime*, 80–82; *semicarbazone*, 222 (*in vacuo*).	2
2,4-Dimethyl-	−9	215–216, 99/10			*Phenylhydrazone*, 88; *semicarbazone*, 225–227.	3,1d
2,5-Dimethyl-		220/738, 100/10			4-*Nitrophenylhydrazone*, 182; *semicarbazone*, 217.	3,1d

(continued)

Table 5 *(continued)*

Benzaldehyde	*m.p. (°C)*	*b.p. (°C/mm)*	n_D°	$d_{\circ}^{\circ}$	*Derivative, m.p. (°C)*	*Reference*
2,6-Dimethyl-	11	226–228/751.5			*2,4-Dinitrophenylhydrazone*, 252; *semicarbazone*, 158.	4
3,4-Dimethyl		225–226			*Phenylhydrazone*, 96; *semicarbazone*, 224 (slow heating).	1,3,5,1d
3,5-Dimethyl-	9	220–222			*Semicarbazone*, 201–202.	6
4-Isopropyl-		235–236, 103–104/10	1.5301^{20}	0.9775^{20}	*Phenylhydrazone*, 137–138; *semicarbazone*, 210–211 (222).	1
2,3,4-Trimethyl-	7–8	121.5/11			*Oxime*, 131–132.	7
2,3,6-Trimethyl-		113–114/10			*Oxime*, 124–126; *semicarbazone*, 167–169.	8
2,4,5-Trimethyl-	43.5	243, 121/10			*Phenylhydrazone*, 127; *semicarbazone*, 243–244.	9
2,4,6-Trimethyl-		237–240, 192/50	192/50		*Phenylhydrazone*, 74; *semicarbazone*, 185–188.	3,5,10,11,11a
2,4,5,6-Tetramethyl-						11b
2,4,6-Tri-isopropyl-						10
4-*tert*-Butyl-		246				12,12a
2,4-Dimethyl-6-*tert*-butyl	60					13

References to Table 5
1 See text.
1a *L. E. Smith* and *J. Nichols*, J. org. Chem., 1941, **6**, 489.
1b *W. F. Beech*, J. chem. Soc., 1954, 1297.
1c *W. S. Trahanovsky, L. B. Young* and *G. L. Brown*, J. org. Chem., 1967, **32**, 3865.
1d *F. Weygand* and *R. Mitgau*, Ber., 1955, **88**, 301.
2 *O. Brunner, H. Hofer* and *R. Stein*, Monatsh., 1933, **63**, 93.
3 *L. E. Hinkel, E. E. Ayling* and *W. H. Morgan*, J. chem. Soc., 1932, 2797.
4 *G. Lock* and *K. Schmidt*, J. prakt. Chem., 1934, **140**, 229.
5 *A. Rieche, H. Gross* and *E. Höft*, Ber., 1960, **93**, 88; Org. Synth., 1967, **47**, 1.
6 *M. Weller*, Ber., 1900, **33**, 465.
7 *L. I. Smith* and *C. L. Agre*, J. Amer. chem. Soc., 1938, **60**, 651.
8 *L. I. Smith* and *J. Nichols*, J. org. Chem., 1941, **6**, 501.
9 *E. L. Niedzielski* and *F. F. Nord*, *ibid.*, 1943, **8**, 148.
10 *R. C. Fusen et al.*, Org. Synth., Coll. Vol. 3, 1955, p. 549.
11 *R. P. Barnes*, *ibid.*, p. 551.
11a *K. Kulka*, Amer. Perfumer, 1957, **70**, 37.
11b *G. A. Olah* and *S. J. Kuhn*, J. Amer. chem. Soc., 1960, **82**, 2380.
12 *A. Tschitschibabin, S. Elgasine* and *V. Lengold*, Bull. Soc. chim. Fr., 1928, [iv], **43**, 239.
12a *J. W. Baker, W. S. Nathan* and *C. W. Shopper*, J. chem. Soc., 1935, 1847.
13 *A. Baur-Thurgau* and *A. Bischler*, Ber., 1899, **32**, 3647.

Ph Ph Ph Ph
N N N O N
H_2C CH_2 H_2C PhCHO ‖ CH_2
C=C ⇌ C: ⟶ PhC—CH
H_2C CH_2 H_2C CH_2
N N N N
Ph Ph Ph Ph

⟶ PhCO·CHO

(d) Individual arenecarbaldehydes and their derivatives

(i) Benzaldehyde and its homologues

Physical constants and selected derivative m.p. data are summarised in Table 5.

(*1*) **Benzaldehyde,** bitter almond oil, PhCHO, is a colourless liquid with a characteristic odour of bitter almonds. It occurs free in the leaves of the peach and some species of *Prunus*, and as the glucoside *amygdalin* in bitter almonds and in the kernels of some stone fruit such as apricot, peach and cherry. It was isolated and studied by *F. Wöhler* and *J. Liebig* (Ann., 1837, **22,** 1) who also recognised its chemical relationship to benzyl alcohol and benzoic acid. It is sparingly soluble in water and miscible with most organic solvents. It is used to some extent as a flavouring agent, in the formulation of perfumes, but mainly as an intermediate in chemical synthesis.

Benzaldehyde is formed by the hydrolysis of amygdalin, a glucoside of benzaldehyde cyanohydrin, by dilute acid or by the enzyme emulsin which occurs together with amygdalin in bitter almonds:

$$PhCH(CN)OC_{12}H_{21}O_{10} \xrightarrow{H_2O} \underset{\text{Glucose}}{2\,C_6H_{12}O_6} + PhCHO + HCN$$

Benzaldehyde is manufactured from toluene either by direct oxidation or by conversion into benzylidene chloride followed by hydrolysis. The direct vapour-phase oxidation of toluene is the method of choice, since a chlorine-free product is obtained. A gaseous mixture of toluene and air is passed over a catalyst consisting of oxides of uranium and molybdenum or related metals. High temperatures and short contact times are essential for maximum yields. In the liquid phase, toluene is oxidised with lead dioxide or manganese dioxide and diluted sulphuric acid. In the manufacture of benzaldehyde *via* the side-chain chlorination of toluene, benzyl chloride and benzotrichloride are obtained as by-products. Hydrolysis of benzylidene dichloride to benzaldehyde can be effected with either aqueous alkali or sulphuric acid (for a discussion of the manufacture of benzaldehyde see *W. L. Faith, D. B. Keyes* and *R. L. Clark*, in "Industrial Chemicals", Wiley, New York, 3rd Edn., 1965).

Benzaldehyde-d, is conveniently prepared by decarboxylation of phenylglyoxylic acid-d, by refluxing in benzene with benzoic anhydride and pyridine (*T. Cohen* and *I. H. Song*, J. Amer. chem. Soc., 1965, **87,** 3780):

$$PhCOCO_2D \rightarrow PhCDO + CO_2$$

Reactions. Benzaldehyde is oxidised by air to benzoic acid and benzaldehyde containing benzoic acid is purified by washing with 10% aqueous sodium carbonate, drying over anhydrous sodium carbonate, and vacuum distillation from a trace of zinc dust in an oxygen-free atmosphere. In the oxidation of benzaldehyde to benzoic acid, perbenzoic acid has been identified as an intermediate; it arises by a radical-initiated process:

$$PhCHO + R^{\bullet} \longrightarrow Ph\dot{C}{=}O + RH \quad \text{(Initiation)}$$

$$Ph\dot{C}{=}O + O_2 \longrightarrow PhC({=}O){-}O{-}O^{\bullet} \quad \text{(Propagation)}$$

$$PhC({=}O){-}O{-}O^{\bullet} + PhCHO \longrightarrow PhC({=}O){-}O{-}OH + Ph\dot{C}{=}O$$

A heterolytic mechanism for the formation of benzoic acid from perbenzoic acid is as follows:

$$PhCHO + PhC({=}O){-}O{-}OH \longrightarrow PhC(H)(OH){-}O{-}O{-}C({=}O)Ph \longrightarrow 2\ PhCO_2H$$

Benzaldehyde is conveniently oxidised to benzoic acid by permanganate (see p. 75) or aqueous acid dichromate. On irradiation in a sealed tube, it slowly polymerises to *tribenzaldehyde*, m.p. 250°, and *tetrabenzaldehyde*, m.p. 160–170° (*G. Ciamician* and *P. Silber*, Ber., 1909, **42,** 1386).

Dry chlorine reacts with benzaldehyde to give benzoyl chloride, also formed by an excess of sulphuryl chloride (*H. Durrans*, J. chem. Soc., 1922, **121,** 45) and by ethyl hypochlorite in carbon tetrachloride (*S. Goldschmidt et al.*, Ber., 1925, **58,** 576). Bromine similarly gives benzoyl bromide. Chlorination of benzaldehyde with phosphorus pentachloride or thionyl chloride gives benzylidene chloride (*F. Loth* and *A. Michaelis*, *ibid.*, 1894, **27,** 2548; *cf.* however, *P. Höring* and *F. Baum*, *ibid.*, 1908, **41,** 1918), with chlorine and excess of aluminium chloride, 3-chlorobenzaldehyde is formed (*Pearson, Pope* and *Hargrove*, *loc. cit.*).

Hydrazoic acid in the presence of sulphuric acid reacts with benzaldehyde to yield benzonitrile and some formanilide (*K. F. Schmidt*, Ber., 1924, **57,** 705; *Schmidt* and *P. Zutaverin*, U.S.P. 1,637,661/1927):

$$\mathrm{PhNHCHO + N_2 \xleftarrow{HN_3} PhCHO \xrightarrow{HN_3} PhCH + N_2 + H_2O}$$

Other reactions of benzaldehyde have been given on pp. 74 *et seq.* as illustrative of the general reactions of aromatic aldehydes.

(*2*) **Tolualdehydes.** 2-*Methylbenzaldehyde* has been prepared by the reaction of *o*-xylyl bromide and the sodium salt of 2-nitropropane (*H. B. Hass* and *M. L. Bender*, Org. Synth., Coll. Vol. 4, 1963, p. 932), alternatively from *o*-toluanilide by the Sonn–Müller procedure (*J. W. Williams, C. H. Witten* and *J. A. Krynitsky, ibid.*, Coll. Vol. 3, 1955, p. 818), or from *o*-tolunitrile by reaction with formic acid and Raney nickel (*B. Staskun* and *O. G. Backeberg*, J. chem. Soc., 1964, 5880).

2-, 3-, and 4-Methylbenzaldehydes are obtained from the corresponding xylenes by oxidation with chromyl chloride, for example, 3-*methylbenzaldehyde* is prepared from *m*-xylene in 80% yield (*W. H. Hartford* and *M. Darrin*, Chem. Reviews, 1958, **58,** 25; *E. Bornemann*, Ber., 1884, **17,** 1462).

4-*Methylbenzaldehyde* is prepared by the reaction of toluene with carbon monoxide and hydrogen chloride in the presence of aluminium chloride (*G. H. Coleman* and *D. Craig*, Org. Synth., Coll. Vol. 2, 1943, p. 583), or by diazotisation of 4-methylbenzylamine in dimethyl sulphoxide (*K. H. Scheit* and *W. Kampe*, Angew. Chem., internat. Edn., 1965, **4,** 787), or by oxidation of 4-methylbenzyl chloride with sodium dichromate in the presence of sodium carbonate (*K. Kulka*, Amer. Perfumer, 1957, **70,** 37). It is reported to have a pepper-like odour.

4-**Isopropylbenzaldehyde,** *cuminal*, occurs together with cymene (4-isopropyltoluene) in caraway oil from *Cuminium cyminium* and in hemlock oil from *Cicuta virosa*. It is also present in many eucalyptus oils (*A. R. Penfold*, J. chem. Soc., 1922, **121,** 268). It has been prepared by reaction of 4-isopropylphenylmagnesium bromide with orthoformic ester (*L. Bert*, Compt. rend., 1928, **186,** 700), from 4-isopropylbenzyl chloride by reaction with hexamethylenetetramine (*Fabr. Prod. de Chimie organique de Laire*, G.P. 268,786/1913) or by boiling with lead nitrate (*G. Errara*, Gazz., 1884, **14,** 278), and from isopropylbenzene by reaction with dichloromethyl methyl ether and titanium tetrachloride (*A. Rieche, H. Gross* and *E. Höft*, Ber., 1960, **93,** 88).

(ii) Functional derivatives of benzaldehyde

(*1*) *Acetals*, $PhCH(OR)_2$, the ethers of the hydrated aldehyde, are formed from the aldehyde by the action of the appropriate alcohol containing a small amount of dry hydrogen chloride (*E. Fischer* and *G. Giebe*, *ibid.*, 1898, **31,** 545; *R. D. Haworth* and *A. Lapworth*, J. chem. Soc., 1922, **121,** 80; see also *D. P. Roelofsen* and *H. van Bekkum*, Synthesis, 1972, 419); or by reaction with an orthoformic ester in the presence of a catalytic amount of strong acid (*L. Claisen*, Ber., 1907, **40,** 3907; see also *H. Adkins* and *R. Adams*, J. Amer. chem. Soc., 1925, **47,** 1370; *W. H. Hartung* and *Adkins*, *ibid.*, 1927, **49,** 2518):

$$PhCHO + HC(OR)_3 \rightarrow PhCH(OR)_2 + HCO_2R$$

Alternative methods involve the use of orthosilicates, $Si(OR)_4$ (*B. Helferich* and *J. Hausen*, Ber., 1924, **57,** 795); of alkyl sulphites (*W. Voss*, Ann., 1931, **485,** 293);

$$PhCHO + (MeO)_2S{=}O \xrightarrow{HCl} PhCH(OMe)_2 + SO_2$$
(81%)

the reaction of the arylmagnesium bromide with an orthoformic ester (*A. E. Tschitschibabin*, Ber., 1904, **37,** 188; *L. I. Smith* and *J. Nichols*, J. org. Chem., 1941, **6,** 489), the reaction of benzalaniline with an alcohol in the presence of either sulphuric or hydrochloric acid (*H. Meerwein*, in *Houben–Weyl*, "Methoden der Organischen Chemie", Thieme, Stuttgart, 1965, 4th Edn., Vol. 6/3, p. 220), and the reaction of benzylidene dichloride with alcoholic solutions of alkoxides (*J. E. Mackenzie*, J. chem. Soc., 1901, **79,** 1213). The electrolysis of α-methoxyphenylacetic acid in absolute methanol affords 62% of benzaldehyde dimethyl acetal and similarly the electrolysis of α-ethoxyphenylacetic acid in methanol gives 71% of benzaldehyde ethyl methyl acetal (*B. Wladislaw* and *A. M. J. Ayres*, J. org. Chem., 1962, **27,** 281).

Reduction of triethyl orthobenzoate with diisobutylaluminium hydride gives 95% of benzaldehyde diethyl acetal (*L. I. Zakharin* and *I. M. Khorlina*, Izvest. Akad. Nauk S.S.S.R., Otdel. khim. Nauk, 1959, 2255).

Benzaldehyde dimethyl acetal, *benzylidene dimethyl ether*, has b.p. 199°; the *diethyl acetal* has b.p. 221° and the *diisopropyl acetal*, b.p. 234°.

The acetals are readily hydrolysed by warm aqueous acid into the aldehyde and alcohol. The diethyl acetal on heating with acetic anhydride, yields α-ethoxybenzyl acetate, $Ph{\cdot}CH(OEt)O_2C{\cdot}CH_3$ (*Claisen*, Ber., 1898, **31,** 1019). The dimethyl acetal reacts with acetyl chloride containing thionyl chloride to give α-methoxybenzyl chloride, PhCH(OMe)Cl (*F. Straus* and *H. Heinze*, Ann., 1932, **493,** 191). This is decomposed by water: on heating it gives methyl chloride and benzaldehyde.

Benzaldehyde reacts with 1,2- and 1,3-diols under the influence of dry hydrogen chloride or similar condensing agents to give cyclic acetals. The *dimethylene acetal* from ethylene glycol, b.p. 106–107°/11 mm, is readily oxidised, and the resulting peroxide readily transformed to a benzoate ester (*A. Rieche*, *E. Schmitz* and *E. Beyer*, Ber., 1958, **91,** 1935):

$$PhCH\begin{matrix}O-CH_2\\ |\\ O-CH_2\end{matrix} \xrightarrow{O_2} \begin{matrix}Ph\\ HO{\cdot}O\end{matrix}C\begin{matrix}O-CH_2\\ |\\ O-CH_2\end{matrix} \xrightarrow{Na_2SO_3} PhCOOCH_2{\cdot}CH_2OH$$

The *trimethylene acetal* has m.p. 50°, b.p. 125°/14 mm (*Fischer*, *ibid.*, 1894, **27,** 1537). With glycerol, benzaldehyde forms a mixture of the 1,2- and 1,3-benzylidene derivatives (*Fischer*, *loc. cit.; J. Irvine et al.*, J. chem. Soc., 1915, **107,** 344; *H. Hibbert et al.*, J. Amer. chem. Soc., 1928, **50,** 2237). The preparation of 1,3-dibenzylidene-glycerol is described by *Hibbert* and *N. M. Carter*, *ibid.*, 1929, **51,** 1601; see also *B. F. Stimmel* and *C. G. King*, *ibid.*, 1934, **56,** 1724). For the cyclic derivatives of benzaldehyde with sugars see Vol. I F, pp. 351 *et seq.*

(2) *Benzylidene diacyl esters* are prepared by reaction of benzaldehyde with acyl anhydrides in the presence of various catalysts such as sulphuric acid, copper sulphate

or zinc chloride (*E. Knoevenagel*, Ann., 1914, **402,** 117; *E. Späth*, Monatsh, 1915, **36,** 38), or by the action of the salts of the carboxylic acids on benzylidene dichloride (*F. Bodroux*, Bull. Soc. chim. Fr., 1899, [iii], **21,** 331).

Benzylidene diacetate, $PhCH(OAc)_2$, m.p. 46°, b.p. 154°/20 mm, is also formed by the oxidation of toluene by chromium trioxide in acetic anhydride containing sulphuric acid. *Benzylidene dipropionate* has b.p. 158°/10 mm.

Some acid chlorides combine additively with benzaldehyde to give chloroacyl derivatives. Thus oxalyl chloride gives *bis*(*α-chlorobenzyl*)*oxalate*, $PhCHClO_2C\cdot CO_2CHClPh$, m.p. ~ 212° (dec.); the corresponding *bromo* compound, m.p. 130° (dec.) is also known (*H. Staudinger* and *E. Anthes*, Ber., 1913, **46,** 1436; *Adams* and *E. H. Vollweiler*, J. Amer. chem. Soc., 1918, **40,** 1741).

(*3*) *Additive compounds with sulphur acids.* Although an addition compound of benzaldehyde with sulphurous acid is unknown, the *bisulphites* of the alkali metals readily form crystalline addition compounds of the general type, $ArCH(OH)SO_2OM$. The crystalline, hydrated, sodium bisulphite adduct of benzaldehyde is used for separating benzaldehyde from mixtures with other organic compounds. The aldehyde can be regenerated from the adduct by distillation with water or more easily by the action of acids or alkalis.

On shaking benzaldehyde with an alkaline solution of *sodium dithionite*, $Na_2S_2O_4$, the air-sensitive sodium salt, $PhCH(OH)SO_2Na$, is obtained. The free acid is unknown; barium, zinc and other salts have been described (*M. Bazlen*, Ber., 1909, **42,** 4634; *A. Binz et al.*, *ibid.*, 1924, **57,** 1399).

(*iii*) *Sulphur and selenium analogues*

(*1*) **Thiobenzaldehyde.** Benzaldehyde reacts in alcoholic solution with hydrogen sulphide to give an amorphous white powder $(PhCHS)_x$ which softens on heating at about 85°. If the reaction is brought about in the presence of hydrogen chloride or zinc chloride, crystalline cyclic trimers are formed; the product is a mixture of two stereoisomeric 2,4,6-*triphenyl*-s-*trithianes:*

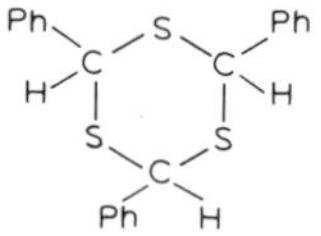

The labile *α-form*, m.p. 167°, is converted into the more stable *β-form*, m.p. 226°, by heating with iodine in benzene or by treatment with an acid catalyst (*E. Baumann* and *E. Fromm*, Ber., 1891, **24,** 1436; *J. F. Suyver*, Rec. Trav. chim., 1905, **24,** 398). Both forms on oxidation with hydrogen peroxide in glacial acetic acid give the same trisulphoxide $(PhCHSO)_3$, and the same trisulphone $(PhCHSO_2)_3$.

(*2*) **Benzaldehyde dimethyl dithioacetal,** $PhCH(SMe)_2$, is formed by reaction of methanethiol with benzaldehyde in the presence of hydrogen chloride. Oxidation gives the corresponding disulphone, *bismethylsulphonylphenylmethane*, $PhCH(SO_2Me)_2$, m.p. 163° (*J. Bongartz*, Ber., 1888, **21,** 487). The corresponding *bisethylsulphonyl* deriv., m.p. 134°, and the *bis*-n-*butylsulphonyl* deriv., m.p. 86°, are prepared similarly (*Baumann*, ibid., 1885, **18,** 885; *T. C. Witner* and *E. E. Reid*, J. Amer. chem. Soc., 1921, **43,** 639).

(*3*) **Selenobenzaldehyde,** PhCHSe, has been isolated in three forms. The *α-form*, m.p. 85°, is obtained by reaction of benzaldehyde and hydrogen selenide in alcohol. The *β-form*, m.p. 218°, and *γ-form*, m.p. 167°, are obtained by reaction of benzaldehyde and hydrogen selenide in alcohol saturated with hydrogen chloride (*L. Vanino* and *A. Schinner*, J. prakt. Chem., 1915, [ii], **91,** 26).

(iv) Reaction of benzaldehyde with ammonia and amines

(*1*) *Ammonia.* A concentrated alcoholic solution of benzaldehyde reacts with ammonia at −20° to give **benzaldehyde ammonia,** $[PhCH(OH)]_2NH$, m.p. 45°. It decomposes rapidly into water, benzaldehyde and **hydrobenzamide,** $(PhCH{:}N)_2CHPh$, m.p. 110° (*F. Francis*, Ber., 1909, **42,** 2216) which is the normal product of the action of concentrated aqueous ammonia on benzaldehyde. On heating at 120–130°, it yields the imidazoline (amarin) (IV) and this on further heating is converted into 2,4,5-triphenylimidazole (lophine) (V) (*H. H. Strain*, J. Amer. chem. Soc., 1927, **49,** 1558):

$$(PhCH{=}N)_2CHPh \xrightarrow{120-130^\circ} \text{(IV)} \xrightarrow{\Delta} \text{(V)}$$

(IV): PhCH—N / PhCH CPh / N–H ring (amarin); (V): PhC—N / PhC CPh / N–H ring (lophine)

Hydrobenzamide reacts with alcoholic hydrogen chloride in benzene to give *benzylideneamine hydrochloride*, $Ph{\cdot}CH{=}\overset{\oplus}{N}H_2\overset{\ominus}{Cl}$, which is decomposed by water into benzaldehyde and ammonium chloride, and with alcohol gives benzaldehyde diethyl acetal (*M. Busch*, Ber., 1896, **29,** 2144). The *stannichloride*, $(PhCH{=}\overset{\oplus}{N}H_2)_2SnCl_6^{2-}$, forms colourless needles (*A. Sonn* and *W. Meyer*, *ibid.*, 1925, **58,** 1097; *H. Stephen*, J. chem. Soc., 1925, **127,** 1876).

(*2*) *Amines.* **Benzylidenemethylamine,** PhCH=NMe, b.p. 90–91°/30 mm, is prepared by reaction of benzaldehyde with concentrated aqueous methylamine (*C. K. Ingold* and *C. W. Shoppee*, *ibid.*, 1929, 1204). *Benzylidene-ethylamine*, b.p. 193°; *-propylamine* and *-isobutylamine* are prepared similarly (*H. Zaunschirm*, Ann., 1888, **245,** 279). *N*-Alkylimines can be used as intermediates in the preparation of aliphatic secondary amines by alkylation and hydrolysis.

$$PhCH{=}NR \xrightarrow{MeI} PhCH{=}\overset{\oplus}{N}MeR\ \overset{\ominus}{I} \xrightarrow{H_2O/NaOH} PhCHO + MeNHR + NaI$$

Benzylideneaniline, PhCH=NPh, m.p. 54°, is formed by the condensation of benzaldehyde with aniline (*L. A. Bigelow* and *H. Eatough*, Org. Synth., Coll. Vol. 1, 2nd Edn., 1947, p. 80). Benzylidene derivatives of the homologues of aniline are prepared similarly. Many anils give methanolamine hydrochlorides, $PhCH(OH)\overset{\oplus}{N}H_2Ar\ Cl^{\ominus}$, when treated with cold fuming hydrochloric acid; the free bases cannot usually be isolated (*O. Dimroth*, Ber., 1902, **35,** 984).

Benzylideneaniline is hydrolysed on boiling with dilute hydrochloric acid. Nitration in sulphuric acid yields benzylidene-4-nitroaniline; in acetic acid some *ortho* isomer is also produced (*C. Schwalbe*, *ibid.*, 1902, **35,** 3302; *F. Arnall* and *T. Lewis*, J. Soc. chem. Ind., 1929, **48,** 160). Reduction under a variety of conditions gives benzylaniline (see *W. S. Emerson*, Org. Reactions, 1948, **4,** 206). Grignard compounds combine with benzylideneaniline to give magnesyl compounds which on hydrolysis yield *C*-alkylbenzylanilines (*Busch*, Ber., 1905, **38,** 1761). The addition of acetoacetic ester, malonic ester and similar compounds is described by *E. Knoevenagel* (*ibid.*, 1898, **31,** 2596), *R. Schiff* (*ibid.*, 1899, **32,** 332) and *F. E. Francis* (*ibid.*, 1903, **36,** 937). *N,α*-**Diphenylnitrone,** N-*phenylbenzaldoxime*, $PhCH{=}\overset{\oplus}{N}(O^{\ominus})Ph$, m.p. 112°, the *N*-oxide of benzylideneaniline, is formed either from benzaldehyde and *N*-phenylhydroxylamine or by oxidation of benzylphenylhydroxylamine, $PhCH_2N(OH)Ph$ (*E. Bamberger* and *A. Voss*, *ibid.*, 1894, **27,** 1556; *R. Huisgen et al.*, Org. Synth., 1966, **46,** 127). It undergoes 1,3-dipolar addition with styrene to give 2,3,5-triphenylisoxazolidine:

$$\text{PhCH}{=}\overset{\oplus}{\text{N}}(\text{O}^{\ominus})\text{Ph} + \text{H}_2\text{C}{=}\text{CHPh} \longrightarrow \text{2,3,5-triphenylisoxazolidine}$$

Benzylidene-o-*toluidine* is a liquid, b.p. 176°/15 mm, *benzylidene*-m- and -p-*toluidines* have m.p. 32° and 35°, respectively.

N-*Benzylidene*-o-*phenylenediamine*, m.p. 60–61°, and N,N′, *dibenzylidene*-o-*phenylenediamine*, m.p. 106° are obtained by reaction of benzaldehyde with *o*-phenylenediamine. The dibenzylidene derivative is converted into 1-benzyl-2-phenylbenzimidazole on keeping, on heating in alcohol, or by the action of hydrochloric acid (*O. Hinsberg* and *P. Koller*, Ber., 1896, **29,** 1499):

$$C_6H_4(N{=}CHPh)_2 \longrightarrow \text{1-benzyl-2-phenylbenzimidazole } (N\text{-}CH_2Ph)$$

N,N′-*Dibenzylidene*-m-*phenylenediamine* has m.p. 104–105° and N,N′-*dibenzylidene*-p-*phenylenediamine*, m.p. 140°.

Benzylideneglycine, $PhCH{:}NCH_2{\cdot}CO_2H$, is obtained as a brownish powder, m.p. ~207°, by condensation of benzaldehyde and glycine in the presence of barium or calcium hydroxides (*M. Bergmann et al.*, *ibid.*, 1925, **58,** 1037). For the benzylidene derivatives of other α-amino acids see *O. Gerngross* and *E. Zühkle* (*ibid.*, 1924, **57,** 1486); *Bergmann et al.*, (*loc. cit.* and Z. physiol. Chem., 1925, **142,** 288).

(*3*) *Hydrazines.* **Benzaldehyde hydrazone,** *benzylidenehydrazine*, $PhCH{:}N{\cdot}NH_2$, m.p. 16°, b.p. 140°/14 mm, is formed by reaction of benzaldehyde and hydrazine, or by heating benzalazine (below) with hydrazine hydrate. Oxidation with yellow mercuric oxide gives phenyldiazomethane (*Staudinger* and *A. Gaule*, Ber., 1916, **49,** 1906). The free amino group can react with carbonyl groups; it may also be acylated. *Benzaldehyde acetylhydrazone*, $PhCH{:}N{\cdot}NHCOMe$, has m.p. 134° (*T. Curtius* and *H. Franzen*, Ber., 1902, **35,** 3234).

Benzalazine, *dibenzylidenehydrazine*, $PhCH{:}N{\cdot}N{:}CHPh$, m.p. 93°, is formed by action of benzaldehyde on a deficiency of hydrazine (*H. H. Hatt*, Org. Synth., Coll.

Vol. 2, 1943, p. 395). On heating at 300°, benzalazine decomposes with loss of nitrogen; stilbene is the main organic product (*H. E. Zimmerman* and *S. Somasckhara*, J. Amer. chem. Soc., 1960, **82,** 5865). It undergoes photochemical cleavage in the presence of benzophenone to give benzonitrile in high yield (*J. E. Hodgkins* and *J. A. King*, *ibid.*, 1963, **85,** 2679). Oxidation of benzalazine with peracetic acid gives the mono-*N*-oxide (*L. Horner*, *W. Kirmse* and *H. Fernekess*, Ber., 1961, **94,** 279). Reduction may give benzaldehyde benzylhydrazone, *N,N'*-dibenzylhydrazine, benzylamine or dibenzylamine, according to the conditions (*K. W. Rosenmund* and *E. Pfankuch*, *ibid.*, 1923, **56,** 2261; *Curtius*, J. prakt. Chem., 1889, [ii], **39,** 47; 1900, [ii], **62,** 99; *J. Thiele*, Ann., 1910, **376,** 261; *H. Mazourevitch*, Bull. Soc. chim. Fr., 1925, [iv], **37,** 1163). It combines additively with bromine to give a tetrabromide (*Curtius* and *E. Quedenfeldt*, J. prakt. Chem., 1898, [ii], **58,** 385; *G. Reddelien*, *ibid.*, 1915, [ii], **91,** 221). For the addition of chlorine and hydrogen bromide see *R. Stollé* (*ibid.*, 1912, [ii], **85,** 390). It combines also with methyl sulphate to give a salt which is hydrolysed by water to benzaldehyde and methylhydrazine sulphate (*Hatt*, *loc. cit.*, p. 396).

Benzaldehyde phenylhydrazone, benzylidenephenylhydrazine, PhCH=N·NHPh, forms colourless crystals which become pink on exposure to light. In the oxidation of benzaldehyde phenylhydrazone, the initial step appears to be formation of the radical (VIa,b) which may dimerise by C–N or N–N bond formation:

$$\text{PhCH=N-NHPh} \rightarrow \underset{\text{(VIa)}}{\text{PhCH=N-}\dot{\text{N}}\text{Ph}} \leftrightarrow \underset{\text{(VIb)}}{\text{Ph}\dot{\text{C}}\text{H-N=NPh}}$$

Oxidation with oxygen in alcoholic alkali gives *anti*-benzil osazone (VII) as the principal product (*H. Biltz* and *O. Amme*, Ann., 1902, **321,** 1):

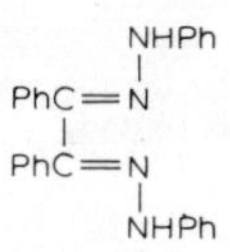

NHPh | PhC=N | PhC=N | NHPh (VII)

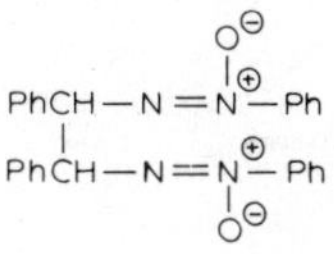

PhCH–N=N⁺(O⁻)–Ph | PhCH–N=N⁺(O⁻)–Ph (VIII)

Peroxyacids convert benzaldehyde phenylhydrazone into a 1,2-bisazoxyethane VI which may exist in a *meso* or *racemic* form (*P. A. S. Smith*, in "Open Chain Nitrogen Compounds", Benjamin, New York, 1966, Vol. 2, p. 168). Reduction of benzaldehyde phenylhydrazone gives *N*-benzyl-*N'*-phenylhydrazine, products arising from N–N bond fission are also formed. *C*-Nitrosation occurs on treatment of benzaldehyde phenylhydrazone with amyl nitrite and base (*Bamberger* and *W. Pemsel*, Ber., 1903, **36,** 62):

$$\text{Ph-CH=N-NH-Ph} \longrightarrow \text{Ph-}\underset{\underset{\text{NOH}}{\|}}{\text{C}}\text{-N=N-Ph}$$

(*4*) *Hydroxylamine.* **Benzaldoximes.** Benzaldehyde forms two stereoisomeric oximes α-(or syn)-*benzaldoxime*, m.p. 34° and β-(or anti)-*benzaldoxime**, m.p. 127°. The α-form

*Recently adapted nomenclature rules designates the α- and β-forms, *E*- and *Z*-, respectively.

is readily isolated from the reaction of benzaldehyde and hydroxylamine in alkaline solution. Treatment of the α-oxime in ether with dry hydrogen chloride at 0° gives an unstable hydrochloride which on crystallisation is converted into the hydrochloride of the β-form. This same product is obtained by the action of hydrogen chloride on the α-oxime without cooling. Treatment of the β-oxime hydrochloride with sodium carbonate gives the β-oxime (*E. Beckmann*, Ber., 1890, **23**, 1684). The β-oxime has also been obtained by reacting benzaldehyde with hydroxylamine hydrochloride and treating the resulting hydrochloride with water (*G. Zvilichovsky* and *L. Heller*, Synthesis, 1972, 563). Assignment of configuration is based on the observation that α-benzaldoxime gives an *acetate*, m.p. 14°, from which the parent oxime is regenerated on treatment with sodium carbonate solution. β-Benzaldoxime, however, yields an *acetate*, m.p. 56°, which on similar treatment is converted to benzonitrile. This facile elimination reaction is indicative of *trans* disposition of H and AcO about the carbon–nitrogen double bond. Supporting chemical evidence for the assignment of configuration, is the observation that α-benzaldoxime gives formanilide on treatment with polyphosphoric acid, whereas β-benzaldoxime yields benzamide (*E. C. Horning* and *V. L. Stromberg*, J. Amer. chem. Soc., 1952, **74**, 5151).

α - or *syn*- benzaldoxime (*E*-form) — aq. Na_2CO_3 ⇅ $(AcO)_2O$ — PhCH=N–OAc

β - or *anti* - benzaldoxime (*Z*-form) — $(AcO)_2O$ → PhCH=N–OAc — aq. Na_2CO_3 → PhC≡N

The configurations of the 4-chlorobenzaldoximes assigned on chemical evidence have been confirmed by X-ray analysis of the crystal structure (for a summary of the physical methods used for assignment of configuration see *E. L. Eliel*, "Stereochemistry of Carbon Compounds", McGraw-Hill, New York, 1962, p. 336).

α- and β-Benzaldoximes yield isomeric phenylcarbamates, PhCH=NOCONHPh, on treatment with phenyl isocyanate. *Ethoxycarbonyl-α-benzaldoxime*, m.p. 35°, is formed from α-benzaldoxime and ethyl chloroformate (*O. L. Brady* and *G. P. McHugh*, J. chem. Soc., 1923, **123**, 1190).

α-Benzaldoxime O-*methyl ether*, PhCH=NOMe, b.p. 192°, is formed from the α-oxime by reaction with methyl iodide or methyl sulphate in alkaline solution or by the action of diazomethane (*M. O. Forster* and *F. P. Dunn*, *ibid.*, 1909, **95**, 425). It is hydrolysed by acids into benzaldehyde and *O*-methylhydroxylamine. *β-Benzaldoxime*-O-*methyl ether*, b.p. 190°, is formed when the silver salt of the β-oxime is treated with methyl iodide (see *Brady et al.*, *ibid.*, 1926, 2386).

N-*Methylbenzaldoxime*, $PhCH=\overset{\oplus}{N}(-\overset{\ominus}{O})Me$, m.p. 82°, is formed together with some of the isomeric *O*-methyl derivative by the action of methyl iodide on the β-aldoxime in alcoholic methoxide, or by keeping the α-oxime with methyl iodide (*idem*, *ibid.*, 1926, 2396). It is also formed by the reaction of benzaldehyde and *N*-methylhydroxylamine (*idem*, *ibid.*, 2390).

(v) Halogenobenzaldehydes

The more common methods for the preparation of halogen-substituted benzaldehydes include (*1*) the hydrolysis of halogen-substituted benzylidene dihalides (see preparation of 4-chlorobenzaldehyde, *W. L. McEwen*, Org. Synth., Coll. Vol. 2, 1943, p. 133, and preparation of 4-bromobenzaldehyde, *G. H. Coleman* and *G. E. Honeywell*, *ibid.*, p. 89); (*2*) the Sommelet method (p. 57) from halogen-substituted benzyl chlorides; (*3*) the oxidation of halogen-substituted toluenes (see preparation of 4-bromobenzaldehyde, *S. V. Liebermann* and *R. Connor*, *ibid.*, p. 441); (*4*) the diazo reaction from the appropriate aminobenzaldehyde (see preparation of 3-chlorobenzaldehyde, *J. S. Buck* and *W. S. Ide*, ibid., p. 130).

The reactions of the halogenobenzaldehydes are similar to those of benzaldehyde. Selected physical constants and derivative m.p. data for the halogeno-benzaldehydes are summarised in Table 6.

Iodoso- and iodoxy-benzaldehydes are described by *C. Willgerodt* (J. prakt. Chem., 1912, [ii], **86,** 276) and *T. S. Patterson* (J. chem. Soc., 1896, **69,** 1002).

(vi) Nitrobenzaldehydes

Convenient methods for the preparation of the nitrobenzaldehydes include the regulated oxidation of nitrotoluenes with chromium trioxide in acetic anhydride and sulphuric acid and subsequent hydrolysis of the resulting diacetate, and the regulated reduction of the corresponding nitro-acid chlorides. Specific preparations are noted under the individual aldehydes. Di-*ortho*-substituted aldehydes are best prepared from the corresponding benzyl halides by Kröhnke's method (see p. 57). Physical constants and selected derivative m.p. data are summarised in Table 7.

2-**Nitrobenzaldehyde,** is also known in a labile form, m.p. 38° (*Brady* and *S. Harris*, J. chem. Soc., 1923, **123,** 484). It is synthesised from 2-nitrotoluene either by intermediate preparation of 2-nitrobenzylidene diacetate (*S. M. Tsang*, *E. H. Wood* and *J. R. Johnson*, Org. Synth., Coll. Vol. 3, 1955, p. 641; *T. Nishimura*, *ibid.*, Coll. Vol. 4, 1963, p. 713), or by a route which involves the preparation of 2-nitrobenzyl bromide and its pyridinium salt (*A. Kalir*, *ibid.*, 1966, **46,** 81). The *syn*-oxime is converted into the *anti*-oxime by boiling in benzene. The change is reversed by irradiation in benzene. On irradiation in inert solvents or by the action of ammoniacal cyanide solution 2-nitrobenzaldehyde is isomerised to 2-nitrosobenzoic acid (*G. Ciamician* and *P. Silber*, Ber., 1901, **34,** 2041; *E. J. Bowen et al.*, J. chem. Soc., 1924, **125,** 1219). Irradiation of an alcoholic solution of 2-nitrobenzaldehyde gives an ester of 2-nitrosobenzoic acid. In the presence of hot alkali, 2-nitrobenzaldehyde and acetone react to give indigo; this reaction can be used as a test for 2-nitrobenzaldehydes (see *R. M. Acheson*, in "An Introduction to the Chemistry of Heterocyclic Compounds", Interscience, London, 1967, 2nd Edn., p. 265).

3-**Nitrobenzaldehyde** is obtained in high yield by the low temperature nitration

TABLE 6

HALOGENOBENZALDEHYDES

Benzaldehyde	*m.p. (°C)*	*b.p.(°/mm)*	*Derivative, m.p. (°C)*	*Reference*
2-Fluoro-	–44.5	175, 80.5/36	*Phenylhydrazone*, 89.5; *4-nitrophenylhydrazone*, 205.	1
3-Fluoro-		173, 76/26	*Phenylhydrazone*, 114; *4-nitrophenylhydrazone*, 202.	1
4-Fluoro-	– 10	181.5, 104.5/74	*Phenylhydrazone*, 147; *4-nitrophenylhydrazone*, 212.	1
2-Chloro-	11	213–214	*Phenylhydrazone*, 86.	2,4b
3-Chloro-	17–18	213–214	*Phenylhydrazone*, 134–135.	3
4-Chloro-	47	213/748 mm, 72–75/3	*Phenylhydrazone*, 126–127.	4,4a,4c,4d
2-Bromo-	22	230, 118–119/12	*Semicarbazone*, 214.	5
3-Bromo-		233–236, 66–68/2	*Semicarbazone*, 205.	3
4-Bromo-	67		*Phenylhydrazone*, 112–113; *semicarbazone*, 227–228.	4b,6,7
2-Iodo-	37	129/14	*Phenylhydrazone*, 79; *2,4-dinitrophenylhydrazone*, 215.	8
3-Iodo-	57		*Phenylhydrazone*, 155.	9
4-Iodo-	77–78	264.5/725	*Phenylhydrazone*, 121.	10,10a
2,3-Dichloro-	65–67			11
2,4-Dichloro-	72		2,4-*Dinitrophenylhydrazone*, 224–227.	12
2,5-Dichloro-	58	231–233	*Oxime*, 128.	13
2,6-Dichloro-	71		*Oxime*, 149–150; *4-nitrophenylhydrazone*, 154.	14,15
3,4-Dichloro-	44	247–248	*Oxime*, 118–119; *4-nitrophenylhydrazone*, 276–277.	16,17
3,5-Dichloro-	65		*Oxime*, 112; *phenylhydrazone*, 106.5.	18
2,3,4-Trichloro-	90			19
2,3,5-Trichloro-	56			20
2,3,6-Trichloro-	86–87			15
2,4,5-Trichloro-				19
2,4,6-Trichloro-	58–59		2,4-*Dinitrophenylhydrazone*, 275–277.	15
3,4,5-Trichloro-	90–91		*Phenylhydrazone*, 147; *semicarbazone*, 252–254.	21
2,4-Dibromo-	80			22
2,3,5-Tribromo-	114			23
2,4,6-Tribromo-	99		*Oxime*, 175.	24
3,4,5-Tribromo-	109		*Oxime*, 172; *phenylhydrazone*, 158.	24

References
1 *B. Shoesmith, C. E. Sosson* and *R. H. Slater*, J. chem. Soc., 1926, 2761.
2 *H. Stephen, ibid.*, 1925, **127,** 1877.
3 *J. S. Buck* and *W. S. Ide*, Org. Synth., Coll. Vol. 2, 1943, p. 130.
4 *W. L. McEwen*, p. 133.
4a *T. van Es* and *B. Staskun*, J. chem. Soc., 1965, 5774.
4b *W. S. Trahanovsky, L. B. Young* and *G. L. Brown*, J. org. Chem., 1967, **32,** 3865.
4c *W. F. Beech*, J. chem. Soc., 1954, 1297.
4d *D. Markees*, J. org. Chem., 1958, **23,** 1490.
5 *R. Adams* and *E. H. Vollweiler*, J. Amer. chem. Soc., 1918, **40,** 1732.
6 *G. H. Coleman* and *G. E. Honeywell*, Org. Synth., Coll. Vol. 2, 1943, p. 89.
7 *S. V. Liebermann* and *R. Connor*, *ibid.*, p. 442.
8 *W. S. Rapson* and *R. G. Shuttleworth*, J. chem. Soc., 1941, 488.
9 *I. R. L. Barker* and *W. A. Waters*, *ibid.*, 1952, 150.
10 *P.P.T. Sah*, J. Amer. chem. Soc., 1942, **64,** 1487.
10a *O. H. Wheeler*, Canad. J. Chem., 1958, **36,** 667.
11 *C. S. Marvel et al.*, J. Amer. chem. Soc., 1946, **68,** 861.
12 *G. Lock* and *E. Böck*, Ber., 1937, **70,** 923.
13 *T. de Grauw*, Rec. Trav. chim., 1931, **50,** 773.
14 *M. S. Reich*, Bull. Soc. chim. Fr., 1917, [iv], **21,** 223.
15 *Geigy*, G.P. 199,943 (Chem. Zentr., 1908, II, 363):
16 *H. H. Hodgson* and *H. G. Beard*, J. chem. Soc., 1927, 25.
17 *G. M. Kraay*, Rec. Trav. chim., 1930, **49,** 1086.
18 *F. Asinger* and *G. Lock*, Monatsh., 1933, **62,** 344.
19 *E. Seelig*, Ann., 1887, **237,** 149.
20 *Hodgson* and *Beard*, J. chem. Soc., 1927, 2382.
21 *C. van der Bunt*, Rec. Trav. chim., 1929, **48,** 131.
22 *J. J. Blanksma*, Chem. Zentr., 1910, I, 261.
23 *Idem*, *ibid.*, 1912, II, 1964.
24 *Lock* and *R. Schreckeneder*, Ber., 1939, **72,** 511.

of benzaldehyde (*R. N. Icke, C. E. Redemann* and *B. B. Wisegarver*, Org. Synth., Coll. Vol. 3, 1955, p. 644). The hazards associated with this preparation are pointed out by *J. Lange* and *T. Urbánski*, Chem.and Ind., 1967, 1424.

4-**Nitrobenzaldehyde** is obtained from 4-nitrotoluene by intermediate preparation of 4-nitrobenzylidene diacetate (*S. V. Liebermann* and *R. Connor*, Org. Synth., Coll. Vol. 2, 1943, p. 441; *T. Nishimura, ibid.*, Coll. Vol. 4, 1963, p. 713) and from 4-nitrobenzoyl chloride by reduction either with hydrogen and a palladium catalyst (*E. Mosettig* and *P. Mozingo*, Org. Reactions, 1948, **4,** 362) or with lithium tri-*tert*-butoxyhydridoaluminate (*H. C. Brown* and *B. C. Subba Rao*, J. Amer. chem. Soc., 1958, **80,** 5377). The *syn*-oxime changes to the *anti*-oxime on fusion.

2,4-**Dinitrobenzaldehyde** is obtained by condensation of 4-nitrosodimethylaniline with 2,4-dinitrotoluene and subsequent fission of the resulting azomethine (*G. M. Bennett* and *E. V. Bell*, Org. Synth., Coll. Vol. 2, 1943, p. 223). It has also been prepared from 2,4-dinitrobenzylaniline by oxidation to the benzylidene compound and subsequent hydrolysis (*F. Sachs* and *W. Everding*, Ber., 1902, **35,** 1237; *P. Cohn* and *Friedländer*, *ibid.*, p. 1266).

2,4,6-**Trinitrobenzaldehyde** is obtained by the condensation of 2,4,6-trinitrotoluene with nitrosodimethylaniline and subsequent acidic fission of the resulting azomethine (*A. Lowy* and *E. H. Balz*, J. Amer. chem. Soc., 1921, **43,** 343). Just as 2-nitrobenzaldehyde isomerises to 2-nitrosobenzoic acid on irradiation, 2,4-di- and 2,4,6-tri-nitro-

TABLE 7

NITROBENZALDEHYDES

Benzaldehyde	*m.p. (°C)*	*Derivative, m.p. (°C)*	*Reference*
2-Nitro-	43–44	Syn-*oxime*, 154; anti-*oxime*, 103; *phenylhydrazone*, 154; *semicarbazone*, 256.	1,1a,1b
3-Nitro-	58	Syn-*oxime*, 74; anti-*oxime*, 123; *semicarbazone*, 246.	1,1a
4-Nitro-	106	Syn-*oxime*, 184; anti-*oxime*, 133; *phenylhydrazone*, 153–154; *semicarbazone*, 211.	1,1a,1b
2-Chloro-4-nitro-	74 (79)	*Phenylhydrazone*, 154; 2,4-*dinitrophenylhydrazone*, 247.	2
2-Chloro-5-nitro-	80	*Phenylhydrazone*, 182.	3
2-Chloro-6-nitro-	70–71	*Oxime*, 156–157; *phenylhydrazone*, 161.	4
4-Chloro-2-nitro-	67–68	*Oxime*, 172.	4,5
4-Chloro-3-nitro-	64–65	*Oxime*, 146; *phenylhydrazone*, 148	6
5-Chloro-2-nitro-	77–78	*Oxime*, 112.	7
2-Bromo-6-nitro-	82	*Phenylhydrazone:* two forms, m.p. 142 and 146.	4
3-Bromo-5-nitro-	100		8
4-Bromo-2-nitro-	97–98	*Oxime*, 164; *phenylhydrazone*, 181–182.	9
4-Bromo-3-nitro-	106	*Oxime*, 154; *phenylhydrazone:* two forms m.p. 146–147 and 148–149.	6
5-Bromo-2-nitro-	73–74	*Oxime*, 113; *phenylhydrazone*, 180 (decomp.).	10
2,4-Dinitro-	72	*Oxime*, 127–128; *phenylhydrazone*, 227–228.	1
2,6-Dinitro-	123	*Oxime*, 115; *phenylhydrazone*, 159.	4
3,4-Dinitro-	64	*Phenylhydrazone*, 184–186.	11
3,5-Dinitro-	85	*Oxime*, 159; *phenylhydrazone*, 210.	12
2,4,6-Trinitro-	119	*Oxime*, 158; *phenylhydrazone*, 202.	1

References
1 See text.
1a *W. S. Trahanovsky, L. B. Young* and *G. I. Brown*, J. org. Chem., 1967, **32,** 3865.
1b *D. G. Markees, ibid.*, 1958, **23,** 1490.
2 *L. Chardonnes* and *P. Heinrich*, Helv., 1940, **23,** 292.
3 *H. Erdmann*, Ann., 1892, **272,** 153.
4 *K. Clarke*, J. chem. Soc., 1957, 3808.
5 *D. P. Spalding et al.*, J. Amer. chem. Soc., 1946, **68,** 1598.
6 *H. H. Hodgson* and *H. G. Beard*, J. chem. Soc., 1927, 24.
7 *A. Eichengrün* and *A. Einhorn*, Ann., 1891, **262,** 137.
8 *J. J. Blanksma*, Chem. Weekblad, 1913, **9,** 862; C.A., 1913, **7,** 770.
9 *H. J. Barber* and *C. E. Stickings*, J. chem. Soc., 1945, 167.
10 *C. Mettler*, Ber., 1905, **38,** 2811.
11 *H. Goldstein* and *R. Voegeli*, Helv., 1943, **26,** 1125.
12 *Hodgson* and *E. W. Smith*, J. chem. Soc., 1933, 315.

benzaldehydes also yield nitrosobenzoic acids on exposure to light (for summary see *J. G. Calvert* and *J. N. Pitt Jr.*, in "Photochemistry", Wiley, New York, 1966, p. 478).

When the phenylhydrazone of 2,6-dinitrobenzaldehyde is treated with aqueous-ethanolic potassium hydroxide, one *ortho* nitro group is eliminated and 1-phenyl-4-nitroindazole is formed (*S. Reich* and *G. Gaigailian*, Ber., 1913, **46,** 2380):

NO_2, CH=N·NHPh, NO_2 ⟶ NO_2, N, N, Ph

(vii) Hydroxyimino-, nitroso-, azoxy- and azo-benzaldehydes

The regulated reduction of nitrobenzaldehydes by zinc dust or electrolytically to hydroxylamines in general leads to condensation products unless the aldehyde group is protected. Thus 3- and 4-nitrobenzaldehydes give nitrones of the general formula:

$$NO_2C_6H_4\cdot CH=\overset{\oplus}{N}(\overset{\ominus}{O})C_6H_4\cdot CHO$$

2-Nitrobenzaldehyde gives anthranil (IX). If, however, the reduction of 2-nitrobenzaldehyde by zinc dust is brought about in the presence of ethyl nitrite, the *N*-nitroso-derivative of 2-hydroxyiminobenzaldehyde (X) is formed (*E. Bamberger*, Ber., 1909, **42,** 2574). This compound is also obtained by the action of nitrous acid on anthranil. It is decomposed by acids into 2-diazobenzaldehyde and 2-**nitrosobenzaldehyde,** m.p. 110°:

CH, O, N (IX) CHO, NOH, NO (X) CH=NOH, NHOH (XI)

If the aldehyde group is protected from further condensation by conversion into the oxime, the nitro group can be reduced to the corresponding hydroxyimino group.

TABLE 8

AMINOBENZALDEHYDES

Benzaldehyde	*m.p. (°C)*	*Derivative, m.p. (°C)*	*Reference*
2-Amino-	39–40	*Oxime*, 134–135; *4-nitrophenylhydrazone*, 218–220; N-*acetyl*, 70–71.	1
3-Amino-	28–30	*4-Nitrophenylhydrazone*, 225–226; N-*acetyl*, 122.	1
4-Amino-	70–72	*Oxime*, 124; *phenylhydrazone*, 175 (156); N-*acetyl*, 156.	1
4-Methylamino-	60–61 (57–58)		1
3-Dimethylamino-	b.p. 112–114°/3 mm	*Oxime*, 75–76; *4-nitrophenylhydrazone*, 188; *methiodide*, 185–186 (decomp.).	1
4-Dimethylamino-	74	*Oxime*, 144; *phenylhydrazone*, 148; *4-nitrophenylhydrazone*, 182.	1
4-Diethylamino-	41	*Oxime*, 93; *phenylhydrazone*, 121 (103)	1
4-Benzylmethylamino-	63	*4-Nitrophenylhydrazone*, 179.	1
2-Amino-4-nitro-	124	*Oxime*, 193; *anil*, 147.	2
2-Amino-5-nitro-	200.5	*Oxime*, 203; N-*acetyl*, 160–161.	3
4-Amino-2-nitro		*Oxime*, 177–178; *phenylhydrazone*, 163.	4
4-Amino-3-nitro	191	*Oxime*, 207; N-*acetyl*, 155.	1

References
1 *See text.*
2 *F. Sachs* and *E. Sichel*, Ber., 1904, **37,** 1862.
3 *P. Cohn* and *L. Springer*, Monatsh., 1903, **24,** 98.
4 *Sachs* and *R. Kempf*, Ber., 1902, **35,** 1234.

Thus 2-*hydroxyiminobenzaldoxime* (XI), m.p. 128°, is formed by suitable reduction of 2-nitrobenzaldoxime or by the action of hydroxylamine on anthranil. It is oxidised by air to *azoxybenzene-2,2′-dicarbaldehyde dioxime*, $HON{=}CH{\cdot}C_6H_4\overset{\oplus}{N}$—$\overset{\ominus}{O}){=}NC_6H_4{\cdot}CH{=}NOH$, m.p. 211°. *Azoxybenzene-2,2-dicarbaldehyde*, m.p. 119°, is prepared conveniently by the reduction of 2-nitrobenzaldehyde diacetate. *Azoxybenzene*-3,3′- and -4,4′-*dicarbaldehydes*, m.p. 129° and 195°, respectively, are obtained from their oximes which result from the reduction of 3- and 4-nitrobenzaldoximes. Azoxybenzene-4,4′-dicarbaldehyde is also formed as its dianil by the action of potassium hydroxide on 4-nitrobenzylaniline. The acetals of azobenzenedicarbaldehydes are obtained by the reduction of the acetals of the corresponding nitrobenzaldehydes with zinc and alkali. Hydrolysis of the azo-acetal with dilute sulphuric acid yields the indazole (XIII) presumably *via* intramoleculer disproportionation of azobenzene-2,2′-dicarbaldehyde (XII) (*R. C. Elderfield*, in "Heterocyclic Compounds", Wiley, New York, 1957, Vol. 5, p. 175):

(XII) → (XIII) [2]

(viii) Aminobenzaldehydes

Physical constants and selected m.p. data are summarised in Table 8.

2-**Aminobenzaldehyde,** $NH_2C_6H_4CHO$, is prepared by the reduction of 2-nitrobenzaldehyde with ferrous sulphate and ammonia (*F. G. Mann* and *A. J. Wilkinson*, J. chem. Soc., 1957, 3346; *L. I. Smith* and *J. W. Opie*, Org. Synth., Coll. Vol. 3, 1955, p. 56). It changes on storage, and more rapidly in the presence of dilute acid, to anhydrotris-2-aminobenzaldehyde, $C_{21}H_{17}N_3O$ (*A. Albert* and *H. Yamamoto*, J. chem. Soc., B, 1966, 956).

Self-condensation of 2-aminobenzaldehyde in the presence of nickel(II) or copper(II) ions produces complexes of the following type

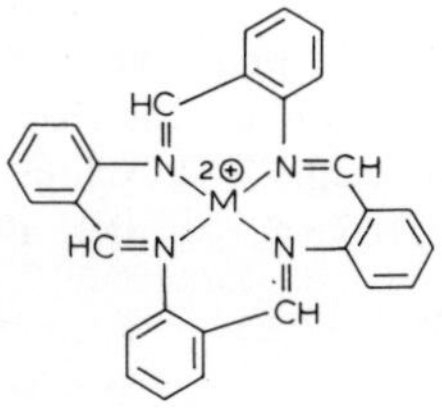

(*G. A. Melson* and *D. H. Busch*, J. Amer. chem. Soc., 1965, **87,** 1706). It is converted to anthranil on oxidation with hydrogen peroxide; with permonosulphuric acid, 2-hydroxyformanilide is formed as well as anthranil (*Bamberger*, Helv., 1924, **7,** 815; Ber., 1903, **36,** 2042).

Formation of heterocyclic compounds. 2-Aminobenzaldehyde condenses with compounds having the grouping $-CH_2-CO-$ to give quinoline derivatives (Friedländer Synthesis). The condensation is generally brought about by alkali or a base such as piperidine:

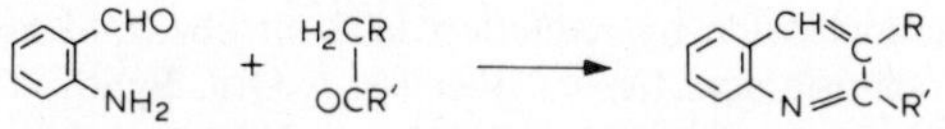

It has been carried out with a wide range of compounds including acetaldehyde, acetone, acetylacetone, benzoylacetone, dibenzoylmethane, acetylpyruvic acid ester ($R = Ac$, $R' = CO_2Et$), oxaloacetic ester and malonic acid (for review see *Elderfield* in "Heterocyclic Compounds", Wiley, London and New York, 1952, Vol. 4, p. 45). Condensation of *o*-aminobenzaldehyde with methazonic acid, $NO_2CH_2{\cdot}CH{=}NOH$, proceeds in the same way to give 3-nitroquinoline (*B.A.S.F.*, G.P. 335,197/1919); with acetonesulphonic acid, quinaldine-3-sulphonic acid is formed.

2-*Acetamidobenzaldehyde* (XIV) cyclises under the influence of alkali to give 2-quinolone (XV), and if ammonia is used in place of alkali, 2-methylquinazoline (XVI) is formed (*A. Bischler*, Ber., 1891, **24,** 507; 1895, **28,** 297):

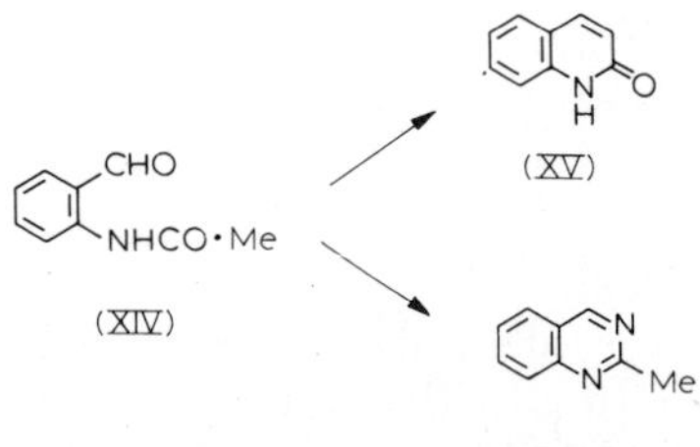

(XIV) (XV) (XVI)

Cyclisation of 2-anilinobenzaldehyde (XVII) in the presence of sulphuric acid gives acridine (XVIII) (see *A. Albert*, in "The Acridines", Arnold, London, 1966, p. 78):

(XVII) (XVIII)

2-**Azidobenzaldehyde** (XIX), m.p. 37°, prepared from 2-aminobenzaldehyde by the action of sodium azide on the diazonium chloride, yields anthranil (XX) when it is heated alone or with water. Its oxime XXI cyclises on heating with aqueous alkali to form 2-hydroxyindazole (XXII) (*Bamberger*, Ber., 1902, **35,** 1885):

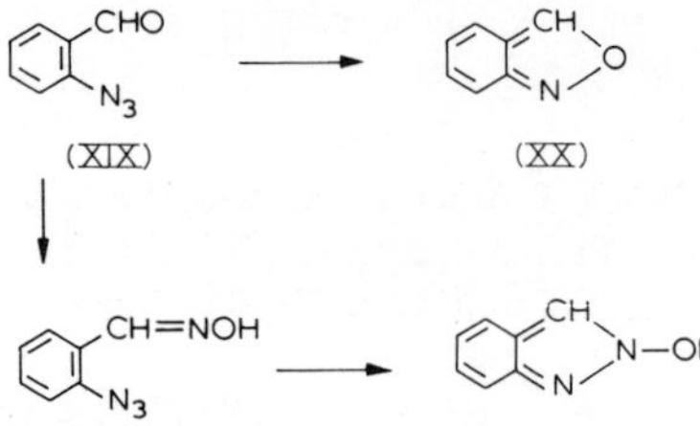

(XIX) (XX) (XXI) (XXII)

3-**Aminobenzaldehyde** is unstable (*W. Borsche* and *F. Sell*, Ber., 1950, **83,** 78) and is isolated as a stannichloride by reduction of 3-nitrobenzaldehyde with stannous chloride and hydrochloric acid (*R. B. Woodward*, Org. Synth., Coll. Vol. 3, 1955, p. 453). The *dimethyl acetal*, b.p. 123–124°/4 mm, is prepared by reduction of the corresponding nitro compound with hydrogen and Raney nickel (*R. N. Icke et al.*, *ibid.*, p. 59). Catalytic reduction of 3-nitrobenzaldehyde in the presence of formaldehyde in alcoholic solution gives 3-dimethylaminobenzaldehyde (*V. M. Ingram*, J. chem. Soc., 1950, 2247).

4-**Aminobenzaldehyde,** which readily undergoes acid-catalysed self-condensation, is prepared by reaction of 4-nitrotoluene with sodium sulphide and sulphur in aqueous sodium hydroxide (*E. Campaigne*, *W. M. Budde* and *G. F. Schaefer*, Org. Synth., Coll. Vol. 4, 1963, p. 31).

4-**Methylaminobenzaldehyde** and 4-**dimethylaminobenzaldehyde** are formed by reaction of mono- and di-methylaniline respectively with alloxan and subsequent acid hydrolysis of the product (*J. R. Geigy and Co.*, G.P. 103,578/1897; *C. F. Boehringer u. Söhne*, G.P. 108,026/1898).

$$\mathrm{RR^1NPh} + \mathrm{CO}\begin{matrix}\diagup \mathrm{CO-NH} \diagdown \\ \diagdown \mathrm{CO-NH} \diagup\end{matrix}\mathrm{CO} \longrightarrow \mathrm{RR^1NC_6H_4C(OH)}\begin{matrix}\diagup \mathrm{CO-NH} \diagdown \\ \diagdown \mathrm{CO-NH} \diagup\end{matrix}\mathrm{CO}$$

4-Dimethylaminobenzaldehyde is prepared in 80–84% yield by treating dimethylaniline with the dimethylformamide–phosphoryl chloride complex (*Campaigne* and *W. L. Archer*, Org. Synth., Coll. Vol. 4, 1963, p. 331).

A modified Vilsmeier formylation procedure involves using dimethylformamide and triphenylphosphine dibromide and gives 72% of the aldehyde (*H. J. Bestmann, J. Lienert* and *L. Mott*, Ann., 1968, **718,** 24). Alternatively it can be obtained in 56–59% yield from reaction of dimethylaniline, formaldehyde and 4-nitrosodimethylaniline (*R. Adams* and *G. H. Coleman*, Org. Synth., Coll. Vol. 1, 2nd Edn., 1947, p. 214). 4-Dimethylaminobenzaldehyde is used as a reagent in colour tests for pyrroles having at least one unsubstituted CH grouping indole, skatole and tryptophan (see *F. Feigl*, in "Spot Tests in Organic Analysis", 7th Edn., Elsevier, Amsterdam, 1966, p. 381). It reacts with dichloromethylenetriphenylphosphorane, $Ph_3P{=}CCl_2$, to give β,β-dichloro-4-dimethylaminostyrene, $[4]Me_2NC_6H_4{\cdot}CH{=}CCl_2$ (*A. J. Speziale, K. W. Ratts* and *D. E. Bissing*, Org. Synth., 1965, **45,** 33).

4-**Diethylaminobenzaldehyde** has been prepared by heating a mixture of diethylaniline, hexamethylenetetramine, formic and acetic acids (*J. C. Duff*, J. chem. Soc., 1945, 276). A number of 4-dialkylaminobenzaldehydes have been prepared in this way including benzylmethylaminobenzaldehyde.

4-**Amino**-3-**nitrobenzaldehyde,** is obtained from the nitration of 4-acetamidobenzaldehyde in acetic anhydride and subsequent hydrolysis of the product (*H. H. Hodgson* and *H. G. Beard*, *ibid.*, 1927, 20).

(ix) Benzaldehydesulphonic acids

Benzaldehyde-2-**sulphonic acid,** 2-*formylbenzenesulphonic acid*, is prepared by heating 2-chlorobenzaldehyde with sodium sulphite at 190–200° (*J. R. Geigy and Co.*, G.P. 88,952/1896). It is also obtained by the oxidation of toluene-2-sulphonic acid with manganese dioxide and sulphuric acid (*Chem. Fabr. v. Sandoz in Basel*, G.P. 154,528/1902) or of stilbene-2,2′-disulphonic acid with permanganate (*Levinstein Ltd.*, G.P. 119,163/1897). Its *chloride* has m.p. 114–115° (*I. Goldberger*, Monatsh., 1916, **37,** 125). 4-**Nitrobenzaldehyde**-2-**sulphonic acid** is obtained by oxidation of the corresponding dinitrostilbenedisulphonic acid (**A. G. Green** and **A. R. Wahl,** Ber., 1897, **30,** 3101).

Benzaldehyde-3-**sulphonic acid** is obtained conveniently by the sulphonation of benzaldehyde (*F. J. Moore* and *R. Thomas*, J. Amer. chem. Soc., 1922, **44,** 368). **Benzaldehyde**-4-**sulphonic acid** and **benzaldehyde-2,4-disulphonic acid** are prepared by oxidation of the corresponding toluenesulphonic acids by manganese dioxide in sulphuric acid. *Benzaldehyde*-4-*sulphonamide*, m.p. 122–124°, has been obtained by the oxidation of 4-toluenesulphonamide with chloramine (*H. D. Dakin*, Biochem. J., 1917, **11,** 94), alternatively from 4-toluenesulphonyl chloride *via* the benzylidene diacetate (*T. P. Sycheva* and *M. N. Shchukina*, C.A., 1955, **49,** 932).

(e) Aryl-substituted aliphatic aldehydes

Many homologous aryl-substituted acetaldehydes have been prepared from the appropriate aralkylmagnesium chloride and orthoformic ester (*L. Bert*, Compt. rend., 1928, **186,** 700) also from the appropriate aralkylglycidic ester (*P. Chuit* and *J. Bolle*, Bull. Soc. chim. Fr., 1924, [iv], **35,** 201).

Physical constants and selected derivative m.p. data for the aldehydes discussed in this section are summarised in Table 9.

TABLE 9

ARYL-SUBSTITUTED ALIPHATIC ALDEHYDES

	m.p. (°C)	*b.p. (°C/mm)*	n_D°	$d_\circ^\circ$	*Derivative, m.p. (°C)*	*Reference*
$PhCH_2 \cdot CHO$	33–34	195,88/10	$1.5255^{19.6}$	$1.0272_4^{19.6}$	*Semicarbazone,* 158–159; *2,4-dinitrophenyl-hydrazone,* 189–190.	1
$PhCH_2 \cdot CH_2 \cdot CHO$		221–224/744			*Semicarbazone,* 127; *2,4-dinitrophenyl hydrazone,* 144–145.	1
$PhCHMe \cdot CHO$		202–205,	1.5176^{20}	1.0089_4^{20}	*Semicarbazone,* 153–154; *2,4-dinitrophenyl hydrazone,* 136–137.	1
$PhCMe_2 \cdot CHO$		215–218, 105–108/14		0.9912°	*Semicarbazone,* 176–177.	1,2
$Ph(CH_2)_3CHO$					*2,4-Dinitrophenyl hydrazone,* 112–115.	3

References
1 See text.
2 *J. Cantacuzène* and *J.-M. Normant*, Tetrahedron Letters, 1970, 2947.
3 *A. Goosen* and *H. A. Laue*, J. chem. Soc., C, 1969, 383.

Phenylacetaldehyde, *α-toluic aldehyde,* $Ph \cdot CH_2 \cdot CHO$, has a strong hyacinth-like odour and is used in perfumery. It is prepared commercially by (*1*) passing styrene oxide vapour at 350–400° over hydrated magnesium silicate (*A. R. Sexton* and *R. C. Britton*, U. S. P. 2,628,255/1951), (*2*) the oxidation and rearrangement of cyclo-octatetraene (*L. E. Craig*, U.S.P. 2,856,431/1951):

$$C_8H_8 \xrightarrow[\text{conc. } H_2SO_4]{30\%\ H_2O_2,\ AcOH} PhCH_2 \cdot CHO$$

(*3*) a route involving glycidic ester formation from benzaldehyde and ethyl chloroacetate (*A. Knorr*, *A. Weisenborn* and *E. Laage*, U.S.P. 1,899,340/1933):

$$PhCHO + ClCH_2 \cdot CO_2Et \longrightarrow PhCH\text{—}CH \cdot CO_2Et\ (\text{epoxide, }O) \xrightarrow[(2)\ \text{Decarboxylation}]{(1)\ \text{Hydrolysis}} PhCH_2 \cdot CHO$$

Reference has already been made (pp. 61 *et seq.*) to various laboratory preparations of phenylacetaldehyde. Application of the Schmidt reaction to cinnamic acid (*M. Oesterlin*, Angew. Chem., 1932, **45,** 536) and the Hofmann reaction to cinnamamide (*E. S. Wallis* and *J. F. Lane*, Org. Reactions, 1946, **3,** 285) gives phenylacetaldehyde. Aniline is formed as a by-product in the Schmidt reaction.

Phenylacetaldehyde is also obtained from the reaction of 2-iodo-1-phenylethanol (styrene iodohydrin) with aqueous silver nitrate (*S. Winstein* and *L. L. Ingram*, J. Amer. chem. Soc., 1955, **77,** 1738).

Phenylacetaldehyde polymerises on keeping at room temperature; with 20% aqueous sulphuric acid it yields a trimeride (*H. Stobbe* and *A. Lippold*, J. prakt. Chem., 1914, [ii], **90,** 280). On treatment of phenylacetaldehyde with potassium hydroxide in ethanol, it is converted into *trans*-1,3-diphenylpropene (*E. K. Raunio* and *W. A. Bonner*, J. org. Chem., 1966, **31,** 396):

$$2\ PhCH_2 \cdot CHO \longrightarrow (H)(PhCH_2)C{=}C(Ph)(H)$$

Treatment of phenylacetaldehyde with dimethylamine in the presence of potassium hydroxide gives β-dimethylaminostyrene (*J. R. Geigy A. G.*, B.P. 832,078/1960; C.A., 1960, **54,** 20877):

$$PhCH_2 \cdot CHO + Me_2NH \xrightarrow{KOH} PhCH{=}CHNMe_2$$

Reaction of phenylacetaldehyde with *N,N'*-dimethylhydrazine in acetic acid gives 51% of 1-methyl-3,4-diphenylpyrrole. A bis-ene-hydrazine is formed intermediately (*W. Sucrow* and *G. Chondromatidis*, Ber., 1970, **103,** 1759):

$$PhCHO \xrightarrow{MeNH \cdot NHMe} (Ph)(H)C{=}C(H)\text{—}N(Me)\text{—}N(Me)\text{—}C(H){=}C(Ph)(H) \xrightarrow{-MeNH_2} \text{3,4-diphenyl-1-methylpyrrole (Ph, Ph, N–Me)}$$

Phenylacetaldehyde reacts with primary and secondary amides and lactams in the presence of toluene-4-sulphonic acid (*H. Moehrle* and *R. Kilian*, Arch. Pharm., 1967, **300,** 917) *e.g.*:

$$PhCH_2 \cdot CHO \xrightarrow{AcNH_2} PhCH_2 \cdot CH(NHAc)_2\ (91\%)$$

Phenylacetaldehyde is readily converted into its *dimethyl acetal*, b.p. 219–220°, n_D^{20} 1.4940, d_4^{20} 1.005, on treatment with methanol and hydrogen chloride, alternatively by the action of methanol and thallium(III) acetate on styrene (*H.-J. Kabbe*, Ann., 1962, **656,** 204). This and the cyclic acetal of ethylene glycol are important in perfumery.

β-**Phenylpropionaldehyde,** *3-phenylpropanal, hydrocinnamaldehyde,* $PhCH_2{\cdot}CH_2{\cdot}CHO$, is present in some samples of oil of cinnamon. It is best prepared by hydrolysis of its acetal which is formed by the reduction of cinnamaldehyde diethyl acetal (*E. Fischer* and *E. Hoffa,* Ber., 1898, **31,** 1992). *β*-Phenylacetaldehyde has also been prepared by reaction of cinnamonitrile with dicobalt octacarbonyl in acidic methanol (*H. Wakanatsu* and *K. Sakamaki,* Chem. Comm., 1967, 1140):

$$PhCH{=}CH{\cdot}CN \quad + \quad Co_2(CO)_8 \quad + \quad 5\ HCl \quad + \quad H_2O \xrightarrow{CH_3OH}$$

$$\underset{(50\%)}{PhCH_2{\cdot}CH_2{\cdot}CHO} \quad + \quad 2\ CoCl_2 \quad + \quad NH_4Cl \quad + \quad 8\ CO$$

α-**Phenylpropionaldehyde,** *2-phenylpropanal, hydratropic aldehyde,* $PhMeCH{\cdot}CHO$, has an odour similar to phenylacetaldehyde and is also used in perfumery. Hydratropic aldehyde does not tend to polymerise and is widely used in hyacinth type and other floral perfumes. It is manufactured from acetophenone by Darzen's synthesis

$$PhCO{\cdot}CH_3 \quad + \quad ClCH_2{\cdot}CO_2Et \longrightarrow PhC(CH_3){-}CHCO_2Et \text{ (epoxide, bridging O)}$$

$$\xrightarrow[(2)\ Decarboxylation]{(1)\ Hydrolysis} PhCH(CH_3){\cdot}CHO$$

(see *Knorr, Weisenborn* and *Laage, loc. cit.*); detailed laboratory instructions for these reactions are given by *C. F. H. Allen* and *J. Van Allan* in Org. Synth., Coll. Vol. 3, 1955, pp. 727 and 733. Alternatively hydratropic aldehyde may be prepared by the following route:

$$PhMgX \quad + \quad CH_3{\cdot}CO{\cdot}CH_2Cl \longrightarrow PhC(CH_3)(OH){\cdot}CH_2Cl \xrightarrow{Alkali} PhC(CH_3){-}CH_2 \text{ (epoxide, bridging O)}$$

$$\xrightarrow{H_2O/H^{\oplus}} PhCH(CH_3){\cdot}CHO$$

This is illustrative of a general reaction for preparing aldehydes of this type (*M. S. Malinowskii* and *A. G. Yudasina,* Zhur. obshcheĭ Khim., 1960, **30,** 1831). In the presence of dicobalt octacarbonyl, styrene reacts with carbon monoxide and hydrogen to give hydratropic aldehyde (*H. Adkins* and *G. Krsek,* J. Amer. chem. Soc., 1949, **71,** 3051). For a discussion of the hydroformylation reaction see *C. W. Bird,* "Transition Metals in Organic Synthesis", Logos Press, London, 1967. Reaction of norephedrin with nitrous acid gives (+)-hydratropic aldehyde which can be isolated as a *semicarbazone,* m.p. 128–130° (*A. McKenzie, E. M. Luis* and *A. G. Mitchell,* Ber., 1932, **65,** 798). Reaction of the aldehyde with molecular oxygen in the presence of a cupric salt gives 78% of acetophenone. This reaction is illustrative of a general synthesis of ketones from *α*-branched aldehydes (*V. Van Rheenen,* Tetrahedron Letters, 1969, 985).

α-**Phenylisobutyraldehyde,** *2-methyl-2-phenylpropanal,* $PhCMe_2{\cdot}CHO$, is prepared from *α,α*-dimethyl-*α′*-phenylethylene glycol, $Me_2C(OH){\cdot}CPhHOH$ (*M. Tiffeneau* and

H. Dorlencourt, Ann. Chim. Fr., 1909, [viii], **16,** 248; *J. Lévy*, Bull. Soc. chim. Fr., 1921, [iv], **29,** 824; *Tiffeneau* and *Lévy*, *ibid.*, 1923, [iv], **33,** 749); from the corresponding epoxide (*Tiffeneau* and *A. P. Orékhov*, *ibid.*, 1921, [iv], **29,** 817; *Tiffeneau et al.*, Compt. rend., 1924, **179,** 979; *Lévy* and *A. Tabart*, *ibid.*, 1929, **188,** 402); or from the iodohydrin (*Tiffeneau*, Ann. Chim. Fr., 1907, [viii], **10,** 363). On treatment with concentrated sulphuric acid at room temperature it rearranges to 3-phenylbutan-2-one:

$$PhC(CH_3)_2 \cdot CHO \rightarrow PhCH(CH_3) \cdot CO \cdot CH_3$$

2-**Methyl**-3-**phenylpropionaldehyde,** 2-**methyl**-3-**phenylpropanal,** $PhCH_2 \cdot CHMe \cdot CHO$, has been prepared by the following route (*R. F. Heck*, Org. Synth., 1972, **51,** 17):

$$PhHgOAc + Pd(OAc)_2 \longrightarrow [PhPdOAc] \xrightarrow{CH_2{=}CMe \cdot CH_2OH}$$

$$\left[\begin{array}{c} PhCH_2 \cdot \underset{\displaystyle PdOAc}{\underset{|}{C}Me} \cdot CH_2OH \end{array}\right] \longrightarrow PhCH_2 \cdot CHMe \cdot CHO$$

2-**Ethyl**-2-**phenylbutanal,** $PhCEt_2 \cdot CHO$, undergoes a similar rearrangement to that of 2-methyl-2-phenylpropanal to give 4-phenylhexan-3-one (*Orékhov* and *Tiffeneau*, Compt. rend., 1926, **182,** 68).

4. Monoketones in the benzene series

In this section two types of aromatic monoketone containing one aryl radical are described, (*1*) those in which the aryl residue is linked directly to the oxo group, as in acetophenone, PhCOMe, and (*2*) the mono-aryl derivatives of aliphatic ketones, of which benzyl methyl ketone, $PhCH_2 \cdot$ COMe, is the simplest representative. Many of the methods of preparation given for aliphatic ketones (Vol. I C, p. 53) are applicable to these ketones.

(*a*) *Methods of formation and preparation*

(i) From alkylbenzenes and secondary alcohols

Ethylbenzene has been oxidised to acetophenone with aqueous silver nitrate–ammonium peroxodisulphate, $(NH_4)_2S_2O_8$, in 73% yield (*F. A. Daniher*, Org. Prep. and Procedures, 1970, **2,** 207) and also by using argentic picolinate (*J. B. Lee* and *T. G. Clarke*, Tetrahedron Letters, 1967, 415). Oxidation of *n*-propylbenzene with chromyl chloride gives propiophenone, $PhCO \cdot CH_2 \cdot CH_3$, and benzyl methyl ketone, $PhCH_2 \cdot CO \cdot CH_3$. The latter compound is the main product and arises from a rearrangement reaction (*K. B. Wiberg*, *B. Marshall* and *G. Foster*, *ibid.*, 1962, 345). The oxidation of secondary alcohols to ketones has been achieved by chromates, permanganate, transition metal ions, aluminium alkoxides, and non-metallic oxidants such as dinitrogen tetroxide (*J. Grundy*, J. chem. Soc., 1957, 5087), *N*-bromoamides (*R. A. Corral* and *O. O. Orazi*, Chem. Comm., 1965, 5) and diethyl azodicarboxylate (*F. Yoneda*, *K. Suzuki* and *Y. Nitta*, J. org. Chem., 1967, **32,** 727) or by dehydrogenation over hot copper or other catalysts. Secondary alcohols which undergo oxidative cleavage on oxidation with chromic acid, lead(IV) compounds or cerium(IV) compounds, may be oxidised to the corresponding ketones using dimethyl sulphoxide in acetic anhydride at room temperature (*W. H. Clement*, *T. J. Dangieri* and *R. W. Tuman*, Chem. and Ind., 1969, 755):

$$PhCH(OH)Bu^t \rightarrow PhCOBu^t \quad (98\%)$$

Carbonyl-forming oxidation reactions have been reviewed by *C. F. Cullis* and *A. Fish*, in "The Chemistry of the Carbonyl Group", ed. *S. Patai*, Interscience, London, p. 79.

(ii) From alkyl halides

(*1*) Alkylation of pyridinium ylids containing an activated methylene group, followed by reductive cleavage of the pyridine residue provides a route to ketone synthesis (*C. A. Henrick*, *E. Ritchie* and *W. C. Taylor*, Austral. J. Chem., 1967, **20,** 2441):

$$PhCH_2Br \xrightarrow[(2)\ Zn,\ AcOH,\ H_2O]{(1)\ C_5H_5\overset{\oplus}{N}-CH=C(O^{\ominus})-CH_3,\ DMF} PhCH_2\cdot CH_2\cdot CO\cdot CH_3 \quad (85\%)$$

(*2*) Unsymmetrical ketones may be prepared by reaction of alkyl halides with acyltetracarbonylferrates (obtained from the corresponding alkyl- or aryl-lithium compounds and iron pentacarbonyl (see *Y. Sawa*, *M. Ryans* and *S. Tsutsumi*, Tetrahedron Letters, 1969, 5189; J. org. Chem., 1970, **35,** 4183):

$$\overset{\oplus}{Li}[n\text{-}BuCOFe(CO)_4]^{\ominus} + ICH_2Ph \xrightarrow{Ether,\ -40^\circ} n\text{-}BuCO\cdot CH_2Ph \quad (50\%)$$

(iii) From arylamines

Condensation of diazonium salts with acetaldoxime, acetaldehyde semicarbazone or propionaldoxime and subsequent hydrolysis gives alkyl aryl ketones (*W. F. Beech*, J. chem. Soc., 1954, 1297):

$$p\text{-}ClC_6H_4NH_2 \longrightarrow p\text{-}ClC_6H_4N_2^{\oplus} \xrightarrow[(2)\ HCl\ -\ H_2O]{(1)\ EtCH=NOH,\ CuSO_4\ -\ AcONa,\ Na_2SO_3\ -\ H_2O} p\text{-}ClC_6H_4COEt$$

(iv) From carboxylic acids and their derivatives

(*1*) *From carboxylic acids.* On heating mixtures of the iron salts of a benzoic acid and an aliphatic acid, alkyl aryl ketones are formed in satisfactory yields. A mixture of the appropriate acids and iron powder is heated at 250° until evolution of hydrogen ceases. On further heating at 280–300°, carbon dioxide is evolved and alkyl aryl ketones are formed (*C. Granito* and *H. P. Schultz*, J. org. Chem., 1963, **28,** 879).

Ketones are formed by passing a mixture of acid vapours over a heated catalyst such as thoria, zirconia and manganous oxide. For example, benzyl methyl ketone is obtained in 55–65% yield by passing a mixture of phenylacetic acid and acetic acid over thoria at 430–450° (*R. M. Herbst* and *R. H. Manske*, Org. Synth., Coll. Vol. 2, 1943, p. 389). Similar treatment of a mixture of *o*-ethylbenzoic acid and acetic acid gives *o*-ethylacetophenone in 74% yield (*W. Winkler*, Ber., 1948, **81,** 256).

Reaction of a carboxylic acid or its lithium salt with an organolithium compound gives ketones (*H. Gilman* and *P. R. van Ess*, J. Amer. chem. Soc., 1933, **55,** 1258; see also *C. Tegner*, Acta Chem. Scand., 1952, **6,** 782):

$$Pr^n{\cdot}CO_2Li + PhLi \longrightarrow (Pr^n)(Ph)C(OLi)_2 \longrightarrow Pr^nCOPh \ (62\%)$$

The reaction of methyl-lithium with the lithium salts of a variety of substituted benzoic acids has been successfully used for acetophenone synthesis. For example, anthranilic acid has been converted in 45% yield to *o*-aminoacetophenone by this method (*J. Itier* and *A. Casadevall*, Bull. Soc. chim. Fr., 1969, 2342). A review of the reactions of organolithium reagents with carboxylic acids has appeared (*M. J. Jorgenson*, Org. Reactions, 1970, **18,** 1).

(*2*) *Carboxylic chlorides and anhydrides* react with Grignard compounds in ether at −30° to give ketones almost exclusively (*H. Fournier*, Bull. Soc. chim. Fr., 1904, [iii], **31,** 483; 1906, [iii], **35,** 19; 1910, [iv], **7,** 836; *M. S. Newman* and *W. T. Booth*, J. Amer. chem. Soc., 1945, **67,** 154; *Newman* and *A. G. Smith*, J. org. Chem., 1948, **13,** 592). The mixed anhydrides prepared as shown below react with organocadmium reagents to give ketones (*D. S. Tarbell* and *J. R. Price*, *ibid.*, 1956, **21,** 144; 1957, **22,** 245):

$$R^1CO_2H \xrightarrow[NEt_3]{ClCO_2Et} R^1CO_2CO_2Et \xrightarrow{R_2Cd} R^1COR$$

Aliphatic carboxylic acid chlorides and anhydrides react with benzene under the conditions of the Friedel–Crafts reaction to give alkyl aryl ketones (*R. Adams* and *C. R. Noller*, Org. Synth., Coll. Vol. 1, 2nd Edn., 1947, p. 109) *e.g.*:

$$PhBr \quad + \quad (CH_3 \cdot CO)_2O \quad \xrightarrow{AlCl_3} \quad [4]CH_3 \cdot CO \cdot C_6H_4Br \quad + \quad CH_3 \cdot CO_2H$$
(69–79%)

Friedel–Crafts acetylation of both *tert*-butylbenzene and 1,4-di-*tert*-butylbenzene gives 4-*tert*-butylacetophenone (*G. F. Hennion* and *S. F. de C. McLeese*, J. Amer. chem. Soc., 1942, **64,** 2421). For reviews see *P. H. Gore*, in "Friedel–Crafts and Related Reactions", ed. *G. A. Olah*, Interscience, New York, 1964, Vol. 3, Pt. 1, p. 1 and *D. P. N. Satchell* and *R. S. Satchell*, in "The Chemistry of the Carbonyl Group", ed. *S. Patai*, Interscience, London, 1966, p. 233). Carboxylic acid chlorides and anhydrides can be converted into ketones by reaction with organocadmium compounds (*D. A. Shirley*, Org. Reactions, 1954, **8,** 28; *M. Cais* and *A. Mandelbaum*, in "The Chemistry of the Carbonyl Group", ed. *S. Patai*, Interscience, London, 1966, p. 303):

$$2\ R^1COCl \quad + \quad R_2Cd \quad \rightarrow \quad 2\ R^1COR \quad + \quad CdCl_2$$

Alternatively organomercury compounds react smoothly with acid chlorides in the presence of aluminium bromide (*A. L. Kurts et al.*, J. organometal. Chem., 1969, **17,** P21) *e.g.:*

$$PhCOCl \xrightarrow{n\text{-}Bu_2Hg\ -\ AlBr_3\ -\ CH_2Cl_2} PhCOBu^n$$
(73%)

Methyl and *n*-alkyl ketones have been prepared from acid chlorides and organocopper reagents:

$$R^1COCl \xrightarrow{R_2CuLi} R^1COR$$

Ethereal solutions of organocopper reagents are prepared from the reaction of an alkyl-lithium with cuprous iodide (*G. H. Posner* and *C. E. Whitten*, Tetrahedron Letters, 1970, 4647; see also *Posner*, *Whitten* and *P. E. McFarland*, J. Amer. chem. Soc., 1972, **94,** 5106).

A general method for the conversion of acid chlorides to methyl ketones is illustrated by the following equations (*C. A. Reynolds* and *C. R. Hauser*, Org. Synth., Coll. Vol. 4, 1963, p. 709):

$$CH_2(CO_2Et)_2 \quad + \quad Mg \quad + \quad EtOH \quad \longrightarrow \quad EtOMgCH(CO_2Et)_2$$

$$\xrightarrow{ArCOCl} ArCO \cdot CH(CO_2Et)_2 \xrightarrow{H_2SO_4\ -\ AcOH} ArCO \cdot CH_3$$

Reaction of alkylidenephosphoranes with acid chlorides gives an intermediate which on hydrolysis, or on reduction with zinc and acetic acid, is converted into a ketone (*H. J. Bestmann* and *B. Arnason*, Ber., 1962, **95,** 1513). The corresponding esters have also been employed as the acylating agent:

$$RCOCl + R^1CH{=}PPh_3 \rightarrow RCO{\cdot}C(R^1){=}PPh_3 \rightarrow RCO{\cdot}CH_2R^1$$

Acid chlorides react with triethyl phosphite to give acylphosphonates, the latter compounds on successive reaction with Grignard reagents (or organolithium reagents) and aqueous alkali are converted into ketones (*I. Shahak* and *E. D. Bergmann*, Israel J. Chem., 1966, **4,** 225):

$$PhCOCl + P(OEt)_3 \longrightarrow PhCOP(O)(OEt)_2 \xrightarrow{n\text{-}BuLi} PhC(OH)(Bu^n)P(O)(OEt)_2 \xrightarrow{OH^{\ominus}} PhCOBu^n \ (61\%)$$

Alkyl aryl ketones have also been prepared from acid chlorides and anhydrides by aluminium chloride-promoted reaction with disilanes (*E. Frainnet, R. Calas* and *P. Gerval*, Compt. rend., 1965, **261,** 1329), *e.g.*:

$$Ph_3Si{\cdot}SiMe_3 \xrightarrow{+ (EtCO)_2O \text{ or } + EtCOCl} PhCOEt \ (90\%)$$

(*3*) *From esters.* Carboxylic esters are converted into methyl ketones by the following route (*E. J. Corey* and *M. Chaykovsky*, J. Amer. chem. Soc., 1964, **86,** 1639; 1965, **87,** 1345):

$$RCOOR^1 \xrightarrow{\overset{\ominus}{C}H_2SOMe} RCO{\cdot}CH_2SOMe \xrightarrow{Al/Hg} RCOMe$$

A solution of methylsulphinyl carbanion is obtained by adding sodium hydride to an excess of dimethyl sulphoxide. Analogous ketone syntheses from carboxylic esters use the sodium salts of *N,N*-dimethylmethanesulphonamide, or dimethyl sulphone (*H. O. House* and *J. K. Lawson*, J. org. Chem., 1968, **33,** 61) or the dilithium salts of sulphinamides (*Corey* and *T. Durst*, J. Amer. chem. Soc., 1966, **88,** 5656):

$$RCO_2R^1 + LiCH_2SONLiC_7H_7 \rightarrow RCO{\cdot}CH_2SONLiC_7H_7 \xrightarrow{H_2O} RCO{\cdot}CH_3$$

The transformations of β-oxosulphoxides ($RCO{\cdot}CH_2SOCH_3$) including their alkylation and reduction to ketones have also been investigated by *G. A. Russell* and *G. J. Mikol* (*ibid.*, 1966, **88,** 5498).

Benzene forms stearophenone when treated with isopropenyl stearate and aluminium chloride (*E. S. Rothman* and *G. G. Moore*, J. org. Chem., 1970, **35,** 2351):

$$C_6H_6 \xrightarrow{C_{17}H_{35}COOC(CH_3){=}CH_2 - AlCl_3} C_6H_5COC_{17}H_{35} \ (60\%)$$

On heating ethyl phenylacetate with 1-ethoxy-*N*,*N*-dimethylvinylamine, and subsequent hydrolysis of the product, phenylacetone is obtained (*T. Oishi et al.*, Chem. Pharm. Bull. Japan, 1969, **17,** 2314):

$$PhCH_2\cdot CO_2Et + (CH_3)_2NC(OEt){=}CH_2 \longrightarrow Ph{-}C({=}C(CH_3)N(CH_3)_2){-}CO_2Et \xrightarrow{HCl - H_2O} PhCH_2\cdot CO\cdot Me$$

The azides derived from disubstituted malonyl esters are rearranged to the corresponding carbamates in alcohol and these are hydrolysed to ketones by mineral acids (*P. A. S. Smith*, Org. Reactions, 1946, **3,** 345):

$$R^1R^2C(CO_2Et)_2 \xrightarrow{NH_2NH_2} R^1R^2C(CONH\cdot NH_2)_2 \xrightarrow{HNO_2} R^1R^2C(CON_3)_2 \xrightarrow{ROH}$$

$$R^1R^2C(NHCO_2R)_2 \xrightarrow{H^{\oplus} - H_2O} R^1R^2C{=}O$$

(*4*) *From nitriles.* Nitriles give ketones by reaction with Grignard compounds (*H. D. Zook* and *W. L. Rellahan*, J. Amer. chem. Soc., 1957, **79,** 881):

$$RCN + R^1MgX \longrightarrow RR^1C{=}NMgX \xrightarrow{H^{\oplus} - H_2O} RR^1C{=}O$$

Enhanced yields of ketones are obtained from acetonitrile when the reaction is carried out in benzene rather than ether (*C. A. Beuhler* and *D. E. Pearson*, in "Survey of Organic Syntheses", Wiley, New York, 1970, p. 717). The following modification of the reaction of nitriles with organometallic reagents can be used for the synthesis of ketones possessing a branched alkyl chain (*T. Cuvigny* and *H. Normant*, Bull. Soc. chim. Fr., 1968, 4990):

$$R^1C{\equiv}N + R^2{-}CH_2M \longrightarrow R^1{-}C({=}NM){-}CH_2{-}R^2 \xrightarrow{R^2{-}CH_2M} R^1{-}C({=}NM){-}CHM{-}R^2$$

$$\xrightarrow[(2)\ H_2O/H^{\oplus}]{(1)\ Hal - R^3} R^1{-}C({=}O){-}CH(R^3){-}R^2 + R^1{-}C({=}O){-}CH_2{-}R^2$$

(where M = MgCl, MgBr or Li)

(v) From β-oxo-esters

Hydrolysis of β-oxo-esters using sulphuric acid (*A. Baeyer* and *W. H. Perkin*, Ber., 1883, **16,** 2129):

$$PhCO{\cdot}CH_2{\cdot}CO_2Et \quad \rightarrow \quad PhCO{\cdot}CH_3$$

or water at 200° (*H. Meerwein*, Ann., 1913, **398,** 248) gives ketones. This synthesis has been extended to the preparation of ketones of type $ArCO{\cdot}CHR_2$, the intermediate oxoester being prepared by reaction of the acid chloride with the enolate of the appropriate ester (*B. E. Hudson* and *C. R. Hauser*, J. Amer. chem. Soc., 1941, **63,** 3163):

$$PhCOCl \xrightarrow[Ph_3CNa\ -\ Et_2O]{Et_2CH{\cdot}CO_2Et\ -} PhCO{\cdot}CEt_2(CO_2Et) \xrightarrow{HI\ -\ AcOH\ -\ H_2O} PhCO{\cdot}CHEt_2$$

The yields of β-oxo-esters varied from 51–74% and the yields of ketones obtained on hydrolysis varied from 69–89%. The dianion of methyl acetoacetate, generated by successive treatment of the β-oxo-ester with sodium hydride and *n*-butyl-lithium readily undergoes alkylation at the γ-carbon.

$$CH_3{\cdot}CO{\cdot}CH_2{\cdot}CO_2Me \xrightarrow{NaH} CH_3{\cdot}CO{\cdot}\overset{\ominus}{C}H{\cdot}CO_2Me \xrightarrow{Bu^nLi}$$

$$\overset{\ominus}{C}H_2{\cdot}CO{\cdot}\overset{\ominus}{C}H{\cdot}CO_2Me \xrightarrow{PhCH_2Cl} PhCH_2{\cdot}CH_2{\cdot}CO{\cdot}CH_2{\cdot}CO_2Me \longrightarrow PhCH_2{\cdot}CH_2{\cdot}COMe$$

A similar technique has been used to generate dianions from substituted β-oxo-esters of the type $RCH_2{\cdot}CO{\cdot}CH_2{\cdot}CO_2Me$ and hence alkylated products of the type $RR^1CH{\cdot}CO{\cdot}CH_2{\cdot}CO_2Me$ (*L. Weiler*, J. Amer. chem. Soc., 1970, **92,** 6702).

(vi) From β-diketones

A general method for ketone synthesis from β-diketones is illustrated in the following reaction scheme (*S. Boatman, T. M. Harris* and *C. R. Hauser*, J. org. Chem., 1965, **30,** 3321):

$$CH_3{\cdot}CO{\cdot}CH_2{\cdot}CO{\cdot}CH_3 \xrightarrow{RX,\ K_2CO_3,\ EtOH} [CH_3{\cdot}CO{\cdot}CHR{\cdot}CO{\cdot}CH_3] \longrightarrow CH_3{\cdot}CO{\cdot}CH_2R \quad \text{(yields 52-79\%)}$$

(vii) From dihydro-1,3-oxazines

The use of dihydro-1,3-oxazines for ketone synthesis is illustrated by the following preparation of 1-phenyl-3-pentanone (*A. I. Meyers* and *E. M. Smith*, J. Amer. chem. Soc., 1970, **92,** 1084):

The corresponding 2-isopropenyloxazine has also been utilised for ketone synthesis (*Meyers* and *A. C. Kovelsky*, *ibid.*, 1969, **91,** 5887), and α-phenylketones have been prepared from 2-(1-phenylvinyl)-4,4,6-trimethyl-4*H*,5*H*,6*H*-1,3-oxazine (*idem*, Tetrahedron Letters, 1969, 4809). The following transformation of the 2-isopropyloxazine (I) to the unsymmetrical ketone (II) illustrates a general synthesis of unsymmetrical ketones (*Meyers*, *Smith* and *A. J. Jurjevich*, J. Amer. chem. Soc., 1971, **93,** 2314):

(I) —2 PhLi→ ... —(1) EtI (2) H_2O - $H^{\oplus}$→ PhCO·CEt(Me)Me (II) (50%)

1,3-Dithianes are used in a related synthesis, these are readily prepared from the corresponding aldehyde. Sequential alkylation, followed by ring opening in the presence of mercuric ion then gives the ketone (*D. Seebach*, Synthesis, 1969, **1,** 17):

$HS(CH_2)_3SH$ + RCHO → ... —(1) BuLi (2) R′Br→ ... —$HgCl_2HgO$→ RR′C=O

(viii) From 1,2-glycols and epoxides

Ketones can be prepared from epoxides, the corresponding 1,2-glycols or halogenohydrins (*E. Fourneau* and *M. Tiffeneau*, Compt. rend., 1905, **141,** 663) *e.g.*:

PhCH—CHMe (epoxide) → $PhCH_2$·COMe

(*Triffeneau*, Ann. Chim. Fr., 1907, [viii], **10,** 345, 368):

PhCH(OH)—CH(OH)—Me —dil. H_2SO_4→ $PhCH_2$·COMe

The secondary mono-esters of open-chain, trisubstituted 1,2-glycols are converted into ketones on heating with zinc dust (*E. Ghera*, J. org. Chem., 1970, **35,** 660):

$$\mathrm{PhC(OH)(Me)-CH(OAc)Me} \xrightarrow[170^\circ]{\mathrm{Zn}} \mathrm{PhCH(Me)\cdot COMe}\ (53\%)$$

The epoxy-nitrile, prepared as shown, has been converted into the ketone by successive treatment with zinc bromide and methylmagnesium bromide (*J. Cantacuzène* and *J.-M. Normant*, Tetrahedron Letters, 1970, 2947):

$$\mathrm{PhCH(Me)\cdot CHO} \xrightarrow{\mathrm{Br_2}} \mathrm{PhC(Br)(Me)-CHO} \xrightarrow{\mathrm{CN^\ominus}} \mathrm{PhC(Me)\overset{O}{-}CH\cdot CN} \xrightarrow[(2)\ \mathrm{MeMgBr}]{(1)\ \mathrm{ZnBr_2}} \mathrm{PhCH(Me)\cdot COMe}$$

Treatment of the epoxide of α-methylstyrene with alkaline hydrogen peroxide gives acetophenone in 85–90% yield (*J. Hoffman*, J. Amer. chem. Soc., 1957, **79,** 503):

$$\mathrm{PhC(Me)\overset{O}{-}CH_2} \longrightarrow \mathrm{PhCOMe}$$

The formation of ketones by heating arylglycidic acids is a convenient method for making aryl-substituted dialkyl ketones (*G. Darzens*, Compt. rend., 1906, **142,** 214; *M. S. Newman* and *B. J. Magerlein*, Org. Reactions, 1949, **5,** 413):

$$\mathrm{Ph(Me)C\overset{O}{-}CMe\cdot CO_2H} \longrightarrow \mathrm{Ph(Me)CH\cdot COMe}$$

(ix) From olefins

Oxidative cleavage of *trans-n*-butylstilbene with osmium tetroxide and sodium metaperiodate gives valerophenone in 71% yield (*M. E. Mulvaney et al.*, J. Amer. chem. Soc., 1966, **88,** 476):

$$\mathrm{PhC{\equiv}CPh} \longrightarrow \mathrm{Ph(H)C{=}C(Bu^n)Ph} \longrightarrow \mathrm{PhCHO} + \mathrm{PhCOBu^n}$$

Alternatively photochemical cleavage of α-methylstyrene has been achieved with *N*-nitrosopiperidine (*Y. L. Chow*, *ibid.*, 1965, **87,** 4642):

$$\underset{\displaystyle Me}{PhC{=}CH_2} \longrightarrow \underset{\displaystyle Me}{PhC{=}NOH}$$

(91%)

Reaction of 1,1-diphenyl-1-butene with hydrazoic acid gives 64% of butyrophenone and 57% of aniline (*L. P. Kuhn* and *J. Di Domenico*, *ibid.*, 1950, **72**, 5777):

$$Ph_2C{=}CH{\cdot}CH_2{\cdot}CH_3 \xrightarrow{NaN_3 - H_2SO_4} PhCO{\cdot}CH_2{\cdot}CH_2{\cdot}CH_3$$

Styrenes are converted quantitatively into ketones with aryl migration on treatment with thallium(III) nitrate (TTN) (*E. C. Taylor* and *A. McKillop*, Accounts of Chem. Research, 1970, **3**, 338; *McKillop et al.*, Tetrahedron Letters, 1970, 5275):

$$PhCMe{=}CH_2 \xrightarrow{TTN/MeOH} PhCH_2{\cdot}COMe$$

(97%)

(x) From acetylenes

Hydration of arylacetylenes gives ketones (*W. Manchot* and *J. Haas*, Ann., 1913, **399**, 130; *M. Miscque, N. M. Hung* and *V. Q. Yen*, Ann. Chim. Fr., 1963, **8**, 157):

$$PhC{\equiv}CH \quad + \quad H_2O \xrightarrow{HgCl_2H^{\oplus}} PhCO{\cdot}CH_3$$

(xi) From aldehydes and ketones

Benzaldehyde has been converted into acetophenone by the following two methods, both of which are in principle general to aromatic aldehydes (*J. Bastús*, Tetrahedron Letters, 1963, 955):

$$PhCHO \xrightarrow[EtOH\ -\ H_2O]{CH_2(CN)_2\ -\ glycine} PhCH{=}C(CN)_2 \xrightarrow[OEt_2]{CH_2N_2}$$

$$\underset{\displaystyle Me}{PhC{=}C(CN)_2} \xrightarrow{NaOH\ -\ H_2O} PhCOMe$$

(100%)

(*K. D. Berlin, B. S. Rathore* and *M. Peterson*, J. org. Chem., 1965, **30**, 226):

$$PhCHO \xrightarrow{HOCH_2{\cdot}CH_2OH} PhCH\langle O{-}CH_2{-}CH_2{-}O\rangle \xrightarrow[(2)\ H_2O]{(1)\ Bu^nLi} PhCOBu^n$$

(87%)

Alternatively oxidation of the adducts from aldehydes and Grignard reagents gives ketones (*T. Mukaiyama, K. Tatsuji* and *I. Kuwajima*, Bull. chem. Soc. Japan, 1968, **41**, 1491):

$$PhCHO + Bu^nMgBr \longrightarrow \underset{Bu^n}{PhCHOMgBr} \xrightarrow{EtO_2C{-}N{=}N{-}CO_2Et} \underset{(50\%)}{PhCOBu^n}$$

Anodic methylation of benzaldehyde giving acetophenone has been reported using platinum electrodes (*A. Takeda, S. Torii* and *H. Oka*, Tetrahedron Letters, 1968, 1781):

$$PhCHO \xrightarrow[MeOH\ -\ H_2O]{AcOH\ -\ KOH} \underset{(18\%)}{PhCOMe}$$

Alkyl benzyl ketones can be made from an aromatic aldehyde and a primary aliphatic nitro compound by way of the oxime:

$$ArCHO + RCH_2NO_2 \longrightarrow ArCH{=}CRNO_2 \xrightarrow{Fe,\ HCl}$$

$$ArCH{=}CRNHOH \rightleftharpoons ArCH_2{\cdot}CR{=}NOH \xrightarrow{HCl\ -\ H_2O} ArCH_2{\cdot}COR$$

The isomerisation of aldehydes into ketones has been carried out by a variety of reagents, for example, mercury(II) chloride, bromide or sulphate (*S. N. Danilov*, Zhur. obshcheĭ Khim., 1948, **18**, 2000):

$$\underset{Me}{PhCH{\cdot}CHO} \xrightarrow[EtOH\ -\ H_2O]{HgCl_2} PhCH_2{\cdot}COMe$$

Some aryl-substituted aldehydes rearrange to ketones under the influence of cold concentrated sulphuric acid (*A. Orekhov* and *M. Tiffeneau*, Compt. rend., 1926, **182**, 68; *S. Danilow* and *E. Venus-Danilova*, Ber., 1927, **60**, 1063):

$$R_2CPh{\cdot}CHO \rightarrow RCHPh{\cdot}COR \quad (R = Me\ or\ Et)$$

Alkylation of a ketone can be achieved by a strong base such as sodamide, a sodium alkoxide or sodium triphenylmethide and an alkyl halide (*H. D. Zook* and *W. L. Rellahan*, J. Amer. chem. Soc., 1957, **79**, 881):

$$PhCO{\cdot}CH_2Et \xrightarrow[(2)\ EtBr]{(1)\ Ph_3CNa} \underset{(62\%)}{PhCO{\cdot}CHEt_2}$$

β-tert-Aminoketone hydrochlorides, prepared by the Mannich reaction, are de-aminated by treatment with hydrogen and Raney nickel (*E. M. Schultz* and *J. B. Bicking*, *ibid.*, 1953, **75**, 1128) *e.g.*:

$$PhCH_2{\cdot}CO{\cdot}CH_3 \xrightarrow{CH_2O\ -\ (CH_3)_2\overset{\oplus}{N}H_2\overset{\ominus}{Cl}} PhCH(CH_2\overset{\oplus}{N}HMe_2)\cdot CO\cdot CH_3\ Cl^{\ominus} \xrightarrow[\text{pressure}]{H_2\ -\ \text{Raney Ni}} PhCH(Me)\cdot CO\cdot CH_3\ (56\%)$$

Addition of acetophenone to 1-octene in the presence of di-*tert*-butyl peroxide gives decanophenone (*J. C. Allen, J. I. G. Cadogan* and *D. H. Hey*, J. chem. Soc., 1965, 1918):

$$PhCOMe \quad + \quad CH_2{=}CH{\cdot}C_6H_{13} \quad \rightarrow \quad PhCO{\cdot}CH_2{\cdot}CH_2{\cdot}CH_2{\cdot}C_6H_{13}\ (10\%)$$

The reaction of ketones in alcohol with diazomethane results in the introduction of a methylene group, epoxides are also formed (*H. Meerwein et al.*, Ber., 1928, **61**, 1840; 1929, **62**, 999). Diethyl methylthiomethylphosphonate can be alkylated by successive treatment with *n*-butyl-lithium and an *n*-alkyl iodide to give the corresponding alkyl(1-methylthio)phosphonate esters. The lithium derivatives of these compounds form adducts with carbonyl compounds which decompose on heating to methyl vinyl sulphides. The mercuric chloride-promoted hydrolysis of the latter compounds gives ketones in good yield (*E. J. Corey* and *J. I. Shulman*, J. org. Chem., 1970, **35**, 777):

$$MeSCH_2P({=}O)Et_2 \xrightarrow[(2)\ RI]{(1)\ Bu^nLi} MeSCHRP({=}O)Et_2 \xrightarrow[(2)\ R^1R^2CO\ (3)\ \text{Heat}]{(1)\ Bu^nLi} R^1R^2C{=}C(R)SMe \quad [+Et_2POO^{\ominus}Li^{\oplus}] \xrightarrow{HgCl_2/H_2O} R^1R^2CH{\cdot}COR$$

A new method for the transposition of the ketonic function is illustrated by the conversion of propiophenone to phenylacetone (*Corey* and *J. E. Richman*, J. Amer. chem. Soc., 1970, **92**, 5276):

$$PhCO{\cdot}CH_2{\cdot}CH_3 \xrightarrow[(2)\ NaBH_4\ (3)\ Ac_2O]{(1)\ RONO} PhCH(OAc){\cdot}C({=}NOAc){\cdot}CH_3 \xrightarrow[THF\ -\ H_2O]{Cr(OAc)_4} PhCH_2{\cdot}CO{\cdot}CH_3$$

The selective reduction of α,β-unsaturated ketones is illustrated by the reduction of benzylideneacetone to 4-phenyl-2-butanone (*R. Cornubert* and *H. G. Eggert*, Bull. Soc. chim. Fr., 1954, 522):

$$PhCH{=}CH{\cdot}CO{\cdot}CH_3 \xrightarrow{H_2-\ Ni\ -\ CH_2Cl_2} PhCH_2{\cdot}CH_2{\cdot}CO{\cdot}CH_3\ (93-96\%)$$

Reactions of α-bromoketones with organoboranes provide a general route to ketones. Thus reaction of bromoacetone with 9-phenyl-9-borabicyclo-[3.3.1]nonane under the influence of potassium 2,6-di-*tert*-butylphenoxide provides phenylacetone (*H. C. Brown*, *H. Nambu* and *M. M. Rogic*, J. Amer. chem. Soc., 1969, **91,** 6852):

$$PhLi + (C_8H_{14})B{-}H \longrightarrow [(C_8H_{14})BHPh]^{\ominus}\,Li^{\oplus} \xrightarrow{MeSO_3H}$$

$$(C_8H_{14})B{-}Ph \xrightarrow{BrCH_2\cdot COMe} \underset{(76\%)}{PhCH_2\cdot COMe}$$

Phenylacyl bromide reacts with a number of trialkylboranes under the influence of potassium *tert*-butoxide to give the corresponding ketones, alternatively 9-alkyl-9-borabicyclo[3.3.1]nonanes may be used (*Brown*, *Rogic* and *M. W. Rathke*, *ibid.*, 1968, **90,** 6218; see also *Brown et al.*, *ibid.*, 1969, **91,** 2147):

$$PhCO\cdot CH_2Br + Et_3B + Bu^tOK \rightarrow \underset{(100\%)}{PhCO\cdot CH_2Et} + KBr + Bu^tOBEt_2$$

Diphenylphosphine converts α-halogeno- or α-mesyloxyketones into the dehalogenated or desmesylated ketone, respectively (*I. J. Borowitz et al.*, J. org. Chem., 1969, **34,** 2687) *e.g.*:

$$PhCO\cdot CH_2Br \quad + \quad Ph_2PH \quad \rightarrow \quad \underset{(100\%)}{PhCO\cdot CH_3} \quad + \quad Ph_2PBr$$

Methyl ketones have been prepared in good to excellent yield from α-chloroethyl methyl ether by an extension of the Wittig olefin synthesis (*D. R. Coulson*, Tetrahedron Letters, 1964, 3323):

$$MeCHCl(OMe) \longrightarrow (H)(OMe)C(Me)(\overset{\oplus}{P}Ph_3)\;Cl^{\ominus} \xrightarrow[(2)\ PhCHO]{(1)\ Base} (Me)(OMe)C{=}C(H)(Ph)$$

$$\xrightarrow{H_2O\ -\ H^{\oplus}} CH_3\cdot CO\cdot CH_2Ph$$

Phenylpropan-2-one

(88%)

(xii) Miscellaneous reactions

Reaction of certain formamides with Raney nickel gives ketones as illustrated by the following preparation of acetophenone (*M. Metayer* and *P. Mastagli*, Compt. rend., 1947, **225,** 457):

TABLE 10.

ALKYL AEYL KETONES

	b.p.(°C/mm)	n_D°	$d_{\circ}^{\circ}$	*Derivative, m.p.(°C)*	*Reference*
Acetophenone (PhCOMe)	202	1.5363^{20}	1.0281_4^{20}	*Oxime*, 59; *semicarbazone*, 198.	1
Propiophenone (PhCOEt)	218	1.5270^{20}	1.0087_{15}^{25}	*Oxime*, 53; *semicarbazone*, 174.	2,2a,2b
Butyrophenone (PhCOPr)	231	1.5203^{20}	0.988_{20}^{20}	*Oxime*, 50; *semicarbazone*, 188.	2,2c
Isobutyrophenone (PhCO·CHMe$_2$)	222	1.5196^{16}	0.9871_4^{16}	*Oxime*, 96 (*anti* phenyl); *semicarbazone*, 181.	3,3a
Valerophenone (PhCOBu^n)	240	1.5146^{20}	0.988_{20}^{20}	*Oxime*, 52; *semicarbazone*, 166.	2,2d
Isovalerophenone (Ph CO·CH$_2$·CHMe$_2$)	236.5	$1.5139^{16.3}$	$0.9701_4^{16.4}$	*Oxime*, 74; *semicarbazone*, 210.	4
tert-Butyl phenyl ketone	221	$1.5086^{18.8}$	0.963^{26}	*Oxime*, 167.	5,5a
n-Amyl phenyl ketone (caprophenone)	265	1.5116^{20}	0.9576^{20}	*semicarbazone*, 132 (124)	2
n-Hexyl phenyl ketone (oenanthophenone)	155/15	1.5020^{20}		*Oxime*, 55.	6
2-Methylacetophenone (methyl *o*-tolyl ketone)	216	1.535^{13}	1.0201_4^{13}	*Oxime*, 61; *semicarbazone*, 206.	7,7a
3-Methylacetophenone	220	1.533^{15}	$1.0106_4^{15.5}$	*Oxime*, 57; *semicarbazone*, 200.	7,8
4-Methylacetophenone	224	1.5335^{20}	1.0051^{20}	*Oxime*, 86; *semicarbazone*, 205.	1
2,4-Dimethylacetophenone	228	1.5340^{20}	1.0121_4^{15}	*Oxime*, 64; *semicarbazone*, 187.	9,10
2,5-Dimethylacetophenone	231	1.5291^{20}	$0.9963_4^{18.5}$	*Oxime*, 58; *semicarbazone*, 169.	11
3,4-Dimethylacetophenone	250	1.5413^{15}	$1.0090_4^{14.4}$	*Oxime*, 85; *semicarbazone*, 234.	12
3,5-Dimethylacetophenone	237			*Oxime*, 114.5; *semicarbazone*, 219.	10
2,4,6-Trimethyl-acetophenone (acetomesitylene)	240.5/735	1.5175^{20}	0.9754_4^{20}		13
4-*tert*-Butylacetophenone	133–134/11	1.5195^{20}	0.9635^{20}	*Semicarbazone*, 220–221.	14

References

1 See text.
2 *C. R. Hauser, W. J. Humphlett* and *M. J. Weiss*, J. Amer. chem. Soc., 1948, **70,** 426.
2a *J. Cason, ibid.*, 1946, **68,** 2078.
2b *G. A. Olah et al., ibid.*, 1962, **84,** 2733.
2c *H. C. Brown et al., ibid.*, 1969, **91,** 2147.
2d *K. D. Berlin, B. S. Rathore* and *M. Peterson*, J. Org. Chem., 1965, **30,** 226.
3 *H. M. Kissman* and *J. Williams*, J. Amer. chem. Soc., 1950, **72,** 5323.

3a *B. E. Hudson* and *C. R. Hauser*, *ibid*., 1941, **63,** 3163.
4 *J.-B. Senderens*, Compt. rend., 1910, **150,** 1337.
5 *A. Favorsky*, Bull. Soc. chim. Fr., 1936, [v], **3,** 239.
5a *M. J. Jorgenson*, Org. Reactions, 1970, **18,** 59.
6 *F. Krafft*, Ber., 1886, **19,** 2987.
7 *K. Auwers*, Ann., 1915, **408,** 242.
7a *N. B. Chapman, K. Clarke* and *D. J. Harvey*, J. chem. Soc., C, 1971, 1202.
8 *F. Mauthner*, J. prakt. Chem., 1922, [ii], **103,** 394.
9 *W. H. Perkin* and *J. F. S. Stone*, J. chem. Soc., 1925, **127,** 2283.
10 *Auwers, M. Lechner* and *H. Bundesmann*, Ber., 1925, **58,** 36.
11 *M. Freund* and *K. Fleischer*, Ann., 1918, **414,** 5.
12 *A. Claus*, J. prakt. Chem., 1890, [ii], **41,** 409.
13 *C. R. Noller* and *R. Adams*, J. Amer. chem. Soc., 1924, **46,** 1893.
14 *G. F. Hennion* and *S. F. de C. McLeese*, *ibid*., 1942, **64,** 2421.

$$PhCH(NHCHO)\cdot CH_3 \xrightarrow{Ni\ -\ 200^\circ} PhCO\cdot CH_3$$

(b) Alkyl aryl ketones

(i) General properties

Some physical constants and selected m.p. data for the derivatives of the commoner alkyl aryl ketones are given in Table 10.

Alkyl aryl ketones have weakly basic properties. Acetophenone is 90% protonated in 82% sulphuric acid, it is therefore more strongly basic than benzaldehyde and this is attributed to electron release from the methyl group. An electron-withdrawing nitro group in the *para* position decreases the basicity of the carbonyl group, whereas an electron-releasing *para*-methoxy group increases the basicity of the carbonyl group. The respective p*K*a values of acetophenone, 4-nitroacetophenone and 4-methoxyacetophenone are −6.15, −7.94 and −4.81 (*D. D. Perrin* in "Dissociation Constants of Organic Bases in Aqueous Solution", Butterworth, London, 1965). The carbonyl-stretching absorption in alkyl aryl ketones is in the region 1670–1700 cm^{-1} (*W. M. Schubert* and *W. A. Sweeney*, J. Amer. chem. Soc., 1955, **77,** 4172; *N. Fuson, M.-L. Josien* and *E. M. Shelton*, *ibid*., 1954, **76,** 2526). The ultraviolet spectrum of acetophenone in *n*-heptane shows λ_{max} 238 mμ ($\log_{10}\varepsilon$ 4.1). *Ortho*-substitution causes a marked decrease in the intensity of this band (*G. D. Hedden* and *W. G. Brown*, *ibid*., 1953, **75,** 3744) confirming steric interference to coplanarity. In the p.m.r. spectrum of acetophenone in $CDCl_3$ the methyl resonance occurs at 7.4 τ, the *ortho* protons are deshielded by the carbonyl group and appear at 2.1 τ. The molecular ion of acetophenone fragments by loss of methyl radical to give an ion which by loss of carbon monoxide gives the phenyl cation (*m/e* 77). Alternative α-fission also occurs as shown by the occurrence of the $CH_3CO^{\oplus}$ ion (*m/e* 43) in the mass spectrum. The mass spectra of acetophenones and of alkyl benzyl ketones are discussed by *H. Budzikiewicz, C. Djerassi* and *D. H. Williams*, in "Interpretation of Mass Spectra of Organic Compounds", Holden-Day, San Francisco, 1964, p. 192.

(ii) Reactions of alkyl aryl ketones

Alkyl aryl ketones undergo many of the same general reactions as aliphatic ketones (see Vol. I C, p. 58 *et seq*.). The carbonyl group reacts normally with hydrogen cyanide, Grignard reagents, hydroxylamine, hydrazine and substituted hydrazines.

Pure acetophenone hydrazone is prepared by first treating acetophenone with *N,N*-dimethylhydrazine, followed by reaction of the resulting *N,N*-dimethylhydrazone with hydrazine (*G. R. Newkome* and *D. L. Fishel*, Org. Synth., 1970, **50,** 102). The azine is prepared conveniently in polyphosphoric acid (*D. B. Mobbs* and *H. Suschitzky*, J. chem. Soc., C, 1971, 175). It has also been prepared by treatment of acetophenone with hydrogen peroxide, ammonia, and methyl cyanide (*J.-P. Schirmann* and *F. Weiss*, Tetrahedron Letters, 1972, 635). Carbonyl compounds may be regenerated from their azines by treatment with periodic acid (*A. J. Fatiadi*, Chem. and Ind., 1971, 64) *e.g.*:

$$\mathrm{Ph(Me)C{=}N{-}N{=}C(Me)Ph} \xrightarrow[\text{AcOH, room temperature}]{HIO_4} \mathrm{PhCOMe}$$

The regeneration of ketones from their oximes has been achieved with thallium nitrate in methanol (*A. McKillop et al.*, J. Amer. chem. Soc., 1971, **93,** 4918), with titanium chloride in buffered aqueous dioxane (*G. H. Timms* and *E. Wildsmith*, Tetrahedron Letters, 1971, 195) and by the action of chromous acetate on the oxime acetate (*E. J. Corey* and *J. E. Richardson*, J. Amer. chem. Soc., 1970, **92,** 5276). The first of these procedures appears to be the method of choice.

Alkyl aryl ketones do not form sodium bisulphite addition compounds. Ammonia is reported to give a condensation product, $[C_6H_5(CH_3)C]_3N_2$, with acetophenone (*C. Thomae*, Arch. Pharm., 1906, **244,** 643). Condensation with primary aromatic amines requires more drastic conditions than are needed with aromatic aldehydes, a catalyst such as zinc chloride is required (*G. Reddelien*, Ann., 1921, **388,** 165).

Condensation with ammonia or primary and secondary amines under reducing conditions leads to the formation of an amine:

$$\mathrm{PhCO{\cdot}CH_3 \quad + \quad NH_3 \quad \xrightarrow{2H} \quad PhCH(NH_2){\cdot}CH_3}$$

In the Leukart reaction (*M. L. Moore*, Org. Reactions, 1949, **5,** 301) the reduction is accomplished with formic acid or a derivative of formic acid. The ketone is heated with ammonium formate or the formic acid salt or formyl derivative of the amine. Alternatively, reduction is brought about by hydrogen and a metal catalyst (*W. S. Emerson*, *ibid*., 1948, **4,** 174).

Other standard reactions of the carbonyl group include the formation of acetals, $>C(OAlk)_2$, by reaction with orthoformic ester (*L. Claisen*, Ber., 1907, **40,** 3908) and dithioacetals, $>C(SAlk)_2$, by the action of thiols and a condensing agent.

Some special points regarding the reactivity of alkyl aryl ketones are discussed below.

(*1*) *Oxidation and reduction.* Vigorous oxidation converts alkyl aryl ketones into aromatic acids. Chromic acid in aqueous sulphuric acid is very suitable for this purpose, sometimes acid permanganate is used:

$$ArCOR \rightarrow ArCO_2H$$

Methyl ketones are conveniently oxidised to aromatic acids by hypochlorite or hypobromite; the reaction depends on the halogenation of the methyl group:

$$ArCO \cdot CH_3 \xrightarrow{NaOX} ArCO \cdot CX_3 \xrightarrow{H_2O,\,NaOH} ArCO_2Na + CHX_3$$

Alkaline permanganate has been used for the oxidation of methyl ketones to α-oxo-acids (*K. Glücksmann*, Monatsh., 1889, **10,** 770; 1890, **11,** 248; *A. Claus et al.*, Ber., 1885, **18,** 1857; and later papers) and alkaline ferricyanide (*K. v. Buchka* and *P. H. Irish*, *ibid.*, 1887, **20,** 386, 1762).

Alkyl aryl ketones are oxidised by selenium dioxide to the corresponding dicarbonyl compounds, thus acetophenone gives phenylglyoxal, $PhCO \cdot CHO$, in 70% yield and propiophenone gives methylphenylglyoxal $PhCO \cdot CO \cdot CH_3$ in 50% yield (*N. Rabjohn*, in Org. Reactions, 1949, **5,** 331). The oxidation of alkyl aryl ketones with peracids leads to ester formation (Baeyer–Villiger reaction) (*C. H. Hassall*, *ibid.*, 1957, **9,** 73) *e.g.*:

$$PhCO \cdot CH_3 \xrightarrow{PhCO_3H} PhOCO \cdot CH_3 \quad (63\%)$$

Alkyl aryl ketones are reduced to the corresponding secondary alcohols by (a) sodium in moist ether or benzene, (b) lithium tetrahydridoaluminate and (c) Meerwein–Ponndorf reduction (*A. L. Wilds*, *ibid.*, 1944, **2,** 178). Thiourea dioxide is an excellent reagent for the reduction of ketones to secondary alcohols, for example, acetophenone is reduced to methylphenylmethanol in 94% yield (*K. Nakagawa* and *K. Minami*, Tetrahedron Letters, 1972, 343):

$$(Ph)(Me)C{=}O + (H_2N^{\oplus}{=})(H_2N)C{-}SO_2^{\ominus} + 2\,OH^{\ominus} \longrightarrow (Ph)(Me)CHOH + (H_2N)_2C{=}O + SO_3^{2\ominus}$$

Two other reagents which effect this change are sodium cyanotrihydridoborate (*R. O. Hutchins, B. E. Maryanoff* and *C. A. Milewski*, J. chem. Soc., D, 1971, 1097) and sodium dihydridotrithioborate, $NaBH_2S_3$ (*J. M. Lalancette* and *A. Frêche*, Canad. J. Chem., 1970, **48,** 2366).

Alkyl aryl ketones are reduced to the corresponding pinacols either electrolytically or by the use of aluminium amalgam (*K. Sisido* and *H. Nozaki*, J. Amer. chem. Soc., 1948, **70,** 776; *J. H. Stocker* and *R. M. Jenevein*, J. org. Chem., 1968, **33,** 294) *e. g.*:

$$PhCO \cdot CH_3 \xrightarrow{Al/Hg} Ph-\underset{OH}{\overset{CH_3}{C}}-\underset{OH}{\overset{CH_3}{C}}-Ph$$

(73% of stereoisomeric mixture)

Photochemical reduction also leads to mixtures of diastereoisomeric pinacols (*Stocker* and *D. H. Kern*, *ibid.*, 1968, **33,** 1270). An alternative procedure for pinacol formation involves treatment of the ketone with chlorotrimethylsilane and magnesium in hexamethylphosphoramide (HMPA) (*T. H. Chan* and *E. Vinokur*, Tetrahedron Letters, 1972, 75).

Reduction of acetophenone with excess of potassium in liquid ammonia in the presence of *tert*-butyl alcohol and potassium *tert*-butoxide gives ethylbenzene, but reduction with lithium and methylamine gives the cyclohexenylcarbinol shown (*E. E. Kaiser*, Synthesis, 1972, 412):

$$PhCOMe \xrightarrow[MeNH_2]{Li} (cyclohexenyl)-CHOHMe$$

Lithium in liquid ammonia and THF converts aromatic ketones to aromatic hydrocarbons (*S. D. Lipsky, F. J. McEnroe* and *A. P. Bartle*, J. org. Chem., 1971, **36,** 2588). Reduction to the corresponding hydrocarbon is effected by the Clemmensen method (*E. L. Martin*, Org. Reactions, 1942, **1,** 155) using amalgamated zinc and hydrochloric acid, the Wolff–Kishner method (*D. Todd*, *ibid.*, 1948, **4,** 379), the hydrogenolysis of dithioacetals with Raney nickel (*M. L. Wolfrom* and *J. V. Karabinos*, J. Amer. chem. Soc., 1944, **66,** 909), or by interaction with the mixed hydride prepared from equimolar quantities of lithium tetrahydridoaluminate and aluminium chloride (*R. F. Nystrom, C. R. A. Berger*, *ibid.*, 1958, **80,** 2896). The 2,4-dinitrophenylhydrazone is the most suitable derivative for the hydrogenolysis of acetophenone to ethylbenzene (*J. W. Burnham* and *E. J. Eisenbraun*, J. org. Chem., 1971, **36,** 737). Acetophenone is smoothly reduced to cyclohexylmethylmethanol on absorption of four moles of hydrogen in the presence of a rhodium-on-carbon catalyst (*E. Breitner, E. Roginski* and *P. N. Rylander*, *ibid.*, 1959, **24,** 1855). The reductive acylation of ketones may be achieved by treatment with an acid chloride and triphenyltin hydride (*L. Kaplan*, J. Amer. chem. Soc., 1966, **88,** 4970). For example, acetophenone is quantitatively converted into α-phenylethyl acetate on reaction with acetyl chloride and triphenyltin hydride:

$$PhCO\cdot CH_3 \xrightarrow[Ph_3SnH]{AcCl} PhCH(OAc)\cdot CH_3$$

(*2*) *Condensation reactions.* Aryl ketones having a methylene group adjacent to the carbonyl group undergo acid-promoted self-condensation to give 1,3,5-triarylbenzenes (*K. Bernhauer, P. Müller* and *F. Neiser*, J. prakt. Chem., 1936, [2], **145,** 301); they condense with aldehydes to give α,β-unsaturated ketones, ArCO·CR=CHR′, and with alkyl nitrites to give isonitroso(hydroxyimino) compounds, ArCO·CR=NOH. Aryl ketones of this type undergo *O*-acetylation on heating with isopropenyl acetate, for example, acetophenone is converted into α-acetoxystyrene (*H. J. Hagemeyer* and *D. C. Hull*, Ind. Eng. Chem., 1949, **41,** 2921):

$$PhCO\cdot CH_3 + CH_3\cdot C(OAc){=}CH_2 \rightarrow \underset{(88\%)}{PhC(OAc){=}CH_2} + CH_3\cdot CO\cdot CH_3$$

On treatment of acetophenone with acetic anhydride and boron trifluoride, *C*-acetylation occurs and benzoylacetone is formed (*H. Meerwein* and *D. Vossen*, J. prakt. Chem., 1934, [ii], **141,** 149):

$$PhCO\cdot CH_3 + Ac_2O \xrightarrow{BF_3} \underset{(83\%)}{PhCO\cdot CH_2\cdot CO\cdot CH_3} + CH_3\cdot CO_2H$$

Benzoylacetone is also obtained by base-promoted reaction of acetophenone and ethyl acetate and this is illustrative of a general reaction for the formation of β-diketones (*Claisen*, Ber., 1905, **38,** 695):

$$PhCO\cdot CH_3 + CH_3\cdot CO_2Et \xrightarrow{NaNH_2} PhCO\cdot CH_2\cdot COMe + EtOH$$

β-Oxo-esters can be obtained by condensation with alkyl carbonates, the alcohol being removed as soon as it is formed (*V. H. Wallingford, A. H. Homeyer* and *D. M. Jones*, J. Amer. chem. Soc., 1941, **63,** 2252):

$$ArCO\cdot CH_2R + (EtO)_2CO \xrightarrow{NaOEt} ArCO\cdot CHR\cdot CO_2Et + EtOH$$

β-Oxo-acids are similarly obtained by treatment of ketones with magnesium methyl carbonate; this reagent is prepared by the reaction of magnesium methoxide with carbon dioxide in dimethylformamide:

$$Mg(OMe)_2 + CO_2 \rightleftharpoons MeOMgOCO_2Me$$

$$ArCO{\cdot}CH_2R + MeOMgOCO_2Me \longrightarrow ArC(O^{\ominus}){=}C(R){-}C(=O)O^{\ominus}\ Mg^{2\oplus} \xrightarrow{(1)\,R'X;\ (2)\,H_3O^{\oplus},\ -CO_2} ArCO{\cdot}CHRR'$$

$$\xrightarrow{H_3O^{\oplus}} ArCO{\cdot}CHR{\cdot}CO_2H$$

Treatment of the intermediate magnesium salt with alkyl halide followed by acid hydrolysis and decarboxylation affords a convenient method for alkylation (*M. Stiles, ibid.*, 1959, **81,** 2598). Alternatively alkylation of ArCO·CH_2R to ArCO·CHRR′ and ArCO·CRR'_2 is brought about by the action of suitable alkyl halides in the presence of sodamide or sodium in an inert solvent. Benzylation of acetophenone can be achieved by reaction with benzyl alcohol in the presence of a catalytic amount of lithium benzyloxide (*E. F. Pratt* and *A. P. Evans, ibid.*, 1956, **78,** 4950):

$$PhCO{\cdot}CH_3 + PhCH_2OH \xrightarrow[PhCH_2OLi,\ \text{Dean-Stark apparatus}]{\text{Xylene}} PhCO{\cdot}CH_2{\cdot}CH_2Ph + H_2O \quad (70\%)$$

The α-carboxymethylation of ketones of this type has been achieved with lithium bromoacetate and lithium amide (*W. H. Puterbaugh* and *R. L. Readshaw*, Chem. and Ind., 1959, 255).

Acetophenone can be converted by "directed aldol condensation" into β-methylcinnamaldehyde by the following route (*G. Wittig* and *H. Reiff*, Angew. Chem. internat. Edn., 1968, **7,** 7):

$$PhCO{\cdot}CH_3 + LiCH_2{\cdot}CH{=}NC_6H_{11} \rightarrow PhC(CH_3)(OLi){\cdot}CH_2{\cdot}CH{=}NC_6H_{11}$$

$$\xrightarrow{H_2O/H^{\oplus}} PhC(CH_3){=}CH{\cdot}CHO + PhC(CH_3){=}CH_2$$

The terminal hydrolysis step produces 71% of the aldehyde and 19% of α-methylstyrene. The feasibility of this reaction depends on protecting the aldehyde group of acetaldehyde by Schiff base formation, lithiation of the Schiff base is carried out with lithium disopropylamide.

An analogous principle is employed in the nucleophilic aminomethylation of ketones, as illustrated in the following scheme (*T. Kauffmann, E. Köppelmann* and *H. Berg, ibid.*, 1970, **9,** 163):

$$PhCO{\cdot}CH_3 + LiCH_2N{=}CPh_2 \rightarrow \underset{\displaystyle OLi}{PhC(CH_3)CH_2N{=}CPh_2} \xrightarrow{2N\ HCl} \underset{\displaystyle OH}{PhC(CH_3){\cdot}CH_2NH_2}$$

Acetophenone can participate in the Mannich reaction (*C. E. Maxwell*, Org. Synth., Coll. Vol. 3, 1955, p. 305; see also *W. Back*, Arch. Pharm., 1970, **303,** 491) *e.g.*:

$$PhCO{\cdot}CH_3 + HCHO + (CH_3)_2\overset{\oplus}{N}H_2\overset{\ominus}{Cl} \rightarrow PhCO{\cdot}CH_2{\cdot}CH_2\overset{\oplus}{N}H(CH_3)_2\overset{\ominus}{Cl} + H_2O \quad (68\text{–}72\%)$$

(*3*) *Reactions with ylids.* The reactions of ketones and diazoalkanes have been reviewed by *C. D. Gutsche*, in Org. Reactions, 1954, **8,** 364. Acetophenone reacts with diazomethane in methanol and ether to give 12–16% of oxirane an 5.5% of homologous ketone:

$$PhCO{\cdot}CH_3 \xrightarrow{CH_2N_2} Ph{-}\overset{CH_3}{C}{-}CH_2\ (\text{–O–}) + PhCH_2{\cdot}CO{\cdot}CH_3$$

The oxirane is formed in 87% yield on reaction of acetophenone with dimethylsulphonium methylide, $(CH_3)_2S{=}CH_2$ (*E. J. Corey* and *M. Chaykovsky*, J. Amer. chem. Soc., 1962, **84,** 3782).

Alkyl aryl ketones react with phosphorus ylids less readily than aromatic aldehydes. Acetophenone fails to react with fluorenylidene tris(n-butyl)phosphorane (III, X = n-C_4H_9), with 3- and 4-nitroacetophenones and 4-chloroacetophenone the yields of fluorenylidene-ketone (IV) are 67, 60 and 9%, respectively:

$$\underset{(III)}{C_{13}H_8{=}PX_3} + ArCO{\cdot}CH_3 \longrightarrow \underset{(IV)}{C_{13}H_8{=}C(Ar)CH_3}$$

These substituted acetophenones fail to react with fluorenylidene triphenylphosphorane (III, X = Ph) (*A. W. Johnson* and *R. B. LaCount*, Tetrahedron, 1960, **9,** 130).

(*4*) *Darzens reaction.* Alkyl aryl ketones undergo base-promoted condensation with α-halogeno-esters to give glycidic esters (*C. F. H. Allen* and *J. Van Allan*, Org. Synth. Coll. Vol. 3, 1955, p. 727) *e.g.*:

$$PhCO{\cdot}CH_3 + ClCH_2{\cdot}CO_2Et \xrightarrow{NaNH_2} \overset{Ph}{\underset{CH_3}{C}}{-}CH{\cdot}CO_2Et\ (\text{–O–})$$

The scope of this reaction has been reviewed by *M. S. Newman* and *B. J. Magerlein*, in Org. Reactions, 1949, **5**, 413).

(5) *Fission reactions.* Ketones of the general type ArCO·CRR′R″ (Ar = aryl; R, R′, R″ = alkyl) undergo fission on heating with sodamide in benzene with the formation of an aromatic hydrocarbon and the amide of an aliphatic carboxylic acid (*A. Haller* and *E. Bauer*, Compt. rend., 1909, **148,** 127; **149,** 5):

$$PhCO{\cdot}C(CH_3)_3 \xrightarrow{NaNH_2} C_6H_6 + (CH_3)_3C{\cdot}CONH_2$$

The stepwise alkylation of an alkyl aryl ketone with an alkyl halide and sodamide, followed by rupture of the alkylated product with sodamide to the amide of a trisubstituted acetic acid is known as the Halle–Bauer reaction (*K. E. Hamlin* and *A. W. Weston*, Org. Reactions, 1957, **9,** 1).

A method for the fission of ketones which appears to be of general application consists in the reaction with hydrazoic acid in the presence of sulphuric acid.

$$PhCO{\cdot}CH_3 \xrightarrow{HN_3} \underset{(77\%)}{PhNHCO{\cdot}CH_3}$$

The scope of this reaction (the Schmidt reaction) has been reviewed by *H. Wolff* (*ibid.*, 1946, **3,** 307).

Aromatic ketones of the type $C_6H_5{\cdot}CH_2CO{\cdot}CH_2R$ are transformed by alkyl azide and sulphuric acid treatment into benzaldehyde and an aldehyde RCHO (*J. H. Boyer* and *L. R. Morgan Jr.*, J. Amer. chem. Soc., 1959, **81,** 3369):

$$PhCO{\cdot}CH_3 \xrightarrow[(2)\ H_2O]{(1)\ C_6H_{11}N_3/H_2SO_4} \underset{(85\%)}{PhCHO} + CH_2O + \underset{(50\%)}{C_6H_{11}NH_2}$$

(6) *Willgerote reaction.* This reaction involves the conversion of alkyl aryl ketones to amides by ammonium polysulphide (*M. Carmack* and *M. A. Spielman*, Org. Reactions, 1946, **3,** 83):

$$ArCO{\cdot}(CH_2)_n{\cdot}CH_3 + 2\,(NH_4)_2S + S \rightarrow Ar(CH_2)_{n+1}{\cdot}CONH_2 + 3\,NH_4SH$$

If the ketone is heated with equimolecular amounts of sulphur and an anhydrous amine instead of aqueous ammonium polysulphide *(Kindler modification)* the thioamide is formed:

$$ArCO{\cdot}CH_3 \xrightarrow{(CH_3)_2NH+S} ArCH_2{\cdot}CSN(CH_3)_2$$

If morpholine is used as the amine, the reaction can be carried out at atmospheric pressure. A simple, but mild, conversion of acetophenones to methyl phenylacetates has been achieved with thallium trinitrate (TTN), the reaction is thought to proceed by the following mechanism: (*A. McKillop, B. P. Swann* and *E. C. Taylor*, J. Amer. chem. Soc., 1971, **93,** 4919):

$$ArCO\cdot CH_3 \rightleftharpoons ArC(OH){=}CH_2 \xrightarrow{TTN-MeOH} H{-}O{-}C(Ar)(OMe){-}CH_2{-}Tl(ONO_2){-}ONO_2$$

$$\longrightarrow ArCH_2\cdot CO_2Me + TlONO_2 + HNO_3$$

(7) *Halogenation.* Predominantly side chain or nuclear halogenation of acetophenone can be achieved depending on the choice of conditions. Reaction of equimolecular amounts of acetophenone and bromine in ether at 0° and in the presence of a catalytic amount of aluminium chloride gives phenacyl bromide, $PhCO\cdot CH_2Br$, in 64–66% yield (*R. M. Cowper* and *L. H. Davidson*, Org. Synth., Coll. Vol. 2, 1943, p. 480). Reaction with two equivalents of bromine gives the corresponding dibromide, $PhCO\cdot CHBr_2$. Treatment of the acetophenone–aluminium chloride complex with bromine at 80–85° and in the presence of excess of aluminium chloride gives 3-bromoacetophenone in 70–75% yield (*D. E. Pearson, H. W. Pope* and *W. W. Hargrove*, Org. Synth., 1960, **40,** 7). This procedure has been used to prepare a number of 3-bromoacetophenones. Reaction of acetophenones with sodium benzenesulphonochloroamide, $PhSO_2NClNa$, gives aryl trichloromethyl ketones and sodium benzenesulphonamide, which interact to give sodium *N*-aroylbenzenesulphonamide $ArCON(Na)SO_2Ph$ and chloroform (*M. M. Kremlev* and *M. T. Plotnikova*, Zhur. org. Khim., 1969, **5,** 279).

(8) *Photolysis of* α-N-*alkylamidoacetophenones.* Compounds of this type are transformed in high yield to the corresponding *N*-substituted 3-azetidinols on irradiation (*E. H. Gold*, J. Amer. chem. Soc., 1971, **93,** 2794):

$$PhCO\cdot CH_2N(CHR^1R^2)(SO_2C_6H_4Me[4]) \xrightarrow[Et_2O]{h\nu} \text{3-Ph-3-OH-2-}R^1\text{-2-}R^2\text{-1-}(SO_2C_6H_4Me[4])\text{azetidine}$$

Cyclisation rather than pinacol formation, occurs on irradiation of 2,4,6-trimethylacetophenone:

$$\text{2,4,6-}Me_3C_6H_2COMe \xrightarrow[Me_2CHOH]{h\nu} \text{(benzocyclobutenol: Me, Me, OH, Me)}$$

As the alkyl group becomes more bulky then irradiation causes fragmentation with fission of the alkyl–carbonyl bond (*T. Matsuura* and *Y. Kitaura*, Tetrahedron, 1969, **25**, 4487).

(iii) Individual alkyl aryl ketones and their analogues

Acetophenone, $C_6H_5 \cdot CO \cdot CH_3$, is the main constituent of oil of *Stirlingia latifolia* and is found in cistus and labdanum oils and in oil of castoreum, an animal product. It is used in perfumery and has been used in medicine as a soporific under the name of hypnone.

Acetophenone is produced commercially by two main processes, the oxidation of ethylbenzene and the peroxidation of cumene (isopropylbenzene). The latter process is operated on a considerable scale for the manufacture of phenol and acetone, but valuable amounts of acetophenone are formed as a by-product. A practical commercial process for the oxidation of ethylbenzene involves passing oxygen through a solution of ethylbenzene at 135° in the presence of iron octaphenylporphyrazine (*R. D. Offenhauer* and *A. J. Silvestri*, U.S.P. 3,073,867/1960). Alternative commercial routes to acetophenone are the Friedel–Crafts reaction of benzene, aluminium chloride and acetic anhydride (*P. H. Groggins* and *R. H. Nagel*, Ind. Eng. Chem., 1934, **26,** 1313), the oxidation of α-methylstyrene with oxygen and in the presence of nickel or cobalt catalysts (*W. C. Keith, E. H. Burk Jr.* and *C. D. Keith*, U.S.P. 2,974,161/1958), and the hydrolysis of α-chlorostyrene with 80% aqueous sulphuric acid (*W. S. Emerson*, U.S.P., 2,372,562/1939), or by passing a 1:5-mixture of benzoic and acetic acids over an aluminium oxide–thorium oxide catalyst at 420° (*V. Martello* and *S. Ceccotti*, Chim. Ind. Milan, 1956, **38,** 289; C.A., 1956, **50,** 15454). For a more detailed discussion of the technology of acetophenone, see *J. Dorsky, F. G. Eichel* and *M. Luthy*, in the *Kirk–Othmer* "Encyclopedia of Chemical Technology", 2nd Edn., Wiley, New York, 1963, Vol. 1, p. 167.

An interesting laboratory preparation of acetophenone, potentially of wide scope for the preparation of methyl ketones, is illustrated in the following reaction sequence:

$$PhCHO + CH_2(CN)_2 \xrightarrow{\text{glycine}} PhCH{=}C(CN)_2 \xrightarrow{CH_2N_2}$$

$$PhC(CH_3){=}C(CN)_2 \xrightarrow[40-50^\circ]{33\%\ \text{aq. NaOH}} PhCO \cdot CH_3$$

Yields are nearly quantitative (*J. Bastús*, Tetrahedron Letters, 1963, 955). Reaction of acetophenone and di-*n*-butylamine in the presence of toluene-4-sulphonic acid gives a mixture of the enamine V and the *N*-butylimine VI. Compound V undergoes acid-catalysed rearrangement to compound VI (*P. Wittig* and *R. Mayer*, Z. Chem., 1967, **7,** 306):

$$PhCO \cdot CH_3 + Bu_2NH \rightarrow \underset{(V)}{PhC(=CH_2)NBu_2} \rightleftharpoons \underset{(VI)}{PhC(=NBu) \cdot CH_2Bu}$$

Acetophenone has been converted into many characteristic ketonic derivatives. Experimental details for conversion into its cyanohydrin are given by *E. L. Eliel* and *J.*

P. Freeman in Org. Synth., Coll. Vol. 4, 1963, p. 58. Its *oxime*, m.p. 60°, is converted by strong acids into acetanilide (Beckmann change, Vol. IB, p. 119). A kinetic study showed that rearrangement is considerably faster in polyphosphoric acid than in sulphuric acid (*D. E. Pearson* and *R. M. Stone*, J. Amer. chem. Soc., 1961, **83,** 1715). The formation of acetanilide indicates that the acetophenone oxime has the configuration shown:

$$\mathrm{Ph(CH_3)C{=}N{-}OH} \longrightarrow \mathrm{CH_3CONHPh}$$

If forms a *hydrazone*, m.p. 22° (*G. R. Newkome* and *D. L. Fishel*, Org. Synth., 1970, **50,** 102), *phenylhydrazone*, m.p. 106°, 2,4-*dinitrophenylhydrazone*, m.p. 237°, and *semicarbazone*, m.p. 203° (decomp.). Treatment of the phenylhydrazone with polyphosphoric acid affords a very convenient route to 2-phenylindole, and reaction of the benzoylhydrazone with peracetic acid gives *N,N'*-dibenzoylhydrazine (*L. Horner* and *H. Fernekess*, Ber., 1961, **94,** 712):

$$\mathrm{2\ PhCONH{\cdot}N{=}C(CH_3)Ph} \xrightarrow{\mathrm{CH_3{\cdot}CO_3H}} \underset{(89\%)}{\mathrm{PhCONH{\cdot}NHCOPh}} + \mathrm{N_2} + \underset{(93\%)}{\mathrm{2\ PhCO{\cdot}CH_3}}$$

Various references have already been made to the oxidation and reduction of acetophenone (see p. 123); detailed directions for the selenium dioxide oxidation of acetophenone to phenylglyoxal are given by *H. A. Riley* and *A. R. Gray* (Org. Synth., Coll. Vol. 2, 1943, p. 509). Acetophenone undergoes self-condensation in the presence of aluminium *tert*-butoxide to give phenyl β-phenylpropenyl ketone, dypnone (*W. Wayne* and *H. Adkins*, *ibid.*, Coll. Vol. 3, 1955, p. 367):

$$\mathrm{PhCO{\cdot}CH_3} \rightarrow \underset{(77\text{–}82\%)}{\mathrm{PhC(CH_3){=}CH{\cdot}COPh}} \rightarrow \mathrm{H_2O}$$

1,3,5-Triphenylbenzene is formed when acetophenone is heated with a mixture of aniline and aniline hydrochloride at 170–175° for 1 hour in an atmosphere of dry carbon dioxide (*D. Vorländer*, *E. Fischer* and *H. Wille*, Ber., 1929, **62,** 2836):

$$\mathrm{3\ PhCO{\cdot}CH_3} \longrightarrow \mathrm{1,3,5\text{-}Ph_3C_6H_3} + \mathrm{3\ H_2O}$$

Acetophenone is readily condensed with nitrous acid, with carboxylic acid esters, and with aldehydes:

$$\mathrm{PhCO{\cdot}CH_3} + \mathrm{PhCO_2Et} \xrightarrow{\mathrm{NaOEt}} \underset{(62\text{–}71\%)}{\mathrm{PhCO{\cdot}CH_2{\cdot}COPh}} + \mathrm{EtOH}$$

(*A. Magnani* and *S. M. McElvain*, Org. Synth., Coll. Vol. 3, 1955, p. 251);

$$\mathrm{PhCO{\cdot}CH_3} + \mathrm{PhCHO} \xrightarrow{\mathrm{NaOH}} \underset{(85\%)}{\mathrm{PhCO{\cdot}CH{=}CHPh}} + \mathrm{H_2O}$$

(*E. P. Kohler* and *H. M. Chadwell*, *ibid.*, Coll. Vol. 1, 2nd Edn., 1947, p. 78).

Low temperature ($-5° \rightarrow 0°$) nitration of acetophenone with a mixture of nitric and sulphuric acids yields 3-nitroacetophenone in 55% yield (*R. B. Carson* and *R. K. Hasen*, *ibid.*, Coll. Vol. 2, 1943, p. 434). Treatment of acetophenone with dioxane–sulphur trioxide, followed by neutralisation with sodium hydroxide, gives sodium benzoylmethanesulphonate, $PhCO \cdot CH_2SO_2ONa$, in approximately 70% yield (*W. E. Truce* and *C. C. Alfieri*, J. Amer. chem. Soc., 1950, **72,** 2740). The conditions for the side-chain and nuclear bromination of acetophenone have been noted on p. 129. ω-Halogenoacetophenones can also be obtained by the Friedel–Crafts reaction from benzene and halogenoacetyl halides.

Acetophenone reacts with dichloromethyl-lithium to give 52% of α-chloro-α-phenylpropionaldehyde (*G. Köbrich* and *W. Werner*, Tetrahedron Letters, 1969, 2181):

$$PhCO \cdot CH_3 + LiCHCl_2 \longrightarrow Ph(CH_3)C(OLi)CHCl_2 \longrightarrow Ph(CH_3)\overset{O}{C-CHCl} \longrightarrow Ph(CH_3)C(Cl)CHO$$

Acetophenone and 4-substituted acetophenones react with 1-bromomethyl-2,2-dimethyloxirane in liquid ammonia and in the presence of sodium amide to give 1-benzoyl-2-hydroxyisopropylcyclopropanes (*B. A. Ershoo et al.*, Zhur. org. Khim., 1969, **5,** 1588; English Edn., p. 1547).

$$PhCO \cdot CH_3 + BrCH_2 \cdot \underset{O}{CH-C(CH_3)_2} \xrightarrow{NaNH_2/NH_3} PhCO \cdot \underset{\underset{H_2}{C}}{CH-CH}-C(OH)(CH_3)_2$$

Acetophenone dimethyl acetal, $Ph \cdot C(OMe)_2 \cdot CH_3$, b.p. 90°/20 mm, is prepared by the reaction of acetophenone and methyl orthoformate in methanol in the presence of a little hydrochloric acid (*M. T. Bogert* and *P. P. Herrera*, J. Amer. chem. Soc., 1923, **45,** 243), by keeping the ketone in methanol containing methyl formimidate hydrochloride ($CH_3OCH{:}NH \cdot HCl$) (*Claisen*, Ber., 1898, **31,** 1012) or methyl acetimidate hydrochloride [$CH_3OC(CH_3){=}NH \cdot HCl$] (*K. Alder* and *H. Niklas*, Ann., 1954, **585,** 97). *Acetophenone diethyl acetal*, b.p. 114°/23 mm, is formed similarly (*H. E. Carswell* and *H. Adkins*, J. Amer. chem. Soc., 1928, **50,** 235). On distillation at atmospheric pressure, or preferably by treatment with acetyl chloride and pyridine, it loses the elements of ethyl alcohol to give α-ethoxystyrene.

ω-**Chloroacetophenone,** *phenacyl chloride*, $PhCO \cdot CH_2Cl$, m.p. 58°, is made from benzene and chloroacetyl chloride by a Friedel–Crafts reaction (*C. Friedel* and *J. M. Craft*, Ann. chim. Fr., 1884, [vi], **1,** 507). It is formed by the chlorination of acetophenone in acetic acid (*R. Scholl* and *H. Korten*, Ber., 1901, **34,** 1902) and by reaction of the ketone with *N*-chlorourea (*A. Béhal* and *A. Detoeuf*, Compt. rend., 1911, **153,** 1231) or with ethyl hypochlorite (*S. Goldschmidt et al.*, Ber., 1925, **58,** 576), and from benzoyl chloride and diazomethane (*F. Arndt* and *J. Amende*, *ibid.*, 1928, **61,** 1122). Like phenacyl bromide, it is lachrymatory and has been used as a tear gas. It is used as a denaturant for industrial alcohol.

ω,ω-Dichloroacetophenone, $PhCO \cdot CHCl_2$, m.p. 20°, b.p. 122°/10 mm, is formed from acetophenone by the action of chlorine (*J. G. Aston et al.*, Org. Synth., Coll.

Vol. 2, 1943, p. 538) or an excess of sulphuryl chloride (*T. H. Durrans*, J. chem. Soc., 1922, **121,** 46). It is also formed by the Friedel–Crafts reaction from benzene and dichloroacetyl chloride (*H. Gautier*, Ann. Chim. Fr., 1888, [iv], **14,** 345, 385).

ω,ω,ω-Trichloroacetophenone, $PhCO \cdot CCl_3$, b.p. 145°/25 mm, results from the reaction of benzene and trichloroacetyl chloride in the presence of aluminium chloride (*Gautier, ibid.*, p. 396) or the aluminium chloride-promoted reaction of benzene and trichloroacetonitrile (*J. Houben* and *W. Fischer*, J. prakt. Chem., 1929, [ii], **123,** 318).

ω-**Bromoacetophenone,** m.p. 50°, is made by the bromination of acetophenone (*Cowper* and *Davidson, loc. cit.*). It is used to identify organic acids with which it forms well defined crystalline salts; on oxidation with dimethyl sulphoxide it is smoothly converted into phenylglyoxal (*N. Kornblum et al.*, J. Amer. chem. Soc., 1957, **79,** 6562). The two stereoisomeric forms of ω-bromoacetophenone oxime are known.

ω,ω-Dibromoacetophenone, $PhCO \cdot CHBr_2$, m.p. 37°, is prepared by the bromination of ω-bromoacetophenone (*W. L. Evans* and *B. J. Brooks, ibid.*, 1908, **30,** 406); on treatment with dilute potassium hydroxide it is converted into the potassium salt of mandelic acid, $PhCHOH \cdot CO_2K$ (*C. Engler* and *E. Wöhrle*, Ber., 1887, **20,** 2201).

ω-**Iodoacetophenone,** m.p. 34°, is prepared from the bromide and sodium iodide (*H. Rheinboldt* and *M. Perrier*, J. Amer. chem. Soc., 1947, **69,** 3149).

Hydroxyiminoacetophenone, *isonitrosoacetophenone*, $PhCO \cdot CH{=}NOH$, m.p. 126°, is derived from acetophenone by reaction with methyl nitrite in ethereal hydrogen chloride (*W. K. Slater*, J. chem. Soc., 1920, **117,** 589) or by treatment with amyl nitrite in alcoholic ethoxide (*L. Claisen*, Ber., 1887, **20,** 656, 2194). It is dehydrated by thionyl chloride to benzoyl cyanide, $PhCO \cdot CN$ (*W. Borsche* and *C. Walter, ibid.*, 1926, **59,** 463).

ω-**Nitroacetophenone,** $PhCO \cdot CH_2NO_2$, m.p. 108°, is formed from ω-iodoacetophenone and silver nitrite (*A. Lucas, ibid.*, 1899, **32,** 602).

ω-**Diazoacetophenone,** $PhCO \cdot CHN_2$, m.p. 49°, results from the action of an excess of diazomethane on benzoyl chloride (*W. Bradley* and *R. Robinson*, J. chem. Soc., 1928, 1316, 1546; *F. Arndt* and *J. Amende*, Ber., 1928, **61,** 1123). It reacts with hydrogen halides to give ω-halogenoacetophenones, $PhCO \cdot CH_2X$, and with acetic acid to give ω-acetoxyacetophenone.

ω-Methylsulphinylacetophenone, $PhCO \cdot CH_2SOMe$, is representative of an increasingly important class of intermediates, the β-oxosulphoxides, ω-Methylsulphinylacetophenone is prepared as shown and it has been converted into ω-bis- and ω-tris-(methylthio)acetophenones (*G. A. Russell* and *L. A. Ochrymowyez*, J. org. Chem., 1969, **34,** 3618):

$$PhCO_2Et + \overset{\ominus}{C}H_2SOMe \longrightarrow PhCO \cdot CH_2SOMe \xrightarrow{SOCl_2}$$

$$PhCO \cdot CHClSMe \xrightarrow{MeSH - CH_2Cl_2} PhCO \cdot CH(SMe)_2 \xrightarrow[(2)\ MeSCl]{(1)\ NaH}$$

$$PhCO \cdot C(SMe)_3$$

Oxidation of ω-tris-(methylthio)acetophenone with iodine and sodium hydrogen carbonate gives *S*-methyl benzoylthioformate, $PhCO \cdot CO(SCH_3)$. As mentioned previously (p. 111), reduction of β-oxosulphoxides affords methyl ketones.

Thioacetophenone, $PhCS \cdot CH_3$, is formed when acetophenone reacts with hydrogen sulphide in alcohol and in the presence of hydrogen chloride (*E. Baumann* and *E. Fromm*, Ber., 1895, **28,** 897) or with magnesium bromohydrosulphide, HSMgBr (*Q. Mingoia*, Gazz., 1926, **56,** 841). The monomer exists as an unstable blue liquid which readily trimerises to give *trithioacetophenone*, a colourless crystalline solid, m.p. 122° (see *E. E. Reid*, in "Organic Chemistry of Bivalent Sulphur", Chemical Publishing Company, New York, 1960, Vol. 3, p. 161). *Acetophenone diethyl dithioacetal*, $PhC(SEt)_2 \cdot CH_3$, is a colourless oil formed by reaction of acetophenone with ethanethiol and hydrogen chloride (*T. Posner*, Ber., 1900, **33,** 3166). It yields *α,α-bis(ethylsulphonyl) ethylbenzene*, m.p. 120°, on oxidation *Acetophenone di-*n*-butyl dithioacetal* has b.p. 168°/3 mm (*T. C. Witner* and *Reid*, J. Amer. chem. Soc., 1921, **43,** 639).

Selenoacetophenone is formed when acetophenone reacts with hydrogen selenide in concentrated hydrochloric acid. It forms a red oil which is the dimeride (*R. E. Lyons* and *W. E. Bradt*, Ber., 1927, **60,** 825).

(iv) Nuclear substituted acetophenones

(*1*) *Alkylacetophenones.* 4-**Methylacetophenone** is used as a perfume in soaps and cosmetics. It is manufactured by the Friedel–Crafts reaction using toluene and acetonitrile (*J. P. Luvisi*, U.S.P. 2,974,172/1957). A laboratory procedure for the preparation of this ketone by the aluminium chloride-promoted reaction of toluene and acetyl chloride is described by *C. R. Noller* and *R. Adams* (J. Amer. chem. Soc., 1924, **46,** 1889).

(*2*) *Halogenoacetophenones.* 4-Halogenoacetophenones are conveniently prepared by the Friedel–Crafts reaction with the appropriate halogenobenzene and acetyl chloride or acetic anhydride (*idem*, Org. Synth., Coll. Vol. 1, 2nd Edn., 1947, p. 109). 2,4- and 3,4-Dichloroacetophenones are similarly prepared (*E. Roberts* and *E. E. Turner*, J. chem. Soc., 1927, 1846). Halogen-substituted acetophenones have been obtained from the corresponding cyanides and methylmagnesium iodide (*R. E. Lutz et al.*, J. org. Chem., 1947, **12,** 617); they have also been prepared by the diazo-reaction from the corresponding aminoacetophenones (*J. Meisenheimer*, *P. Zimmerman* and *U. von Kummer*, Ann., 1926, **446,** 219; *W. J. Bruining*, Rec. Trav. chim., 1922, **41,** 665), by the oxidation of the corresponding arylmethylcarbinols (*K. von Auwers et al.*, Ber., 1925, **58,** 49; *N. B. Chapman*, *K. Clarke* and *D. J. Harvey*, J. chem. Soc., C, 1971, 1202), the hydrolysis of halogenobenzoylacetoacetic acid esters (*L. Thorp* and *E. R. Brunskill*, J. Amer. chem. Soc., 1915, **37,** 1260; *A. Wahl* and *J. Rolland*, Ann. Chim. Fr., 1928, [x], **10,** 27) and the hydrolysis of the appropriate diethyl halogenobenzoylmalonate (*H. G. Walker* and *C. R. Hauser*, J. Amer. chem. Soc., 1946, **68,** 1386). Reaction of 4-iodobenzoyl chloride with lithium dimethyl cuprate(I) gives

4-iodoacetophenone in 98% yield. This type of reaction has also been used to prepare acetophenone homologues (*G. H. Posner, C. E. Whitten* and *P. E. McFarland, 1972*, **94,** 5106).

2-**Chloro-,** 2-**bromo-** and 2-**iodo-acetophenones** are liquids, b.p. 113°/18 mm, 113°/11 mm and 140°/12 mm, respectively. The *para*-isomers are solids; 4-*chloro*-, m.p. 20°, b.p. 237°; 4-*bromo*-, m.p. 54°, b.p. 256°, and 4-*iodo-acetophenone*, m.p. 85°, b.p. 153°/18 mm. Derivatives of 4-iodosoacetophenone are described by *E. A. Werner* and *W. Caldwell* (J. chem. Soc., 1906, **89,** 1625; 1907, **91,** 240).

(*3*) *Nitroacetophenones.* Detailed directions for the conversion of acetophenone into 3-nitroacetophenone are given by *R. B. Carson* and *R. K. Hasen*, in Org. Synth., Coll. Vol. 2, 1943, p. 434. 2- and 4-Nitroacetophenones have been prepared starting from either methylphenylcarbinol or ethylbenzene (*A. H. Ford-Moore* and *H. N. Rydon*, J. chem. Soc., 1946, 679) and by hydrolysis of the appropriate diethyl nitrobenzoylmalonate (*C. A. Reynolds* and *Hauser*, Org. Synth., Coll. Vol. 4, 1963, p. 709). 4-Nitroacetophenone has been obtained by the hydration of 4-nitrophenylacetylene or 4-nitrophenylpropiolic acid (*V. B. Drewson*, Ann., 1882, **212,** 160).

2-**Nitroacetophenone,** b.p. 179°/30 mm, (*oxime*, m.p. 115°); 3-*nitroacetophenone*, m.p. 80°, b.p. 167°/18 mm, (*oxime*, m.p. 131°); 4-*nitroacetophenone*, m.p. 81°, (*oxime*, m.p. 172°).

3,5-*Dinitroacetophenone*, m.p. 82°; 2,4,6-*trinitroacetophenone*, m.p. 90° (*A. Sonn* and *W. Bülow*, Ber., 1925, **58,** 1697).

(*4*) *Aminoacetophenones.* Transfer-hydrogenation with cyclohexene as donor and palladium as catalyst provides an efficient method for the reduction of 2-, 3-, and 4-nitroacetophenones to the corresponding aminoacetophenones (*E. A. Braude, R. P. Linstead* and *K. R. H. Wooldridge*, J. chem. Soc., 1954, 3586). Alkyl nitroaryl ketones have also been reduced to the corresponding alkyl aminoaryl ketones by catalytic hydrogenation (*N. J. Leonard* and *S. N. Boyd*, J. org. Chem., 1946, **11,** 405) and with tin and hydrochloric acid (*B. Camps*, Arch. Pharm., 1902, **240,** 1).

Reduction of 2-nitroacetophenone by zinc dust and ammonium chloride or tin and acetic acid yields *C*-methylanthranil, the anhydride of 2-hydroxyaminoacetophenone (*E. Bamberger*, Ber., 1903, **36,** 1611). 3-Hydroxyaminoacetophenone and the corresponding azo and azoxy compounds are described by *Bamberger* (*loc. cit.*).

2-**Aminoacetophenone,** m.p. 20°, b.p. 135°/17 mm, readily yields cyclic derivatives under suitable conditions and many of its derivatives react similarly. Thus by boiling

2-aminoacetophenone with aqueous alcoholic alkali, 4-methyl-2-(2′-aminophenyl)-quinoline (VII) is formed (*Camps*, Ber., 1899, **32**, 3231):

(VII)

The participation of 2-aminoacetophenone in the Friedländer synthesis is illustrated by its conversion into the tetrahydroacridine (VIII) (*G. Kempter et al.*, Ber., 1964, **97**, 16):

(VIII)

2-Aminoacetophenone condenses with cyanoacetic ester to give 3-cyano-4-methyl-2-quinolone (IX) (*Camps*, Arch. Pharm., 1902, **240**, 144):

(IX)

(X)

(XI)

Diazotised 2-aminoacetophenones yield cinnolones (*e.g.*, X) on keeping in acid solution (*J. C. E. Simpson*, in "Condensed Pyridazine and Pyrazine Rings", Interscience, New York, 1953, p. 17). By the successive action of nitrous acid and sodium sulphite, on a solution of 2-aminoacetophenone hydrochloride, the sodium salt of 3-methylindazole-2-sulphonic acid (XI) is formed (*E. Fischer* and *J. Tafel*, Ann., 1885, **227**, 316). Reaction of 2-*aminoacetophenone oxime*, m.p. 109°, with nitrous acid in acid solution gives 4-methylbenzo-1,2,3-triazine-3-oxide (XII) (*J. Meisenheimer et al.*, Ber., 1927, **60**, 1745):

(XII)

2-*Formamidoacetophenone*, m.p. 79°, cyclises on heating with aqueous ethanolic alkali to 4-quinolone (*Camps*, Ber., 1901, **34**, 2709), similar treatment of 2-*acetamido*-

acetophenone, m.p. 79°, gives 4-methyl-2-quinolone (XIII) (*idem, ibid.*, 1899, **32,** 3230):

aqueous-ethanolic NaOH → (XIII); ethanolic NH_3 → (XIV)

Reaction of 2-acetamidoacetophenone with ethanolic ammonia at 130–150° gives 2,4-dimethylquinazoline (XIV) (*A. Bischler* and *E. Buckhart, ibid.*, 1893, **26,** 1350).

3-*Aminoacetophenone*, m.p. 99°; *acetyl*, m.p. 129°; *oxime*, m.p. 194°; *semicarbazone*, m.p. 196°.

4-*Aminoacetophenone*, m.p. 106°, *oxime*, m.p. 147–148°, *semicarbazone*, m.p. 250°. The *acetyl* derivative, m.p. 167°, is obtained by heating aniline with acetic anhydride and zinc chloride (*J. Klingel, ibid.*, 1885, **18,** 2691). 4-*Dimethylaminoacetophenone*, m.p. 105°, is formed in poor yield by heating dimethylaniline, acetic anhydride, and zinc chloride together (*H. Staudinger* and *N. Kon*, Ann., 1911, **384,** 111).

2-*Aminopropiophenone*, m.p. 47°, b.p. 93°/0.8 mm; *acetyl*, m.p. 71°; *oxime*, m.p. 89° (*L. A. Elson et al.*, J. chem. Soc., 1930, 1131; *B. L. Zenitz* and *W. H. Hartung*, J. org. Chem., 1946, **11,** 444). 3-*Aminopropiophenone*, m.p. 42°; *oxime*, m.p. 113°. 4-*Aminopropiophenone*, m.p. 140°; *acetyl*, m.p. 173°; *oxime*, m.p. 156°. 2-*Aminobutyrophenone*, m.p. 45° (*Elson et al., loc. cit.*). 3-*Aminobutyrophenone*, m.p. 28°, b.p. 189°/16 mm (*G. T. Morgan* and *W. J. Hickinbottom*, J. chem. Soc., 1921, **119,** 1134). 4-*Aminobutyrophenone*, m.p. 84° (*F. Kunckell*, Ber., 1900, **33,** 2643).

(c) Monoaryl derivatives of dialkyl ketones

Benzyl methyl ketone, *phenylacetone*, 1-*phenylpropan-2-one*, $PhCH_2 \cdot COMe$, b.p. 100°/14 mm, *oxime*, m.p. 70°; *semicarbazone*, m.p. 198°. The most satisfactory methods for the preparation of this ketone are (*a*) the acid hydrolysis of α-phenylacetoacetonitrile, PhCH(CN)·COMe (*P. L. Julian* and *J. J. Oliver*, Org. Synth., Coll. Vol. 2, 1943, p. 391) and (*b*) the passage of a mixture of phenylacetic and acetic acids over thoria heated at 430–450° (*R. M. Herbst* and *R. H. Manske, ibid.*, p. 389). On treatment of benzyl methyl ketone with carbon disulphide and base, the desaurin (XV) is obtained (*P. Yates* and *D. L. Moore*, J. Amer. chem. Soc., 1958, **80,** 5577):

$PhCH_2 \cdot CO \cdot CH_3$ → (XV)

Benzyl ethyl ketone, 1-*phenylbutan-2-one*, $PhCH_2 \cdot CO \cdot CH_2 \cdot Me$, b.p. 111°/16 mm; *semicarbazone*, m.p. 135.5° (146°), is obtained by heating the adduct of ethylmagnesium bromide and phenacyl chloride (*M. Tiffeneau*, Ann. Chim. phys., 1907, **10,** 368):

$$PhCO \cdot CH_2Cl + EtMgBr \longrightarrow Ph \cdot \underset{OMgBr}{\overset{Et}{C}} \cdot CH_2Cl \longrightarrow PhCH_2 \cdot COEt$$

Benzyl chloromethyl ketone, $PhCH_2 \cdot CO \cdot CH_2Cl$, b.p. 96–98°/1 mm, is obtained by reaction of diazomethane with phenylacetyl chloride followed by treatment of the resulting diazoketone with hydrogen chloride (*W. D. McPhee* and *E. Klinsberg*, Org. Synth., Coll. Vol. 3, 1955, p. 119).

Benzylacetone, 4-*phenylbutan*-2-*one*, $Ph \cdot CH_2 \cdot CH_2 \cdot COMe$, b.p. 115°/13 mm; *oxime*, m.p. 87°; *semicarbazone*, m.p. 142°, is conveniently prepared by the alkaline hydrolysis of ethyl benzylacetoacetate (*I. H. Mattar*, *J. J. H. Hastings* and *T. K. Walker*, J. chem. Soc., 1930, 2455); alternatively by reduction of benzylideneacetone (*R. Cornubert* and *H. G. Eggert*, Bull. Soc. chim. Fr., 1954, 522; from methylenetriphenylphos-phosphorane, $CH_2{=}PPh_3$, by acylation with β-phenylpropionyl chloride, followed by hydrolysis (*H. J. Bestmann* and *B. Arnason*, Ber., 1962, **95,** 1513) or from benzyl bromide and the ylid generated by treatment of acetonylpyridinium bromide, followed by reduction (*C. A. Henrick*, *E. Ritchie* and *W. C. Taylor*, Austral. J. Chem., 1967, **20,** 2441).

3-*Phenylbutan*-2-*one*, $PhCHMe \cdot CO \cdot CH_3$, b.p. 106–107°/17 mm, *semicarbazone*, m.p. 172°, is obtained by the rearrangement of 2-methyl-1-phenylpropane-1,2-diol or of 2-methyl-1-phenyl-1,2-epoxypropane (*J. Lévy*, Bull. Soc. chim. Fr., 1921, [iv], **29,** 824; *E. Ghera*, J. org. Chem., 1970, **35,** 660), or by the rearrangement of dimethyl(phenyl)-acetaldehyde, $PhMe_2C \cdot CHO$ (*A. Orékhov* and *M. Tiffeneau*, Compt. rend., 1926, **182,** 68), or from methylphenylacetaldehyde (*J. Cantacuzène* and *J.-M. Normant*, Tetrahedron Letters, 1970, 2947).

1-*Phenylpentan*-2-*one*, *benzyl propyl ketone*, $PhCH_2 \cdot COPr^n$, b.p. 244°, *semicarbazone*, m.p. 120° (*Lévy* and *F. Gombinska*, Compt. rend., 1929, **188,** 712). 1-*Phenylpentan*-3-*one*, b.p. 126°/12 mm, *semicarbazone*, m.p. 132° (*Herbst* and *Manske*, *loc. cit.; N. Maxim*, Ann. Chim. Fr., 1928, [x], **9,** 69). 1-*Phenylpentan*-4-*one*, b.p. 132°/17 mm, *oxime*, m.p. 52°, *semicarbazone*, m.p. 128° (*Lévy* and *M. Sfiras*, Compt. rend., 1927, **184,** 1337). 2-*Phenylpentan*-3-*one*, b.p. 228°, *semicarbazone*, m.p. 136° (*Tiffeneau* and *Lévy*, Bull. Soc. chim. Fr., 1923, [iv], **33,** 768). 3-*Phenylpentan*-2-*one*, b.p. 220°, *semicarbazone*, m.p. 188° (*Lévy* and *P. Jullien*, *ibid.*, 1929, [iv], **45,** 942).

1-*Phenylhexan*-5-*one*, b.p. 150°/17 mm, *semicarbazone*, m.p. 141° (*Lévy* and *Sfiras*, *loc. cit.*). 3-*Phenylhexan*-4-*one*, b.p. 114°/13 mm, *oxime*, m.p. 58°, *semicarbazone*, m.p. 140° (*Tiffeneau* and *Lévy*, *loc. cit.*, p. 742; *Orékhov* and *Tiffeneau*, *loc. cit.*). 2-*Methyl*-2-*phenylpentan*-4-*one*, b.p. 134°/22 mm, *oxime*, m.p. 52°, *semicarbazone*, m.p. 164°, from mesityl oxide, benzene, and aluminium chloride (*A. Hoffmann*, J. Amer. chem. Soc., 1929, **51,** 2543).

Curcumone, 4-p-*tolylpentan*-2-*one*, $[4]MeC_6H_4 \cdot CHMe \cdot CH_2 \cdot COMe$, b.p. 121°/9 mm, *semicarbazone*, m.p. 125°, is a liquid with a strong ginger-like smell, obtained by heating curcuma oil with strong alkali. It is dextrorotatory, $[\alpha]_D$ + 80°. The inactive form has been synthesised by the reaction of zinc dimethyl on β-*p*-tolylbutyryl chloride (*H. Rupe* and *Fr. Wiederkehr*, Helv., 1924, **7,** 667). For the preparation of other ketones of this type see *Rupe et al.*, *ibid.*, 1926, **9,** 992; 1931, **14,** 687.

(d) Monoaryl ketenes

Phenylketene, PhCH:CO:O, is known only in solution and has been prepared by the action of zinc on an ethereal solution of α-chlorophenylacetyl chloride (*H. Staudinger*, Ber., 1911, **44,** 536). It dimerises on keeping to give 1,3-*diphenylcyclobutane-*2,4-*dione*, m.p. 73°, converted to the keto-enol form, m.p. 160°, by warming with alkali.

Phenylketene dimethyl acetal, PhCH:C(OMe)$_2$, b.p. 124–126°/14 mm, is formed when methyl orthophenylacetate is distilled with aluminium methoxide at 210° (*S. M. McElvain* and *J. T. Venerable*, J. Amer. chem. Soc., 1950, **72,** 1661). Reaction of the dimethyl acetal with benzylidene dichloride in the presence of potassium *tert*-butoxide gives the dimethyl acetal of diphenylcyclopropenone, and this on acid hydrolysis yields the parent ketone (*R. Breslow et al.*, J. Amer. chem. Soc., 1963, **87,** 1320):

MeO, OMe / C=C / Ph, H —($PhCHCl_2$, $KOBu^t$)→ MeO, OMe / C / C=C / Ph, Ph —($H_2O/H^{\oplus}$)→ O=C / C=C / Ph, Ph

The *diethyl acetal*, PhCH:C(OEt)$_2$, b.p. 136°/12 mm, is formed when the triethylortho ester of phenylacetic acid is distilled under reduced pressure (*Staudinger* and *G. Rathsam*, Helv., 1922, **5,** 652).

2,4,6-*Trimethylphenylketene*, is formed by a method similar to that for phenylketene. It is isolated as a *dimer*, m.p. 197–200° (*R. C. Fuson, L. J. Armstrong* and *W. J. Shenk*, J. Amer. chem. Soc., 1944, **66,** 964).

Methylphenylketene, PhCMe:C:O, b.p. 74°/12 mm, is obtained by the action of zinc on α-chloro-α-phenylpropionyl dichloride in ethereal solution (*Staudinger* and *L. Ruzicka*, Ann., 1911, **380,** 298) or by the action of diphenylketene on phenylmethylmalonic acid in ether (*Staudinger et al.*, Helv., 1923, **6,** 302). With water it gives α-phenylpropionic acid; aniline gives the anilide. It slowly dimerises and it reacts with oxygen at −80° forming a peroxide, $C_9H_8O_3$.

Chapter 15

Phenolic Aralkylamines, Monohydric Alcohols, Monocarbaldehydes, Monoketones and Monocarboxylic Acids

J. GRIMSHAW*

In this chapter are described the phenolic derivatives of monohydric aliphatic alcohols and related alkylamines together with their oxidation products, the corresponding monoaldehydes, monoketones and monocarboxylic acids. Lignin and lignans, hydrolysable tannins and depsides are large classes of related naturally occurring compounds and they will be discussed together in Chapter 16.

1. Phenolic aralkylamines

(a) Methods of preparation

Hydroxybenzylamines and other hydroxyaralkylamines with a longer side-chain may be prepared by methods already given for the aralkylamines (p. 2 *et seq.*). Often it is convenient to prepare the corresponding methoxy compound which can be demethylated with hydrobromic acid.

Phenol, formaldehyde, amine condensations. The high reactivity of the phenolic ring towards electrophilic substitution allows condensation with formaldehyde and a dialkylamine to give benzylamine derivatives. Usually the position *ortho* to the phenolic group is the most reactive but both *ortho*- and *para*-substitution products are obtained. Some factors, not always steric in nature, bring about *para*-substitution in certain cases. If the quantities of reagents are restricted monosubstitution occurs. A useful preparative route to methylphenols results from reduction of the appropriate benzylamine with hydrogen and copper chromite (*W. I. Caldwell* and *T. R. Thompson*, J. Amer. chem. Soc., 1939, **61**, 765, 2354):

* Text revised June 1975.

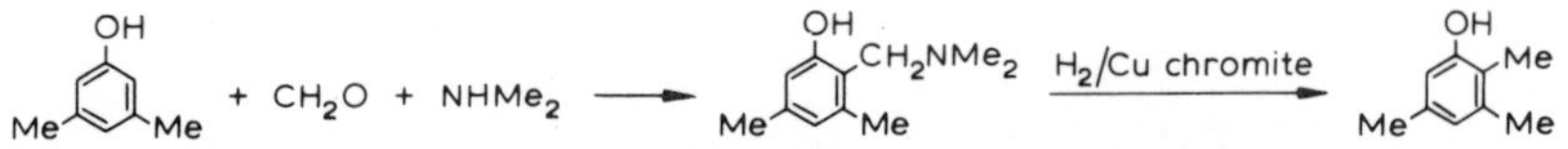

Excess of formaldehyde and amine leads to polysubstitution products (*H. A. Bruson* and *C. W. MacMullen*, *ibid.*, 1941, **63,** 270). In an alternative procedure, which gives mainly *para*-substitution, the phenol is condensed in benzene solution with *N*-chloromethylphthalimide and zinc chloride catalyst to give the hydroxybenzylphthalimide which can be hydrolysed to the benzylamine (*W. Herzberg*, G.P. 442,774/1925).

Phenols with one free *ortho*-position condense with formaldehyde and ammonia to give dibenzylamines (*A. Zincke* and *S. Pucher*, Monatsh., 1948, **79,** 26). Complex reactions occur when such bis(hydroxybenzyl)amines are heated with an excess of phenol resulting in the formation of diphenylmethanes and tris(hydroxybenzyl)amines (*G. Zigeuner* and *W. Schaden*, *ibid.*, 1950, **81,** 1018; *Zincke et al.*, *ibid.*, 1950, **81,** 999). Such reactions have been studied because of their importance in explaining resin formation from phenols, formaldehyde and ammonia:

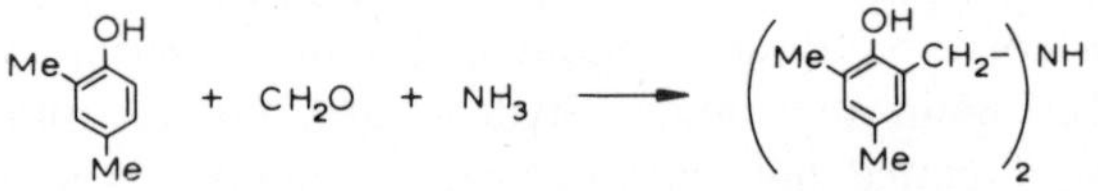

(b) Monohydroxyphenylalkylamines

2-**Hydroxybenzylamine,** *salicylamine*, $HOC_6H_4{\cdot}CH_2NH_2$, m.p. 125°, is prepared by reduction of salicylaldoxime either catalytically with palladium in ethanolic hydrogen chloride (*E. Ott* and *K. Zimmermann*, Ann., 1921, **425,** 328) or with sodium amalgam in ethanol and acetic acid (*L. C. Raiford* and *E. P. Clarke*, J. Amer. chem. Soc., 1923, **45,** 1740). Salicylamine dissolves in aqueous alkali and aqueous acid; with ferric chloride it gives a violet-blue coloration.

2-**Methoxybenzyldimethylamine,** b.p. 113°/20 mm, from dimethylamine and 2-methoxybenzyl bromide (*E. Stedman*, J. chem. Soc., 1927, 1904).

2-**Hydroxy**-3-**methylbenzyldimethylamine,** b.p. 91–95°/2 mm, prepared by the action of formaldehyde and dimethylamine on *o*-cresol (*R. B. Carlin* and *H. P. Landerl*, J. Amer. chem. Soc., 1950, **72,** 2762).

3-**Methoxybenzylamine,** b.p. 131°/26 mm, from hydrogenation of 3-methoxybenzonitrile over Adam's catalyst in acetic anhydride followed by hydrolysis of the resulting *N*-acetyl derivative (*W. H. Carothers et al.*, *ibid.*, 1929, **49,** 2912).

4-**Hydroxybenzylamine,** m.p. 109°, is the chief product of the reaction between phenol and *N*-chloromethylphthalimide (*Herzberg*, *loc. cit.*), and may be prepared by reduction of 4-hydroxybenzaldoxime (*E. C. S. Jones* and *F. L. Pyman*, J. chem. Soc., 1925, **127,** 2592).

4-*Methoxybenzylamine*, b.p. 134°/33 mm, can be prepared from anisaldehyde by the Leuckart reaction (*K. G. Lewis*, J. chem. Soc., 1950, 2249).

1-(4-**Methoxyphenyl)ethylamine,** b.p. 126°/16 mm, is conveniently prepared from 4-methoxyacetophenone by the Leuckart reaction (*A. W. Ingersoll et al.*, J. Amer. chem. Soc., 1936, **58,** 1808).

2-(4-**Hydroxyphenyl)ethylamine, tyramine,** $HOC_6H_5 \cdot CH_2 \cdot CH_2NH_2$, m.p. 164°, is present, together with a number of other bases, in the ergot of rye (*G. Barger*, J. chem. Soc., 1909, **95,** 1128). It is found in many other plant and animal juices including the urine of patients with Parkinson's disease (*A. A. Boulton et al.*, Nature, 1967, **215,** 132), and is formed by the bacterial degradation of tyrosine and of proteins containing tyrosine. It is prepared most conveniently by the thermal decarboxylation of tyrosine (*E. Waser*, Helv., 1925, **8,** 761). Other methods include reduction of 4-hydroxymandelonitrile (*J. S. Buck*, J. Amer. chem. Soc., 1933, **55,** 3388). [α-^{14}C]Tyramine has been prepared (*E. Leete et al.,*) Canad. J. Chem., 1952, **30,** 749). Tyramine alone or mixed with histamine is used in obstetrics. It has the effect of raising the blood pressure and stimulating respiration.

Amides of tyramine and its *O*- and *N*-methyl derivatives are found in plants. These include the 4-hydroxycinnamoylamide (I), m.p. 252°, isolated from *Evodia belahe* B. (*J. Rondest, M. B. C. Das* and *J. Polonsky*, Bull. Soc. chim. Fr., 1968, 2411), the *cinnamoylamide* (II), m.p. 75°, isolated from *Zanthoxylum clavaherculis* L. (*L. Crombie*, J. chem. Soc., 1952, 2997) and a mixture of fatty acid amides (III) isolated from *Anacyclus pyrethrum* D.C. (*R. S. Burden* and *Crombie*, *ibid.*, 1969, 2477):

$$[4]HOC_6H_4 \cdot CH{=}CH \cdot CONHCH_2 \cdot CH_2 \cdot C_6H_4OH[4] \qquad (I)$$

$$PhCH{=}CH \cdot CONMeCH_2 \cdot CH_2 \cdot C_6H_4OMe[4] \qquad (II)$$

$$Me(CH_2)_n \cdot CH{=}CH \cdot CH{=}CH \cdot CONHCH_2 \cdot CH_2 \cdot C_6H_4OH[4] \qquad (III)$$

$$n = 4,\ 6 \text{ and } 8$$

2-(4-*Methoxyphenyl)ethylamine*, b.p. 135–138°/18 mm, is prepared from β-nitro-4-methoxystyrene either by catalytic hydrogenation over platinum in ethanolic hydrogen chloride (*A. Skita*, G.P. 406,149/1922) or by reduction with lithium tetrahydridoaluminate (*C. B. Clarke* and *A. R. Pinder*, J. chem. Soc., 1958, 1967).

2-(4-*Hydroxyphenyl)ethylmethylamine*, m.p. 130°, has been isolated from barley and synthesised by demethylation of the methyl ether using hydrobromic acid (*S. Kirkwood* and *L. Marion*, J. Amer. chem. Soc., 1950, **72,** 2522).

2-(4-**Hydroxyphenyl)ethyldimethylamine, hordenine, anhaline,** $HOC_6H_4 \cdot CH_2 \cdot CH_2NMe_2$, m.p. 118° is present in sprouting barley and other cereal embryos (*G. O. Gaebel*, Arch. Pharm., 1906, **244,** 436) and in cacti of the genus *Anhalonium* (*E. Spath*, Monatsh., 1919, **40,** 129). Hordenine was first synthesised from 2-(4-nitrophenyl)-ethyldimethylamine by reduction and diazotisation (*Barger*, J. chem. Soc., 1909, **95,** 2194). It can be prepared by demethylation of the methyl ether (*K. Kindler*, Arch. Pharm., 1927, **265,** 413).

2-(4-*Methoxyphenyl)ethyldimethylamine*, b.p. 253°, has been prepared by the electro-reduction of *N,N*-dimethyl-4-methoxyphenylthioacetamide at a lead cathode in alcoholic hydrogen chloride (*Kindler, loc. cit.*).

2-(4-**Hydroxyphenyl**)**ethyltrimethylammonium chloride, candicine chloride,** $HOC_6H_4 \cdot CH_2 \cdot CH_2N^{\oplus}Me_3$, $Cl^{\ominus}$, m.p. 285°, hygroscopic, has been isolated from barley and other plants (*L. Reti*, Compt. rend. Soc. biol., 1933, **114,** 811; *V. Erspamer*, Arch. Biochem. Biophys., 1959, **82,** 431); *picrate*, m.p. 165°.

(c) Di- and tri-hydroxyphenylalkylamines

3,4-**Dihydroxybenzylamine,** *hydrochloride*, m.p. 186°, is obtained by demethylation of the dimethyl ether (*R. Douetteau*, Bull. Soc. chim. Fr., 1911, [iv], **9,** 936) which itself is obtained by reduction of 3,4-dimethoxybenzaldoxime with sodium amalgam (*L. Rugheimer* and *P. Schon*, Ber., 1908, **41,** 18).

4-**Hydroxy**-3-**methoxybenzylamine,** m.p. 145°, is obtained from vanillin oxime by hydrogenation (*E. Ott* and *K. Zimmermann*, Ann., 1921, **425,** 328) or by reduction with sodium amalgam in ethanolic acetic acid (*Jones* and *Pyman*, J. chem. Soc., 1925, **127**, 2592). It occurs as the pungent tasting amide *capsaicin*, (IV), m.p. 65°, in fruits of the genus *Capsicum* (*A. Lapworth* and *F. A. Royle*, *ibid*., 1919, **115,** 1109). Capsaicin has been synthesised from the appropriate acid chloride and amine (*E. K. Nelson*, J. Amer. chem. Soc., 1920, **42,** 597; *Spath* and *S. F. Darling*, Ber., 1930, **63,** 737):

$$HO-C_6H_3(OMe)-CH_2NHCO(CH_2)_4CH{=}CH \cdot CHMe_2$$

(IV)

2-(3,4-**Dihydroxyphenyl**)**ethylamine** is a solid which darkens rapidly in air; *hydrochloride*, m.p. 237° (decomp.) (*Barger* and *L. Ewins*, J. chem. Soc., 1910, **97,** 2257). Its various benzyl and benzylmethyl ethers have been employed in synthesis, for example of isoquinoline alkaloids, and the benzyl groups subsequently removed by hydrogenolysis or acid treatment to afford phenolic products. The various ethers may be obtained by reduction of the ω-nitrostyrene with lithium tetrahydridoaluminate; 2-(3,4-*dibenzyloxyphenyl*)*ethylamine*, m.p. 133° (*I. Baxter*, *L. T. Allan* and *G. A. Swan*, *ibid*., 1965, 3645), 2-(4-*benzyloxy*-3-*methoxyphenyl*)*ethylamine*, m.p. 171° (*D. H. Hey* and *A. L. Palluel*, *ibid*., 1957, 2926). 2-(3,4-*dimethoxyphenyl*)*ethylamine*, b.p. 173°/26 mm, has been prepared by hydrogenation of the benzyl cyanide (*J. C. Robinson* and *H. R. Snyder*, Org. Synth., 1943, **23,** 72), by hydrogenation of the ω-nitrostyrene (*Barger et al*., Ber., 1933, **66,** 451) and by the action of hypochlorite on the phenylpropionamide (*Buck* and *W. H. Perkin*, J. chem. Soc., 1924, **125,** 1679).

2-(3,4,5-**Trimethoxyphenyl**)**ethylamine, mescaline,** b.p. 180°/12 mm, *hydrochloride*, m.p. 181°, needles soluble in water, *picrate*, m.p. 219–220° (*Spath*, Monatsh., 1919, **40,** 144; 1921, **42,** 105) occurs in the dried flowering heads (Mezcal buttons) of the cactus *Lohpophora Willia nsii* (Lemaire) Coult. (*A. Heffter*, Ber., 1896, **29,** 221) together with other bases. It produces hallucinations and intoxication and the crude alkaloid mixture is used in native Mexican religious rites. Mescaline is obtained by reduction of 3,4,5-trimethoxy-ω-nitrostyrene over Adam's catalyst (*G. Hahn* and *F. Rumpf*, *ibid*., 1938, **71,** 2141).

(d) Hydroxybenzyl isothiocyanates and their parent glucosides

One of the simplest isothiocyanates or mustard oils of natural occurrence is allyl isothiocyanate which was described in Vol. I C, p. 372. This is formed by hydrolysis of the glycoside sinigrin. *p*-**Hydroxybenzyl isothiocyanate,** a yellow oil, is formed during enzymic hydrolysis of the glucoside **sinalbin** (V), *pentahydrate*, m.p. 83°, $[\alpha]_D$ − 8.4° (water) which occurs in white mustard *Sinapis alba* L. (*W. Schneider*, *H. Fischer* and *W. Specht*, Ber., 1930, **63,** 2789). Sinalbin is the salt of the ionic glycoside with the cation sinapine. Enzymic hydrolysis of the glycosidic link is followed by a Lossen rearrangement to give the isothiocyanate:

$$\left[\begin{array}{l} HO\text{-}C_6H_4\text{-}CH_2C(=NOSO_3^{\ominus})SC_6H_{11}O_5 \\ HO\text{-}C_6H_4\text{-}CH{=}CH\cdot CO_2CH_2\cdot CH_2\overset{\oplus}{N}Me_3 \end{array}\right] \xrightarrow{\text{enzyme}} HO\text{-}C_6H_4\text{-}CH_2NCS$$

Acid hydrolysis affords 4-hydroxyphenylacetic acid and hydroxylamine (*M. G. Ettlinger* and *A. J. Lundeen*, J. Amer. chem. Soc., 1956, **78,** 4172). 4-*Methoxybenzyl isothiocyanate*, *aubrietin*, b.p. 133°/0.7 mm, is an enzymic fission product of glucoaubrietin present in various *Aubrietta* species (*A. Kjaer*, *R. Gmelin* and *R. B. Jensen*, Acta chem. Scand., 1956, **10,** 26).

3-**Methoxybenzyl isothiocyanate, limnanthin,** b.p. 105°/0.45 mm, is present in enzymic hydrolysis products from the seeds of *Limnanthes douglasii* R.Br. (*Ettlinger* and *Lundeen*, J. Amer. chem. Soc., 1956, **78,** 1952). With ammonia it gives 3-*methoxybenzylthiourea*, m.p. 101°. It is synthesised by the action of carbon disulphide on 3-methoxybenzyl-amine.

2. Hydroxyaryl alcohols

(a) Methods of preparation

The general preparative routes to alcohols and phenols are described respectively in Chapters 13 (pp. 30 *et seq.*) and 4 (Vol. III A, pp. 290 *et seq.*). Special methods are available for the preparation of 2- and 4-hydroxy-benzyl alcohols.

Phenol formaldehyde condensation. Phenols condense readily with formal-dehyde and under mild conditions good yields of the 2- and 4-hydroxybenzyl alcohols can be obtained with one mole of aqueous formaldehyde present (*O. Manasse*, Ber., 1894, **27,** 2409; *S. van der Meer*, Rec. Trav. chim., 1944, **63,** 147). The polymeric resinous products which are formed under more vigorous conditions have been discussed in Vol. III A, p. 304. Substitution occurs both *ortho*- and *para*- to the phenolic hydroxyl group and the ratio of monosubstitution products obtained from phenol depends on the base catalyst used, strongly chelating bases give a higher *ortho:para* ratio (*H. G. Peer*, Rec. Trav. chim., 1959, **78,** 851):

$$C_6H_5OH + CH_2O \longrightarrow 2\text{-}HOC_6H_4CH_2OH + 4\text{-}HOC_6H_4CH_2OH$$

(*b*) *Properties and reactions*

Substances in this class show the properties expected as the sum of properties of an isolated phenol and alcohol function. In addition, 2- and 4-hydroxybenzyl alcohols show special reactions which will be described here.

(*i*) *Acidity*

pK_a values for the hydroxybenzyl alcohols are 2- 9.92, 3- 9.83, 4- 9.83. 2-Hydroxybenzyl alcohol forms an intramolecular hydrogen bond between the oxygen atoms which can be detected by i.r. spectroscopy in dilute solution (*R. E. Richards* and *H. W. Thompson*, J. chem. Soc., 1947, 1260).

(*ii*) *Removal of side-chain*

2-Hydroxybenzyl alcohol couples with diazonium salts in the usual manner *para* to the phenolic group. When this position is blocked by a Br or Me substituent then reaction with 4-nitrobenzenediazonium salts occurs with displacement of the hydroxymethyl group to give I (*E. Ziegler* and *G. Zigeuner*, Monatsh., 1948, **79,** 358). A 2- or 4-hydroxymethyl group is displaced also on vigorous nitration (*Ziegler* and *K. Garther*, Monatsh., 1949, **80,** 634), bromination (*W. Authenrieth* and *F. Beuttel*, Arch. Pharm., 1910, **248,** 121) and chlorination (*R. Piria*, Ann., 1845, **56,** 42).

$$2\text{-}HO\text{-}5\text{-}X\text{-}C_6H_3\text{-}N{=}N\text{-}C_6H_4\text{-}NO_2$$

(I)

(*iii*) *Carbonium ion stabilisation*

Electron donation from *ortho* or *para* HO and MeO groups stabilises the carbonium intermediates and transition states formed from the corresponding benzyl alcohols. Thus on heating alone, 2-hydroxybenzyl alcohol yields 2,2′-dihydroxydibenzyl ether together with a tar. Heating with aniline yields 2-hydroxybenzylaniline (*C. Paal et al.*, Ber., 1894, **27,** 1802). 4-Methoxybenzyl alcohol forms the ether on keeping (*E. Spath*, Monatsh., 1913, **34,** 2000) and the chloroformate ester [4]$MeOC_6H_4 \cdot CH_2OCOCl$, has been recommended for generating an amine-protecting group in peptide synthesis which is readily removed on acid treatment (*F. C. McKay* and *N. F. Albertson*, J. Amer. chem. Soc., 1957, **79**, 4686; *F. Weygand* and *K. Hunger*, Ber., 1962, **95**, 1).

Racemisation occurs during the hydrolysis of esters of optically active 1-(4-methoxyphenyl)ethanol (*M. P. Balfe et al.*, J. chem. Soc., 1946, 803). Reaction through a carbonium ion or ion pair intermediate and accompanied by racemisation competes with the normal mechanism of alkaline hydrolysis of an ester. The non-racemised alcohol is obtained by hydrolysis in concentrated alkali.

Sulphonate esters of 2-(4-methoxyphenyl)ethanols solvolyse very rapidly. An intermediate bridged phenonium ion is involved in these solvolyses and the phenomium ion II has been characterised in highly acidic solvents by its n.m.r. spectrum (*G. A. Olah, M. B. Comisarow* and *E. Namanworth*, J. Amer. chem. Soc., 1967, **89**, 5259). The dieneone III has been prepared from 2-(4-hydroxyphenyl)ethyl bromide (*R. Baird* and *S. Winstein*, *ibid.*, 1963, **85**, 567):

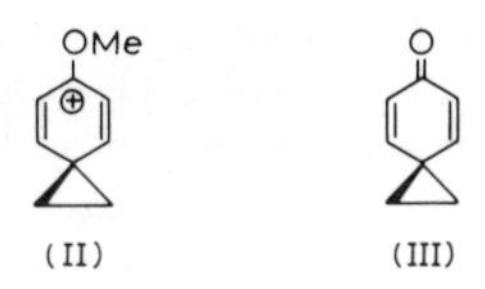

(II) (III)

(c) Monohydroxyaryl alcohols

2-**Hydroxybenzyl alcohol, saligenin,** m.p. 80°, was first isolated from the glucoside salicin by hydrolysis with emulsin or dilute acids (*R. Piria*, Ann., 1845, **56**, 37). It is prepared from salicylaldehyde by catalytic reduction (*L. W. Covert, R. Connor* and *H. Adkins*, J. Amer. chem. Soc., 1932, **54**, 1651) and by reduction of salicylic acid with lithium tetrahydridoaluminate (*R. F. Nystrom* and *W. G. Brown*, *ibid.*, 1947, **69**, 2548). It is formed together with 4-hydroxybenzyl alcohol by reaction of phenol with formaldehyde (*Manasse, loc. cit.*).

Saligenin is readily soluble in alcohol, ether and hot water. Its solutions give a deep blue colour with ferric chloride. It resinifies on vigorous acid treatment and can be oxidised to salicylaldehyde with cold chromic acid or silver oxide. The *diacetyl* derivative, b.p. 103–104°/1 mm, is formed by the action of acetic anhydride (*R. Barthel*, J. pr. Chem., 1942, [2], **161**, 77).

When heated with benzaldehyde containing some benzoic acid saligenin yields 2-phenylbenzo-1,3-dioxin (IV) (*R. Adams et al.*, J. Amer. chem. Soc., 1922, **44**, 1131) and other aldehydes react in a similar manner. Cyclic derivatives V are obtained with benzeneboronic acid (*R. A. Bowie* and *O. C. Musgrave*, J. chem. Soc., 1963, 3945) and by reaction with phosphate derivatives $ROPOCl_2$ and pyridine. The cyclic phosphates VI are insecticides and useful in a synergistic association with malathion (*E. Morifusa*, *E. Todako* and *O. Yasuyoshi*, Agr. Biol. Chem., 1962, **26**, 630):

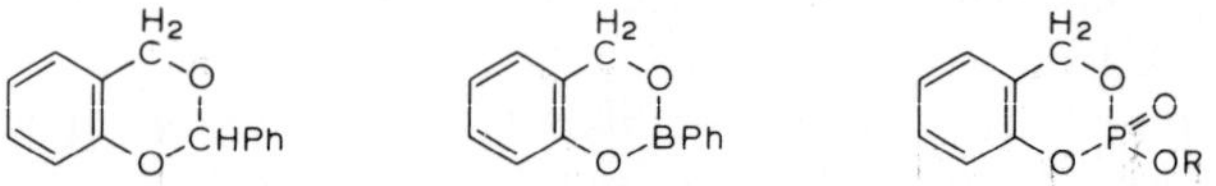

(IV) (V) (VI)

Reaction with bromine water yields 5-bromo- and 3,5-dibromo-2-hydroxybenzyl alcohol (*K. Auwers* and *G. Buttner*, Ann., 1898, **302**, 138). More vigorous halogenation results in the displacement of the hydroxymethyl group with the formation of trihalogenophenol (*Authenrieth* and *Beuttel*, *loc. cit.*).

Salicin, 2-*hydroxymethylphenyl-β-D-glucoside*, m.p. 201°, $[\alpha]_D -62.6°$ (water) is found in the leaves, stems and bark of many species of willow and poplar in quantity which varies with the season (*H. A. D. Jowett* and *C. E. Potter*, Pharm. J., 1902, [4], **15**, 157; *I. A. Pearl et al.*, Tappi, 1961, **44**, 475). Characterisation of the glycosides of the *Salicaceae* has been reviewed by *H. Thieme* (Pharm., 1964, **7**, 471). It can be prepared by hydrolysis of populin and reduction of helicin. The enzymic glucosidation of saligenin (*E. Bourquelot* and *M. Herissey*, Compt. rend., 1913, **156**, 1790) affords 2-*hydroxybenzyl-β-D-glucoside*, $[\alpha]_D -37.5$ (water).

Salicin tetra-acetate, m.p. 120°, $[\alpha]_D$ −14° (chloroform); *penta-acetate*, m.p. 130°, $[\alpha]_D$ −18° (chloroform) (*A. Kunz*, J. Amer. chem. Soc., 1926, **48**, 262).

Populin, 2-*hydroxymethyl-β-D-(6-benzoyl)glucoside*, m.p. 178°, $[\alpha]_D$ −2.0° (pyridine) is present in the bark and leaves of *Populus tremulata* L. and *P. alba* L. (*H. Braconnot*, Ann. chim., 1830, [2], **44**, 311). It has been prepared by the action of benzoyl chloride on salicin (*L. Dobbin* and *A. D. White*, Pharm. J., 1904, [iv], **19**, 233) and the structure was determined by *N. K. Richtmyer* and *E. H. Yeakel* (J. Amer. chem. Soc., 1934, **56**, 2495).

3-**Hydroxybenzyl alcohol,** m.p. 71°, is obtained from 3-hydroxybenzaldehyde by reduction with copper chromite catalyst (*D. Nightingale* and *H. D. Radford*, J. org. Chem., 1949, **14**, 1089) or using sodium tetrahydridoborate (*S. W. Chaikin* and *Brown*, J. Amer. chem. Soc., 1949, **71**, 122).

4-**Hydroxybenzyl alcohol,** m.p. 127°, is formed together with saligenin from the reaction of formaldehyde with phenol. It can be prepared by reduction of ethyl 4-hydroxybenzoate over copper chromite catalyst (*R. Mozingo* and *K. Folkers*, *ibid.*, 1948, **70**, 229).

4-**Methoxybenzyl alcohol,** m.p. 25°, b.p. 102°/2 mm, occurs in Tahiti vanilla beans. It is prepared from anisaldehyde by reduction with sodium tetrahydridoborate (*Chaikin* and *Brown*, *loc. cit.*), by catalytic reduction, or by the crossed Cannizzarro reaction with formaldehyde (*T. F. Dankowa et al.*, J. gen. Chem. U.S.S.R., 1948, **18**, 1724). It is easily converted to the dibenzyl ether. The chloroformate has been recommended as a blocking group in peptide synthesis.

1-(4-**Methoxyphenyl)ethanol,** b.p. 138°/12 mm, has been obtained from the corresponding ketone by catalytic reduction using copper chromite (*D. T. Mowry*, *M. Renoll* and *W. F. Huber*, J. Amer. chem. Soc., 1946, **68**, 1105). Resolution of the racemate is achieved by crystallisation of the cinchonidine salt of the hydrogen phthalate (*Balfe et al., loc. cit.*).

2-(4-**Hydroxyphenyl)ethanol, tyrosol,** m.p. 93°, is formed from tyrosine by the action of lactic acid-bacteria (*W. Grimmer et al.*, Milchwirtsch. Forsch., 1940, **20,** 110).

(d) Di- and tri-hydroxyphenyl alcohols

2,5-**Dihydroxybenzyl alcohol, gentisyl alcohol, salirepol,** m.p. 101°, is obtained by the hydrogenation of the corresponding aldehyde (*J. H. Birkinshaw, A. Bracken* and *H. Raistrick*, Biochem. J., 1943, **37,** 726). It has been isolated from *Penicillium divergens* (*J. Barta* and *R. Mecir*, Experientia, 1948, **4,** 277) and shows antibacterial activity; the 2-β-D-6′-benzoylglucopyranoside, *salireposide* has been found in the bark of *Salix* and *Populus* species and its structure was elucidated by *H. Thieme* (Naturwissenschaften, 1964, **51,** 291).

3,4-**Dihydroxybenzyl alcohol,** m.p. 137–138°, is obtained from the corresponding aldehyde by catalytic reduction (*H. Kaemmerer* and *A. Casaeuberta*, Makromol. Chem., 1963, **67,** 167).

4-*Hydroxy*-3-*methoxybenzyl alcohol*, m.p. 115° (*W. H. Carothers* and *R. Adams*, J. Amer. chem. Soc., 1924, **46,** 1080) is obtained from the action of formaldehyde on guaiacol (*Manasse*, Ber., 1894, **27,** 2411). 3-*Hydroxy*-4-*methoxybenzyl alcohol*, m.p. 132° (*N. Mauthner*, J. pr. Chem., 1941, [2], **158,** 321). 3,4-*Dimethoxybenzyl alcohol*, b.p. 188–192°/20 mm, is obtained from veratraldehyde by a crossed Cannizzarro reaction with formaldehyde (*C. A. Fetscher* and *M. T. Bogert*, J. org. Chem., 1939, **4,** 71).

1-(4-*Hydroxy*-3-*methoxyphenyl)ethanol, apocynol* (VII), m.p. 101°, is related to the ketone apocynin which occurs in Canadian hemp. The 4-benzoyl derivative is obtained by the action of methylmagnesium iodide on benzoylvanillin and hydrolysis gives apocynol (*H. Finnemore*, J. chem. Soc., 1908, **93,** 1522).

1-(4-*Hydroxy*-3-*methoxyphenyl)butan*-3-*ol* (VIII), b.p. 197°/17 mm, is obtained by reduction of zingerone (p. 170) and has a ginger-like taste:

(VII) (VIII)

4-**Hydroxy**-3,5-**dimethoxybenzyl alcohol.** m.p. 135°, prepared from 2-hydroxy-1,3-dimethoxybenzene and formaldehyde (*J. W. Cook et al.*, J. chem. Soc., 1944, 322) is formed during the enzymic oxidation of 4-hydroxy-3,5-dimethoxytoluene (*H. Richtzenhain*, Ber., 1944, **77,** 409).

1-(3,4,5-*Trimethoxyphenyl)ethanol* is obtained from methylmagnesium iodide and 3,4,5-trimethoxybenzaldehyde (*F. Mauthner*, J. pr. Chem., 1915, [ii], **92,** 195).

3. Hydroxyarylcarbaldehydes

(a) Methods of preparation

Methods *(i)–(vi)* for the introduction of an aldehyde group into a phenol are effective because the hydroxyl group activates the benzene ring towards electrophilic substitution and the products are 2- and 4-hydroxybenzaldehydes. Methods *(i)–(v)* are compared in Table I. The general methods for preparation of benzaldehydes are also applicable.

(i) The Reimer–Tiemann reaction

The reaction of phenols with chloroform and alkali to give hydroxybenzaldehydes has been reviewed (*H. Wynberg*, Chem. Reviews, 1960, **60,** 169). It was discovered in 1876 (*K. Reimer* and *F. Tiemann*, Ber., 1876, **9,** 824). Industrial application of the reaction has been limited to the preparation of salicylaldehyde and vanillin. The aldehyde group is introduced *ortho* or *para* to the phenolic group. Considerable amounts of tar are formed during the reaction so that the yields of products isolated are often limited by difficulties of isolation. Recorded yields and *ortho:para* ratios vary widely but in general greater yields of the 2-hydroxybenzaldehyde can be isolated because this is usually steam-volatile and so is readily separated from the reaction mixture. The reaction mixture which contains water and chloroform, is heterogeneous. Addition of methanol was found to improve the yields of products from phenol (*D. F. Pontz*, U.S.P. 3,365,500/1964; C.A., 1968, **69,** 27041) but addition of ethanol or pyridine reduced the yields. The chloroform has been replaced by trichloroacetic acid (*D. E. Armstrong* and *D. H. Richardson*, J. chem. Soc., 1933, 496) which is soluble in water, and also by bromoform (*H. H. Hodgson*, *ibid.*, 1929, 1639).

The reaction involves formation of an intermediate dichlorocarbene from chloroform under alkaline conditions. This attacks the phenol as an electrophile (*Wynberg*, J. Amer. chem. Soc., 1954, **76,** 4998; *J. Hine* and *A. M. Dowell*, *ibid.*, 1954, **76,** 2688; *Hine* and *J. M. van der Veen*, J. org. Chem., 1961, **26,** 1406). Cyclohexadienones are isolated as byproducts from the reaction of 2- and 4-alkylphenols:

$$CHCl_3 + OH^{\ominus} \longrightarrow H_2O + Cl^{\ominus} + :CCl_2$$

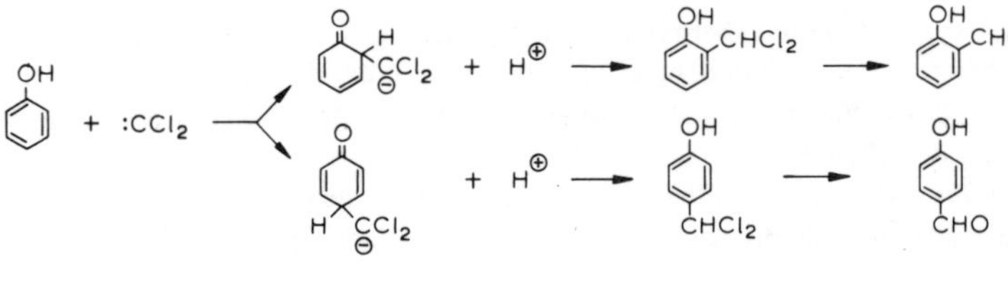

(I) (II)

Thus I has been isolated in 8% yield from *o*-cresol and II in 30% yield from 2,4-dimethylphenol (*K. Auwers* and *G. Keil*, Ber., 1902, **35,** 4207; 1903, **36,** 1861).

(ii) Gattermann reaction

Hydroxybenzaldehydes and their ethers are obtained by the action of dry hydrogen chloride and anhydrous hydrogen cyanide on a phenol or a phenyl ether in some inert solvent such as ether or benzene. Monohydric phenols and phenyl ethers generally require the addition of aluminium chloride. Resorcinol, and phloroglucinol derivatives, will react in the absence of aluminium chloride though in some reactions zinc chloride has been used as catalyst. In place of hydrogen cyanide, zinc cyanide and hydrogen chloride may be used to formylate resorcinol and phloroglucinol derivatives (*R. Adams et al.*, J. Amer. chem. Soc., 1923, **45,** 2373). A mixture of zinc cyanide, hydrogen chloride and aluminium chloride will convert anisole into 4-methoxybenzaldehyde (*Adams* and *E. Montgomery*, *ibid.*, 1924, **46,** 1518). Liquid hydrogen cyanide has also been replaced by sodium cyanide (*E. L. Niedzielsky* and *F. F. Nord*, *ibid.*, 1941, **63,** 1463; J. org. Chem., 1943, **8,** 147), by cyanogen bromide (*P. Karrer*, Helv., 1919, **2,** 89) and by *symm*-triazine (*A. Krentzberger*, Arch. Pharm., 1969, **302,** 828). The reaction has been reviewed (*W. E. Truce*, Organic Reactions, 1957, **9,** 37; *G. A. Olah* and *S. J. Kuhn*, "Friedel–Crafts and Related Reactions", Ed. Olah, Interscience, London, 1964, Vol. III, p. 1191).

An aldimine hydrochloride is the product of electrophilic substitution on the phenol. It is hydrolysed to the aldehyde by addition of water. Few of the nitrogen-containing intermediates have been isolated; in some cases the intermediate is thought to be a polymeric form of the imine. The Gattermann reaction usually gives more *para* than *ortho* substitution with a phenol or phenyl ether:

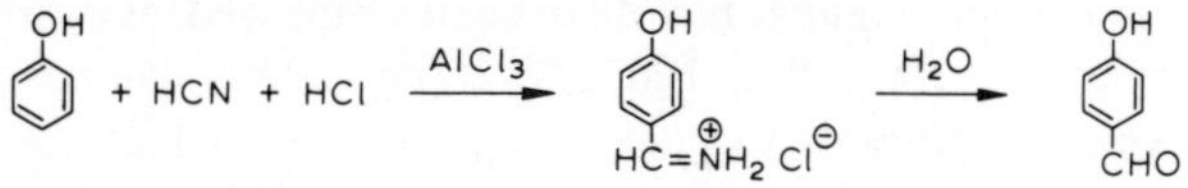

$$C_6H_5OH + HCN + HCl \xrightarrow{AlCl_3} HOC_6H_4HC{=}\overset{\oplus}{N}H_2\ Cl^{\ominus} \xrightarrow{H_2O} HOC_6H_4CHO$$

(iii) Vilsmeier reaction

The formylation of phenols with a mixture of *formanilide* and phosphoryl chloride was first achieved by *O. Dimroth* and *R. Zoppritz* (Ber., 1902, **35,** 995). This reaction was examined extensively by Vilsmeier who introduced the use of N-*methylformanilide* and phosphoryl chloride for the formylation of benzene compounds reactive to electrophilic substitution. Other formamide derivatives have also been used and *dimethylformamide* has become the reagent now most frequently employed (*N. P. Buu-Hoï et al.*, Bull. Soc. chim. Fr., 1955, 1594):

$$\text{1,3-}(MeO)_2C_6H_4 + PhN(Me)CHO + POCl_3 \longrightarrow [HC{=}\overset{\oplus}{N}(Me)Ph]\text{-}C_6H_3(OMe)_2 \xrightarrow{H_2O} OHC\text{-}C_6H_3(OMe)_2$$

For a comparison of these reagents see *J. P. Lambooy* (J. Amer. chem. Soc., 1956, **78,** 771). *Dimethylthioformamide* is suggested as being a superior reagent to dimethylformamide (*J. G. Dingwall, O. H. Reid* and *K. Wade*, J. chem. Soc., C, 1969, 913). The reaction may be carried out without a solvent or with added benzene or *o*-dichlorobenzene and usually proceeds readily at near room temperatures. An aldimine salt is the first product and the aldimine has to be hydrolysed in a second step. Formylation usually gives one isomer with the formyl group entering *para* to the hydroxy or methoxy function. The Vilsmeier reaction has been reviewed (*V. I. Minkin* and *G. N. Dorofeenko*, Usp. Khim., 1960, **29,** 599; *Olah* and *Kuhn*, *op. cit.*, p. 1121).

(iv) Alternative formylating agents

Other formic acid derivatives will condense with phenols to yield hydroxybenzaldehydes (*H. Gross, A. Rieche* and *G. Matthey*, Chem. Reviews, 1963, **96,** 308). *Dichloromethyl methyl ether* reacts with the phenol in dichloromethane as solvent and titanium tetrachloride as catalyst. Hydrolysis of the products affords a mixture of the 2- and 4-hydroxybenzaldehydes. Phenols will also react with *ethyl orthoformate* and aluminium chloride in an inert solvent to give the 2- and 4-hydroxybenzaldehydes. Excellent yields of the aldehydes from 1,2,3-, 1,2,4- and 1,3,5-trihydroxybenzenes can be obtained by these methods.

Chloromethylene dibenzoate, $(PhCO_2)_2CHCl$ formylates anisole to anisaldehyde in good yield (*F. Wenzel* and *L. Bellak*, Monatsh., 1914, **35,** 965). *Dichloromethyl methyl sulphide* has also been employed as a formylating reagent (*Gross* and *Matthey*, Ber., 1967, **97,** 2606). *Diphenylformamidine* condenses with reactive phenols to give a benzanil from which the aldehyde can be obtained by hydrolysis (*J. B. Shoesmith* and *J. Haldane*, J. chem. Soc., 1923, 2704; 1924, 2405):

$$\text{Me, OH, Pr}^i\text{-}C_6H_3 \xrightarrow[\text{or } CH(OEt)_3,\ AlCl_3]{Cl_2CHOMe,\ TiCl_4} \text{Me, CHO, OH, Pr}^i\text{-}C_6H_2 + \text{OHC, Me, OH, Pr}^i\text{-}C_6H_2$$

(v) Duff reaction

Hydroxybenzaldehydes are obtained when a phenol is heated at 150–165° with hexamethylenetetramine in a mixture of glycerol and boric acid (*J. C. Duff, ibid.*, 1941, 547; 1951, 1512; *L. M. Liggett* and *H. Diehl*, Proc. Iowa Acad. Sci., 1945, **52,** 191). The phenol is rapidly converted to the aminomethyl compound which is then slowly dehydrogenated to the imine and this is rapidly hydrolysed to the hydroxybenzaldehyde (*Y. Ogata, A. Kawasaki* and *F. Sugiura*, Tetrahedron, 1968, **24,** 5001). Tarry byproducts may be formed and the reaction is more suitable to the preparation of 2-hydroxybenzaldehydes since these can be isolated from the mixture by

OH, Me + $C_6H_{12}N_4$ $\xrightarrow[150-165°]{H_3BO_3}$ OH, CHO, Me

steam distillation. 4-Hydroxy-3,5-dimethoxybenzaldehyde has been prepared by the Duff reaction (*C. F. H. Allen* and *G. W. Leubner*, Org. Synth. 1963, Coll. Vol. IV, p. 866).

In a related reaction, the condensation of a phenol with formaldehyde in the presence of an oxidising agent yields a hydroxybenzaldehyde and many commercial syntheses of vanillin have been proposed on this basis. Vanillin can be obtained in good yields by the action of formaldehyde on guaiacol in alkaline solution in the presence of nitrobenzene-3-sulphonic acid (G.P. 105,798/1898).

(vi) Use of organometallic reagents

Aryl halides are converted to the Grignard reagent or the related aryllithium, which is reacted with ethyl orthoformate, *N*-methylformanilide or ethyl *N*-phenylformimidate, PhN=CHOEt (*L. Gattermann*, Ann., 1912, **393,** 215; *L. I. Smith* and *M. Bayliss*, J. org. Chem., 1941, **6,** 441). The metalation of phenyl ethers with butyl-lithium takes place *ortho* to the ether group (*H. Gilman* and *J. W. Morton*, Organic Reactions, 1954, **8,** 258) and this reaction can be used in the preparation of otherwise difficultly accessible benzaldehyde derivatives such as 2,6-dimethoxybenzaldehyde (*Adams* and *J. Mathieu*, J. Amer. chem. Soc., 1948, **70,** 2120):

OMe, OMe $\xrightarrow{BuLi}$ OMe, Li, OMe $\xrightarrow[(2)\ hydrolysis]{(1)\ PhN(Me)CHO}$ OMe, CHO, OMe

(vii) By the diazonium reaction

The diazonium reaction has been used to obtain hydroxybenzaldehydes from aminobenzaldehydes and is particularly useful for the preparation of 3-hydroxybenzaldehyde.

(viii) Side-chain oxidations

Reactions are available for the oxidation of hydroxytoluenes to the corresponding hydroxybenzaldehyde. The free phenols are oxidised with pentyl nitrite and good yields of the aldehyde result when electrophilic substitution of the phenol is blocked by alkyl groups (*J. Thiele* and *H. Eichwede*, Ann., 1900, **311,** 363):

$$\text{2,6-Me}_2\text{-4-Me-C}_6\text{H}_2\text{OH} \xrightarrow{C_5H_{11}ONO} \text{2,6-Me}_2\text{-4-CHO-C}_6\text{H}_2\text{OH}$$

Manganese dioxide can also be used (*C. M. Orlando*, J. org. Chem., 1970, **35,** 3714). When the phenolic group is protected as the acetate or the ether, oxidation of the side chain can be accomplished with chromium trioxide in acetic anhydride; the aldehyde diacetate is formed (*Thiele* and *E. Winter*, Ann., 1900, **311,** 353). Phenolic and other benzyl alcohols are oxidised in good yields to the aldehyde by the chromium trioxide–pyridine complex in dichloromethane (*J. C. Collins, W. W. Hess* and *F. J. Frank*, Tetrahedron Letters, 1968, 3363). Manganese dioxide and sulphuric acid is also used for oxidation of the phenyl ethers (*W. M. Easter* and *T. F. Wood*, J. org. Chem., 1951, **16,** 586):

$$\text{4-Me-C}_6\text{H}_4\text{OAc} \xrightarrow[Ac_2O]{CrO_3} \text{4-CH(OAc)}_2\text{-C}_6\text{H}_4\text{OAc} \qquad \text{4-Me-2-Bu}^t\text{-C}_6\text{H}_3\text{OMe} \xrightarrow[H_2SO_4]{MnO_2} \text{4-CHO-2-Bu}^t\text{-C}_6\text{H}_3\text{OMe}$$

Free and protected phenols which possess an olefin side-chain can be oxidised with ozone to the corresponding carbaldehyde. This reaction is of particular use for the preparation of some hydroxyphenylacetaldehydes where the required 2-allylphenol is available by the Claisen rearrangement (*R. Aneja, S. K. Mukergee* and *T. R. Sheshadri*, Tetrahedron, 1958, **2,** 203; 1958, **4,** 256).

The oxidation of styrenes with thallium(III) nitrate yields phenyl acetaldehyde (*A. McKillop et al.*, Tetrahedron Letters, 1970, 5275). Hydroxyphenylstyrenes are oxidised to the benzaldehyde with nitrobenzene and alkali and this process is used commercially to prepare vanillin from isoeugenol (p. 161). Wood shavings and lignin-containing extracts yield vanillin and syringaldehyde on heating with nitrobenzene and aqueous alkali (*R. H. J. Creighton, R. D. Gibbs* and *H. Hibbert*, J. Amer. chem. Soc., 1944, **66,** 32).

The Sommelet reaction (*S. J. Angyl*, Organic Reactions, 1954, **8,** 197) is available for oxidation of benzyl halides to the aldehyde. Hydroxybenzyl alcohols can be oxidised to the aldehyde by the reagents discussed in *(v)*.

(ix) Side-chain reduction

The Rosenmund reduction of acid chlorides to the corresponding aldehyde using hydrogen and a poisoned palladium catalyst is available. Reaction of alkoxybenzoylsulphonohydrazides with alkali (McFadyen–Stevens reaction) gives the aldehyde. Catalytic reduction of β-nitrostyrenes over palladium/charcoal gives the corresponding phenylacetaldehyde oximes (*B. Reichert* and *W. Koch*, Arch. Pharm., 1935, **273,** 265):

$CONH{\cdot}NHSO_2C_6H_4Me[4]$ (OMe) $\xrightarrow{base}$ CHO (OMe) $CH{=}CHNO_2$ (OMe) $\xrightarrow{H_2/Pd}$ $CH_2{\cdot}CH{=}NOH$ (OMe)

(b) Properties and reactions

(i) Chelation and intramolecular hydrogen bonding

2-Hydroxybenzaldehydes form an intramolecular hydrogen bond which can be detected in dilute solution by infrared spectroscopy (*C. J. W. Brooks* and *J. F. Morgan*, J. chem. Soc., 1961, 3373). Because of this intramolecular hydrogen bonding 2-hydroxybenzaldehydes are in general steam-volatile while the *meta*- and *para*-isomers are not. The phenolic p*K*a values for hydroxybenzaldehydes are 2- 8.34, 3- 9.02, 4- 7.62. The *para*-isomer is rendered a relatively strong acid because of the mesomeric effect of the CHO group while hydrogen bonding weakens the *ortho*-isomer p*K*a value (*L. B. Magnusson, C. Postmus* and *C. A. Craig*, J. Amer. chem. Soc., 1963, **85,** 1711).

Selective alkylation of some polyhydric aldehydes is possible since because of hydrogen bonding, a hydroxyl group *ortho* to CHO forms the corresponding ether only slowly under mild conditions. Suitable reagents are an alkyl halide with potassium carbonate in acetone (*A. M. Robinson* and *R. Robinson*, J. chem. Soc., 1932, 1439; *T. Reichstein et al.*, Helv., 1935, **18,** 816) or diazomethane (*H. A. Offe* and *H. Jatzkewitz*, Ber., 1947, **80,** 469):

CHO, OH (OH) $\xrightarrow[acetone]{MeI,\ K_2CO_3}$ CHO, OH (OMe)

2-Hydroxybenzaldehydes form chelates with transition metal ions. The copper(II) complex of salicylaldehyde is soluble in chloroform. Salicylaldehyde imine and oxime form more stable complexes and the oxime has been used as a reagent in the gravimetric analysis of some metal ions.

(ii) Oxidation

Oxidation to the hydroxycarboxylic acid requires a careful choice of reagent. It can be effected with manganese dioxide (*M. Z. Barakat, M. F. Abdel-Wahab* and *M. M. El-Sadr*, J. chem. Soc., 1956, 4685) with silver oxide in alkaline suspension or by fusion with sodium hydroxide (*I. A. Pearl*, Org. Synth., 1963, Coll. Vol. IV, p. 972). Oxidation with hydrogen peroxide in alkaline solution (the Dakin reaction) replaces the CHO group by OH in such cases where hydroxyl groups are present in *ortho*- or *para*-positions (*H. Dakin*, J. Amer. chem. Soc., 1909, **42**, 486; *A. R. Surrey*, Org. Synth., 1955, Coll. Vol. III, p. 759). *m*-Hydroxybenzaldehyde does not react. 2- and 4-Methoxybenzaldehydes under these conditions give a mixture of the methoxyphenol and methoxybenzoic acid (*E. Spath, M. Pailer* and *G. Gergeley*, Ber., 1940, **73**, 935).

(iii) Hydrogenation

Catalytic hydrogenation of the aldehyde function gives either the benzyl alcohol or the toluene derivative depending on the choice of catalyst and conditions (*R. L. Augustine*, "Catalytic Hydrogenation", Arnold, London, 1965). Reduction to methyl has also been accomplished by the Clemmensen method. Electroreduction gives the corresponding hydrobenzoin (*J. H. Stocker* and *R. M. Jenevein*, J. org. Chem., 1968, **33**, 294).

(iv) Cyclisation reactions

Many reactions of 2-hydroxybenzaldehydes lead to cyclised products. These are discussed under the reactions of salicylaldehyde.

2-Hydroxyphenylacetaldehydes cyclise easily to benzofuran derivatives (*Aneja, Mukergee* and *Sheshadri, loc. cit.*):

$H_2C{\cdot}CHO$ OC O OH ⟶ OC O O

(c) Monohydroxybenzaldehydes and their homologues

A comparison of some methods of preparation is given in Table 1.

TABLE 1

METHODS FOR THE FORMATION OF HYDROXYBENZALDEHYDES AND THEIR ETHERS

Starting benzene derivative	*Benzaldehyde formed*	*Method, yeild and ref.*
Phenol	1-HO	R–T (25–57%)[1,2]; D (20%)[3]
	4-HO	R–T (8–27%)[1,2]; G (30%)[4]; V (85%)[5]
Anisole	4-MeO	G (94%)[6]; V (10%)[5]
Catechol	3,4-$(HO)_2$	Gr (45%)[7]; R–T[8]
1-HO-2-MeO	2-HO-3-MeO	R–T[8]
	4-HO-3-MeO	R–T[8]
1,2-$(MeO)_2$	3,4-$(MeO)_2$	G (60%)[4]; V (30–40%)[5]
Resorcinol	2,4-$(HO)_2$	R–T (23%)[8]; G (56–97%)[9,10]; V (60%)[11]; Gr (64%)[7]
1-HO-3-MeO	2-HO-4-MeO	R–T[13]; G (45%)[12]
	4-HO-2-MeO	R–T (25%)[13]; G (75%)[4,9]; G (50%)[12]
1,3-$(MeO)_2$	2,4-$(MeO)_2$	G (80%)[4,6]; V (85%)[14]
Quinol	2,5-$(HO)_2$	R–T[15]
1-HO-4-MeO	2-HO-5-MeO	R–T (50–65%)[16]; D[20]
Pyrogallol	2,3,4-$(HO)_3$	G (45–50%)[4,10]; Gr (92%)[7]
1,3-$(HO)_2$-2-MeO	2,4-$(HO)_2$-3-MeO	G (93%)[17]
1,2,4-$(HO)_3$	2,4,5-$(HO)_3$	G (100%)[4,18]; Gr (89%)[7]
Phloroglucinol	2,4,6-$(HO)_3$	G (70%)[4,18]; Gr (96%)[7]
1,3-$(HO)_2$-5-EtO	2,4-$(HO)_2$-6-EtO	G (97%)[19]

R–T: Reimer–Tiemann reaction; G: Gattermann reaction and Gr: Gross modification of Gattermann reaction; V: Vilsmeier reaction; D: Duff reaction.

References

1 *H. Wynberg*, Chem. Reviews, 1960, **60**, 169.
2 *D. F. Pontz*, U.S.P. 3,365,500/1964.
3 *J. C. Duff*, J. chem. Soc., 1941, 547.
4 *L. Gattermann*, Ann., 1907, **357**, 313.
5 *N. P. Buu-Hoï et al.*, Bull. Soc. chim. Fr., 1955, 1594.
6 *R. Adams* and *E. Montgomery*, J. Amer. chem. Soc., 1924, **46**, 1518.
7 *H. Gross*, *A. Rieche* and *G. Matthey*, Ber., 1963, **96**, 308.
8 *F. Tiemann* and *P. Koppe*, *ibid.*, 1881, **14**, 2015.
9 *Gattermann* and *W. Berchelmann*, *ibid.*, 1898, **31**, 1765.
10 *Adams* and *I. Levine*, J. Amer. chem. Soc., 1923, **45**, 2373.
11 *O. Dimroth* and *R. Zoepritz*, Ber., 1902, **35**, 995.
12 *T. E. de Kiewiet* and *H. Stephen*, J. chem. Soc., 1931, **133**, 84.
13 *Tiemann* and *A. Parrisius*, Ber., 1880, **13**, 2354.
14 *A. R. H. Sommers*, *R. J. Michaelis* and *A. W. Weston*, J. Amer. chem. Soc., 1952, **74**, 5546.
15 *Tiemann* and *W. H. M. Muller*, Ber., 1881, **14**, 1986.
16 *H. B. Gillespie*, Biochem. Preparations, 1953, **3**, 79.
17 *E. Spath* and *H. Schmid*, Ber., 1941, **74**, 193.
18 *Gattermann* and *M. Kobner*, Ber., 1899, **32**, 278.
19 *A. Robertson* and *T. S. Subramanian*, J. chem. Soc., 1937, 288.
20 *H. Diehl et al.*, Iowa Coll. J. Sci., 1947, **22**, 91.

Salicylaldehyde, 2-hydroxybenzaldehyde, m.p. + 1.6°, b.p. 93°/25 mm, is present in the essential oil of various species of *Spiraea*. The *β-D-glucopyranoside, helicin,* m.p. 176°, $[\alpha]_D$ −60° (water) does not occur naturally but has been obtained by the oxidation of salicin. It has been synthesised (*A. Robertson* and *R. B. Waters*, J. chem. Soc., 1930, 2732). *Helicin tetra-acetate*, m.p. 142°, $[\alpha]_D$ −37° (acetone).

(i) Methods of preparation

(*1*) Salicylaldehyde is manufactured largely by the Reimer–Tiemann reaction. It is purified by distillation and through the solid sodium bisulphite compound which is collected and decomposed with sulphuric acid. Purification can also be effected through the copper(II) salt (*E. C. Horning, M. G. Horning* and *D. A. Dimmig*, Org. Synth., 1955, Coll. Vol. III, p. 166).

(*2*) Saligenin is oxidised to salicylaldehyde by a variety of reagents including air and palladium catalyst (*P. A. R. Marchand* and *J. B. Grenet*, U.S.P. 3,321,526/1967).

(*3*) Aluminium phenoxide reacts with carbon monoxide under pressure to yield 2- and 4-hydroxybenzaldehydes (*L. Schmerling*, U.S.i. 3,098,875/1963).

(*4*) A number of processes have been used for the oxidation of *o*-cresol. Side-chain chlorination of tri-*o*-cresyl phosphate followed by hydrolysis of the benzylidene dichloride gives salicylaldehyde (*J. Terry* and *R. M. Tusskin*, U.S.P. 3,314,998/1967; *F. C. Buckley*, Ger. Offen., 1,925,111/1969).

(*5*) Salicylic acid can be reduced to salicylaldehyde electrochemically in the presence of boric acid which protects the aldehyde from further reduction (*H. V. K. Udupa*, Bull. Acad. Polon. Sci., Ser. Sci. chim., 1961, **9,** 51). Salicylaldehyde is obtained from salicylic acid *N*-benzenesulphonohydrazide by the McFadyen–Stevens reaction (*J. S. McFadyen* and *T. S. Stevens*, J. chem. Soc., 1936, 584).

(ii) Reactions

(*1*) *Chelation.* Chelates are formed with some metal ions and stability constant measurements have been made (*L. E. Maley* and *D. P. Mellor*, Austral. J. scient. Res., [A], 1949, **2,** 92). Salicylaldehyde gives a deep purple colour with ferric chloride.

(*2*) *Substitution.* Electrophilic substitution occurs readily in the 3- and 5-positions. 5-*Bromo-2-hydroxybenzaldehyde* (III), m.p. 105°, and 3,5-*dibromo-2-hydroxybenzaldehyde*, m.p. 83°, are obtained by bromination (*L. C. Raiford* and *L. K. Tanzer*, J. org. Chem., 1941, **6,** 722) and the corresponding chloro compounds are obtained by chlorination (*H. Biltz* and *K. Stepf*, Ber., 1904, **37,** 4022). The prolonged action of bromine yields 2,4,6-tribromophenol (*A. W. Francis* and *A. J. Hill*, J. Amer. chem. Soc., 1924, **46,** 2498). Nitration gives as the first products a mixture of 3-, m.p. 109°, and 5-*nitro-2-hydroxybenzaldehyde*, m.p. 126° (*W. von Miller*, Ber., 1887, **20,** 1928) and subsequently 3,5-*dinitro-2-hydroxybenzaldehyde*, m.p. 58° (*A. B. E. Lovelt* and *E. Roberts*, J. chem. Soc., 1928, 1978):

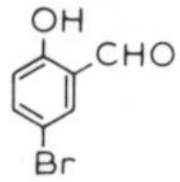

OH CHO Br

(III)

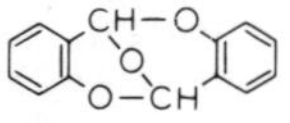

CH—O O O—CH

(IV)

(3) *Cyclisations.* By the action of acid chlorides, phosphorus trichloride, or by the addition of a small amount of sulphuric acid to a solution of salicylaldehyde in acetic anhydride, *disalicylaldehyde*, m.p. 130° (sublimes), is formed and to which structure IV has been given (*R. Adams*, J. Amer. chem. Soc., 1922, **44,** 1126). Substituted salicylaldehydes give similar products (*W. P. Bradley* and *F. B. Dains*, Amer. chem. J., 1892, **14,** 295).

Salicylaldehyde with acetic anhydride and sodium acetate gives coumarin (V, R = H) along with 2-acetoxycinnamic acid (*W. H. Perkin*, Ber., 1875, **8,** 1599). Sodium acetate can be replaced with a salt of phenylacetic acid (*A. Oglialoro*, Gazz., 1879, **9,** 428) or methoxyacetic acid (*I. M. Heilbron, D. W. Hill* and *H. N. Walls*, J. chem. Soc., 1931, 1701) and others to give the corresponding coumarin (V, R = Ph or MeO).

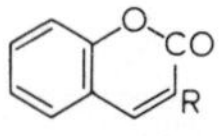

(V) 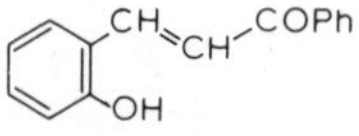 (VI) 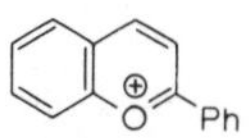 (VII)

Salicylaldehyde undergoes the aldol condensation catalysed by alkali with acetophenone and its derivatives to give the corresponding 2-hydroxychalkone (VI). Such chalkones cyclise under acidic conditions to give the 2-phenylchromenylium salt (VII) (*H. Decker* and *T. von Fellenberg*, Ann., 1907, **356,** 302).

2-*Acetoxybenzaldehyde*, m.p. 38°, is prepared by the acetylation of salicylaldehyde in pyridine (*A. Neuberger*, Biochem. J., 1948, **43,** 599). The *triacetate* $CH_3 \cdot CO_2 \cdot C_6H_4 \cdot CH(OAc)_2$, m.p. 103°, is formed by the action of acetic anhydride and mineral acid (*R. Wegscheider* and *E. Spath*, Monatsh., 1909, **30,** 851).

Salicylaldehyde-imine is known as its complexes with transition metal ions (*G. N. Tyson* and *S. C. Adams*, J. Amer. chem. Soc., 1940, **62,** 1228). The *copper*(II) *complex* (VIII) is formed by reaction of salicylaldehyde with ammonia and copper(II) acetate (*L. Hunter* and *J. A. Marriott*, J. chem. Soc., 1937, 2000). Similar complexes are formed with primary aliphatic amines in place of ammonia. Salicylaldehyde and ethylenediamine form crystalline *ethylene bis*(*salicylidene-imine*) (IX), m.p. 125°, which also gives complexes with transition metal ions.

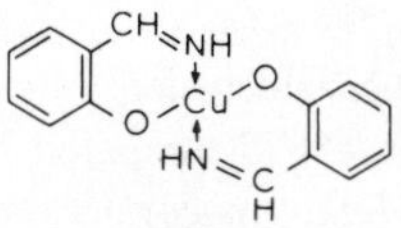

(VIII) 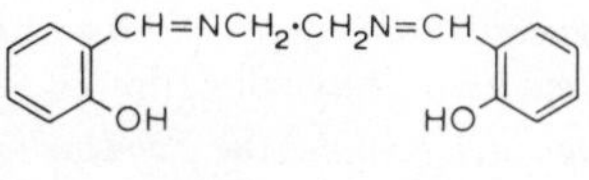 (IX)

Salicylaldoxime, m.p. 59°, forms complexes with metal ions of the type (VIII) and has been used in quantitative analysis (*F. I. Welcher*, "Organic Analytical Reagents", Princetown, 1962, Vol. 3); 2,4-*dinitrophenylhydrazone*, m.p. 248–252°.

2-*Methoxybenzaldehyde*, m.p. 39°, b.p. 127°/20 mm, is prepared by methylation of salicylaldehyde with methyl sulphate and alkali (*N. V. Sidgwick* and *N. S. Bayliss*, J. chem. Soc., 1930, 2027); 2,4-*dinitrophenylhydrazone*, m.p. 253°.

3-**Hydroxybenzaldehyde,** m.p. 104°, occurs as the cyanhydrin-β-D-glucoside, *zierin*, m.p. 156°, $[\alpha]_D$ −29.5° in *Zieria laevigata* Sm. (*H. Finnemore* and *J. M. Cooper*,

J. Proc. roy. Soc. N.S.W., 1936, **70,** 175). It is obtained from 3-nitrobenzaldehyde by a diazonium reaction of the corresponding amine (*R. B. Woodward*, Org. Synth., 1955, Coll. Vol. III, p. 453). It undergoes the Cannizzarro reaction (*G. Lock*, Ber., 1929, **62,** 1177) under the usual conditions whereas 2- and 4-hydroxybenzaldehydes only react in the presence of finely divided silver (*I. A. Pearl*, J. org. Chem., 1947, **12,** 85). Oxidation by air in alkaline solution or by fusion with potassium hydroxide gives 3-hydroxybenzoic acid. Reaction with bromine in chloroform gives 4,6-dibromo- and 2,4,6-tribromo-3-hydroxybenzaldehyde (*Lock*, Monatsh., 1930, **55,** 307). *Oxime*, m.p. 87°; 4-*nitrophenylhydrazone*, m.p. 221°.

3-*Methoxybenzaldehyde*, b.p. 105°/8 mm, is obtained by methylation of 3-hydroxybenzaldehyde with dimethyl sulphate and alkali or by catalytic reduction of 3-methoxybenzoyl chloride (*J. L. Hartwell* and *S. R. L. Kornberg*, J. Amer. chem. Soc., 1945, **67,** 1606).

4-**Hydroxybenzaldehyde,** m.p. 116°, (see Table 1 for preparation) is not volatile in steam. It has been found in *Papaver somniferum* L. and in the waste liquors from paper manufacture from aspen and spruce wood where it is a breakdown product of lignin. *Oxime*, m.p. 65°; 2,4-*dinitrophenylhydrazone*, m.p. 280° (decomp.).

4-*Methoxybenzaldehyde, anisaldehyde*, m.p. + 2°, b.p. 116°/10 mm, is a product of the aerial oxidation of essential oils containing anisole, star anise oil, cassia flower oil and vanilla oil. For preparation see Table 1, or by the hydrogenation of anisoyl chloride (*K. W. Rosenmund* and *F. Zetzche*, Ber., 1923, **56,** 1483) and the McFadyen–Stevens reaction of anisic acid (J. chem. Soc., 1936, 584).

Anisaldehyde has the usual reactions of an aromatic aldehyde, forming anisoin when refluxed with potassium cyanide solution and undergoing the Cannizzarro reaction with aqueous alkali. Electrochemical reduction gives *meso* and (±)-hydroanisoin (*J. Grimshaw* and *J. S. Ramsey*, J. chem. Soc., C, 1966, 653). Demethylation to 4-hydroxybenzaldehyde is effected by heating with pyridine hydrochloride (*V. Prey*, Ber., 1941, **74,** 1219). The nitration of anisaldehyde yields 3-*nitroanisaldehyde*, m.p. 86° (*P. Pfeiffer* and *B. Segall*, Ann., 1928, **460,** 129). Under vigorous conditions 3,5-dinitroanisaldehyde and 2,4,6-trinitroanisole are formed (*M. P. de Lange*, Rec. Trav. chim., 1926, **45,** 46).

4-*Ethoxybenzaldehyde*, m.p. 14°, b.p. 139°/20 mm, is made from phenetol by the Gattermann reaction or by ethylation of 4-hydroxybenzaldehyde.

2-*Hydroxyphenylacetaldehyde, semicarbazone*, m.p. 171°, is prepared by ozonolysis of 2-allylphenol (*I. J. Rinkes, ibid.*, 1926, **45,** 823). 2-*Methoxyphenylacetaldehyde oxime*, m.p. 95°, is formed by palladium-catalysed hydrogenation of β-nitro-2-methoxystyrene (*B. Reichert* and *W. Koch*, Arch. Pharm., 1935, **273,** 265).

3-*Methoxyphenylacetaldehyde*, b.p. 117°/13 mm, is obtained by reduction of 3-methoxyacetonitrile with stannous chloride and hydrogen chloride in ether (*K. H. Lin* and *R. Robinson*, J. chem. Soc., 1938, 2005). 4-*Methoxyphenylacetaldoxime*, m.p. 46.5°, is obtained by catalytic reduction of the corresponding β-nitrostyrene.

4-**Methoxyhydratropaldehyde,** $MeOC_6H_4{\cdot}CHMe{\cdot}CHO$, b.p. 135°/15 mm, (*M. Tiffeneau*, Bull. Soc. chim. Fr., 1907, [iv], **1,** 1212) is obtained from anethole through its iodohydrin.

(d) Dihydroxybenzaldehydes and their homologues

A comparison of some methods of preparation is given in Table 1.

(i) Derivatives of catechol

2,3-**Dihydroxybenzaldehyde,** m.p. 108°, is obtained by demethylation of 2-*hydroxy*-3-*methoxybenzaldehyde*, o-*vanillin*, m.p. 44°, with hydrobromic acid (*H. Pauly et al.*, Ann., 1911, **383,** 288). *o*-Vanillin is a byproduct from the manufacture of vanillin from guaiacol. 2,3-*Dimethoxybenzaldehyde*, m.p. 54°, is prepared by methylation of *o*-vanillin (*E. Rupp* and *K. Linck*, Arch. Pharm., 1915, **253,** 35).

Protocatechualdehyde, 3,4-**dihydroxybenzaldehyde,** m.p. 153°, can be prepared from catechol, by demethylation of vanillin with aluminium bromide (*K. W. Mertz* and *J. Fink*, Arch. Pharm., 1956, **289,** 347), or by the action of aluminium bromide (*E. Mosettig* and *A. Burger*, J. Amer. chem. Soc., 1930, **52,** 2988; *P. Pfeiffer* and *W. Loewe*, J. pr. Chem., 1937, [2], **147,** 293) or chloride on piperonal. It has been obtained from 3-bromo-4-hydroxybenzaldehyde with aqueous sodium hydroxide and a copper catalyst (*K. Shinra* and *Y. Shinra*, J. Soc. chem. Ind., Japan, 1944, **47,** 71). On fusion with alkali it yields 2,3-dihydroxybenzoic acid. *Oxime*, m.p. 157°.

Vanillin, 4-**hydroxy**-3-**methoxybenzaldehyde,** m.p. 81°, is the odoriferous principle of the vanilla pod (*Vanilla planifolia* Andr.) and is in demand for flavouring purposes.

Methods of manufacture

(*1*) From eugenol (X) which is available from oil of cloves. Eugenol is isomerised by an acid catalyst to isoeugenol (XI), this is acetylated and the side-chain oxidised with ozone, chromic acid, or nitrobenzene and alkali to yield acetylvanillin which is hydrolysed to give vanillin:

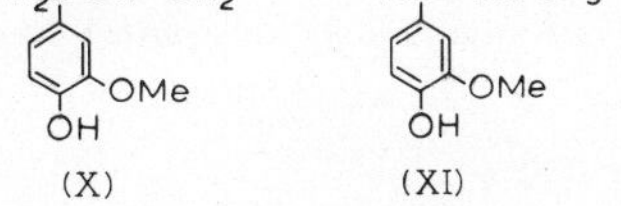

$H_2C{\cdot}CH{=}CH_2$ ... OMe, OH

(X)

$HC{=}CH{\cdot}CH_3$ ... OMe, OH

(XI)

(*2*) From the lignin byproduct of the manufacture of cellulose from wood fibre. The lignin is oxidised with nitrobenzene and alkali and the benzaldehyde products separated. Lignins can yield a mixture of 4-hydroxybenzaldehyde, vanillin and syringaldehyde and a source rich in vanillin has to be chosen.

(*3*) From guaiacol by any of the general methods for formylation of phenols.

Reactions

Vanillin can be oxidised to vanillic acid by silver oxide or by fusion with potassium hydroxide (*Pearl*, Org. Synth., 1963, Coll. Vol. IV, p. 972). Vigorous alkali fusion results in demethylation to protocatechuic acid (*idem, ibid.*, 1955, Coll. Vol. III, p. 745).

Bromination affords 5-bromo- (*R. L. Shriner* and *P. McCutchan*, J. Amer. chem.

Soc., 1929, **51,** 2194), and 5,6-dibromo-4-hydroxy-3-methoxybenzaldehyde. Other electrophiles react in a similar manner. Nitration of *O*-acetylvanillin gives a product which, after hydrolysis, affords 2-*nitrovanillin* yellow needles, m.p. 137°, in 80% yield. Methylation of this gives 2-*nitroveratraldehyde*, m.p. 54° (*D. H. Hey* and *L. C. Lobo*, J. chem. Soc., 1954, 2249). Bromination of *O*-acetylvanillin followed by hydrolysis yields 6-*bromo-4-hydroxy-3-methoxybenzaldehyde*, m.p. 178° (*L. C. Raiford* and *W. C. Stoesser*, J. Amer. chem. Soc., 1927, **49,** 1079).

Vanillin oxime, m.p. 117°.

*Glucovanillin, vanillin β-*D*-glucopyranoside, dihydrate*, m.p. 189°, $[\alpha]_B$ −87° (water), is a product of the oxidation of coniferin (*F. Tiemann*, Ber., 1885, **18,** 1596, 3482); the *tetra-acetate*, m.p. 143°, is obtained from vanillin and acetobromoglucose (*W. V. Thorpe* and *R. T. Williams*, J. chem. Soc., 1937, 494).

3-*Hydroxy-4-methoxybenzaldehyde, isovanillin*, m.p. 116°, is obtained by demethylation of veratraldehyde (*J. Shinoda* and *M. Kawagoye*, J. pharm. Soc. Japan, 1928, **48,** 119) or the careful methylation of protocatechualdehyde (*Lock*, Monatsh., 1934, **64,** 341).

3-*Ethoxy-4-hydroxybenzaldehyde*, m.p. 78°, has a much stronger taste and odour than vanillin and is used commercially under the name *bourbonal* or *ethyl vanillin.*

4-*Ethoxy-3-hydroxybenzaldehyde*, m.p. 126°, from ozonolysis of 6-ethoxy-3-propenylphenol (*N. Hirao*, J. chem. Soc., Japan, 1931, **52,** 26).

3,4-*Dimethoxybenzaldehyde, veratraldehyde*, m.p. 43°; 3-*benzyloxy-4-methoxybenzaldehyde*, m.p. 64°; 4-*benzyloxy-3-methoxybenzaldehyde*, m.p. 64° (*Merz* and *Fink*, *loc. cit.*).

Piperonal, *heliotropin*, 3,4-*methylenedioxybenzaldehyde*, (XII), m.p. 37°, is manufactured from safrole (XIII). It has also been prepared by the oxidation of piperoic acid and by the action of diiodomethane on protocatechualdehyde.

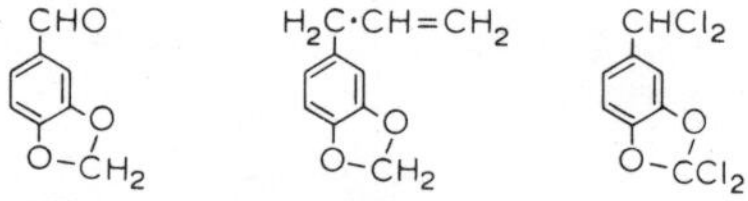

CHO; O–CH_2 (XII) — $H_2C{\cdot}CH{=}CH_2$; O–CH_2 (XIII) — $CHCl_2$; O–CCl_2 (XIV)

Piperonal has an odour of heliotrope and finds application in perfumery. *Oxime*, m.p. 110°.

The methylenedioxy group can be split off by heating with hydrochloric acid at 190° or with aluminium bromide (see 3,4-dihydroxybenzaldehyde). Thionyl chloride or phosphorus pentachloride yield *piperonylidene dichloride*, m.p. 59°, and the further action causes halogenation of the methylenedioxy group to give XIV, m.p. 15°, b.p. 152°/9 mm, which can be hydrolysed to the carbonate of protocatechualdehyde (*G. Barger*, J. chem. Soc., 1908, **93,** 572). Bromination of piperonal affords 6-*bromopiperonal*, m.p. 129°, and further bromination gives 4,5-dibromomethylenedioxybenzene (*A. Orr, R. Robinson* and *M. M. Williams*, *ibid.*, 1917, **111,** 948). The corresponding reaction occurs on chlorination. Nitration affords 6-*nitropiperonal*, m.p. 96°, and 3,4-methylenedioxynitrobenzene (*A. H. Salway*, *ibid.*, 1909, **95,** 1163).

(ii) Derivatives of resorcinol

2,4-**Dihydroxybenzaldehyde,** *β-resorcylaldehyde*, m.p. 135°, see Table 1. *Oxime*, m.p. 192°. Both monomethyl ethers, *2-hydroxy-4-methoxybenzaldehyde*, m.p. 42°, and *4-hydroxy-2-methoxybenzaldehyde* are formed by the Reimer–Tiemann (*F. Tiemann* and *P. Koppe*, Ber., 1881, **14,** 2021) and the Gattermann reactions (*T. E. de Kiewiet* and *H. Stephen*, J. chem. Soc., 1931, **133,** 84).

β-Resorcylaldehyde with an alkyl halide and base yields first the 4-alkoxy compound and then the 2,4-dialkoxide (*A. M. Robinson* and *R. Robinson*, *ibid.*, 1932, 1439).

2,6-**Dihydroxybenzaldehyde,** *γ-resorcylaldehyde*, m.p. 155°, is made by demethylation with aluminium chloride of *2,6-dimethoxybenzaldehyde*, m.p. 98°, obtained by treating 1,3-dimethoxybenzene with phenyl-lithium and then reacting the resultant lithium derivative with *N*-methylformanilide (*G. Wittig*, Angew. Chem., 1940, **53,** 241). It has also been obtained from 2,4-dihydroxy-3-formylbenzoic acid (*R. C. Shah* and *M. C. Laiwalla*, J. chem. Soc., 1938, 1828).

3,5-**Dihydroxybenzaldehyde,** *α-resorcylaldehyde*, m.p. 161° (decomp.), is obtained by reduction of 3,5-diacetoxybenzoyl chloride with a palladium–barium sulphate catalyst and hydrolysis of the product (*E. Spath* and *F. Liebherr*, Ber., 1941, **74,** 869). *3,5-Dimethoxybenzaldehyde*, m.p. 45°, is obtained by treating 3,5-dimethoxybenzoyltoluene-4-sulphonohydrazide with potassium carbonate (*R. Adams*, *S. MacKenzie* and *S. Loewe*, J. Amer. chem. Soc., 1948, **70,** 664); *oxime*, m.p. 120°.

4,6-**Dihydroxy**-2-**methylbenzaldehyde,** *orcylaldehyde*, (XV), m.p. 180°, is prepared from orcinol by the Gattermann reaction (*Adams* and *I. Levine*, *ibid.*, 1923, **45,** 2376); *oxime*, m.p. 200°. Partial methylation with iodomethane and potassium carbonate in acetone gives *6-hydroxy-4-methoxy-2-methylbenzaldehyde*, *everninaldehyde*, (XVI), m.p. 65° (*A. Robertson* and *R. J. Stephenson*, J. chem. Soc., 1932, 1388):

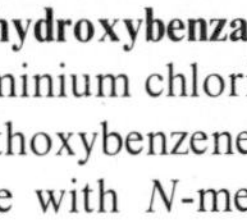

(XV) 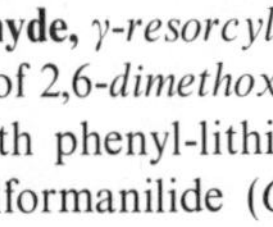 (XVI) 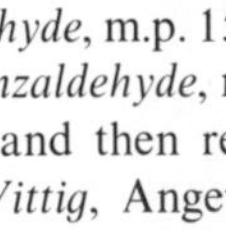 (XVII) 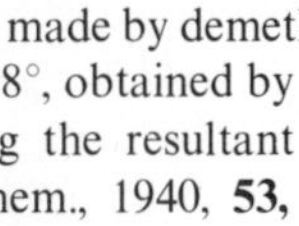 (XVIII)

2,6-**Dihydroxy**-4-**methylbenzaldehyde,** *atranol*, (XVII, R=H), m.p. 124°, is prepared from 3,5-dimethoxytoluene *via* the lithium derivative as for 2,6-dihydroxybenzaldehyde (*Adams* and *J. Mathieu*, J. Amer. chem. Soc., 1948, **70,** 2120). *Chloratranol*, (XVII, R=Cl), m.p. 143° (*A. St. Pfau*, Helv., 1934, **17,** 1319).

4,6-Dihydroxy-2-pentylbenzaldehyde, *olivetolaldehyde*, (XVIII), m.p. 66°, by the Gattermann reaction on 5-pentylresorcinol (*Y. Asahina* and *M. Yasue*, Ber., 1937, **70,** 206).

(iii) Derivatives of hydroquinone

2,5-**Dihydroxybenzaldehyde,** *gentisaldehyde*, m.p. 99°, is prepared from quinol (Table 1, also *I. Amakasu* and *K. Sato*, Bull. chem. Soc., Japan, 1967, **40,** 1428) or by oxidation of salicylaldehyde with persulphate (*H. H. Hodgson*, J. chem. Soc., 1927, 2330). *2-Hydroxy-5-methoxybenzaldehyde*, m.p. 4°, b.p. 133°/15 mm (*H. B. Gillespie*, Biochem. Prepn., 1953, **3,** 79).

Flavoglaucin (XIX), pale yellow, m.p. 103°, and **auroglaucin** (XX), orange red, m.p. 153°, are isolated from species of *Aspergillus* along with other pigments (*H. Raistrick, R. Robinson* and *A. R. Todd*, J. chem. Soc., 1937, 80). The structures shown were proposed by *A. Quilico et al.* (Gazz., 1949, **79,** 89; 1953, **83,** 756) and dihydroflavoglaucin has been synthesised (*L. Panizzi* and *R. Nicolaus*, Gazz., 1953, **83,** 774):

(XIX) (XX)

(e) Tri- and tetra-hydroxybenzaldehydes

A comparison of some methods of preparation is given in Table 1.

(i) Derivatives of pyrogallol

2,3,4-**Trihydroxybenzaldehyde,** *pyrogallolaldehyde*, m.p. 158°, for preparation see Table 1. Methylation with methyl sulphate and alkali gives 2-*hydroxy*-3,4-*dimethoxybenzaldehyde*, m.p. 74° (*Spath* and *F. Boschan*, Monatsh., 1933, **63,** 141) or 2,3,4-*trimethoxybenzaldehyde*, m.p. 37° (*Barger* and *A. J. Ewins*, J. chem. Soc., 1910, **97,** 2258) depending on the conditions used. These aldehydes have been prepared by standard methods from pyrogallol 1,2-dimethyl ether (*W. Baker* and *H. A. Smith*, *ibid.*, 1931, 2545) and the trimethyl ether (*F. Schaaf* and *A. Labouchere* Helv., 1924, **7,** 357), respectively.

3,4,5-**Trihydroxybenzaldehyde,** *gallaldehyde*, m.p. 212° (decomp.) is obtained by reduction of triacetyl- or tri(methoxycarbonyl)-galloyl chloride over palladised barium sulphate and hydrolysis of the product (*K. W. Rosenmund* and *F. Zetzsche*, Ber., 1918, **51,** 598). Obtained by this route from the appropriate derivative of gallic acid are 3,4-*dihydroxy*-5-*methoxybenzaldehyde*, m.p. 139° (*F. Mauthner*, J. pr. Chem., 1928, [ii], **119,** 309), 3-*hydroxy*-4,5-*dimethoxybenzaldehyde*, m.p. 177° (*idem*, Ann., 1926, **449,** 105) and 3,4,5-*trimethoxybenzaldehyde*, m.p. 75°, b.p. 198°/10 mm (*W. Baker* and *R. Robinson*, J. chem. Soc., 1929, 156). 4,5-Dihydroxy-3-methoxybenzaldehyde can be obtained by heating 5-bromovanillin with sodium hydroxide and a copper catalyst (*W. Bradley, Robinson* and *G. Schwarzenbach*, *ibid.*, 1930, 811).

(ii) Derivatives of 1,3,4-trihydroxybenzene

2,4,5-**Trihydroxybenzaldehyde,** m.p. 223°, for preparation see Table 1 and also *Y. Tanase* (J. pharm. Soc., Japan, 1941, **61,** 341). Its trimethyl ether, *asarylaldehyde*, m.p. 114° is a product of the ozonolysis of asarone (*J. van Alphen*, Rec. Trav. chim., 1927, **46,** 195). It has been prepared from 1,3,4-trimethoxybenzene by the Gattermann reaction (*L. Gattermann* and *F. Eggers*, Ber., 1899, **32,** 289). The 4-*methyl ether* has m.p. 209°, 4,5-*dimethyl ether*, m.p. 105° (*F. S. Head* and *Robertson*, J. chem. Soc., 1930, 2441), 5-*methyl ether*, m.p. 152° (*idem, ibid.*, 1931, 1241) are all obtained by the Gattermann reaction.

2,5-*Dihydroxy-3-methoxybenzaldehyde*, m.p. 143°, is obtained by the oxidation of *o*-vanillin with persulphate (*W. Baker, N. C. Brown* and *J. A. Scott, ibid.*, 1939, 1922).

2,3,6-**Trihydroxy-4-methylbenzaldehyde,** *thamnol*, m.p. 185°, is formed by heating thamnolic acid with formic acid at 150° (*Asahina* and *F. Fuziwaka*, Ber., 1932, **65,** 58).

(*iii*) *Derivatives of phloroglucinol*

2,4,6-**Trihydroxybenzaldehyde,** *phloroglucinaldehyde, dihydrate* loses water at 105° and chars without melting, is prepared from phloroglucinol by the action of hydrogen chloride and hydrogen cyanide (*D. G. Pratt* and *Robinson*, J. chem. Soc., 1924, **125,** 194) or zinc cyanide (*T. Malkin* and *M. Neirenstein*, J. Amer. chem. Soc., 1931, **53,** 241). It is also obtained by formylation with ethyl orthoformate (*H. Gross, A. Rieche* and *G. Matthey*, Ber., 1963, **96,** 308) or $MeSCHCl_2$ (*Gross* and *Matthey*, Chem. Reviews, 1964, **97,** 2606). Methylation with methyl sulphate and alkali (*P. Karrer et al.*, Helv., 1927, **10,** 378) or diazomethane (*J. Herzig et al.*, Monatsh, 1903, **24,** 862) gives 2-*hydroxy-4,6-dimethoxybenzaldehyde*, m.p. 70° and finally 2,4,6-*trimethoxybenzaldehyde*, m.p. 118°. 2-*Methoxy-4,6-dihydroxybenzaldehyde*, m.p. 203° (decomp.,), is obtained by the Vilsmeier reaction from *O*-methylphloroglucinol (*Pratt* and *Robinson, loc. cit.*, 1924, **125,** 194). 3,5-*Dimethoxyphenol* in the Gattermann reaction gives principally 6-*hydroxy-2,4-dimethoxybenzaldehyde*, m.p. 70° and a little 4-*hydroxy-2,6-dimethoxybenzaldehyde*, m.p. 222° (decomp.) (*W. Gruber*, Ber., 1943, **76,** 135).

(*iv*) *Derivatives of tetrahydroxybenzene*

2,3,4,5-**Tetrahydroxybenzaldehyde,** *apionolaldehyde*, darkens at 220°, and 6-*hydroxy-2,3,4-trimethoxybenzaldehyde, antiarolaldehyde*, m.p. 65° (*E. Chapman, A. G. Perkin* and *Robinson*, J. chem. Soc., 1927, 3030) are obtained by the Gattermann reaction. *Apiolaldehyde*, 2,5-*dimethoxy-3,4-methylenedioxybenzaldehyde*, m.p. 102°, is obtained from parsley-apiol (XXI) a constituent of parsley seed oil. 5,6-*Dimethoxy-3,4-methylenedioxybenzaldehyde*, m.p. 75°, is obtained from dill-apiol (XXII) a constituent of oil of dill by isomerisation of the olefin and oxidation:

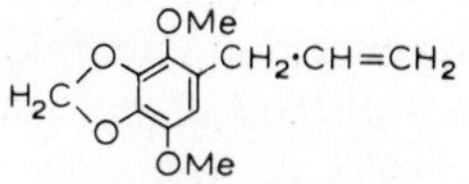

(XXI) 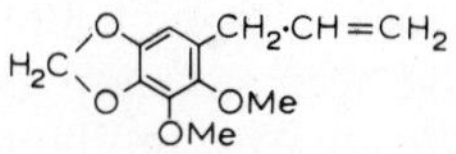 (XXII)

4. Hydroxyphenyl ketones

(*a*) *Methods of preparation*

Many of the general methods for the preparation of aromatic ketones (pp. 107 *et seq.*) are applicable here.

(i) Fries reaction

An ester of a phenol is converted to a mixture of the *o*- and *p*-hydroxyketones by treatment with aluminium chloride, either alone or in a solvent such as carbon disulphide, nitrobenzene or tetrachloroethane. One mole of aluminium chloride is required. The ratio of *ortho*- to *para*-migration depends markedly on the temperature and also on the solvent and catalyst used:

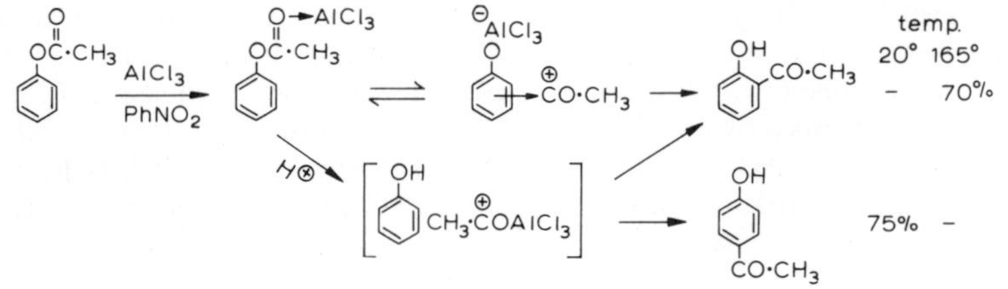

Esters of mono-, di- and tri-hydric phenols are successfully converted to monoketones. For a review see *A. H. Blatt* (Org. Reactions, 1942, **1,** 342) and *A. Gerecs* ("Friedel-Crafts and Related Reactions", Ed. *G. A. Olah*, Interscience, London, 1964, Vol. III, p. 299). The reaction can proceed by two mechanisms. The first of these processes leads exclusively to *ortho*-migration and is probably intramolecular proceeding *via* a π-complex. The second mechanism is also that of the corresponding acylation reaction, leads to *ortho*- and *para*-migration and probably proceeds *via* an ion-pair type intermediate (*Y. Ogata* and *H. Tabuchi*, Tetrahedron, 1964, **20,** 1661; *M. J. S. Dewar* and *L. S. Hart*, Tetrahedron, 1970, **26,** 973).

The rearrangement of phenyl esters can also be effected photochemically by u.v. irradiation in the absorption region of the ester (Photo-Fries reaction). Ethanol and cyclohexane are commonly used solvents. Reaction leads to *ortho*- and *para*-migration. Both intramolecular and free radical-cage mechanisms have been proposed (*D. Bellus* and *P. Hrdlovic*, Chem. Reviews, 1967, **67,** 599; *Bellus*, Adv. in Photochem., 1971, **8,** 109).

(ii) Friedel–Crafts reaction

Reaction of acyl chlorides with phenols or their ethers in the presence of aluminium, stannic or zinc chloride and an inert solvent.

(iii) Hoesch reaction

Nitriles react with phenols in ether solution in the presence of hydrogen chloride to give ketimines from which the ketone is obtained by hydrolysis. In some cases anhydrous zinc chloride is necessary as a catalyst:

$$H_3C{\cdot}CN + H^{\oplus} \longrightarrow H_3C{\cdot}\overset{\oplus}{C}{=}NH \longrightarrow$$

OH, OH, $H_3C-C{=}NH$ $\longrightarrow$ OH, OH, $H_3C-C{=}O$

The reaction proceeds best with resorcinol and phloroglucinol and their ethers. Catechol and quinol do not react. The reaction has been reviewed by *P. E. Spoerri* and *A. S. Dubois* (Org. Reactions, 1942, **1,** 388).

(iv) Diazo reaction

From aminoaryl ketones by the diazo reaction, this reaction is useful for the preparation of 3-hydroxyacetophenone.

(v) Replacement of halogen

Reaction with hydroxide ions in the presence of copper oxide.

(b) Properties and reactions

The phenol ketones have the usual reactions of carbonyl compounds and of phenols.

Acidity and hydrogen bonding. The pK_a values for hydroxyacetophenones are 2- 10.26, 3- 9.19, 4- 7.87 (*L. B. Magnusson, C. Postmus* and *C. A. Craig*, J. Amer. chem. Soc., 1963, **85,** 1711; *F. G. Bordwell* and *G. D. Cooper*, *ibid.*, 1952, **74,** 1058). Intramolecular hydrogen bonding between the hydroxyl and the carbonyl group of 2-hydroxyacetophenone is responsible for the low acidity of the *ortho* isomer. This intramolecular hydrogen bond has been demonstrated by i.r. spectroscopy (*C. J. W. Brooks* and *J. F. Morgan*, J. chem. Soc., 1961, 3373) and it is stronger than the intramolecular hydrogen bond in salicylaldehyde. 2-Hydroxyacetophenone and its homologues are steam-volatile and this property has been used to separate a mixture of *ortho* and *para* isomers from a Fries rearrangement (*C. E. Coulthard et al.*, *ibid.*, 1930, 286). Under mild conditions of methylation, *para* and *meta* hydroxyl groups react at a much faster rate than *ortho* groups, because they are more acidic, so selective reactions can be achieved (*H. Appel et al.*, *ibid.*, 1937, 738; *G. N. Vyas* and *N. M. Shah*, Org. Synth., 1963, Coll. Vol. IV, p. 836).

2-Methoxyacetophenones are readily demethylated by aluminium chloride at or near room temperature (*S. von Kostanecki* and *J. Tambor*, Ber., 1899, **32,** 2262), conditions which do not affect *meta* and *para* methoxyl groups.

(c) Monohydroxyaryl ketones

2-**Hydroxyacetophenone,** $HOC_6H_4 \cdot CO \cdot CH_3$, m.p. 28°, b.p. 263°, is best obtained by heating phenyl acetate with aluminium chloride at 140° (*K. W. Rosenmund* and *W. Schnurr*, Ann., 1928, **460,** 88). It is found in the oil from the bark of *Chione glabra* D.C. (*W. R. Dunstan* and *T. A. Henry*, J. chem. Soc., 1899, **75,** 66).

Condensation with benzaldehyde at pH 9.8 yields flavone (I) and in more alkaline medium pH >10.9 the intermediate 2-hydroxychalkone is stable (*L. Reichel et al.*, Ber., 1941, **74,** 1741; 1802):

OH, CO·CH₃ + PhCHO ⟶ OH, CO·CH=CHPh ⇌ (pH 9.8 / pH >10.9) O, Ph, O (I)

2-*Methoxyacetophenone*, b.p. 122°/16 mm.

3-**Hydroxyacetophenone,** m.p. 95°, is obtained from 3-aminoacetophenone (*J. C. E. Simpson et al.*, J. chem. Soc., 1945, 653). 3-*Methoxyacetophenone*, b.p. 138°/24 mm, by methylation of the phenol or from 3-methoxybenzoyl chloride and ethyl acetoacetate (*R. H. Martin* and *R. Robinson*, *ibid.*, 1943, 497).

4-**Hydroxyacetophenone,** *piceol*, m.p. 110°, occurs as the β-D-glucopyranoside, *picrin*, *salinigrin*, *ameliaroside*, m.p. 193°, $[\alpha]_D$ −88° in the needles of *Picea excelsa* and the bark of some species of *Salix* (*M. Bridel* and *J. Rabate*, Compt. rend., 1929, **189,** 1304). It is prepared by reaction of acetyl chloride and phenol in presence of aluminium chloride (*K. V. Auwers* and *W. Mauss*, Ann., 1928, **460,** 274). 4-*Methoxyacetophenone*, m.p. 39°, b.p. 150°/15 mm.

2-**Hydroxyphenylpropiophenone,** b.p. 110°/6 mm, and 4-**hydroxyphenylpropiophenone,** m.p. 148°, are formed by rearrangement of phenyl propionate by aluminium chloride in carbon disulphide (*E. Miller* and *W. H. Hartung*, Org. Synth., 1943, Coll. Vol. II, p. 543).

4-**Hydroxyphenylacetone,** $HOC_6H_4 \cdot CH_2 \cdot CO \cdot CH_3$, m.p. 35°. The *methyl ether*, *anisylacetone*, b.p. 139°/13 mm, is a constituent of star anise oil.

2-*Hydroxybenzylacetone*, 1-(1′-*hydroxyphenyl*)*butan*-3-*one*, m.p. 48°, is obtained by reduction of salicylideneacetone; its *methyl ether*, b.p. 147°/10 mm, is similarly prepared (*M. Faillebin*, Ann. chim., 1925, [x], **4,** 413) as are 3-*hydroxybenzylacetone*, m.p. 85° and 4-*hydroxybenzylacetone*, m.p. 83°.

2-(4′-**Methoxyphenyl)butan**-3-**one,** b.p. 260°, from 2-(4′-methoxyphenyl)butan-3-one, b.p. 260°, from 2-(4′-methoxyphenyl)-but-2-ene *via* the corresponding epoxide or glycol (*M. Tiffeneau*, Bull. Soc. chim. Fr., 1931, [iv], **49,** 1714).

(d) Di- and tri-hydroxyphenyl ketones

3,4-**Dihydroxyacetophenone,** m.p. 116°, is prepared by the rearrangement of catechol diacetate (*Rosenmund* and *H. Lohfert*, Ber., 1928, **61,** 2603) or by acylation of catechol (*Coulthard*, *J. Marshall* and *F. L. Pyman*, J. chem. Soc., 1930, 280). 4-*Hydroxy*-3-*methoxyacetophenone*, **acetovanillone,** m.p. 115°, occurs in the roots of *Iris pallida* Lamanck and as the glycoside *androsin* in the rhizomes of *Apocynum cannabinum* L. (*H. Finnemore*, *ibid.*, 1908, **93,** 1513). Acetovanillone is obtained from guaiacol and acetic acid in presence of zinc chloride *(Coulthard, Marshall* and *Pyman, loc. cit.);* phosphoryl chloride as catalyst affords largely 3-*hydroxy*-4-*methoxyacetophenone*, m.p. 92°.

2,3-**Dihydroxyacetophenone,** m.p. 97°, is obtained by demethylation of 2,3-*dimethoxyacetophenone*, b.p. 143°/14 mm, prepared from the corresponding benzonitrile and methylmagnesium iodide (*W. Baker* and *A. R. Smith*, J. chem. Soc., 1936, 346) or from the benzoyl chloride and dimethylcadmium (*E. H. Woodruff*, J. Amer. chem. Soc., 1942, **64,** 2859). 2-*Hydroxy*-3-*methoxyacetophenone*, m.p. 54°, by partial demethylation of the dimethyl ether with aluminium chloride in ether *(Baker* and *Smith*, *loc. cit.)*.

2,4-**Dihydroxyacetophenone, resacetophenone,** m.p. 143°, is prepared from resorcinol and acetic acid containing zinc chloride (*R. S. Cooper*, Org. Synth., 1955, Coll. Vol. III, p. 761) or by reaction with acetonitrile, hydrogen chloride and zinc chloride (*K. Hoesch*, Ber., 1915, **48,** 1125; *W. K. Slater* and *H. Stephen*, J. chem. Soc., 1920, **117,** 311; *J. Houben*, Ber., 1926, **59,** 2887). 2-*Hydroxy*-4-*methoxyacetophenone, paeonal*, m.p. 50°, is found in the rhizomes of *Primula auricula* L. It can be obtained from 3-methoxyphenol, acetic acid and zinc chloride *(Coulthard, Marshall* and *Pyman*, *loc. cit.)* and by partial demethylation of 2,4-dimethoxyacetophenone (*J. Shinoda*, *D. Sato* and *M. Kawagoye*, J. pharm. Soc., Japan, 1932, **52,** 766). Methylation of resacetophenone gives paeonal and 2-*hydroxy*-4-*methoxy*-3-*methylacetophenone*, m.p. 83° (*Appel et al., loc. cit.; Robinson* and *R. C. Shah*, J. chem. Soc., 1934, 1491). 4-*Hydroxy*-2-*methoxyacetophenone*, m.p. 138°, is formed together with paeonal from 3-methoxyacetophenone by the Hoesch reaction (*Baker*, *ibid.*, 1934, 1684). 2,4-*Dimethoxyacetophenone*, m.p. 40°, is obtained from 1,3-dimethoxybenzene, acetic anhydride and aluminium chloride (*C. R. Noller* and *R. Adams*, J. Amer. chem. Soc., 1924, **46,** 1892).

2,5-**Dihydroxyacetophenone,** *quinacetophenone*, m.p. 203°, is formed by heating quinol with acetic and zinc chloride (*A. Russell* and *S. F. Clark*, *ibid.*, 1939, **61,** 2651) or from the Fries rearrangement of quinol diacetate (*S. G. Morris*, *ibid.*, 1949, **71,** 2056). The *dimethyl ether*, m.p. 21°, b.p. 141°/10 mm (*L. D. Abbott* and *J. D. Smith*, Biochem. Prepns., 1955, **4,** 6; *G. N. Vyas* and *N. M. Shah*, Org. Synth., 1963, Coll. Vol. IV, p. 836) gives 2-*hydroxy*-5-*methoxyacetophenone*, m.p. 50°, on partial demethylation with aluminium chloride (*W. S. Ide* and *R. Baltzly*, J. Amer. chem. Soc., 1948, **70,** 1084).

3,5-**Dihydroxyacetophenone,** m.p. 147°. The *dimethyl ether*, m.p. 42°, is prepared from 3,5-dimethoxybenzoic acid by reaction of the amide with methylmagnesium iodide (*R. Adams et al.*, *ibid.*, 1949, **71,** 1624) or the acid chloride with dimethyl cadmium (*E. H. Woodruff*, *ibid.*, 1942, **64,** 2859).

2,6-**Dihydroxyacetophenone,** m.p. 157°, is best obtained by the Fries rearrangement of 7-acetoxy-4-methylcoumarin followed by alkali cleavage of the resulting coumarin (*A. Russell* and *J. R. Frye*, Org. Synth., 1955, Coll. Vol. III, p. 281):

$$H_3C\cdot\overset{O}{\overset{\|}{C}}O\text{-(4-Me-coumarin-7-yl)} \xrightarrow{AlCl_3} \text{7-HO-8-}CO\cdot CH_3\text{-4-Me-coumarin} \xrightarrow{NaOH} \text{2,6-(HO)}_2C_6H_3\cdot CO\cdot CH_3 + CO_2 + Me_2CO$$

Methylation gives 6-*hydroxy*-2-*methoxyacetophenone*, m.p. 60° (*Baker*, J. chem. Soc., 1939, 956) and 2,6-*dimethoxyacetophenone*, m.p. 72°.

2,4-**Dihydroxypropiophenone,** m.p. 101°, is obtained by heating resorcinol with propionic acid and zinc chloride (*C. M. Brewster* and *J. C. Harris*, J. Amer. chem. Soc., 1930, **52,** 4866).

1-(4-**Hydroxy**-3-**methoxyphenyl)butan**-3-**one, zingerone,** m.p. 41°, b.p. 186°/13 mm, is the hot-tasting principle of ginger (*Zingiber officinale* Rosc.) (*A. Lapworth et al.*, J. chem. Soc., 1917, **111,** 784). It is prepared by catalytic reduction of vanillylideneacetone over palladised carbon (*K. Kim*, J. pharm. Soc., Japan.

2,3,4-**Trihydroxyacetophenone,** *gallacetophenone*, m.p. 172°, is prepared by heating pyrogallol with acetic anhydride, zinc chloride and acetic acid (*I. C. Badhwar* and *K. Venkataraman*, Org. Synth., 1934, Coll. Vol. II, p. 304). 2,3,4-*Trimethoxyacetophenone*, b.p. 185°/20 mm, is obtained from pyrogallol trimethyl ether and acetyl chloride in presence of aluminium chloride at −10° (*N. P. Buu-Hoï* and *P. Cagniant*, Rec. Trav. chim., 1945, **64,** 214). Reaction at room temperature gives 2-*hydroxy*-3,4-*dimethoxyacetophenone*, m.p. 78°, also formed by partial methylation of gallacetophenone (*Baker*, J. chem. Soc., 1941, 662).

2,4,5-**Trihydroxyacetophenone,** m.p. 200°, from 1,2,4-trihydroxybenzene with acetic anhydride and zinc chloride in acetic acid (*T. C. Chadha* and *Venkataraman*, J. chem. Soc., 1933, 1073). *Trimethyl ether*, m.p. 103° (*Baker*, *ibid*., 1934, 71).

2,4,6-**Trihydroxyacetophenone,** *phloracetophenone*, m.p. 218° (anhydr.), crystallises from water as the monohydrate. It is prepared by the Hoesch reaction from phloroglucinol (*Venkataraman et al.*, Org. Synth., 1943, Coll. Vol. II, p. 522).

Phloracetophenone is converted to phloroglucinol on warming with 2 *N* sodium hydroxide (*H. Brockmann* and *K. Maier*, Ann., 1938, **535,** 156). Methylation with dimethyl sulphate or iodomethane and alkali gives 2-*hydroxy*-4,6-*dimethoxyacetophenone*, m.p. 84° found in the essential oils of some *Geijera* species (*A. R. Penfold*, J. Proc. roy. Soc. N.S.W., 1930, **64,** 264), and 2-*hydroxy*-4,6-*dimethoxy*-3-*methylacetophenone*, m.p. 141° (*F. H. Curd* and *A. Robertson*, J. chem. Soc., 1933, 437; *V. D. N. Sastri* and *T. R. Seshadri*, Proc. Indian Acad., [A], 1946, **23,** 262).

2,4,6-*Trimethoxyacetophenone*, m.p. 103°, from 1,3,5-trimethoxybenzene by the Hoesch reaction (*J. Houben* and *W. Fischer*, J. pr. Chem., 1929, [ii], **123,** 97) or by condensation with acetyl chloride and ferric chloride (*S. von Kostanecki* and *J. Tambor*, Ber., 1899, **32,** 2262). Partial demethylation affords the 4,6-dimethyl ether.

2,4,6-**Trihydroxypropionphenone,** m.p. 207°, and its homologues are prepared by the Hoesch reaction (*J. Shinoda*, J. Pharm., Japan, 1927, 35; Chem. Ztbl., 1927, II, 97; *P. Karrer*, Helv., 1919, **2,** 473; *E. Klarmann* and *W. Figdor*, J. Amer. chem. Soc., 1926, **48,** 804) or from phloroglucinol and the appropriate acid chloride in the presence of a condensing agent (*K. W. Rosenmund* and *H. Lohfert*, Ber., 1928, **61,** 2606).

Aspidinol, 2,6-*dihydroxy*-4-*methoxy*-3-*methylbutyrophenone*, m.p. 116.5°, occurs in extracts of male fern (*R. Bochm*, Ann., 1901, **318,** 247). It has been synthesised from 2-methylphloroglucinol 1-methyl ether (*W. Riedl* and *R. Mittledorf*, *ibid*., 1956, **89,** 2589).

3,4,5-**Trihydroxyacetophenone,** m.p. 188°. 3,4,5-*Trimethoxyacetophenone*, m.p. 79°, is obtained from 3,4,5-trimethoxybenzyl chloride and the ethoxymagnesium salt of diethyl malonate followed by hydrolysis (*E. Haggett* and *S. Archer*, J. Amer. chem. Soc., 1949, **71,** 2255). 4-Hydroxy-3,5-dimethoxyacetophenone is obtained by the Fries rearrangement of 2,6-dimethoxyphenyl acetate (*J. M. Pepper* and *H. Hibbert*, *ibid.*, 1948, **70,** 67).

5. Phenolic monocarboxylic acids: the hydroxybenzoic acids

Many hydroxybenzoic acids occur in plants, particularly as depsides in lichens, and gallic acid is an important constituent of the hydrolysable tannins. Their methyl ethers are oxidation products of many natural compounds and have been important in structure determination. Salicylic acid and aspirin are important pharmaceutical chemicals.

(a) Methods of preparation

(i) Kolbe-Schmitt reaction

This reaction has been reviewed by *A. S. Lindsey* and *H. Jeskey* (Chem. Reviews, 1957, **57,** 583).

Historical. H. Kolbe and *E. Lautermann* (Ann., 1860, **113,** 125) succeeded in preparing salicylic acid by heating sodium phenoxide in a stream of carbon dioxide. Phenol distilled from the reaction mixture and the remaining disodium salt gave salicylic acid (I) on acidification:

ONa + CO_2 → (ONa, CO_2Na) → (OH, CO_2H)

(I)

p-Cresol and thymol were also converted to the corresponding acids. *R. Schmitt* (G.P. 29,939/1884) carried out the reaction in an autoclave under a pressure of carbon dioxide and achieved a high conversion to sodium salicylate. *S. Marasse* (G.P. 78,708/1893) introduced the procedure of heating the phenol with potassium carbonate and carbon dioxide under pressure. This avoids the necessity for making the phenoxide but the procedure is expensive on an industrial scale and the cheaper sodium carbonate is ineffective. More recently *W. H. Meek* has introduced dimethylformamide as a solvent for the sodium phenoxide during carbonation under pressure (F.P. 1,553,473/1969; Ger. Offen. 1,926,063/1969).

Carbonation of monohydric phenols. The carbonation of sodium or potassium phenoxides requires a temperature above 100° and the reaction mixture must be dry. A pressure of carbon dioxide above 4 atm. is needed to ensure that the equilibrium between the phenoxide and carbon dioxide to give the intermediate phenyl carbonate lies in a favourable direction. Carbonation of sodium phenoxide gives salicylic acid at all temperatures. Potassium phenoxide and the Marasse modification give a substantial amount of 4-hydroxybenzoic acid (II) at 100–150° and at higher temperatures 4-hydroxyisophthalic acid (III) is formed. In general the reaction is useful for the *ortho*-carbonation of phenols. The *para*-carbonation process has been exploited only in the production of 4-hydroxybenzoic acid. Electron-donating substituents facilitate the reaction.

Potassium phenoxide, potassium carbonate and carbon monoxide heated under pressure also give 4-hydroxybenzoic acid, after acidification. ^{14}C-labelling experiments indicate that the carboxyl group is derived from the potassium carbonate (*Y. Yasuhara et al.*, J. org. Chem., 1968, **33,** 4512; J. chem. Soc., Japan, 1969, **42,** 2070).

(I) (II) (III)

TABLE 2

YIELDS IN THE KOLBE REACTION OF PHENOL

$M^{\oplus}$	$t°C$	% Yield in acid mixture		
		I	*II*	*III*
$Na^{\oplus}$	100–250	96	2	2
$K^{\oplus}$	100	46	54	0
$K^{\oplus}$	200	79	16	5
$K^{\oplus}$	250	67	1	32

Reference
O. Bain et al., J. org. Chem., 1954, **19,** 510.

Carbonation of polyhydric phenols. Many di- and tri-hydric phenols can be carbonated in aqueous solution in the presence of an alkali metal carbonate and at atmospheric pressure. Alternatively the mixture of phenol and potassium hydrogen carbonate is heated in a stream of carbon dioxide.

Resorcinol and plhoroglucinol derivatives are carbonated by methylmagnesium carbonate (*R. Mechoulam* and *Z. Ben-Zvi*, Chem. Comm., 1969, 343).

Mechanism. Phenol will undergo carbonation only in the form of the phenoxide. Sodium phenoxide and carbon dioxide form sodium phenylcarbonate at temperatures below 100° and when heated in a sealed tube at 120–130° this substance is converted quantitatively to sodium salicylate (*Schmitt*, J. pr. Chem., 1885, [ii], **31,** 397). The dissociation pressure of sodium phenylcarbonate is around 4 atm. at 120° and so this salt could be an intermediate in the Kolbe reaction (*I. A. Davies*, Z. physik. Chem., 1928, **134,** 57). Another reaction scheme suggests that this is a side product and that electrophilic substitution by carbon dioxide on the sodium phenoxide is the reaction course. *ortho*-Substitution is favoured because th sodium ion in sodium phenoxide can form a coordination complex containing both phenoxide and carbon dioxide ligands. Potassium ions form a weaker complex and so allow more *para*-product to be formed (*J. Idris Jones et al.*, J. chem. Soc., 1954, 3145; 1958, 3152).

(ii) Friedel–Crafts reaction

A variety of reagents undergo the Friedel–Crafts reaction with phenyl ethers to yield carboxylic acids or their derivatives. Catechol dichloromethylene ether reacts in the presence of aluminium or stannic chlorides to yield acids (*H. Gross, J. Rusche* and *M. Mirsch*, Ber., 1963, **96,** 1382). Amides, anilides and thioanilides are formed by reaction of phenol ethers with carbamoyl chloride, phenyl isocyanate or phenyl isothiocyanate (*L. Gattermann*, Ann., 1888, **244,** 41; *K. von Auwers* and *C. Beger*, Ber., 1894, **27,** 1733).

(iii) Reimer–Tiemann reaction

A strongly alkaline solution of phenol reacts with carbon tetrachloride in the presence of a copper catalyst to give a mixture of the corresponding 2- and 4-hydroxybenzoic acids (*K. Reimer* and *F. Tiemann, ibid.*, 1876, **9,** 1285). Phenol gives 25% salicylic acid and 35% 4-hydroxybenzoic acid (*J. Zeltner* and *M. Landau*, G.P. 258,887/1912; Friedl., 1915, **11,** 208).

(iv) Direct introduction of hydroxyl

The oxidation of the ammonium salts of benzoic acid and its homologues with hydrogen peroxide gives hydroxybenzoic acid (*H. D. Dakin* and *M. D. Herter*, J. biol. Chem., 1907, **3,** 419). Of more importance is the oxidation of a hydroxybenzoic acid with persulphate ion to introduce a further hydroxyl substituent in a position *para* to the one already present (*S. M. Sethna*, Chem. Reviews, 1951, **49,** 91).

Basic copper benzoate, basic copper toluates and other homologues on heating sometimes with an inert diluent give the corresponding 2-hydroxybenzoic acid (*W. W. Kaeding* and *A. T. Shulgin*, F.P. 1,321,264/1961).

(v) From organometallic compounds

The alkyl ethers of hydroxybenzoic acids are obtained by the action of carbon dioxide on arylmagnesium halides or the lithium aryls. Selective metalation of the benzene ring by butyl-lithium in a position *ortho* to a methoxyl group is useful for the preparation of some 2-methoxybenzoic acids (*H. Gilman* and *J. W. Morton*, Org. Reactions, 1954, **8,** 258).

(vi) By side-chain oxidation

Cresols are oxidised to the hydroxybenzoic acid by fusion with alkali alone or in the presence of lead dioxide, other metal oxides, or potassium chlorate (*C. Graebe* and *H. Kraft*, Ber., 1906, **39,** 796; *C. Rudolph*, Z. angew. Chem., 1906, **19,** 384; *G. Lock* and *F. Stitz*, Ber., 1939, **72,** 77). Phenolic aldehydes can be oxidised by fusion with alkali or by treatment with aqueous alkali in the presence of a silver catalyst (*I. A. Pearl*, J. Amer. chem. Soc., 1945, **67,** 1628; 1946, **68,** 429, 110; J. org. Chem., 1947, **12,** 79). Alkylsubstituted phenyl ethers are oxidised by potassium permanganate to the corresponding alkoxybenzoic acid. Hydroxyacetophenones are oxidised to hydroxybenzoic acids by treatment with iodine in pyridine and subsequent treatment with alkali (*L. C. King et al.*, J. Amer. chem. Soc., 1945, **67,** 2089):

$$HOC_6H_4{\cdot}CO{\cdot}CH_3 \xrightarrow[C_5H_5N]{I_2} [HOC_6H_4{\cdot}CO{\cdot}CH_2NC_5H_5]^{\oplus}\ I^{\ominus} \xrightarrow{NaOH} HOC_6H_4{\cdot}CO_2^{\ominus} + CH_3\overset{\oplus}{N}C_5H_5$$

(b) Properties and reactions

(i) Acidity

The dissociation constants of some hydroxybenzoic acids are given in Table 3. A hydroxy group *ortho* to carboxyl strenghtens the acidity of the latter relative to benzoic acid, pK_a 4.20. This results from hydrogen bonding which is stronger between OH and $CO_2^{\ominus}$ than between OH and CO_2H. As a consequence of this hydrogen bonding, the pK_2 for salicylic acid is weak. The u.v. spectra of 2-, 3- and 4-hydroxybenzoic acids at various pH's have been measured (*L. Doub* and *J. M. Vandenbelt*, *ibid.*, 1947, **69,** 2714; 1949, **71,** 2414). Salicylic acid is steam-volatile and comparatively more soluble in chloroform than the 3- and 4-hydroxybenzoic acids.

(ii) Substitution

Electrophilic substitution proceeds *ortho* and *para* to the hydroxyl groups. An *ortho*- or *para*-carboxyl group can be displaced by the incoming substituent. Thus vigorous bromination of either 2- or 4-hydroxybenzoic acid gives 2,4,6-tribromophenol (*W. Robertson*, J. chem. Soc., 1902, **81,** 1480; *E. Schunck* and *L. Marchlewski*, Ann., 1893, **278,** 348) and vigorous nitration

TABLE 3

DISSOCIATION CONSTANTS OF HYDROXYBENZOIC ACIDS

Benzoic acid	pK_1	pK_2	*temp.*	*ref.*
2-hydroxy	3.03	13.4	18°	1,2
3-hydroxy	4.17	10.0	18°	1,2
4-hydroxy	4.59	9.3	18°	2,3
2,4-dihydroxy	3.22		30°	5
2,5-dihydroxy	3.22	> 9.5	—	4
2,6-dihydroxy	1.22	—	30°	6
3,4,5-trihydroxy	4.33	8.85	30°	7

References

1 *E. Larson* and *B. Adell*, Z. physik. Chem., 1931, **A157,** 342.
2 *Larson*, Z. anorg. Chem., 1929, **183,** 30.
3 *G. Briegleb* and *A. Bieber*, Z. Elektrochem., 1951, **55,** 250.
4 *E. M. Kapp* and *A. F. Coburn*, J. biol. Chem., 1942, **145,** 549.
5 *C. T. Abichandani* and *S. K. K. Jatkar*, J. Indian chem. Soc., 1939, **16,** 385.
6 *Idem*, J. Indian Inst. Sci., 1941, **A23,** 77.
7 *Idem, ibid.*, 1938, **A21,** 417.

gives picric acid. 4-Hydroxybenzoic acid couples with arenediazonium salts by replacing the carboxyl group rather than by entering the vacant *ortho* positions (*E. Grandmougin* and *H. Freimann*, Ber., 1907, **40,** 3453).

(iii) Macrocyclic ester formation

Salicylic acid and some of its substituted derivatives form macrocyclic esters on dehydration. These are discussed as derivatives of salicylic acid.

(c) 2-Hydroxybenzoic acids: Salicylic acid and its derivatives

(i) Salicylic acid

Salicylic acid, 2-*hydroxybenzoic acid*, m.p. 158°, is the most important member of this group being used in medicine and in the dyestuffs industry. It occurs in the free state in the flowers of *Spiraea ulmaria* and as the methyl ester in the essential oil of *Gaultheria procumbens* L. (wintergreen) and of *Betula lenta*. Methyl salicylate is present in these plants as the glycoside *gaultherin, monotropitin* and this is hydrolysed enzymatically when the plant is crushed (*E. Bourquelot*, Compt. rend., 1896, **122,** 1002; *M. L. Duparc*, *ibid.*, 1901, **132,** 1237). The occurrence in *Gaultheria* species has been examined (*G. H. N. Towers*, *A. Tre* and *W. S. G. Maass*, Phytochem., 1966, **5,** 677).

History. R. Piria (Ann., 1838, **30,** 165) obtained salicylaldehyde by the oxidation of salicin with chromic acid and steam distillation of the reaction mixture. Fusion of this salicylaldehyde with potassium hydroxide gave salicylic acid. Salicylic acid was also obtained from coumarin by fusion with potassium hydroxide (*Z. Delalane*, *ibid.*, 1843, **45,** 332) and *A. Cahours* (*ibid.*, 1843, **48,** 60) showed that oil of wintergreen, then recently introduced to Europe from Canada, consisted chiefly of methyl salicylate. *B. W. Gerland* (*ibid.*, 1853, **86,** 147) obtained salicylic acid from anthranilic acid by treatment with nitrous fumes. The synthesis by the action of carbon dioxide on heated dry sodium phenoxide was achieved by *Kolbe* and *Lautermann* (*ibid.*, 1860, **115,** 201).

Manufacture. Most salicylic acid is made by the Kolbe–Schmitt process. Dry sodium phenoxide is heated to 120–140° under a pressure of 5–6 atm. of carbon dioxide until the calculated quantity has been adsorbed. A fluidised–solids reaction system has been described for this operation. Sodium salicylate is formed and is acidified to give salicylic acid which is finally purified by sublimation. Small quantities of 4-hydroxybenzoic acid and 4-hydroxyisophthalic acid are formed as byproducts.

Properties and reactions

Aqueous solutions of salicylic acid give a violet coloration with ferric chloride. Salicylic acid is an antiseptic. The free acid and many of its salts are used in medicine as analgesics and antipyretics and in the treatment of acute rheumatic fever.

(*1*) *Self-condensation.* When heated alone, salicylic acid begins to decarboxylate at 200°. Rapid heating to 250° gives xanthone (IV) along with the decarboxylation product, phenol (*A. Klepl*, J. pr. Chem., 1883, [ii], **28,** 217). Salicyloylsalicylic acid and the salicylides, macrocyclic lactones, are obtained by the action of dehydrating agents:

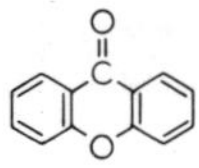

(IV) 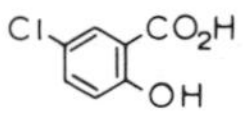 (V) 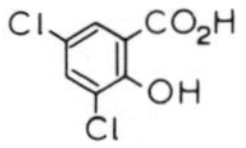 (VI)

(*2*) *Reduction.* Pimelic acid is the chief product from reduction of salicylic acid with sodium and pentyl alcohol (*A. Muller*, Org. Synth., 1943, Coll. Vol. 1, p. 535), the intermediate tetrahydrosalicylic acid undergoing the "acid hydrolysis" characteristics of β-oxo acid derivatives:

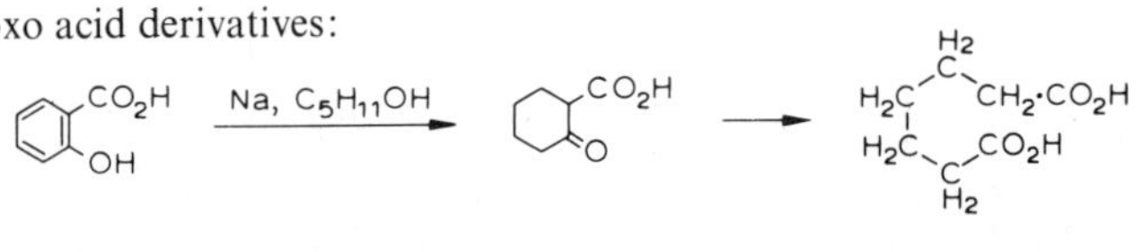

$\xrightarrow{Na,\ C_5H_{11}OH}$ $\longrightarrow$

Hydrogenation of the acid using a platinum catalyst gives *cis-* and *trans-*2-hydroxycyclohexanecarboxylic acids (*J. Pascual et al.*, J. chem. Soc., 1949, 1943; *H. E. Ungnade* and *F. V. Morriss*, J. Amer. chem. Soc., 1948, **70,** 1898). Hydrogenation using a molybdenum sulphide catalyst yields *o*-cresol (*K. Itabashi*, C.A., 1959, **53,** 14918).

(*3*) *Substitution.* Chlorination yields the 5-chloro- (V) and 3,5-dichloro-salicylic acids (VI). Bromination leads to similar products and 2,4,6-tribromophenol is formed by reaction with excess bromine in acetic acid (*Schunck* and *Marchlewski*, *loc. cit.*). Nitration affords a mixture of 3-nitro- and 5-nitro-salicylic acids and further reaction gives 3,5-dinitrosalicylic acid and picric acid is formed by the action of fuming nitric acid (*R. F. Marchand*, J. pr. Chem., 1842, [i], **26,** 397).

(ii) Ethers and esters of salicylic acid

Methyl salicylate, m.p. −8°, b.p. 224°, d_4^{20} 1.184, is the chief constituent of oil of wintergreen. The commercial product is made by esterification of salicylic acid with methanol and sulphuric acid. It has been used for the treatment of gout, neuralgia, pleurisy and rheumatism. In plants it occurs as the glycoside **monotropitin, gaultherin** (*M. Bridel*, Compt. rend., 1923, **177,** 642; 1924, **179,** 991; J. pharm. Chim., 1924, [vii], **30,** 400), m.p. 180°, $[\alpha]_D$ −58° (acetone). The glycoside is hydrolysed by a specific enzyme to methyl salicylate and primeverose, by dilute mineral acids to methyl salicylate, glucose and xylose. It has been synthesised (*A. Robertson* and *R. B. Waters*, J. chem. Soc., 1931, 1881). *Monotropitin hexa-acetate*, m.p. 189°, $[\alpha]_D$ −85.7° (chloroform).

Violutoside, *methyl salicylate vicianoside*, m.p. 169°, $[\alpha]_D$ −36° (water), has been isolated from *Viola cornuta* L. (*P. Picard*, Compt. rend., 1926, **182,** 1167).

Ethyl salicylate, m.p. 1.3°, b.p. 234°, has been used as an antiseptic.

Methoxymethyl salicylate, mesotan, $HOC_6H_4 \cdot CO_2CH_2OMe$, b.p. 153°/32 mm (*Bayer*, G.P. 127,585/1902; Chem. Zent., 1903, I, 112) has been used in the treatment of rheumatism. It is prepared from sodium salicylate and chloromethoxymethane.

Phenyl salicylate, salol, m.p. 43°, b.p. 172°/12 mm, is prepared by treating salicylic acid with phenol and phosphoryl chloride at 110° for 4–5 h (*R. Seifert*, J. pr. Chem., 1885, [ii], **31,** 472). This reaction is general. Salol is also prepared by heating salicylic acid at 200–200°; from polysalicylide by heating with phenol; or by the action of phosgene on sodium salicylate and sodium phenoxide. It is used as an antiseptic. When heated it yields phenol, carbon dioxide and xanthone (*A. F. Holleman*, Org. Synth., 1943, Coll. Vol. I, 537). 4-*Acetamidophenyl salicylate, phenetsal, salophen*, m.p. 187–188°, is prepared by reduction and acetylation of 4-nitrophenyl salicylate (*R. Q. Brewster*, J. Amer. chem. Soc., 1918, **40,** 1136). It is used in veterinary medicine as an antiseptic, antipyretic and antirheumatic.

Salicyloylsalicylic acid, diplosal, $HOC_6H_4 \cdot CO_2C_6H_4 \cdot CO_2H$, m.p. 148°, is the simplest depside (see p. 203). It is obtained by the action of phosgene, phosphoryl chloride or thionyl chloride on salicylic acid in benzene or pyridine (*Boehringer*, G.P. 211,403/1907; Chem. Zent., 1909, II, 319) and has been used as an analgesic, antipyretic and antirheumatic. *Acetylsalicyloylsalicylic acid*, m.p. 161° (*A. Einhorn* and *A. van Bagh*, Ber., 1911, **44,** 437).

Acetylsalicylic acid, *aspirin*, m.p. 135° (varies with rate of heating and quality of the glass capillary used; on long standing the m.p. is lowered; these phenomena are due to partial hydrolysis), is prepared by treating salicylic acid with acetyl chloride, acetic anhydride, or ketene. Ketene gives first a mixed anhydride of salicylic acid and acetic acid (*C. D. Hurd* and *W. Williams*, J. Amer. chem. Soc., 1938, **58,** 962). Aspirin is widely used as an analgesic; "soluble aspirin" is the calcium salt. With thionyl chloride the acid gives *acetylsalicyloyl chloride*, m.p. 60° (*B. Riegel* and *H. Wittcoff*, *ibid.*, 1942, **64,** 1486). Thionyl chloride or phosphoryl chloride in benzene and pyridine give *acetylsalicylic anhydride*, m.p. 85° (*Bayer*, G. P. 201,325/1907; Chem. Zent., 1908, II, 996).

Methoxycarbonylsalicylic acid, $MeOCO_2C_6H_4 \cdot CO_2H$, m.p. 135° (decomp.), is prepared from salicylic acid, methoxycarbonyl chloride and dimethylaniline (*E. Fischer*,

Ber., 1919, **42,** 218). Reaction of sodium salicylate with phosgene yields the cyclic anhydride, 1,3-*benzodioxane*-2,4-*dione*

$$\overline{\text{O[2]}\cdot C_6H_4\cdot\text{[1]CO}\cdot\text{O}\cdot\text{CO}}$$

m.p. 114° (decomp.) which with methanol gives methoxycarbonylsalicylic acid (*A. Tchichibabin*, Compt. rend., 1941, **213,** 355).

2-**Methoxybenzoic acid,** m.p. 100°, is prepared by methylation of methyl salicylate with methyl sulphate and alkali and saponification of the *methyl* 2-*methoxybenzoate*, b.p. 127/11 mm (*E. R. Marshal et al.*, J. org. Chem., 1942, **7,** 444). 2-*Methoxybenzoyl chloride*, b.p. 136°/12 mm, 2-*methoxybenzonitrile*, m.p. 5°, b.p. 135°/12 mm (*H. H. Hodgson* and *F. Heyworth*, J. chem. Soc., 1949, 1131).

2-*Phenoxybenzoic acid*, m.p. 114°, is obtained by heating 2-chlorobenzoic acid with phenol, sodium hydroxide and copper powder at 190° (*Brewster* and *F. Strain*, J. Amer. chem. Soc., 1934, **56,** 117).

Disalicylic acid, $O(C_6H_4 \cdot CO_2H)_2$, m.p. 223°, is prepared from di-*o*-tolyl ether by oxidation with aqueous potassium permanganate (*M. Tomita*, J. pharm. Soc., Japan, 1937, **57,** 391). When treated with acetic anhydride or thionyl chloride in pyridine it forms an *anhydride*, m.p. 248° (*L. Anschutz* and *R. Neher*, J. pr. Chem., 1941, [ii], **159,** 264). With sulphuric acid it gives xanthone-4-carboxylic acid (*Anschutz* and *W. Claasen*, Ber., 1922, **55,** 684).

(iii) Salicyloyl chloride and amides

Salicyloyl chloride, 2-*hydroxybenzoyl chloride*, m.p. 18°, b.p. 90°/11 mm, is obtained from salicylic acid by the action of oxalyl chloride (*R. Adams* and *L. H. Ulich*, J. Amer. chem. Soc., 1920, **42,** 604) or thionyl chloride (*L. McMaster* and *F. F. Ahmann*, *ibid.*, 1928, **50,** 148). It readily loses hydrogen chloride to give salicylides (p. 183) but reacts normally with phenols and alcohols. Phosphorus chlorides give phosphorus derivatives of salicylic acid (p. 179).

Salicyloylamide, $HOC_6H_4 \cdot CONH_2$, m.p. 139°, is obtained by the action of ammonia on methyl salicylate (*E. R. Kline*, J. chem. Education, 1942, **19,** 332). p*K*a 7.95 for the phenol dissociation. The u.v. absorption at pH 6 and pH 11 has been measured (*L. Doub* and *J. M. Vandenbelt*, J. Amer. chem. Soc., 1949, **71,** 2414). It forms a disodium copper(II) complex (*P. Pfeiffer* and *H. Glaser*, J. pr. Chem., 1938, [ii], **151,** 145).

The *O*-acyl salicyloylamides are unstable and rearrange to the *N*-acyl compounds when melted or heated with pyridine. On boiling with acetic acid the *N*-acyl compound reverts to the *O*-acyl compound (*K. Auwers*, Ber., 1907, **40,** 3506; *Anschutz et al.*, Ann., 1925, **442,** 20).

When submitted to the Hofmann reaction with bromine and alkali salicyloyl amide gives benzoxazolone (VII) which undergoes further bromination (*W. van Dam*, Rec. Trav. chim., 1899, **18,** 418). Sodium hypochlorite also gives benzoxazolone (*C. Graebe* and *S. Rostovzeff*, Ber., 1902, **35,** 2751). When treated with phosgene in pyridine it gives salicylonitrile and "*carbonylsalicylamide*" (VIII), m.p. 227° which is also obtained by the action of ethoxycarbonyl chloride and pyridine (*A. Einhorn* and *C. Mettler*, *ibid.*, 1902, **35**, 3746).

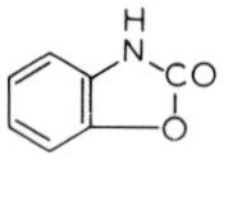

(VII) 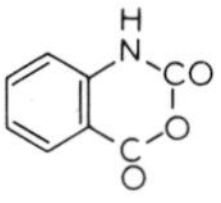 (VIII)

Salicyloylanilide, $HOC_6H_4 \cdot CONHPh$, m.p. 135° (*C. F. H. Allen* and *J. A. van Allen*, Org. Synth., 1955, Coll. Vol. III, p. 765) is used as a fungicide in textiles. When heated it gives acridone.

Salicylonitrile, m.p. 98°, is obtained from salicylaldoxime and acetic anhydride (*W. A. Bone*, J. chem. Soc., 1893, **63,** 1350; *E. Beckmann*, Ber., 1893, **26,** 2621). *Salicylohydrazide*, m.p. 147°, gives *salicyloyl azide*, $HOC_6H_4 \cdot CON_3$, m.p. 27°, with nitrous acid (*A. Struve* and *R. Radenhausen*, J. pr. Chem., 1895, [ii], **52,** 240), which on heating in benzene yields benzoxazolone (VII).

2-**Hydroxyhippuric acid,** m.p. 170°, is found in the urine when salicylic acid has been administered and has been synthesised from methoxycarbonylsalicyloyl chloride and glycine (*E. Fischer*, Ber., 1909, **42,** 219); *ethyl ester*, m.p. 88°.

(iv) Derivatives of salicylic acid containing phosphorus

2-**Chloro-4-oxo-4-*H*-benzo**[1,3,2]**dioxaphosphorin** (IX). The action of phosphorus trichloride on salicylic acid gives a *compound* $C_7H_4ClO_3P$, m.p. 37°, b.p. 127°/11 mm (*Anschutz* and *W. O. Emery*, Ann., 1887, **239,** 301; *J. A. Cade* and *W. Gerrard*, Chem. and Ind., 1954, 402). Replacement of the chlorine in this product by reaction with butanol in pyridine and ether at −10° yields the same compound (X) as can be obtained from butyl phosphorodichloridite and salicylic acid. Thus the Anschutz's compound is formulated as IX.

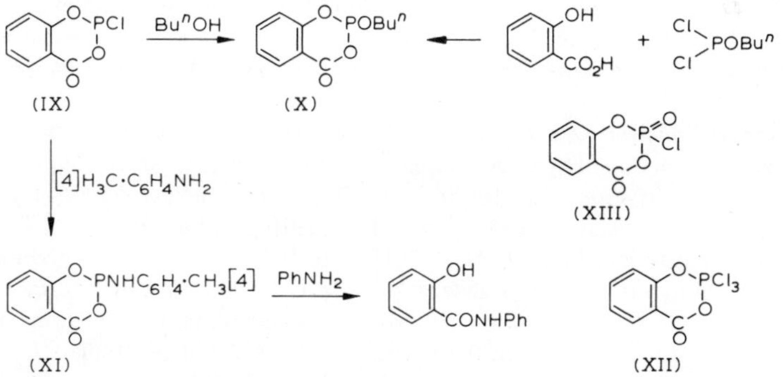

Bu^nOH ; $[4]H_3C \cdot C_6H_4NH_2$; $PhNH_2$

(IX) (X) (XIII) (XI) (XII)

Anschutz's compound gives a *p*-toluidide (XI) on treatment with *p*-toluidine and this on hydrolysis gives salicyloyl *p*-toluidide by some rearrangement which caused confusion over the structures of these compounds. Reaction of XI with aniline gives salicyloylanilide (*R. W. Young*, J. Amer. chem. Soc., 1952, **74,** 1672).

2-**Chloro-4-oxo-4*H*-benzo**[1,3,2]**dioxaphosphorin**-2-**oxide** (XIII). Reaction between salicylic acid or methyl salicylate and phosphorus pentachloride gives a compound $C_7H_4Cl_3O_3P$ (XII), b.p. 168°/11 mm (*A. S. Couper*, Ann., 1859, **109,** 369; *R. Anschutz*, *ibid.*, 1906, **346,** 286) which on partial hydrolysis gives the *monochloro* compound XIII, m.p. 95° (*A. G. Pinkus*, *P. G. Waldrep* and *W. J. Collier*, J. org. Chem., 1961, **26,** 682). The *trichloride* can be obtained from the monochloride (IX) by the action of chlorine or of phosphorus pentachloride and so the trichloride is formulated as XII. Autoxidation of IX in benzene yields the oxide XIII; its structure follows from these reactions. This oxide can be prepared also from salicylic acid and phosphoryl chloride (*H. A. C. Montgomery*, *J. H. Turnbull* and *W. Wilson*, J. chem. Soc., 1956, 4603).

Compounds of uncertain structure. Phosphorus pentachloride reacts with either the trichloro compound XII or salicyloyl chloride to give a pentachloro *compound*, $C_7H_4Cl_5O_2P$, b.p. 179°/11 mm (*L. Anschutz*, Ann., 1927, **454,** 76). Anschutz gave the structure as $ClOC{\cdot}C_6H_4OPCl_4$ on the basis of an incorrect structure for XII. When heated in a sealed tube at 180° the pentachloro compound gives 2-chlorobenzotrichloride. A second *substance* $C_{14}H_8ClO_6P$, m.p. 82°, is obtained from the action of phosphorus pentachloride on salicyloyl chloride.

2-*Carboxyphenyl dihydrogen phosphate*, $HO_2C{\cdot}C_6H_4OPO(OH)_2$, is obtained from salicylic acid by treatment with phosphorus pentachloride followed by hydrolysis with water (*J. D. Chanley et al.*, J. Amer. chem. Soc., 1952, **74,** 4351) or by alkaline hydrolysis of the pyrophosphate which results from the action of phosphorus oxychloride on methyl salicylate (*P. G. Walker* and *E. J. King*, Biochem. J., 1950, **47,** 93).

(*v*) *Sulphur derivatives of salicylic acid*

2-**Hydroxythiobenzoic acid,** m.p. 33°, with an odour reminiscent of phosphorus, is obtained by the action of alcoholic sodium hydrogen sulphide on acetylsalicyloyl chloride (*B. Riegel* and *H. Wittcoff*, J. Amer. chem. Soc., 1942, **64,** 1486) and together with the *para* isomer by the action of carbon oxysulphide on sodium phenoxide at 125–180° under pressure (*D. W. Grisley*, U.S.P. 3,136,800/1964). On oxidation with iodine it gives *salicyloyl disulphide*, $(HOC_6H_4{\cdot}CO)_2S_2$, m.p. 142°; *acetylsalicyloyl disulphide*, m.p. 101°.

2-**Hydroxydithiobenzoic acid,** $HOC_6H_4CS_2H$, m.p. 48–50°, bright orange needles, is formed by the action of hydrogen sulphide and hydrogen chloride on salicylaldehyde in benzene; *methyl ester*, m.p. 10–20°, b.p. 170°/13 mm; *methyl ester methyl ether*, m.p. 43–44° (*F. Hoehn* and *I. Block*, J. pr. Chem., 1910, [ii], **82,** 495).

2-**Mercaptobenzoic acid,** $HSC_6H_4{\cdot}CO_2H$, m.p. 164° (indefinite), is obtained from diazotised anthranilic acid by standard methods (*C. F. H. Allen* and *D. D. McKay*, Org. Synth., 1943, Coll. Vol. II, p. 580), from benzenethiol, potassium carbonate and carbon dioxide at 160–180° under pressure (*P. Saraswati et al.*, Indian J. Pharm., 1968, **30,** 124) and by heating 2-chlorobenzoic acid with alkali metal hydrogen sulphides in the presence of copper powder. 2-Mercaptobenzoic acid is a diacid with pK_1 3.54 and pK_2 8.60 (*V. M. Tarayan* and *A. N. Pogosyan*, Arm. Khim. Zhur., 1969, **22,** 569). It forms complexes with metal ions and has been used in analysis (*M. S. Gresser* and *T. S. West*, Analyst, 1968, **93,** 595), as a photographic fixer and in the production of fine grain photographic emulsions. 2-Mercaptobenzoic acid is readily oxidised to bis(2-*carboxyphenyl) disulphide*, m.p. 289° (*R. List* and *M. Stein*, Ber., 1898, **31,** 1665). It condenses with benzene in the presence of sulphuric acid to form thioxanthone (*W. G. Prescott* and *S. Smiles*, J. chem. Soc., 1911, **99,** 640).

Phenyl 2-mercaptobenzoate, m.p. 91°, is obtained from the acid, phenol and phosphoryl chloride (*E. Mayer*, Ber., 1909, **42,** 1134).

2-*Methylmercaptobenzoic acid*, m.p. 168°, is obtained by methylation of 2-mercaptobenzoic acid (*P. Friedlaender*, Ann., 1907, **351,** 401). 2-*Carboxydiphenyl sulphide*, $PhSC_6H_4 \cdot CO_2H$, m.p. 167° is obtained from 2-chlorobenzoic acid and sodium thiophenoxide in the presence of copper powder (*J. Goldberg*, Ber., 1904, **37,** 4526).

(2-**Carboxyphenylthio)acetic acid,** (XIV), m.p. 213°, is obtained from 2-mercaptobenzoic acid and chloroacetic acid (*C. Hansch* and *H. G. Lindwall*, J. org. Chem., 1945, **10,** 381) or from mercaptoacetic acid and diazotised anthranilic acid (*Kalle and Co.*, G.P. 194,040/1905; Chem. Zent., 1908, I, 1221). It is used in the manufacture of thioindigo dyestuffs. Thus on heating with an aromatic nitro compound as oxidising agent it gives thioindigo (XV) *(Friedlaender, loc. cit.):*

CO_2H, $SCH_2 \cdot CO_2H$ (XIV) → CO, S, C=C, S, CO (XV)

Diphenyl diselenide 2,2′-**dicarboxylic acid,** m.p. 234°, is obtained by the action of diazotised anthranilic acid on potassium selenide (*R. Lesser et al.*, Ber., 1913, **46,** 2640; 1914, **47,** 2505).

(vi) Nuclear-substituted salicylic acids

3-**Fluorosalicylic acid,** m.p. 144°, is obtained from 2-fluorophenol by the Kolbe reaction (*L. N. Ferguson et al.*, J. Amer. chem. Soc., 1950, **72,** 5315); 4-*fluorosalicylic acid*, m.p. 186° (*Hodgson* and *J. Nixon*, J. chem. Soc., 1929, 1632); 5-*fluorosalicylic acid*, m.p. 179°, from 4-fluorophenol by the Kolbe reaction (*K. Kraft* and *F. Dengel*, Ber., 1952, **85,** 577), 5-*fluoro-2-methoxybenzoic acid*, m.p. 87°, is prepared by metalation of 4-fluoroanisole with butyl-lithium followed by reaction with carbon dioxide (*H. Gilman et al.*, J. Amer. chem. Soc., 1941, **63,** 545).

3-**Chlorosalicylic acid,** m.p. 182°, is obtained by the action of *tert*-butyl hypochlorite on salicylic acid (*J. M. Shackelford*, J. org. Chem., 1961, **26,** 4908); 4-*chlorosalicylic acid*, m.p. 211°, from 2,4-dichlorobenzoic acid on heating with aqueous barium hydroxide and copper powder (*J. T. Sheehan*, J. Amer. chem. Soc., 1948, **70,** 1665); 5-*chloro-*, m.p. 173° and 3,5-*dichloro-salicylic acid*, m.p. 220°, are obtained by chlorination of salicylic acid; 6-*chlorosalicylic acid*, m.p. 166°, is obtained by the diazonium reaction from 6-chloro-2-aminobenzoic acid (*P. Cohn*, Chem. Zent., 1901, II, 925).

3-**Bromosalicylic acid,** m.p. 181°, is obtained by the action of bromine on the mercuration product from salicylic acid *(Shackelford, loc. cit.);* 4-*bromosalicylic acid*, m.p. 212° (*H. Ohta*, Nippon Kagaku Zasshi, 1957, **78,** 1608; C.A., 1959, **53,** 21341); 5-*bromo-*, m.p. 164° and 3,5-*dibromo-salicylic acid*, m.p. 223°, are obtained by bromination of salicylic acid.

3-**Iodosalicylic acid,** m.p. 198°, is obtained by the action of iodine on the mercuration product of salicylic acid (*O. Dimroth*, Ber., 1902, **35,** 2873) and is formed together with 5-*iodo-*, m.p. 198° and 3,5-*diiodo-salicylic acid*, m.p. 235°, by iodination (*E. Lautermann*, Ann., 1861, **120,** 301); 4-*iodosalicylic acid*, m.p. 230° (*Ohta, loc. cit.).*

3-**Nitrosalicylic acid,** m.p. 148° when anhydrous, and 5-*nitrosalicylic acid*, m.p. 228°, are obtained by nitration (*A.N. Meldrum* and *N. W. Hirve*, J. Indian chem. Soc., 1928, **5,** 95); 4-*nitrosalicylic acid*, m.p. 235°, is prepared from 2-amino-4-nitrotoluene (*S. S. Sabnis* and *M. V. Shirsat*, J. sci. ind. Research, India, 1959, **18B,** 240); 6-*nitrosalicylic acid*, m.p. 166°, from 3-nitrophthalic acid (*S. Seki et al.*, Nippon Kagaku Zasshi, 1962, **83,** 117; C.A., 1963, **59,** 500).

5-**Nitrososalicylic acid,** bluish green crystals, m.p. 162° (decomp.), is tautomeric with 1,4-benzoquinonemonoxime-2-carboxylic acid (see CCC, 2nd Edn., Vol. IIIA, p. 346); *methyl ester*, blue crystals, m.p. 89°. It is formed by heating nitroso-*N*-methylanthranilic acid with sodium hydroxide or sulphuric acid or by treating sodium salicylate with copper sulphate and sodium nitrite (*W. Gulinov*, Chem. Zent., 1928, II, 759).

3-**Aminosalicylic acid,** 3-*amino-2-hydroxybenzoic acid*, m.p. 235° (decomp.), and its isomers are prepared by reduction of the corresponding nitrosalicylic acid; 4-*aminosalicylic acid*, m.p. 220° (decomp.); 5-*aminosalicylic acid*, decomp. 280° (*E. Grandmougin*, Ber., 1906, **39,** 3930). 4-*Aminosalicylic acid* has been made by the Kolbe–Schmitt process (*Sheehan, loc. cit.).*

The hydrochloride of methyl 5-diethylaminoacetamidosalicylate, *nirvanin*,

TABLE 4

HOMOLOGUES OF SALICYLIC ACID

Derivative	*m.p.*	*ref.*	*Derivative*	*m.p.*	*ref.*
3-Methyl	162°	1,2	3,6-Dimethyl	195°	7
4-Methyl	174°	2	4-Isopropyl[a]	94°	8
5-methyl	149°	2,3	6-Methyl-3-isopropyl[b]	127°	9
6-Methyl	168°	4,5	3-Methyl-6-isopropyl[c]	136°	10
3,5-Dimethyl	179°	6			

[a] Isohydrocumic acid; [b] *o*-Thymotic acid; [c] *o*-Carvacrotic acid

References

1 *K. Brunner*, Ann., 1907, **351,** 319.
2 *A. Engelhardt* and *P. Latschinow*, J. Russian Phys,-chem. Soc., 1869, **1,** 220.
3 *J. Zeltner* and *M. Landau*, G.P. 258,887/1912.
4 *F. Mayer* and *S. Schulze*, Ber., 1925, **58,** 1469.
5 *W. W. Kaeding* and *A. T. Shulgin*, F.P. 1,321,264/1961.
6 *H. Meerwein*, Ann., 1907, **358,** 71.
7 *W. Baker et al.*, J. chem. Soc., 1954, 2044.
8 *O. Jacobsen*, Ber., 1878, **11,** 1061.
9 *J. P. Street, C. E. Georgi* and *P. J. Jannke*, J. Amer. pharm. Assoc., 1948, **37,** 180.
10 *S. Lusting*, Ber., 1886, **19,** 18.

$Et_2NCH_2 \cdot CONHC_6H_3(OH)CO_2Me$, has been used as a local anaesthetic (*A. Einhorn* and *M. Oppenheimer*, Ann., 1900, **311** 154). The diazonium salt from 5-aminosalicylic acid gives Diamond Black F when it is coupled with 2-naphthylamine and the resulting aminoazo compound is diazotised and coupled with 1-naphthol-4-sulphonic acid.

5-**Sulphosalicylic acid,** hydrated, m.p. 113°, is obtained by sulphonation of salicylic acid *A. N. Meldrum* and *M. S. Shah*, J. chem. Soc., 1923, **123,** 1988).

(vii) Homologues of salicylic acid (see Table 4)

The methylsalicylic acids formed by Kolbe–Schmitt carboxylation of *o*-, *m*- and *p*-cresols are known respectively as *o*-, *m*- and *p*-cresotic acids. For a comparison of the variations of this method see *O. Baine et al.*, (J. org. Chem., 1954, **19,** 510).

6-**Methylsalicylic** acid is present in the culture filtrate from some *Penicillium* species (*H. Raistrick et al.*, Biochem. J., 1931, **25,** 39; 1935, **29,** 1102; *A. Grosser* and *W. Friedrich*, Z. Naturforsch., 1948, **3b,** 380). It has been prepared by heating basic copper *o*-toluate (*W. W. Kaeding* and *A. T. Skulgin*, F.P. 1,321,264/1961). *Penicillium griseofulvum* has been shown to utilise acetic acid in forming 6-methylsalicylic acid (see CCC, 2nd Edn., Vol. IIIA, p. 135).

3,5-*Dimethylsalicylic acid* is obtained by the condensation of 2-methylpentenal with diethyl malonate in the presence of sodium ethoxide and hydrolysis of the resulting ester:

(viii) Cyclic anhydrides of salicylic acid and its homologues

The dehydration of salicylic acid with phosphorus pentoxide and phosphoryl chloride was first examined by *C. Gerhardt* (Ann., 1853, **87,** 159). In a careful examination of the reaction products under various conditions, *R. Anschutz et al.*, isolated four different anhydrides from salicylic acid and the analogous compounds from the cresotic acids (Ber., 1892, **25,** 3506; Ann., 1893, **273,** 73, 97; Ber., 1919, **52,** 1883). There were thought to be two isomeric disalicylides, named α- and β-, of which the α-form was more readily hydrolysed, a tetra- and a poly-salicylide. *R. Spallino* and *G. Provenzal* (Gazz., 1909, **39,** II, 325) isolated two anhydrides of *o*-thymotic acid which they considered to be the α- and β-di-*o*-thymolides. In the later literature the isomerism of the α- and β-disalicylides and the reason for the high-energy barrier to their interconversion was discussed.

The correct structures for these anhydrides are given by *W. Baker et al.* (see refs. to Table 5). They are macrocyclic anhydrides of salicylic acid. The readily hydrolysable α-disalicylide is the true disalicylide (XVI), the β-disalicylide is shown by molecular weight measurements to be trisalicylide (XVIII). The tetrasalicylide is XIX as suggested by previous workers and the polymer is shown to be a hexasalicylide XX by molecular weight measurements. The true linear high polymers of salicylic acid can be drawn into fibres or made into films (*E. Simon* and *F. W. Thomas*, U.S.P. 2,698,838/1955;

TABLE 5

MACROCYCLIC ANHYDRIDES OF SALICYLIC ACID AND ITS HOMOLOGUES

Macrocyclic anhydride	*m.p.(°C)*	*ref.*
from *salicylic acid*		
Disalicylide (XVI)	234 (dec.)	1
Trisalicylide (XVIII)	200	1
Tetrasalicylide (XIX)	298–300	1
Hexasalicylide(XX)	375 (dec.)	1
from 3-*methylsalicylic acid*		
Di-*o*-cresotide	240 (dec.)	2
Tri-*o*-cresotide	264–265	2
Tetra-*o*-cresotide	299–300	2
from 4-*methylsalicylic acid*		
Di-*m*-cresotide	255 (dec.)	2
Tri-*m*-cresotide	207	2
Hexa-*m*-cresotide	380 (dec.)	2
from 5-*methylsalicylic acid*		
Di-*p*-cresotide	235 (dec.)	2
Tri-*p*-cresotide	245	2
Tetra-*p*-cresotide	347 (dec.)	2
from o-*thymotic acid*		
Di-*o*-thymotide	207	3
Tri-*o*-thymotide–EtOH complex	172 (dec.)	3
from o-*carvacrotic acid*		
Di-*o*-carvacrotide	174	4
Tri-*o*-carvacrotide	247	4
from 3,6-*dimethylsalicylic acid*		
Tri-*3,6-dimethylsalicylide*	254	4

References

1 *W. Baker*, *W. D. Ollis* and *T. S. Zealley*, J. chem. Soc., 1951, 201.
2 *Baker*, *B. Gilbert*, *Ollis* and *Zealley*, *ibid.*, 1951, 209.
3 *Baker*, *Gilbert* and *Ollis*, *ibid.*, 1952, 1443.
4 *Baker*, *J. B. Harborne*, *A. J. Price* and *A. Rutt*, *ibid.*, 1954, 2043.

C.A., 1955, **49,** 4334; *H. K. Hall* and *A. K. Schneider*, J. Amer. chem. Soc., 1958, **80,** 6409).

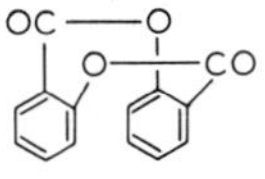

(XVI) 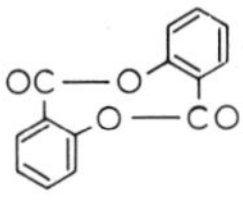 (XVII) 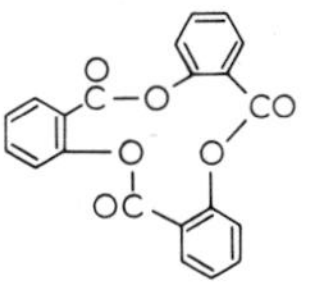 (XVIII)

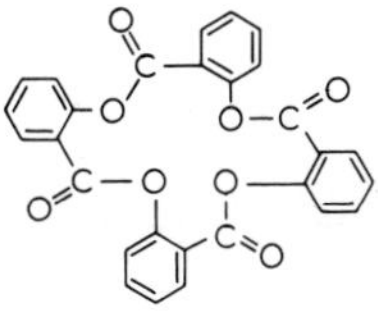

(XIX) 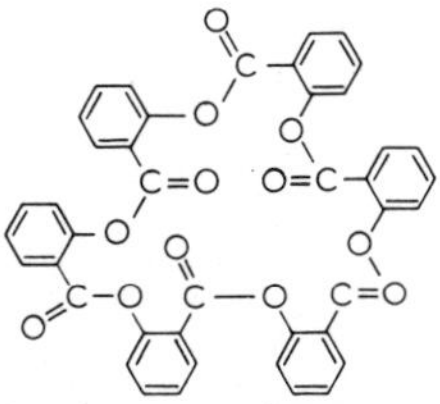 (XX)

Disalicylide is obtained together with trisalicylide by distilling acetylsalicylic acid or the silicic ester of methyl salicylate under reduced pressure (*R. Schwartz* and *W. Kuchen*, Ber., 1952, **85,** 624). Under mild conditions the ring is opened to give salicoylsalicylic acid or its derivatives; one equivalent of sodium hydroxide gives the sodium salt, methanolic hydrogen chloride the methyl ester whilst aniline or benzylamine yield the corresponding *N*-substituted amide of salicoylsalicylic acid.

Two conformations, *cis* (XVI) and *trans* (XVII) have been considered for disalicylide and its homologues. Dipole moment studies (*P. G. Edgerley* and *L. E. Sutton*, J. chem. Soc., 1951, 1069) indicate the conformation to be *cis* (XVI); *cis*-disalicylide has a dipole moment (6.26D) which is close to that calculated for conformation XVI (6.0–6.2D) while th conformation XVII is expected to have a dipole moment near zero. Di-*o*-salicylides of conformation XVI could exist in optically active forms. The temperature dependence of the ^{1}H-n.m.r. methyl signals for the isopropyl groups in di-*o*-thymotide and di-*o*-carvocrotide is attributed to the interchange of the conformation like XVI with its mirror image and for this process $\Delta G^{\ddagger}$ is 18 kcal·mol^{-1} (*W. D. Ollis* and *J. F. Stoddart*, Chem. Comm., 1973,571). In most cases the enantiomeric forms of a di-*o*-salicylide would racemise too rapidly at room temperature to permit of their isolation.

Disalicylide and its homologues have a higher carbonyl-bond stretching frequency (1773 cm^{-1}) in the infrared than the tri- and tetra-salicylides (1740–1749 cm^{-1}) (*L. N. Short*, *ibid.*, 1952, 206). In the dimers the carbonyl group is twisted out of effective conjugation with the benzene ring so that the stretching frequency is like that for phenyl acetate (1766 cm^{-1}). Conjugation with the benzene ring is effective in the case of the tri- and higher salicylides so that the carbonyl-bond stretching frequency is like that for phenyl benzoate (1738 cm^{-1}).

TABLE 6

PROPORTIONS OF THE PROPELLER (XXI) AND HELIX (XXII) CONFORMATIONS PRESENT AT EQUILIBRIUM IN SOLUTIONS OF TRISALICYLIDE DERIVATIVES

Macrocyclic lactone	*Solvent*	*temp. (°C)*	*Propeller conformation (%)*	*Helix conformation (%)*
Tri-*o*-thymotide	C_2HCl_5	68	86	14
Tri-*o*-thymotide	C_2HCl_5	90	80	20
Tri-*o*-carvacrotide	C_5H_5N	20	64	36
Tri-*o*-carvacrotide	C_5H_5N	90	58	42
Tri-3,6-dimethyl-salicylide	$CDCl_3$	−10	67	33

Reference
A. P. Downing, W. D. Ollis and *I. O. Sutherland*, J. chem. Soc., B, 1970, 24.

Trisalicylide also has a non-zero dipole moment (2.95 D) and adopts a nonplanar conformation with three *trans*-ester linkages.

Tri-*o*-thymotide crystallises from a number of solvents to give adducts of the clathrate type (*Baker, B. Gilbert* and *W. D. Ollis, ibid.*, 1952, 1443; *D. Lawton* and *H. M. Powell, ibid.*, 1958, 2339). The adducts with *n*-hexane, benzene and chloroform were found to give crystals belonging to enantiomorphous space groups. When large single crystals of the benzene adduct were obtained it was shown that spontaneous optical resolution had taken place. Each crystal gave a solution showing optical activity, either positive or negative, which diminished rapidly through racemisation (*A. C. D. Newman* and *Powell, ibid.*, 1952, 3747). Thus tri-*o*-thymotide adopts an asymmetric conformation in the clathrate and in solution conformational changes cause racemisation.

An n.m.r. investigation of *tri-o-carvacrotide* (XXIb) *tri-o-thymotide* (XXIa) and *tri-3,6-dimethylsalicylide* (XXIc) at varying temperatures allows the conformational equilibria to be unravelled (*A. P. Downing, Ollis* and *I. O. Sutherland*, J. chem. Soc., B, 1970, 24). In the solid state and predominantly in solution at room temperature tri-*o*-thymotide adopts the propeller conformation of which XXIa is one enantiomer. The benzene rings are arranged like the three blades of a propeller. This conformation is in equilibrium with the helix of which XXIIa is one enantiomer. The benzene rings in the latter conformation are approximately parallel and helically arranged on three different levels. The conformers are interconverted by rotation about the Ar–O and Ar–CO single bonds. This rotation also allows interconversion of the two enantiomeric helical conformations and optically active tri-*o*-thymotide is racemised by this process:

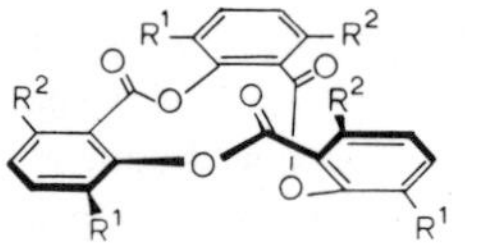

(XXI) 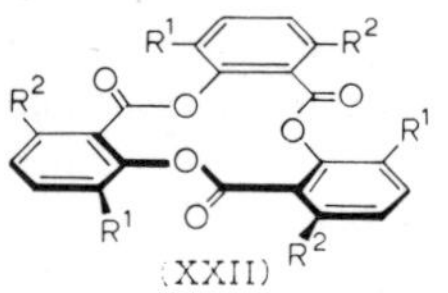 (XXII)

(a) $R^1 = Pr^i$, $R^2 = Me$
(b) $R^1 = Me$, $R^2 = Pr^i$
(c) $R^1 = R^2 = Me$

The circular dichroism of (+)-tri-*o*-thymotide is similar in the solid state and in solution in ether at $-78°$. A calculation of the expected sign for the Cotton effect shows that (+)-tri-*o*-thymotide has the absolute configuration associated with the propeller conformation XXIa (*Downing et al.*, Chem. Comm., 1968, 329).

Tri-*o*-thymotide forms molecular inclusion compounds with a wide range of molecules. If the included molecule does not differ much from the "straight" chain type, the resulting compound belongs to one of two classes. A clathrate type, $2C_{33}H_{36}O_6$, M, is formed with included molecules, M, of greatest length less than 9.5 Å. Longer molecules are enclosed in a different channel-type structure of composition $C_{33}H_{36}O_6$, *x*M, where *x* need not be rational (*Lawton* and *Powell*, *loc. cit.*). Either type of complex when freshly dissolved gives a solution which contains only the propeller conformation of tri-*o*-thymotide. Most of the complexes underwent spontaneous optical resolution on crystallisation indicating that the individual crystals are composed of one enantiomeric propeller conformation. This property has been applied to resolve 2-bromobutane by taking advantage of the fact that the holes in a host crystal are themselves enantiomeric. One crystal plays host preferentially to one enantiomer of 2-bromobutane and the optically active 2-bromobutanes can be recovered from individual crystals of the complex (*Powell*, Nature, 1952, **170**, 155).

(*d*) *Other monohydroxybenzoic acids*

3-**Hydroxybenzoic acid,** m.p. 200° (sublimes), is prepared from 3-aminobenzoic acid by the diazo reaction (*F. M. Beringer* and *S. Sands*, J. Amer. chem. Soc., 1953, **75,** 3319) and can be made by reduction of piperonylic acid with Raney nickel–aluminium alloy and alkali (*E. Schwenk* and *D. Papa*, J. org. Chem., 1945, **10,** 232). Its preparation by aerial oxidation of phthalic acid in the presence of copper(II) phthalate is described in the patent literature (*M. B. Pearlman*, U.S.P. 2,727,924/1955; C.A., 1956, **50,** 10771). Iodination of the ammonium salt in aqueous solution gives 3-*hydroxy*-4-*iodobenzoic acid*, m.p. 225° (decomp.) (*H. R. Frank*, *P. E. Fanta* and *D. S. Tarbell*, J. Amer. chem. Soc., 1948, **70,** 2314). Chlorination gives 2-*chloro*-3-*hydroxybenzoic acid*, m.p. 156° (*E. Plazek*, Roczniki Chem., 1930, **10,** 761). 3-*Methoxybenzoic acid*, m.p. 109°; 3-*acetoxybenzoic acid*, m.p. 132°. 3-(3′-*Hydroxybenzoyl*)*oxybenzoic acid* m.p. 199° (*R. Anschutz et al.*, Ann., 1925, **442,** 45).

Methyl 3-*methoxy*-2-*methylaminobenzoate*, *damascenine*, m.p. 26°, b.p. 147°/60 mm, has a nutmeg like odour. It is the odoriferous principle of the oil from *Nigella damascena* L. seeds and is used in perfumery. It is prepared synthetically from 3-hydroxybenzoic acid (*A. J. Ewins*, J. chem. Soc., 1912, **101,** 544) or by degradation of 8-methoxyquinoline (*A. Kaufmann* and *E. Rothlin*, Ber., 1916, **49,** 589).

3-*Hydroxy*-2-*methylbenzoic acid*, m.p. 141°, is obtained by heating technical sodium 2-naphthylamine-4,8-disulphonate with 50% sodium hydroxide (*O. Baudisch* and *W. H. Perkin*, J. chem. Soc., 1909, **95,** 1883; *L. F. Fieser* and *W. C. Lothrop*, J. Amer. chem. Soc., 1936, **58,** 752). 3-*Hydroxy*-4-*methylbenzoic acid*, m.p. 207° (*A. N. Meldrum* and *Perkin*, J. chem. Soc., 1908, **93,** 1419). 3-*Hydroxy*-5-*methylbenzoic acid*, m.p. 208°, (*idem*, *ibid*., p. 1895). 5-*Hydroxy*-2-*methylbenzoic acid*, m.p. 183–184° from naphthalene-1,3,7-trisulphonic acid (*Kalle and Co.*, G.P. 91,201/1893; *Friedl.*, 1894–1897, **4,** 148), or 2-naphthol-6,8-disulphonic acid (*T. Zincke* and *H. Fischer*, Ann., 1906, **350,** 253) by heating with aqueous sodium hydroxide.

4-**Hydroxybenzoic acid,** m.p. 210°, occurs in *Papaver somniferum* L. and the young leaves of *Populus Balsamifera* L. It is prepared in the laboratory (*C. A. Buehler* and *W. E. Cato*, Org. Synth., 1943, Coll. Vol. II, p. 341) and commercially by heating potassium salicylate. The acid and its esters are used commercially as disinfectants and food preservatives. Bromination at temperatures below −5° gives 3,5-*dibromo*-4-*hydroxybenzoic acid*, m.p. 268°. At temperatures above 20° tribromophenol is formed (*A. W. Francis* and *A. J. Hill*, J. Amer. chem. Soc., 1924, **46,** 2498). *Methyl* 4-*hydroxybenzoate*, m.p. 131°. 4-*Acetoxybenzoic acid*, m.p. 189° (*F. D. Chattaway*, J. chem. Soc., 1931, 2495). 4-(4′-*Hydroxybenzoyloxy*)*benzoic acid*, m.p. 277° (*E. Fischer* and *K. Freudenberg*, Ann., 1909, **372,** 47). *Methyl* 3-*amino*-4-*hydroxybenzoate*, *orthocaine*, *orthoform*, m.p. 143° (*A. Einhorn* and *B. Pfyl*, Ann., 1900, **311,** 46) is used as a local anaesthetic. *Methyl* 4-*amino*-3-*hydroxybenzoate*, m.p. 121°, was previously used under the name of orthoform but is now called orthoform-old.

Anisic acid, 4-*methoxybenzoic acid*, m.p. 184°, can be obtained from anethole, the principal constituent of anise oil and some other essential oils, by oxidation with dilute nitric or chromic acids. It is prepared either by methylation of 4-hydroxybenzoic acid or by the action of carbon dioxide on the Grignard reagent from 4-bromoanisole. *Methyl anisate*, m.p. 49°, b.p. 255°; *ethyl anisate*, m.p. 7°, b.p. 269°. *Anisoyl chloride*, m.p. 24°, b.p. 145°/15 mm (*J. S. Pierce et al.*, J. Amer. chem. Soc., 1942, **64,** 1691); *amide*, m.p. 163°. *Anisonitrile*, m.p. 62°, b.p. 131°/14 mm.

4-*Hydroxy*-2-*methylbenzoic acid*, m.p. 177° and 4-*hydroxy*-3-*methylbenzoic acid*, m.p. 174° are obtained from the appropriate cresol by the action of carbon tetrachloride and alkali (*C. Schall*, Ber., 1879, **12,** 816).

(e) Dihydroxybenzoic acids

(i) Derivatives of catechol

The Kolbe–Schmitt reaction applied to catechol gives both 2,3- and 3,4-dihydroxybenzoic acids. Under the Marasse procedure the 2,3-acid is the major product (*O. Baine et al.*, J. org. Chem., 1954, **19,** 510) while heating catechol, ammonium carbonate and water in a sealed tube gives the 3,4-acid as the principal product (*A. K. Miller*, Ann., 1883, **220,** 116).

2,3-**Dihydroxybenzoic acid,** *pyrocatechuic acid*, m.p. 204° with decomp. to catechol and carbon dioxide. It is also prepared by demethylation of the dimethyl ether (*G. R. Pettit* and *D. M. Piatak*, J. org. Chem., 1960, **25,** 721). 2,3-*Dimethoxybenzoic acid*, m.p. 122°, is obtained from *o*-vanillin (*W. H. Perkin* and *R. Robinson*, J. chem. Soc., 1914, **105,** 2383). Nitration gives 2,3-*dimethoxy*-5-*nitrobenzoic acid*, m.p. 174° (*J. C. Cain* and *J. L. Simonsen*, *ibid.*, 1914, **105,** 159). 2-*Hydroxy*-3-*methoxybenzoic acid*, m.p. 152°, is obtained by the Kolbe reaction with guaiacol (*Baine et al., loc. cit.*). 2,3-*Diacetoxybenzoic acid*, m.p. 148°.

Bromination of *methyl* 2-*acetoxy*-3-*methoxybenzoate*, m.p. 62° (*A. Klemenc*, Monatsh., 1914, **35,** 97) followed by hydrolysis of the product yields 6-*bromo*-2-*hydroxy*-3-*methoxybenzoic acid*, m.p. 150° (*C. Weizmann* and *L. Haskelberg*, J. org. Chem., 1944, **9,** 121). Demethylation with aluminium bromide gives 6-*bromo*-2,3-*dihydroxy*-

benzoic acid, m.p. 182–185°. 6-*Bromo-2,3-dimethoxybenzoic acid*, m.p. 87°, on treatment with hydrobromic acid yields 5-*bromo-2,3-dimethoxybenzoic acid (Pettit* and *Piatak, loc. cit.)* by a rearrangement, no demethylation occurring.

Protocatechuic acid, 3,4-*dihydroxybenzoic acid*, m.p. 199° (anhyd.) is also obtained by fusion of vanillin with alkali (*I. A. Pearl*, Org. Synth., 1955, Coll. Vol. III, p. 745) and crystallises from water as the monohydrate. It has been found in the fruit of *Illicium* species and in *Penicillium* species (*J. Barta* and *R. Mecir*, Experientia, 1948, **4,** 277) and has been identified as the substance responsible for the hardening of cockroach oötheca (*A. R. Todd et al.*, Biochem. J., 1946, **40,** 627). One colleterial gland of the cockroach stores 4-*O*-*β*-D-*glucodisoprotocatechuic acid*, m.p. 180°, $[\alpha]_D$ −59° (methanol), and secretes this with the protein constituents of sclerotin and a polyphenol oxidase. The other gland secretes a *β*-glucosidase such that in the intermingled secretions, protocatechuic acid is liberated in the genital vestibulum. This is oxidised to the *o*-quinone which condenses with the protein to form the resilient sclerotin constituent of the oötheca (*P. W. Kent* and *P. C. J. Brunet*, Tetrahedron, 1959, **7,** 252). Protocatechuic acid is also involved in the sclerotization processes of locusts and silk worms (*B. M. Jones* and *W. Sinclair*, Nature, 1958, **181,** 926; *S. Kawase*, *ibid.*, 1958, **181,** 1350).

Solutions of protocatechuic acid give a green coloration with ferric chloride; on addition of very dilute sodium carbonate solution the colour turns to blue, then red (*F. Tiemann* and *W. Will*, Ber., 1881, **14,** 956).

3,4-*Diacetoxybenzoic acid*, m.p. 162° (*T. Malkin* and *M. Nierenstein*, *ibid.*, 1928, **61,** 797).

Vanillic acid, 4-*hydroxy-3-methoxybenzoic acid*, m.p. 211°, is prepared from vanillin by oxidation with silver oxide or by fusion with potassium hydroxide below 240° (*Pearl*, Org. Synth., 1963, Coll. Vol. IV, p. 972).

Isovanillic acid, 3-*hydroxy-4-methoxybenzoic acid*, m.p. 252°, is obtained by the partial demethylation of veratric acid with hydrobromic acid (*A. Lovecy et al.*, J. chem. Soc., 1930, 818). Reduction with Raney nickel–aluminium alloy in sodium hydroxide solution gives 3-hydroxybenzoic acid (*E. Schwenk et al.*, J. org. Chem., 1944, **9,** 1).

Veratric acid, 3,4-*dimethoxybenzoic acid*, m.p. 179°, occurs in the seeds of *Veratrum Sabadilla* Retz. together with the alkaloid veratrine. It is obtained from many natural products by methylation and oxidation. Veratric acid is prepared by oxidation of veratraldehyde (*L. C. Raiford* and *R. P. Perry*, J. org. Chem., 1942, **7,** 354). *Veratronitrile*, m.p. 67° (*J. S. Buck* and *W. S. Ide*, Org. Synth., 1943, Coll. Vol. II, p. 622).

4-*Ethoxy-3-methoxybenzoic acid*, m.p. 202° (*R. D. Haworth* and *W. Kelly*, J. chem. Soc., 1936, 998), 3-*ethoxy-4-methoxybenzoic acid*, m.p. 165° (*R. Manske*, Canad. J. Res. [B], 1937, **15,** 159), 3,4-*diethoxybenzoic acid*, m.p. 167° (*N. Hirao*, J. chem. Soc., Japan, 1931, **52,** 34), 3-*benzyloxy-4-methoxybenzoic acid*, m.p. 178° and 4-*benzyloxy-3-methoxybenzoic acid*, m.p. 168° *(Lovecy et al., loc. cit.)*, are obtained by oxidation of the corresponding aldehyde with potassium permanganate.

Piperonylic acid, 3,4-*methylenedioxybenzoic acid* (XXIII), m.p. 230°, is obtained by oxidation of piperonal (*R. L. Shriner* and *E. G. Kleiderer*, Org. Synth., 1943, Coll. Vol. II, p. 538). It is reduced to *m*-hydroxybenzoic acid by Raney nickel–aluminium alloy and sodium hydroxide (*Schwenk* and *Papa*, *loc. cit.*). It gives protocatechuic acid on treatment with mild dealkylating agents such as aluminium bromide in benzene (*P. Pfeiffer* and *W. Lowe*, J. pr. Chem., 1937, [ii], **147,** 305). *Piperonyl chloride*, m.p. 80°, b.p. 140°/8 mm, is formed by the action of refluxing thionyl chloride on the acid; thionyl chloride at 180° gives 3,4-*carbonyldioxybenzoyl chloride*, m.p. 68°, b.p. 166°/12 mm, which is hydrolysed in formic acid to give 3,4-*carbonyldioxybenzoic acid*, (XXIV), m.p. 228° (decomp.) (*G. Barger*, J. chem. Soc., 1908, **93,** 567). Phosphorus pentachloride with piperonylic acid gives 3,4-dichloromethylenedioxybenzoyl chloride (XXV). 6-Nitropiperonylic acid, (XXVI), m.p. 172°, is formed by nitration (*J. B. Ekeley* and *M. S. Klemme*, J. Amer. chem. Soc., 1928, **50,** 2713). Chlorination and bromination give 4,5-dichloro- and 4,5-dibromomethylenedioxybenzene respectively (*R. Robinson et al.*, J. chem. Soc., 1917, **111,** 949; 913). *Amide*, m.p. 169°; *nitrile*, m.p. 95°.

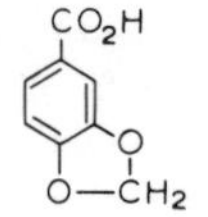

(XXIII) 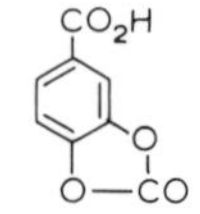 (XXIV) 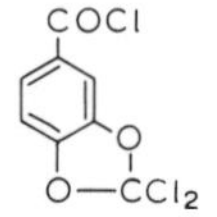 (XXV) 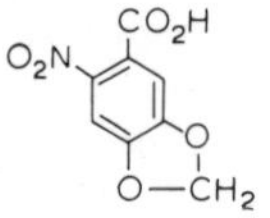 (XXVI)

(ii) Derivatives of resorcinol

The Kolbe reaction can be carried out by passing carbon dioxide into a hot aqueous solution of resorcinol and potassium hydrogen carbonate. Under these conditions 2,4-dihydroxybenzoic acid is the principal product (*Nierenstein* and *D. A. Clibbens*, Org. Synth., 1943, Coll. Vol. II, p. 557). Aqueous solutions of the sodium salt of 2,4-dihydroxybenzoic acid give mixture containing 50% of 2,6-dihydroxybenzoic acid when heated at 200° for some time. Carbonation of resorcinol at higher temperatures and pressures gives a mixture of these two acids (*D. K. Hale et al.*, J. chem. Soc., 1952, 3503). Carbonation of the anhydrous salt of resorcinol in dimethylformamide gives 2,6-dihydroxybenzoic acid (*E. R. Stove*, B.P. 916,548/1963; *F. P. Doyle et al.*, J. chem. Soc., 1962, 1453). Separation of the two resorcylic acids is achieved by elution from a column of ion-exchange resin (*Hale, loc. cit.*) and by taking advantage of the faster reaction of 2,4-dihydroxybenzoic acid with toluene-*p*-sulphonyl chloride (*L. G. Shah* and *G. D. Shah*, J. Sci. Ind. Research, India, 1956, **15B,** 159). The mixture is also separated by conversion to the mixed dimethyl ether methyl esters. Only methyl 2,4-dimethoxybenzoate is hydrolysed by base and can be separated (*Stove, loc. cit.*).

2,4-**Dihydroxybenzoic acid,** *β*-**resorcylic acid,** m.p. 218° (anhydr.), is obtained from resorcinol by the Kolbe reaction. Bromination gives the 5-*bromo* derivative, m.p. 186° (decomp.); nitration gives the 5-*nitro* derivative, m.p. 215° (decomp.), together with styphnic acid (*N. Kaneniwa*, J. pharm. Soc., Japan, 1955, **75,** 791). 2,4-*Dihydroxybenzoyl chloride* (*A. W. Scott* and *W. O. Kearse*, J. org. Chem., 1940, **5,** 600); *nitrile*, m.p. 175°; *methyl ester*, m.p. 118°. 2,4-*Diacetoxybenzoic acid*, m.p. 136°, on hydrolysis with two moles of sodium hydroxide yields 2-*acetoxy*-4-*hydroxybenzoic acid*, m.p. 167° (*M. Bermann* and *P. Dangschat*, Ber., 1919, **52,** 380).

2-*Hydroxy*-4-*methoxybenzoic acid*, m.p. 158°, is obtained by methylation of *β*-resorcylic acid with two moles of dimethyl sulphate and sodium hydroxide (*M. Gornberg* and *L. C. Johnson*, J. Amer. chem. Soc., 1917, **39,** 1687). The *methyl ester*, m.p. 49°, occurs in the roots of *Primula officinalis* Jacq. as the primaveroside, *primaverin*, m.p. 203°, $[\alpha]_D$ −71.5° (water), from which it is released enzymatically when the roots are crushed (*A. Goris* and *M. Mascre*, Compt. rend., 1909, **149,** 947). 2-*Benzyloxy*-4-*methoxybenzoic acid*, m.p. 103° (*V. Bruckner et al.*, J. Amer. chem. Soc., 1948, **70,** 2697).

4-*Hydroxy-2-methoxybenzoic acid*, m.p. 186°, is prepared by oxidation of 4-acetoxy-2-methoxybenzaldehyde and hydrolysis of the acetyl group (*F. Tiemann* and *A. Parrisius*, Ber., 1880, **13,** 2375). 4-*Benzyloxy-2-methoxybenzoic acid*, m.p. 112° (*K. G. Dave et al.*, J. sci. ind. Research, India, 1960, **19B,** 470).

2,4-*Dimethoxybenzoic acid*, m.p. 108° (*Robinson* and *K. Venkataraman*, J. chem. Soc., 1929, 62) is selectively demethylated to 2-hydroxy-4-methoxybenzoic acid by boron trichloride (*F. M. Dean et al.*, Tetrahedron Letters, 1966, 4153); 2,4-*dibenzyloxybenzoic acid*, m.p. 180° (*C. J. Cavallito* and *J. S. Buck*, J. Amer. chem. Soc., 1943, **65,** 2140).

2,6-**Dihydroxybenzoic acid,** γ-**resorcylic acid,** m.p. 166°, can be prepared free from β-resorcylic acid by metalation of resorcinol bistetrahydropyranyl ether with butyllithium, followed by reaction with carbon dioxide and acid hydrolysis (*W. E. Parham* and *E. L. Anderson*, *ibid.*, 1948, **70,** 4187); from 2,6-dihydroxyacetophenone (*D. B. Limaye* and *G. R. Kelkar*, J. Indian chem. Soc., 1935, **12,** 788); by reduction of phloroglucinolcarboxylic acid 4-monotosylate (*Shah* and *Shah*, *loc. cit.*). 2-*Hydroxy-6-methoxybenzoic acid*, m.p. 135°, is found in the tubers of *Gloriosa superba* L. (*H. Clewer et al.*, J. chem. Soc., 1915, **107,** 837) and is prepared by partial methylation of methyl γ-resorcylate (*Doyle et al.*, *loc. cit.*). 2,6-*Dimethoxybenzoic acid*, m.p. 186° (*H. Gilman et al.*, J. Amer. chem. Soc., 1940, **62,** 667); 2-*benzyloxy-6-methoxybenzoic acid*, m.p. 104°; 2,6-*dibenzyloxybenzoic acid*, m.p. 124° (*Doyle et al.*, *loc. cit.*). The esters of 2,6-dialkoxybenzoic acids are only slowly hydrolysed under alkaline conditions and the acids are best esterified by the action of methyl iodide on the silver salt (*Gilman et al.*, *Doyle et al.*, *loc. cit.*).

3,5-**Dihydroxybenzoic acid,** α-**resorcylic acid,** m.p. 237° (anhydr.) is obtained from benzoic acid by sulphonation and alkali fusion (*A. W. Weston* and *C. M. Suter*, Org. Synth., 1955, Coll. Vol. III, p. 288) or from 3,5-dibromobenzoic acid by heating with aqueous calcium hydroxide at 160° in a copper vessel (*C. F. Boehringer u. Sohn*, G.P. 286,226/1912; Friedl., 1914–1916, **12,** 158). Partial methylation with dimethyl sulphate and alkali gives 3-*hydroxy-5-methoxybenzoic acid*, m.p. 203° (*E. Spath* and *K. Kromp*, Ber., 1941, **74,** 1424). 3,5-*Dimethoxybenzoic acid*, m.p. 186°. *Methyl* 3,5-*dihydroxybenzoate*, m.p. 165° (*J. H. Birkinshaw* and *A. Bracken*, J. chem. Soc., 1942, 368).

A number of derivatives of β- and γ-resorcylic acid have been isolated from the hydrolysis of depsides which occur in lichens. Only the acids derived from orcinol are described here, for others see p. 207.

Orsellinic acid, 4,6-*dihydroxy-2-methylbenzoic acid*, m.p. 176° (decomp.), p*K*a 3.90, is obtained by oxidation of orcylaldehyde (*K. Hoesch*, Ber., 1913, **46,** 886) and by condensing ethyl crotonate with ethyl acetoacetate in the presence of alkali, dehydrogenating the resulting compound with bromine, hydrolysing the ester function under acid conditions and finally removing bromine substituents by hydrogenolysis (*A. Sonn*, *ibid.*, 1928, **61,** 926; *R. A. Kloss* and *D. A. Clayton*, J. org. Chem., 1965, **30,** 3566):

$CH_3\cdot CO\cdot CH_2\cdot CO_2Et$ + $CH_3\cdot CH{=}CH\cdot CO_2Et$ $\xrightarrow{NaOEt}$ HO–C(=CH–CO)–CH_2–CHMe–$CH\cdot CO_2Et$ $\xrightarrow{Br_2}$ (Br, Me, CO_2Et, OH, Br, HO benzene) $\xrightarrow{H_2SO_4}$ (Br, Me, CO_2H, OH, Br, HO benzene) $\xrightarrow{H_2/Pd}$ (Me, CO_2H, OH, HO benzene)

Methyl orsellinate, m.p. 142° (*A. Robertson* and *R. J. Stephenson*, J. chem. Soc., 1932, 1388). *Ethyl orsellinate*, m.p. 132° (*Sonn, loc. cit.*) decarboxylates on attempted alkaline hydrolysis (*Kloss* and *Clayton, loc. cit.*).

Everinic acid, 6-*hydroxy*-4-*methoxy*-2-*methylbenzoic acid*, m.p. 171°, is obtained by oxidation of the corresponding benzaldehyde, protecting the phenolic group as its acetate (*Robertson* and *Stephenson, loc. cit.*). Its *methyl ester, sparassol*, m.p. 67°, is isolated from the fungus *Sparassis racemosa* Schaff. (*E. Wedekind* and *K. Fleischer*, Ber., 1924, **57,** 1121). 6-*Hydroxy*-4-*methoxycarbonylbenzoic acid*, m.p. 153°, is obtained from orsellinic acid and methoxycarbonyl chloride (*E. Fischer* and *Hoesch*, Ann., 1912, **391,** 364). Methylation of this gives the *methyl ether, methyl ester*, m.p. 87° (*F. W. Canter et al.*, J. chem. Soc., 1933, 493) from which **isoeverinic acid,** 4-*hydroxy*-6-*methoxy*-2-*methylbenzoic acid*, m.p. 175°, is obtained on hydrolysis (*Y. Asahina* and *F. Fujikawa*, Ber., 1932, **65,** 582).

Paraorsellinic acid, 2,6-*dihydroxy*-4-*methylbenzoic acid*, m.p. 178° (decomp.), is obtained by the Kolbe reaction on orcinol (*A. Bistrzycki* and *S. V. Kostanecki*, Ber., 1885, **18,** 1986; *Robertson* and *R. Robinson*, J. chem. Soc., 1927, 2199) or by oxidation of the corresponding aldehyde (*A. St. Pfau*, Helv., 1926, **9,** 666). p*K*a 1.39. The acid decarboxylates on boiling with water. 2-*Hydroxy*-6-*methoxy*-4-*methylbenzoic acid*, m.p. 176° and 2,6-*dimethoxy*-4-*methylbenzoic acid*, m.p. 178°, are obtained by methylation with diazomethane and hydrolysis of the corresponding esters (*J. Herzig et al.*, Monatsh., 1903, **24,** 896).

(iii) Derivatives of hydroquinone

Gentisic acid, 2,5-*dihydroxybenzoic acid*, m.p. 204°, has been isolated from various *Penicillium* species (*H. Raistrick* and *P. Simonart*, Biochem. J., 1933, **27;** 628; *J. Barta* and *R. Mecir*, Experientia, 1948, **4,** 277). Aspirin is metabolised in humans and secreted in the urine as gentisic acid (*D. R. Boreham* and *B. K. Martin*, Brit. J. Pharmacol., 1969, **37,** 294). It is prepared from hydroquinone by the Kolbe reaction (*K. Brunner*, Ann., 1907, **351,** 321; *M. L. Clemens*, U.S.P. 2,816,137/1957; C.A., 1958, **52,** 6404), or from salicylic acid by oxidation with potassium persulphate (*F. Mauthner*, J. pr. Chem., 1940, [ii], **156,** 1[illegible]0). It gives a deep blue colour with ferric chloride. *Methyl ester*, m.p. 88°; the benzyl ester occurs as the 2-*β*-D-glucoside, *trichocarpin*, in the bark of *Populus trichocarpa* Torr. and Gray (*T. K. Ester* and *I. A. Pearl*, Tappi, 1967, **50,** 318).

2-*Hydroxy*-5-*methoxybenzoic acid*, m.p. 143°, is obtained by partial methylation of gentisic acid (*S. v. Kostanecki* and *J. Tambor*, Monatsh., 1895, **16,** 920; *C. Graebe* and *E. Martz*, Ann., 1905, **340,** 215) or from 4-methoxyphenol by the Kolbe reaction (*G. Korner* and *G. Bertoni*, Ber., 1881, **14,** 848); *methyl ester*, b.p. 235°, occurs as the primaveroside, *primulaverin*, m.p. 163°, $[\alpha]_D$ −66.6° (water), in the roots of *Primula officinalis* Jacq. (*A. Goris*, |. *Mascre* and *Ch. Vischniac*, Bull. Sci. pharmacol, 1912, **19,** 593). The glycoside has been synthesised (*Robertson et al.*, J. chem. Soc., 1948, 2220).

Treatment of gentisic acid with one mole of methoxycarbonyl chloride and alkali gives 2-*hydroxy*-5-*methoxycarbonylbenzoic acid*, m.p. 171° (*Fischer*, Ber., 1909, **42,** 222). Methylation of this with diazomethane followed by hydrolysis gives 5-*hydroxy*-2-*methoxybenzoic acid*, m.p. 155° (*Fischer* and *O. Pfeffer*, Ann., 1912, **389,** 204).

2,5-*Dimethoxybenzoic acid*, m.p. 76°, is obtained by oxidation of the corresponding aldehyde (*F. Tiemann* and *W. Muller*, Ber., 1881, **14,** 1993) or from 2-iodo-1,4-dimethoxybenzene by a Grignard reaction. The *amide*, m.p. 140°, is formed by the Friedel–Crafts reaction of 1,4-dimethoxybenzene with carbamoyl chloride (*H. Kauffmann*, Ann., 1905, **344,** 73).

(*f*) *Trihydroxybenzoic acids*

(*i*) *Derivatives of pyrogallol*

The most important trihydroxybenzoic acid is gallic acid, the 3,4,5-derivative which is readily available from plant sources. Carbonation of pyrogallol is achieved by heating the phenol with potassium bicarbonate and yields 2,3,4-trihydroxybenzoic acid (*W. Baker* and *H. A. Smith*, J. chem. Soc., 1931, 2544; *R. D. Haworth et al.*, *ibid*., 1951, 1323).

2,3,4-**Trihydroxybenzoic acid,** *pyrogallolcarboxylic acid*, m.p. 206° to 220° (decomp.) depending on the rate of heating, is obtained from pyrogallol by the Kolbe reaction. It decarboxylates on heating with water to 60° or with aniline at 40°. *Methyl ester*, m.p. 153° (anhydr.) (*A. R. Penfold*, *G. R. Ramage* and *J. L. Simonsen*, J. chem. Soc., 1938, 756); *triacetyl derivative*, m.p. 164° (*E. Pacsu*, Ber., 1923, **56,** 418).

2,3-*Dihydroxy*-4-*methoxybenzoic acid*, m.p. 222°, and 2-*hydroxy*-3,4-*dimethoxybenzoic acid*, m.p. 170°, are obtained by partial methylation of the methyl ester followed by saponification (*Penfold*, *Ramage* and *Simonsen*, *loc. cit*.); 2-*benzyloxy*-3,4-*dimethoxybenzoic acid*, m.p. 95° (*G. Fodor et al.*, J. Amer. chem. Soc., 1949, **71,** 3694), 4-*benzyloxy*-2,3-*dimethoxybenzoic acid*, is an oil (*W. Augstein* and *C. K. Bradsher*, J. org. Chem., 1969, **34,** 1349).

(*ii*) *Gallic acid and its derivatives*

Gallic acid, 3,4,5-*trihydroxybenzoic acid*, m.p. 239° (anhydr.), crystallises from water as the monohydrate, dehydrated at 120°. It was isolated by *C. Scheele* (Crell's Ann., 1787, I, 3) from fermented oak galls and is of wide occurrence in plants both in the free state and as esters with glucose, the hydrolysable tannins which will be discussed on p. 227. It is obtained technically from gall nuts, the tissue formed by species of *Quercus* L. and *Rhus semialata* Murr. around larvae of the gall wasp. The technical production has been reviewed (Science Library Bibliographical Series No. 678, Science Museum, London, 1949). It has been prepared by fusing 4-bromo-3,5- or 5-bromo-3,4-dihydroxybenzoic acid with alkali.

Properties and uses. Gallic acid crystallises from water in silky needles. It has an astringent taste. It is a powerful reducing agent, reducing gold and silver salts and Fehling's solution. It gives a dark blue precipitate with ferric chloride. Gallic acid is decarboxylated by heating with water at 200° or aniline at 115° to give pyrogallol which is manufactured by this route.

Gallic acid is used in the manufacture of blue-black writing inks. These usually contain either tannin or gallic acid, ferrous sulphate, small amounts of mineral acid, gum and a little phenol to prevent growth of moulds. The acids are neutralised by fillers present in the paper and oxidation then takes place with deposition of the blue-black ferric complex of gallic acid. A little blue dyestuff is incorporated in the ink which would otherwise leave practically no mark on the paper initially. Gallic acid and gallamide are used in the manufacture of gallocyanine mordant dyes. Esters of gallic acid, especially *n*-propyl gallate are used as antioxidants in fats and foodstuffs.

Basic bismuth gallate, *dermatol*, structure probably XXVII (*P. Pfeiffer* and *E. Schmitz*, Pharmazie, 1950, **5**, 517), is used as an odourless antiseptic. Bismuth iodosubgallate, *airol*, which is used as an antiseptic dusting powder, is probably a diphenyl derivative since it can be hydrolysed to ellagic acid (*M. Nierenstein* and *C. W. Webster*, Pharm. J., 1945, **154**, 14). Other workers (*S. Takagi* and *Y. Nagase*, J. pharm. Soc., Japan, 1936, **56**, 228) have claimed that the hydrolysis product is gallic acid and that the structure is analogous to that of dermatol with iodine in place of hydroxyl coordinated to bismuth:

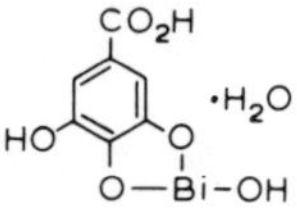

(XXVII)

Oxidation. Oxidation of gallic acid under acid conditions gives diphenyl derivatives (see Table 7). Ellagic acid (XXVIII) and flavellagic acid (XXIX) are obtained by oxida-

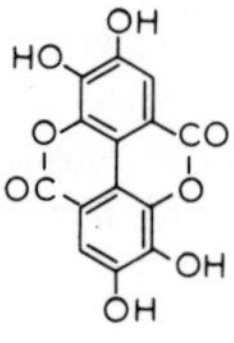

(XXVIII) 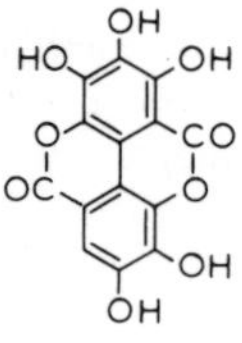 (XXIX) 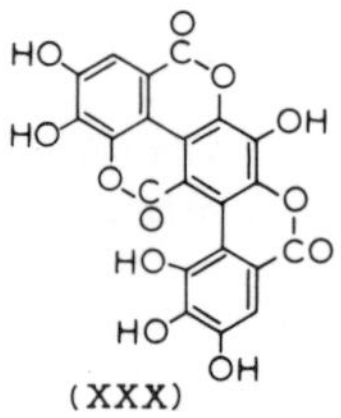 (XXX)

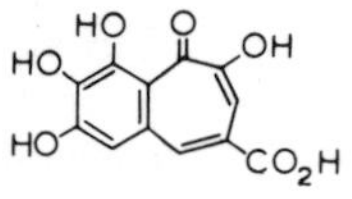

(XXXI) 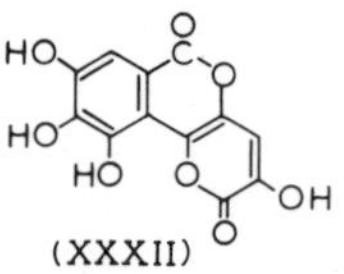 (XXXII)

tion with potassium persulphate and sulphuric acid, depending on the conditions used. Oxidation with arsenic pentoxide gives ellagic acid and flavogallol (XXX). Ellagic acid is also obtained by autoxidation of methyl and ethyl gallate in aqueous ammonia (*J. Herzig, J. Pollak* and *M. v. Bronneck*, Monatsh., 1908, **29,** 278; *D. E. Hathway*, J. chem. Soc., 1957, 519). Purpurogallincarboxylic acid (XXXI) is obtained by electrolytic oxidation in neutral solution or by oxidation of a mixture of pyrogallol and gallic acid (*W. D. Crow* and *Haworth*, J. chem. Soc., 1951, 1325). Autoxidation of gallic acid in aqueous potassium hydroxide followed by acidification of the precipitated potassium salt gives galloflavin (XXXII).

Ellagic acid is an important constituent of hydrolysable tannins and is discussed on p. 237. It gives fluorene on distillation with zinc dust.

Flavellagic acid also gives fluorene on distillation with zinc duct. Its structure follows from reaction of the pentamethyl ether with potassium hydroxide and methyl sulphate to give a heptamethoxydiphenic acid dimethyl ester.

TABLE 7

OXIDATION PRODUCTS OF GALLIC ACID

Substance	*Structure*	*ref.*	*Derivatives*	*Isolation ref.*
Ellagic acid	XXVIII	3	*tetra-acetate*, m.p. 343°; *tetrabenzoate*, m.p. 332°	1,2
Flavellagic acid	XXIX	4,5	*pentamethyl ether*, m.p. 245°; *penta-acetate*, m.p. 317°	4
Flavogallol	XXX	7	*hexa-acetate*, m.p. 278°(dec.); *hexabenzoate*, m.p. 326°	6
Purpurogallin-carboxylic acid	XXXI	9	*trimethyl ether methyl ester*, m.p. 182°; *tetramethyl ether methyl ester*, m.p. 120°	8
Galloflavin	XXXII	11	*tetramethyl ether*, m.p. 235°; *tetra-acetate*, m.p. 232°	10

References
1 *V. Griessmayer*, Ann., 1871, **160,** 55.
2 *A. G. Perkin* and *M. Nierenstein*, J. chem. Soc., 1905, **87,** 1415.
3 *C. Graebe*, Ber., 1903, **36,** 212; *G. Goldschmiedt*, Monatsh., 1905, **26,** 1139; *J. Herzig* and *J. Polak*, *ibid.*, 1908, **29,** 263.
4 *Perkin*, J. chem. Soc., 1906, **89,** 251.
5 *Herzig* and *R. Tscherne*, Ann., 1907, **351,** 34; Monatsh., 1908, **29,** 281.
6 *H. Bleuler* and *Perkin*, J. chem. Soc., 1916, **109,** 529.
7 *J. Grimshaw* and *R. D. Haworth*, *ibid.*, 1956, 4225.
8 *Perkin* and *A. B. Steven*, *ibid.*, 1903, **83,** 199; *A. G. Perkin* and *F. M. Perkin*, 1908, **93,** 1186.
9 *W. D. Crow* and *Haworth*, *ibid.*, 1951, 1325.
10 *R. Bohn* and *C. Graebe*, Ber., 1887, **20,** 2328; *Herzig* and *Tscherne*, Monatsh., 1904, **25,** 607.
11 *Haworth* and *J. M. McLachlan*, J. chem. Soc., 1952, 1583; *Grimshaw* and *Haworth*, *ibid.*, 1956, 418.

Flavogallol gives an oxonium sulphate XXXIII when crystallised from sulphuric acid. Treatment of the hexamethyl ether with methyl sulphate and excess alkali gives the diacid XXXIV which exists in meso and (±)-forms due to restricted rotation about the diphenyl links. Because of this steric hindrance the remaining ester function is not hydrolysed by alkali. The diacid has been synthesised by a series of Ullmann reactions and demethylated to give flavogallol:

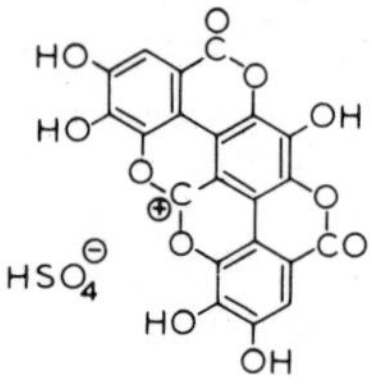

(XXXIII) 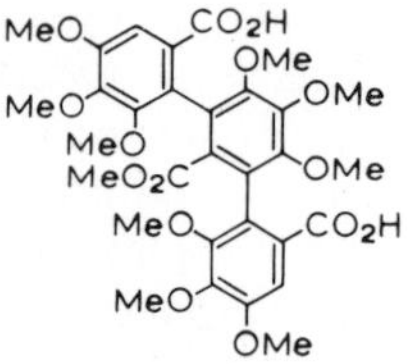 (XXXIV)

Purpurogallincarboxylic acid can be decarboxylated to purpurogallin (CCC, 2nd Edn., Vol. III A, p. 410). Further oxidation of the acid with hydrogen peroxide gives a tropolone derivative XXXV which can be converted to uvitic acid (XXXVI) thus establishing the structure:

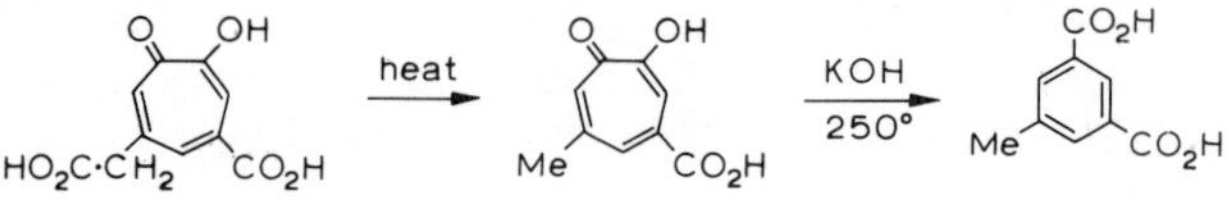

(XXXV) (XXXVI)

Galloflavin is converted to isogalloflavin (XXXVII) on warming with alkali followed by acidification. Tri-*O*-methylisogalloflavin is decarboxylated and the product reduced catalytically to the alcohol XXXVIII. Oxidation of this with chromic acid yields the corresponding acid which has been synthesised:

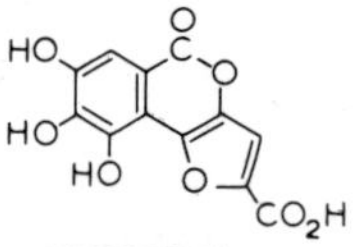

(XXXVII) 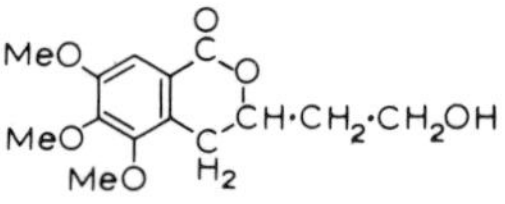 (XXXVIII)

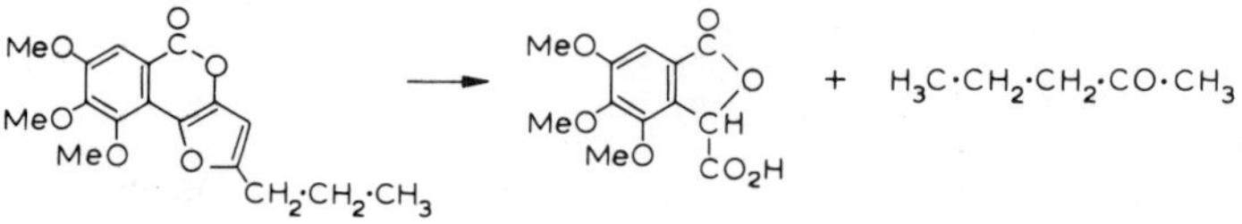

(XXXIX) (XL)

Tri-*O*-methylisogalloflavin acid chloride was reacted with diethyl cadmium and the resulting ketone reduced to XXXIX. On treatment with alkali this underwent rupture of the furan ring to give XL and pentan-2-one thus establishing the position of the side-chain. Thus isogalloflavin is shown to have structure XXXVII and galloflavin is XXXII.

Reduction. Catalytic reduction of gallic acid gives *dihydrogallic acid* (XLI), m.p. 192°, and further hydrogenation of this gives cis-3,4,5-*trihydroxycyclohexane*-1-*carboxylic acid*, m.p. 198°, which is not identical with dihydroshikimic acid or its epimer (*W. Mayer et al.*, Ber., 1955, **88**, 316; 1959, **92**, 213):

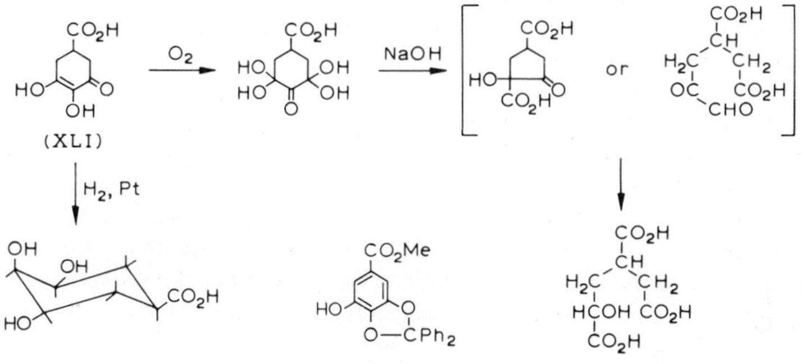

(XLI) (XLIII) (XLII)

Autoxidation of dihydrogallic acid gives 3,4,5-*tri-oxocyclohexanecarboxylic acid*, isolated as the *dihydrate*, m.p. 138°. This is isomerised to gallic acid on warming with aqueous sodium acetate. With sodium hydroxide ring opening occurs to give 1-hydroxybutane-1,3,4-tricarboxylic acid XLII.

Derivatives of gallic acid

Methyl gallate, m.p. 201°; *ethyl gallate*, m.p. 158°; *propyl gallate*, m.p. 147°. Methyl gallate and dichlorodiphenylmethane afford *methyl* 5-*hydroxy*-3,4-*diphenylmethylenedioxybenzoate* (XLIII), m.p. 165° (*R. Robinson et al.*, J. chem. Soc., 1930, **132,** 812).

4,5-*Dihydroxy*-3-*methoxybenzoic acid*, m.p. 220°, is obtained from XLIII by methylation and hydrolysis (*Robinson et al., loc. cit.*). *Methyl* 3,5-*dihydroxy*-4-*methoxybenzoate*, m.p. 136°, is obtained by methylation of methyl gallate with one mole of methyl sulphate (*C. Schopf* and *L. Winterhalder*, Ann., 1940, **544,** 74). 5-*Hydroxy*-3,4-*dimethoxybenzoic acid*, m.p. 193°, is obtained by heating 5-bromo-3,4-dimethoxybenzoic acid with sodium hydroxide and copper catalyst (*R. L. Shriner* and *P. McCutchan*, J. Amer. chem. Soc., 1929, **51,** 2195).

Syringic acid, 4-*hydroxy*-3,5-*dimethoxybenzoic acid*, m.p. 204°, is obtained by degradation of many natural products and occurs in the bark of *Robinia pseudacacia* L. and in cascara sagrada (bark of *Rhamnus purshiana* DC) extract. It is prepared by partial demethylation of tri-*O*-methylgallic acid (*W. Bradley* and *Robinson*, J. chem. Soc., 1928, **130,** 1553). The *β*-D-glucoside, *glucosyringic acid*, m.p. 225°, $[\alpha]_D$ −18° (aqueous solution of sodium salt) is obtained by the oxidation of syringin with potassium permanganate and is a constituent of many woods (*I. A. Pearl et al.*, J. org. Chem., 1959, **24,** 443). 4-*Benzyloxy*-3,5-*dimethoxybenzoic acid*, m.p. 157° (*Bradley* and *Robinson, loc. cit.*).

3,4,5-**Trimethoxybenzoic acid,** m.p. 168° (*F. Mauthner*, Org. Synth., 1932, Coll. Vol. I, p. 522); *methyl ester*, m.p. 82°; *chloride*, m.p. 77°, prepared by the action of thionyl chloride (*J. T. Marsh* and *H. Stephen*, J. chem. Soc., 1925, **127,** 1635); *anhydride*, m.p. 160°; *amide*, m.p. 244°. 3,4,5-*Triethoxybenzoic acid*, m.p. 112°; 3,4,5-*tribenzyloxybenzoic acid*, m.p. 196° (*R. O. Clinton* and *T. A. Geissman*, J. Amer. chem. Soc., 1943, **65,** 85).

Mono-4-acyl and -aroyl, derivatives of gallic acid do not exist. Attempts to prepare them in aqueous solution lead to migration of the group to the 3-position.

3,4,5-**Triacetoxybenzoic acid,** m.p. 172°, on careful alkaline hydrolysis gives 4-*hydroxy-3,5-diacetoxybenzoic acid*, m.p. 174° (*E. Fischer et al.*, Ber., 1918, **51,** 45). Benzoylation of this gives the 4-*benzoyl* derivative, m.p. 183°, which on hydrolysis with ammonia, sodium acetate or dilute hydrochloric acid gives 3-*benzoyloxy*-4,5-*dihydroxybenzoic acid*, m.p. 240° (decomp.) (*Fischer et al., loc. cit.*). 3,4,5-*Triacetoxybenzoyl chloride*, m.p. 106° (*A. Russell* and *W. G. Tebbens*, J. Amer. chem. Soc., 1942, **64,** 2275).

3,4,5-*Tris(methoxycarbonyloxy)benzoic acid*, m.p. 137° (*Fischer*, Ber., 1908, **41,** 2882); reaction between gallic acid and one mole of methoxycarbonyl chloride gives 4,5-*dihydroxy-3-methoxycarbonyloxybenzoic acid*, m.p. 209° (decomp.) (*Fischer* and *K. Freudenberg*, Ber., 1912, **45,** 2716). 3,4,5-*Tris(methoxycarbonyloxy)benzoyl chloride*, m.p. 86°.

(iii) Derivatives of 1,2,4-trihydroxybenzene

1,2,4-Trihydroxybenzene is carbonated by heating with aqueous sodium hydrogen carbonate under one atmosphere of carbon dioxide (*J. Thiele* and *K. Jaeger*, Ber., 1901, **34,** 2840; *F. Wessely et al.*, Monatsh., 1950, **81,** 1071) to give 2,4,5-trihydroxybenzoic acid.

2,4,5-**Trihydroxybenzoic acid,** m.p. 217°. **Asaronic acid,** 2,4,5-*trimethoxybenzoic acid*, m.p. 144°; 2,5-*dihydroxy-4-methoxybenzoic acid*, m.p. 201° (decomp.) is obtained by oxidation of 2-hydroxy-4-methoxybenzoic acid with persulphate (*T. R. Sheshadri et al.*, Proc. Indian Acad. Sci., 1949, **30A,** 265).

Rissic acid (XLIV), m.p. 262° (decomp.) and **derric acid** (XLV), m.p. 171°, are decomposition products of rotenone (*F. B. LaForge*, J. Amer. chem. Soc., 1931, **53,** 3896):

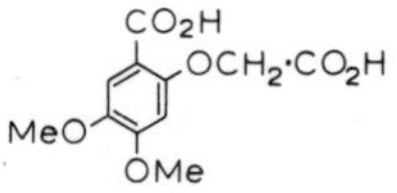

(XLIV) 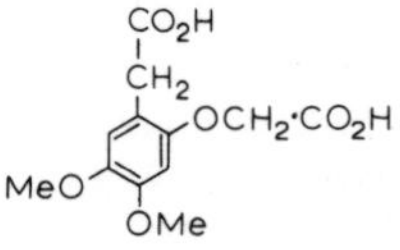 (XLV)

2,3,5-**Trihydroxybenzoic acid,** m.p. 223° (decomp.), is obtained by the oxidation of 3,5-dihydroxybenzoic acid (*A. R. Todd et al.*, J. chem. Soc., 1950, 5) or salicylic acid with persulphate (*R. U. Schock* and *D. L. Tabern*, U.S.P. 2,641,609/1953; C.A., 1954, **48,** 6467). 2,3,5-*Trimethoxybenzoic acid*, m.p. 100°, is prepared by carbonation of the lithium derivative of 2,3,5-trimethoxybromobenzene (*A. Kreuchunas*, J. org. Chem., 1956, **21,** 910).

2,3,6-**Trihydroxybenzoic acid.** Derivatives of this acid are obtained from the Thiele acetylation of methyl 1,4-benzoquinone-2-carboxylate (*H. S. Wilgus* and *J. W. Gates*,

Canad. J. Chem, 1967, 1975). The *methyl ester*, m.p. 138°, is obtained from methyl furoate by the following sequence of reactions (*N. Clauson-Kaas* and *P. Nedenskov*, Acta chem. Scand., 1955, **9,** 27; *M. Murakami* and *J. C. Chen*, Bull. chem. Soc., Japan, 1963, **36,** 263):

CH_3OH, Br_2 → MeO OMe CO_2Me → (Na,MeOAc) MeO OMe $CO{\cdot}CH_2{\cdot}CO_2Me$ → (HCl) HO, OH, OH, CO_2Me

2,3,6-*Trimethoxybenzoic acid*, m.p. 149°, is prepared by metalation of 1,2,4-trimethoxybenzene with butyl-lithium and then carbonation (*H. Gilman* and *J. R. Thirtle*, J. Amer. chem. Soc., 1944, **66,** 858).

(iv) Derivatives of phloroglucinol

2,4,6-**Trihydroxybenzoic acid,** *phloroglucinolcarboxylic acid*, decomposes into carbon dioxide and phloroglucinol at about 100°. It is prepared by heating phloroglucinol with aqueous potassium hydrogen carbonate (*D. E. White et al.*, J. chem. Soc., 1950, 2811). Due to the steric hindrance which results from 2,6-disubstitution the methyl ester cannot be prepared by acid-catalysed esterification. *Methyl ester*, m.p. 174°, by reaction with diazomethane, methyl iodide on the silver or potassium salt (*Herzig et al.*, Monatsh., 1903, **22,** 219; 1904, **23,** 86; *F. H. Curd* and *A. Robertson*, J. chem. Soc., 1933, 442). 2,4,6-*Trimethoxybenzoic acid*, m.p. 144° (decomp.), prepared by oxidation of the corresponding aldehyde, is also esterified by the above methods (*Herzig et al.*, Monatsh., 1904, **23,** 92); *methyl ester*, m.p. 66°.

3-*Chloro*-2-*hydroxy*-4,6-*dimethoxybenzoic acid*, m.p. 220° (decomp.), is a degradation product of griseofulvin (*J. MacMillan*, J. chem. Soc., 1959, 1823).

6. Phenolic monocarboxylic acids: mono-, di- and tri-hydroxyphenylacetic and homologous acids

(a) Hydroxyphenylacetic acids

(i) Preparation

(*1*) All three isomeric hydroxyphenylacetic acids have been obtained by the Willgerodt reaction by heating the corresponding hydroxyacetophenones with sulphur and morpholine and hydrolysis of the resultant phenylthioacetmorpholides (*J. A. King* and *F. H. MacMillan*, J. Amer. chem. Soc., 1946, **68,** 2335; *A. C. Ott et al., ibid.*, 1946, **68,** 2633; *E. Schwenk* and *P. Papa*, J. org. Chem., 1946, **11,** 798). The method is useful for the preparation of methoxyphenylacetic acids and has been reviewed (*M. Carmack* and *M. A. Spielman*, Org. Reactions, 1946, **3,** 83).

(*2*) The methoxyphenylacetic acids are obtained from the corresponding benzaldehyde by the azlactone process. The aldehyde is condensed with hippuric acid to give an azlactone which on hydrolysis affords benzoic acid and an arylpyruvic acid. The latter is separated as its soluble bisulphite compound and then oxidised to the arylacetic acid by alkaline hydrogen peroxide (*R. D. Haworth, W. H. Perkin* and *J. Rankin*, J. chem. Soc., 1924, **125,** 1686; *J. C. Cain, J. L. Simonsen* and *C. Smith*., *ibid*., 1913, **103,** 1035):

$$MeO\text{-}C_6H_4\text{-}CHO + CH_2{\cdot}CO_2H(NHCOPh) \longrightarrow MeO\text{-}C_6H_4\text{-}CH{=}C\text{—}CO\ (N{=}C(Ph)\text{-}O)$$

$$\xrightarrow{\text{base}} MeO\text{-}C_6H_4\text{-}CH_2{\cdot}CO{\cdot}CO_2H \xrightarrow{H_2O_2} MeO\text{-}C_6H_4\text{-}CH_2{\cdot}CO_2H$$

(*3*) Reduction of a benzaldehyde cyanhydrin with hydriodic acid (*S. Czaplicki, S. v. Kostanecki* and *V. Lampe*, Ber., 1909, **42,** 828; *T. S. Stevens*, J. chem. Soc., 1927, 181; *J. Levine et al.*, J. Amer. chem. Soc., 1948, **70,** 1930).

(*4*) Oxidation of acetophenones by thallium(III) nitrate in methanol gives the methyl ester of phenylacetic acid (*A. McKillop et al., ibid.*, 1971, **93,** 4919).

(ii) Individual compounds

2-*Hydroxyphenylacetic acid*, m.p. 147° (*Levine et al., loc. cit.*). When heated it gives the lactone coumaranone and the lactone ring can be reopened with alkali. 2-*Methoxyphenylacetic acid*, m.p. 123° (*F. Mauthner*, Ann., 1909, **370,** 374).

3-*Hydroxyphenylacetic acid*, m.p. 129°, results from the reduction of homopiperonylic acid with Raney nickel–aluminium alloy and sodium hydroxide (*Schwenk* and *Papa*, J. org. Chem., 1945, **10,** 232). 3-*Methoxybenzoic acid*, m.p. 67°.

4-*Hydroxyphenylacetic acid*, m.p. 153°, is found in the urine as a metabolic product of tyrosine. The 4-hydroxyphenylacetamide of 6-aminopenicillanic acid is penicillin-X produced by a mutant of *Penicillium chrysogenum* particularly when fed 4-hydroxyphenylacetic acid. 4-*Methoxyphenylacetic acid*, m.p. 86°.

3,4-*Dihydroxyphenylacetic acid*, m.p. 127°, has been prepared from veratraldehyde cyanhydrin by refluxing with hydriodic acid (*A. Pictet* and *A. Garns*, Ber., 1909, **42,** 2949). 3-*Hydroxy*-4-*methoxy*-, m.p. 125°, 4-*hydroxy*-3-*methoxy*-, m.p. 142°, 3,4-*dimethoxy*-, m.p. 99° (anhydr.) and 3,4-*methylenedioxy-phenylacetic acid*, m.p. 127° are all prepared from the corresponding benzaldehyde by the azlactone synthesis (*E. Spath* and *N. Lang*, Monatsh., 1921, **42,** 280; *Mauthner, loc. cit.; H. R. Snyder et al.*, Org. Synth., 1943, Coll. Vol. II, p. 333).

2,5-*Dihydroxyphenylacetic acid, homogentisic acid*, m.p. 149°, is found in the urine of patients suffering from alkaptonuria. It is prepared from the benzoate of hydroquinone monoallyl ether, which undergoes the Claisen rearrangement to allylhydroquinone monobenzoate; conversion to the dibenzoate and ozonolysis yields the dibenzoate of homogentisic acid (*G. Hahn* and *W. Stenner*, Z. physiol. Chem., 1907, **52,** 375). It is also obtained by hydrolysis of the adduct from ketene diethylacetal and benzoquinone (*S. M. McElvain* and *H. Cohen*, J. Amer. chem. Soc., 1942, **64,** 260).

3,5-*Dihydroxyphenylacetic acid*, m.p. 127° (anhydr.). The triethyl ester (I) is produced by self-condensation of ethyl acetonedicarboxylate in presence of sodium (*H. von Pechmann* and *L. Wolman*, Ber., 1898, **31,** 2014):

$CH_2{\cdot}CO_2Et$ / CO / EtO_2C-CH_2 ; $CH_2{\cdot}CO_2Et$ / CO / EtO_2C / CH_2 → EtO_2C, $CH_2{\cdot}CO_2Et$, CO_2Et, HO, OH (I) → $CH_2{\cdot}CO_2H$, HO, OH

On fusion with alkali it gives 3,5-dihydroxyphenylacetic acid.

(b) Hydroxyphenylpropionic acids

(i) Preparation

(*1*) 3-Phenylpropionic acid derivatives are prepared by reduction of the corresponding cinnamic acid usually catalytically or with sodium amalgam.

(*2*) Oxidation of 1-phenylpropyne with thallium(III) nitrate in methanol gives the methyl ester of the 2-phenylpropionic acid (*A. McKillop et al.*, J. Amer. chem. Soc., 1971, **93,** 7331).

(ii) Individual acids

3-(2-*Hydroxyphenyl)propionic acid, melilotic acid*, m.p. 82°, occurs in *Melilotus officinalis* Lam. When heated or treated with thionyl chloride or hydrobromic acid it gives the lactone, dihydrocoumarin (*G. Lasch*, Monatsh., 1913, **34,** 1653). It is made by hydrogenation of coumarin and alkaline hydrolysis of the lactone. *Methyl ether*, m.p. 87°.

3-(3-*Hydroxyphenyl)propionic acid*, m.p. 111°; *methyl ether*, m.p. 51° (*E. Tiemann* and *R. Ludwig*, Ber., 1882, **15,** 2050; *K. Brand* and *O. Horn*, J. pr. Chem., 1927, [ii], **115,** 374).

3-(4-*Hydroxyphenyl)propionic acid, phloretic acid*, m.p. 129°, occurs in urine as a metabolic product of tyrosine. *Methyl ether*, m.p. 101° (*J. M. Galland et al.*, J. chem. Soc., 1929, 1448).

3-(3,4-*Dihydroxyphenyl)propionic acid, hydrocaffeic acid*, m.p. 139°, the *dimethyl ether*, m.p. 98° (*C. Schopf, H. Perrey* and *I. Jackh*, Ann., 1932, **497,** 52) is obtained by hydrogenation of the cinnamic acid sodium salt over palladium–calcium carbonate. 3-(3-*Hydroxy-4-methoxyphenyl*)-, m.p. 164°, and 3-(4-*hydroxy-3-methoxyphenyl)propionic acid*, m.p. 89°.

3-(2,4-*Dihydroxyphenyl)propionic acid*, m.p. 164° (decomp.) is prepared by reaction of resorcinol with 3-ethoxypropionitrile in the presence of hydrogen chloride and zinc chloride and hydrolysis of the product (*C. D. Hurd* and *G. W. Fowler*, J. Amer. chem. Soc., 1939, **61,** 253).

(c) Hydroxyphenylbutyric and higher acids

(i) Preparation

(*1*) Phenylbutyric acids can be prepared by Clemmensen reduction of the oxo acids obtained by condensing a benzene derivative with succinic anhydride (*R. D. Haworth* and *G. Sheldrick*, J. chem. Soc., 1934, 1950; *Haworth* and *C. R. Mavin*, *ibid.*, 1932, 1486). Reduction of CO to CH_2 is also effected catalytically over palladium charcoal (*W. Borsche*, *P. Hofmann* and *H. Kuhn*, Ann., 1943, **554,** 33; *E. C. Horning* and *D. B. Reisner*, J. Amer. chem. Soc., 1949, **71,** 1036):

(Scheme: anisole (OMe) + succinic anhydride ($H_2C{-}CO$, $H_2C{-}CO$, O) $\xrightarrow{AlCl_3}$ p-MeO-C_6H_4-$CO{\cdot}CH_2{\cdot}CH_2{\cdot}CO_2H$ $\xrightarrow{Zn/Hg,\ HCl}$ p-MeO-C_6H_4-$H_2C{\cdot}CH_2{\cdot}CH_2{\cdot}CO_2H$)

(*2*) 5-Phenylpentanoic acids are prepared from the benzene derivative using glutaric anhydride as in (*1*) (*D. G. Thomas* and *A. H. Nathan*, *ibid.*, 1948, **70,** 331; *D. Caunt et al.*, J. chem. Soc., 1950, 1631).

(*3*) 3-Phenylbutyric acids are obtained from acetophenones and ethyl bromoacetate by the Reformatsky reaction to give a crotonic acid which is reduced electrochemically or catalytically (*E. H. Woodruff*, J. Amer. chem. Soc., 1942, **64,** 2859):

(Scheme: 3,4-(MeO)$_2C_6H_3$-$CO{\cdot}CH_3$ + $BrCH_2{\cdot}CO_2Et$ $\xrightarrow{Zn}$ 3,4-(MeO)$_2C_6H_3$-$C(CH_3){=}HC{-}CO_2Et$ $\longrightarrow$ 3,4-(MeO)$_2C_6H_3$-$HC(CH_3){-}H_2C{\cdot}CO_2H$)

(ii) Individual acid

4-(2-*Methoxyphenyl)butyric acid*, m.p. 39°, is obtained from 2-(2-methoxyphenyl)-ethyl bromide and the sodium derivative of diethyl malonate followed by hydrolysis and decarboxylation (*E. Hardegger et al.*, Helv., 1945, **28,** 632).

4-(3-*Methoxyphenyl)butyric acid*, m.p. 49°, is obtained by treating 3-(3-methoxy-phenyl)propylmagnesium iodide with methoxycarbonyl chloride and hydrolysis of the product (*R. Robinson* and *J. Walker*, J. chem. Soc., 1936, 747).

4-(4-*Methoxyphenyl)butyric acid*, m.p. 63° (*Haworth* and *Sheldrick*, *loc. cit.*). This acid is readily cyclised to 7-methoxy-1-oxo-tetrahydronaphthalene.

5-(3,4-*Dimethoxyphenyl)pentanoic acid*, b.p. 193°/0.01 mm, is cyclised to the benzcycloheptеneone (*Caunt et al.*, *loc. cit.*).

Chapter 16

Depsides, Hydrolysable Tannins, Lignans, Lignin and Humic Acid

J. GRIMSHAW

Two pathways are followed in nature for the formation of phenolic compounds. These have already been discussed (see C.C.C., 2nd Edn., Vol. III A, pp. 135–137). Depsides afford many examples of phenolcarboxylic acids which are elaborated by the head to tail combination of acetate units or of acetyl-coenzyme-A and malonyl-coenzyme-A units. The hydrolysable tannins, lignin and lignans are examples of phenolic substances elaborated by the shikimic acid pathway.

1. Depsides, depsidones and depsones

(a) Occurrence and general structure

Depsides are substances derived from two or more molecules of phenolcarboxylic acids by esterification of the carboxyl group of one molecule (termed the S-component) with the hydroxyl group of a second (the A-component). The name depside was introduced by *E. Fischer* and is derived from the Greek word δέψειν, to tan. The first typical depside to be thoroughly investigated was digallic acid, which occurs as a component of gallotannin and is biosynthesised by the shikimic acid pathway. Most natural depsides occur in lichens and are biosynthesised by the acetate pathway. The term depside has become restricted to the latter group of compounds. In this chapter digallic acid will be discussed along with the tannins.

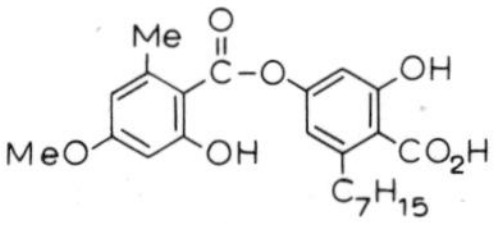

Sphaerophorin
(I)

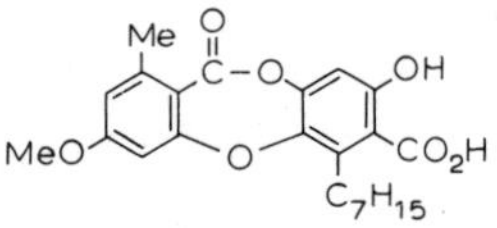

Grayanic acid
(II)

Sphaerophorin (I) is an example of a typical depside isolated from *Sphaerophorus melanocarpus* DC (*Y. Asahina* and *A. Hashimoto*, Ber., 1934, **67,** 416). Depsidones, of which grayanic acid (II) is a typical example, also occur in lichens and can be envisaged as being biosynthesised from the corresponding depside by an oxidative cyclisation to form the diphenyl ether link. Grayanic acid is isolated from the lichen *Cladonia grayi* Merrill (*S. Shibata* and *H.-H. Chiang*, Chem. Pharm. Bull., Tokyo, 1963, **11,** 926; C.A., 1963, **59,** 11318). The range of phenolic acid units which occur in depsides and depsidones is given in Table 1. Some of the most common naturally occurring substances are lecanoric acid (III), gyrophoric acid (IV), barbatic acid (V), atranorin (VI) and psoromic acid (VII):

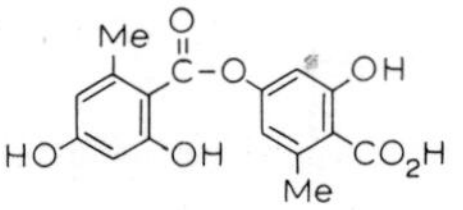

Lecanoric acid
(III)

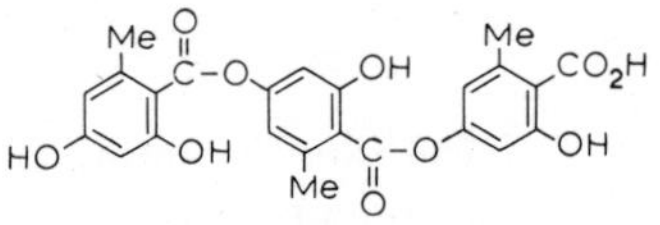

Gyrophoric acid
(IV)

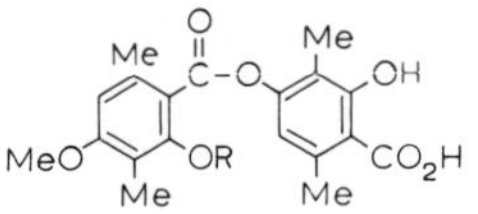

Barbatic acid (V, R=H)
Diffractaic acid (V, R=Me)

Atranorin (VI, R=H)
Chloroatranorin (VI, R=Cl)

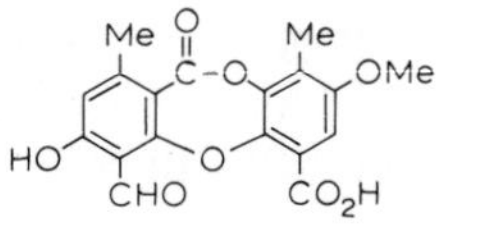

Psoromic acid
(VII)

Lichens are symbiotic associations of algae and fungi (*K. Mosbach*, Angew. Chem. intern. Edn., 1969, **8,** 240). The physiology of the isolated fungal and algal symbionts is discussed by *V. Ahmadjian* ("The Lichen Symbiosis", Baisdell, London, 1967). The taxonomy of many lichen species is rather difficult because only slight differences exist between species, and because the chemical constituents of species often vary more markedly the possibilities of chemotaxonomy are very considerable. Monographs on this subject are available and are reviewed particularly in "The Bryologist". Chemotaxonomy aspects are covered by *C. F. Culberson* ("Chemical and Botanical Guide to Lichen Products", University of N. Carolina Press, Chapel Hill, 1969; supplemented in Bryologist, 1970, **73,** 177). Other books and review articles give more weight to the chemistry of depsides (*Asahina* and *Shibata*,

"Chemistry of Lichen Substances", Asher, Amsterdam, 1971, reprint of 1954 edition; *Shibata* in "Moderne Methoden der Pflanzenanalyse", Eds. *H. F. Linskens* and *M. V. Tracey*, Springer, Berlin, 1963, Vol. 6, p. 155; *S. Huneck* in "Progress in Phytochemistry", Ed. *L. Reinhold* and *Y. Liwschitz*, Interscience, London, 1968, Vol. 1, p. 223; *Hunek*, Fortschr. org. Naturstoffe, 1971, **29,** 209).

Other phenolic substances, anthraquinones, xanthones and dibenzofurans derived by the acetate pathway occur in lichens as do the pulvinic acid and 2,5-(diphenyl) benzoquinone derivatives which are derived by the shikimic acid pathway.

(b) Isolation

For isolating lichen substances the dried pulverised lichens are extracted with organic solvents (benzene, ether, acetone) and in many cases the depsides and depsidones separate out from the extract. An ethereal extract can be shaken consecutively with sodium hydrogen carbonate, sodium carbonate and sodium hydroxide solutions, to separate the components according to acidity, and the products recovered by acidification can be purified by crystallisation. *Asahina* and *Shibata (loc. cit.)* have collected many examples of this method. The depside link is however readily hydrolysed when an *ortho*-OH group is present. Crude products have been purified by column chromatography (*T. R. Seshadri* and *G. B. V. Subramanian*, J. Indian chem. Soc., 1963, **40,** 7) and by thin-layer chromatography (*O. Bachmann*, Österr. botan. Z., 1963, **110,** 103; *J. Santesson*, Acta chem. Scand., 1967, **21,** 1162; *Huneck, loc. cit.; Culberson* and *H. Kristinsson*, J. Chromat., 1970, **46,** 85). Paper chromatography has also been used for the analysis of mixtures (*C. A. Wachtmeister*, Acta chem. Scand., 1952, **6,** 818; Botan. Notiser, 1956, **109,** 313; *J. L. Ramant*, Bull. Soc. roy. botan. Belg., 1961, **93,** 27; *H. A. Borthwick*, Amer. Naturalist, 1964, **97,** 347; *Shibata, loc. cit.*).

(c) Biosynthesis

The orsellinic acid structural unit is present in so many depsides and this fact was part of the evidence put forward by *A. J. Birch* and *F. W. Donovan* (Austral. J. Chem., 1953, **6,** 361) to support the hypothesis that molecules are elaborated in nature, at least in part, by the head to tail linkage of acetate units. The acids (VIII, R=Pr^n, C_5H_{11}, C_7H_{15}, $CH_2{\cdot}CO{\cdot}C_5H_{11}$) with longer alkyl side chains can also be built up by the head to tail linkage of acetate units.Others such as atraric acid (XII) have an extra C-1 unit at position 5 and in a few examples (X, XI) further hydroxylation of the orcinol group has occurred.

TABLE 1

PHENOLCARBOXYLIC ACIDS FROM THE HYDROLYSIS OF DEPSIDES

Acid	*Structure*	*m.p.*	*Derivatives*	*Ref.*
Orsellinic acid	VIII, R = Me	+ 1.H_2O, 177°	4-OMe, **everinic acid,** m.p. 174° (decomp.); 4,6-*(OMe)*$_2$, m.p. 145° (decomp.)	1
Divaric acid	VIII, R = Pr^n	179° (decomp.)	4-OMe, **divaricatinic acid,** m.p. 157° (decomp.); 4,6-*(OMe)*$_2$, m.p. 64°	2
Olivetolcarboxylic acid	VIII, R = C_5H_{11}	145°	4-*OMe*, m.p. 126°; 4,6-*(OMe)*$_2$, m.p. 52°	3
Sphaerophorol-carboxylic acid	VIII, R = C_7H_{15}	140°		4
Olivetonic acid	VIII, R = $CH_2 \cdot CO$–C_5H_{11} III	159°	4,6-*(OMe)*$_2$, m.p. 93°. **Olivetonide** (IX), m.p. 110°; 8-*OMe*, m.p. 146°; 6-*OMe*, m.p. 57°; 6,8-*(OMe)*$_2$, m.p. 94°	5
2,3,4-Trihydroxy-6-methylbenzoic acid	X	decarboxylates	*Me ester*, m.p. 153°	6
3,4,6-Trihydroxy-2-methylbenzoic acid	XI, R = H	194° (decomp.)	*Me ester*, m.p. 205°	7
Atraric acid	XII	184° (decomp.)	*Me ester*, m.p. 145°; 4-*(OMe)*, **rhizoninic acid,** m.p. 215° (decomp.); 2,4-*(OMe)*$_2$, m.p. 105°	8
Haematommic acid	XIII	173°	*Me ester*, m.p. 147°	9
2,6-Dihydroxy-4-methyl-isophthalic acid	XIV	—	6-*OMe*, m.p. 207° and *Me ester*, m.p. 168°	10
Thamnolcarboxylic acid	XI, R = CHO	decarboxylates	**thamnol,** m.p. 185°; *thamnol anil*, m.p. 128°	11
Barbatolcarboxylic acid	XV	234° (decomp.)	7-*methoxy*-5,6-*dimethylphthalide* (XVI), m.p. 173.5°	12

References

1 See p. 191.
2 *A. Sonn*, Ber., 1928, **61,** 2480.
3 *Y. Asahina* and *M. Yasue*, *ibid.*, 1937, **70,** 206; *idem*, J. pharm. Soc. Japan, 1937, **57,** 553.
4 *A. Hashimoto*, *ibid.*, 1938, **58,** 776; C.A., 1939, **33,** 2504.
5 *Asahina* and *F. Fujikawa*, Ber., 1935, **68,** 2022; *Asahina* and *H. Nogami*, Bull. chem. Soc. Japan, 1942, **17,** 221.
6 *G. Koller* and *H. Hamburg*, Monatsh., 1935, **65,** 367; *Asahina* and *Yasue*, Ber., 1936, **69,** 2327.
7 *T. Kusaka*, J. pharm. Soc. Japan, 1941, **61,** 355.
8 *Fujikawa*, *ibid.*, 1936, **56,** 237; *Asahina* and *Fujikawa*, Ber., 1932, **65,** 175.
9 *A. St. Pfau*, Helv., 1933, **16,** 285; *F. H. Curd*, *A. Robertson* and *R. J. Stephenson*, J. chem. Soc., 1933, 131.
10 *Asahina* and *M. Yanagita*, Ber., 1933, **66,** 393; *Asahina* and *Z. Simosato*, *ibid.*, 1938, **71,** 2561.
11 *Asahina* and *S. Ihara*, *ibid.*, 1929, **62,** 1196.
12 *E. E. Suominen*, Suomen Kem., 1939, **12,** 26; Chem. Zentr., 1940, I, 385.

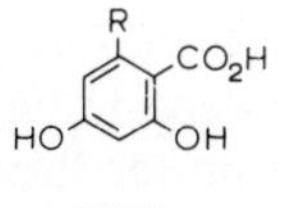

(VIII) 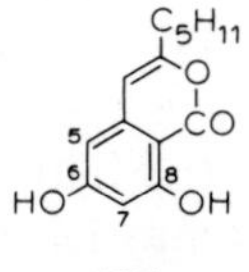 (IX) 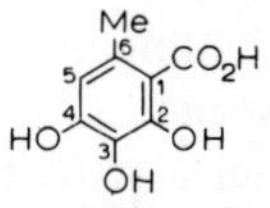 (X)

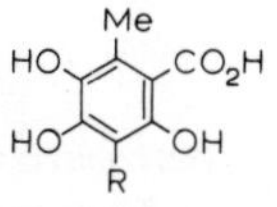

(XI) 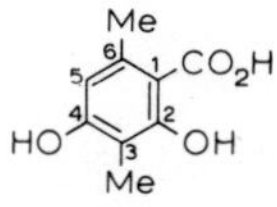 (XII) 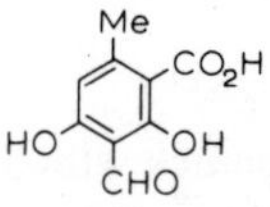 (XIII)

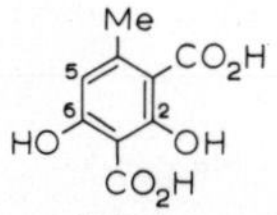

(XIV) 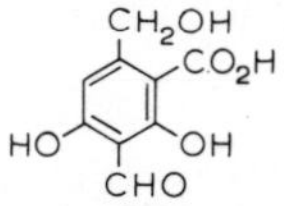 (XV) 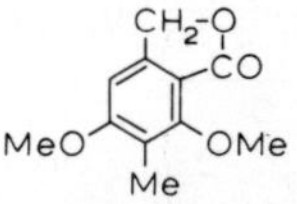 (XVI)

Confirmation of this route for the biosynthesis of some depsides has been obtained by ^{14}C-labelling experiments. *Mosbach* (Acta chem. Scand., 1964, **18,** 329) incubated *Umbilicaria pustulate* (L.) Hoffm. with diethyl [1-^{14}C]malonate and determined the labelling pattern in the gyrophoric acid (IV) so formed. The individual units of orsellinic acid were isolated as their methylation products by hydrolysis of fully *O*-methylated gyrophoric acid and each had the same specific activity. Orsellinic acid, obtained by hydrolysis, was then degraded to determine the ^{14}C-labelling pattern as shown below which indicates the formation from units of acetylcoenzyme-A and malonylcoenzyme-A:

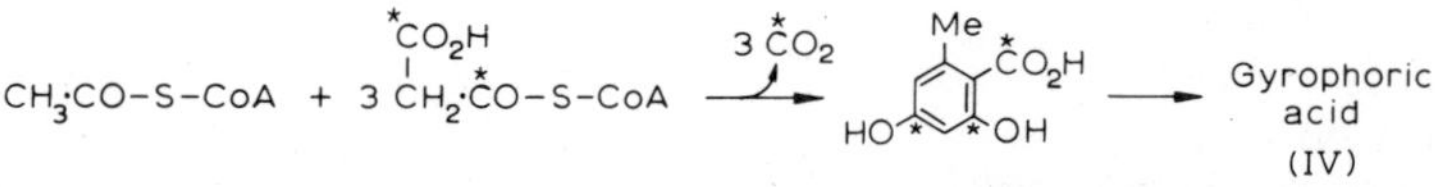

$CH_3CO{-}S{-}CoA + 3\,\overset{*}{C}O_2H{-}CH_2{\cdot}\overset{*}{C}O{-}S{-}CoA \xrightarrow{3\,\overset{*}{C}O_2}$ [orsellinic acid, labelled] $\longrightarrow$ Gyrophoric acid (IV)

Feeding of *Parmelia tinctorum* Despr. with [1-^{14}C]acetate and [^{14}C]formate resulted in incorporation of the labelled acetate into lecanoric acid (III), atranorin (VI, R=H) and chloroatranorin (VI, R=Cl) whereas formate was incorporated only into atranorin and chloroatranorin thus revealing that it participates in forming the extra C-1 groups. The distribution of the label from acetate was in full agreement with the acetate hypothesis (*M. Yamazaki et al.*, Chem. Pharm. Bull., Tokyo, 1965, **13,** 1015). Tritium labelled orsellinic acid was not interconverted with atraric acid (XII) and *vice versa* (*idem, ibid.*, 1966, **14,** 96).

It is the algal partner in the lichen symbiotic association which photosynthesises sugars from carbon dioxide. These are then used by the fungal partner to elaborate the depsides. The two symbionts are very closely bound so that the labelling process from $^{14}CO_2$ is rapid (*C. H. Fox* and *Mosbach*, Acta chem. Scand., 1967, **21,** 2327). Many of the depsides and depsidones have antibacterial activity which may be the reason for their elaboration.

(d) Depsides

(i) Degradation

All the depsides which possess an *ortho*-hydroxybenzoate ester function are hydrolysed by 5% methanolic potassium hydroxide over 2–3 h at 40° and decompose to the A-portion and a methyl ester of the S-portion (*Asahina* and *H. Akagi*, Ber., 1935, **68,** 1130):

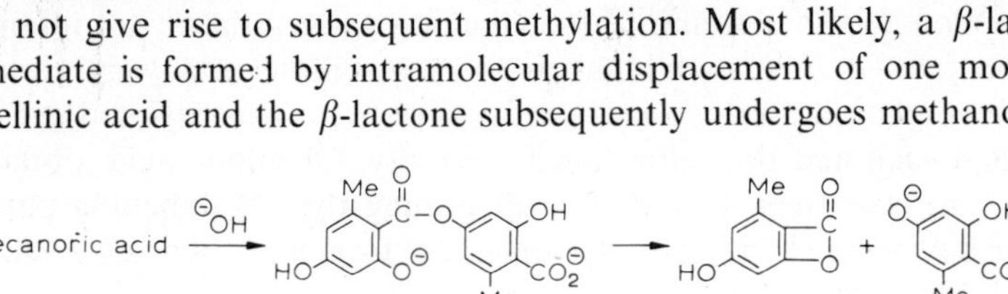

Lecanoric acid $\xrightarrow{MeOH,\ KOH}$ (methyl orsellinate) + (VIII, R=Me)

The normal mechanism of ester hydrolysis cannot operate here as this would not give rise to subsequent methylation. Most likely, a β-lactone intermediate is formed by intramolecular displacement of one molecule of orsellinic acid and the β-lactone subsequently undergoes methanolysis:

Lecanoric acid $\xrightarrow{^{\ominus}OH}$ (dianion) $\longrightarrow$ (β-lactone) + (orsellinate dianion)

Rupture of the depside link can be achieved by heating with methanol alone at 150° and again the S-portion is isolated as the methyl ester while the A-portion now undergoes decarboxylation (*G. Koller* and *H. Hamburg*, Monatsh., 1935, **65,** 373):

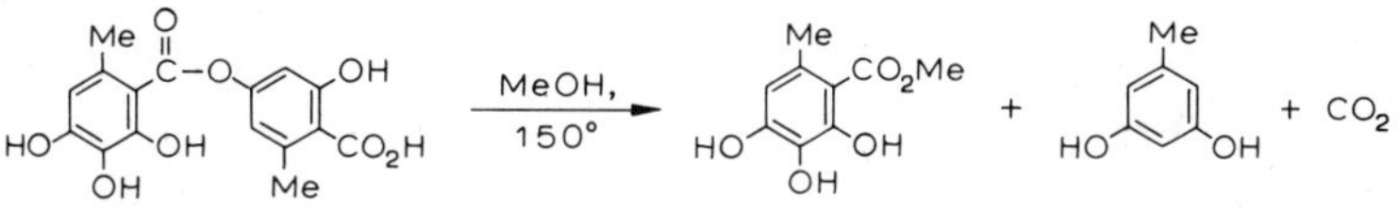

Diploschistesic acid
(XVII)

Boiling potassium carbonate solution is also effective for rupture of the depside link (*Asahina* and *M. Yanagita*, Ber., 1933, **66,** 36). Depsides such as diffractaic acid (V, R = Me) which do not have an OH group *ortho* to the depside carbonyl are difficult to hydrolyse under alkaline conditions (*Asahina* and *F. Fusikawa*, *ibid.*, 1932, **65,** 583).

As an alternative process, the depside link is broken by treatment with acids. Refluxing, formic acid (*Asahina* and *J. Asano*, *ibid.*, 1932, **65,** 475) is employed although this frequently causes decarboxylation of the A-portion:

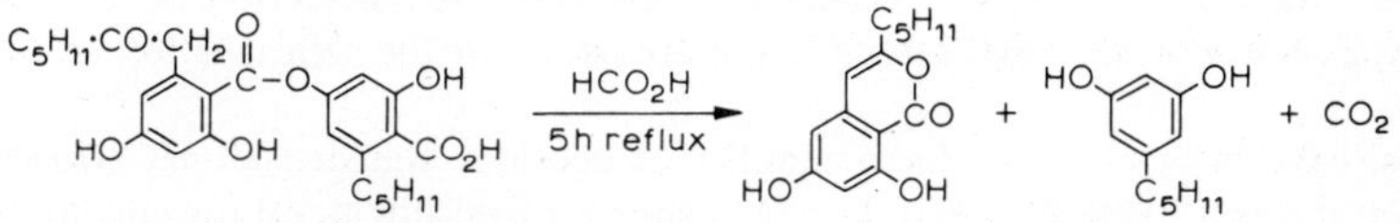

Olivetoric acid
(XVIII)

Cold concentrated sulphuric acid (*K. Fujii* and *S. Osumi*, J. pharm. Soc. Japan, 1936, **56,** 101; *Asahina et al.*, Ber., 1941, **74,** 827) gives quantitative recovery of the intact S- and A-portions and usually only slowly hydrolyses the methyl ester of the A-portion. These reactions probably involve protonation of the depside ester function followed by acyl-oxygen cleavage of the depside link to the phenol and a carbonium ion which reacts with the solvent.

Evidence for the structure of squamatic acid (XIX) is presented in Scheme 1 to illustrate the application of these techniques to a depside and its fully *O*-methylated derivative (*idem*, *ibid.*, 1933, **66,** 36; 1937, **70,** 62).

(*ii*) *Spectral properties*

The u.v. and i.r. spectra of many depsides have been recorded and are useful for diagnostic purposes (*Huneck*, Progress in Phytochemistry, *loc. cit.*; *Seshadri et al.*, Proc. Indian Acad. Sci., Sect. A, 1967, **66,** 1). The carbonyl stretching frequency in

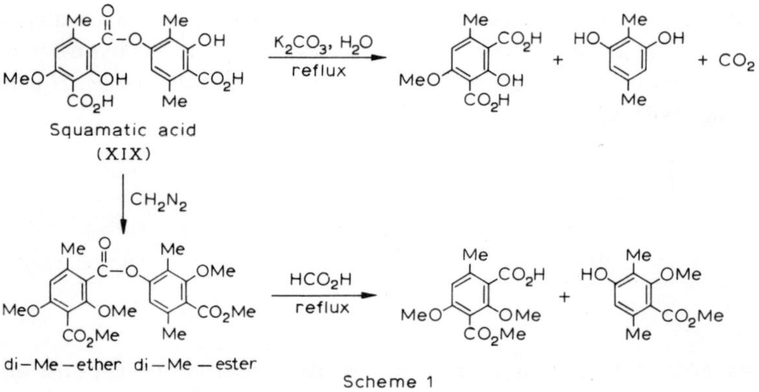

Squamatic acid (XIX)

K_2CO_3, H_2O reflux

CH_2N_2

di-Me-ether di-Me-ester

HCO_2H reflux

Scheme 1

O-methylated depsides is 1755–1760 cm^{-1} (carbon tetrachloride solution) which indicates that the carbonyl group is twisted out of the plane of the benzene ring. This carbonyl stretching frequency is close to that for phenyl acetate (1766 cm^{-1}) and higher than that for phenyl benzoate (1738 cm^{-1}). The hydroxylated depsides show a carbonyl stretching frequency around 1720–1736 cm^{-1} and intramolecular hydrogen bonding is considered responsible for this lowering of frequency. However the spectra were only measured in KBr pellets and not in solution.

N.m.r. data for depsides and depsidones have been tabulated (*Huneck* and *P. Lisscheid*, Z. Naturforsch, 1968, **23b,** 717) and are useful for the identification of substituents.

M.s. have been recorded for a number of depsides and depsidones (*Huneck et al.*, Tetrahedron, 1968, **24,** 2707). This technique is especially useful for making structural assignments to the S- and A-parts of the depside molecules because the most abundant fragments are formed by cleavage of the depside link. The fragmentation of methyl evernate (XX) molecular ion is given as an example:

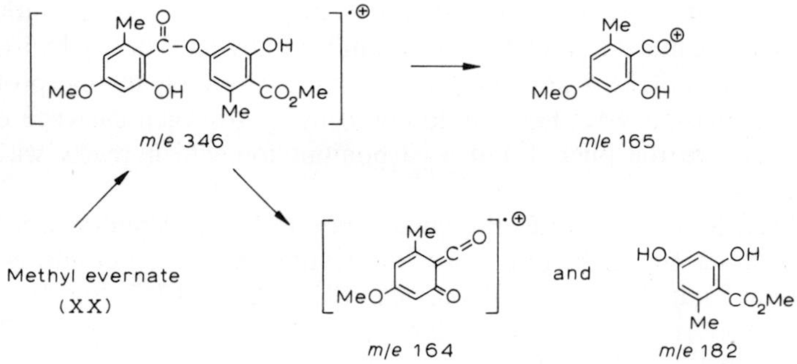

m/e 346 *m/e* 165

Methyl evernate (XX)

m/e 164 and *m/e* 182

(iii) Synthesis

E. Fischer (Ber., 1913, **46,** 1138) achieved the first synthesis of a lichen depside, lecanoric acid (III) by the condensation of bis (*O*-methoxycarbonyl)-

orsellinic acid chloride and orsellinic acid in dilute potassium hydroxide and subsequent removal of the phenol protecting groups. The method is not generally suitable because the acid anhydride can be formed, especially when pyridine is used as the base:

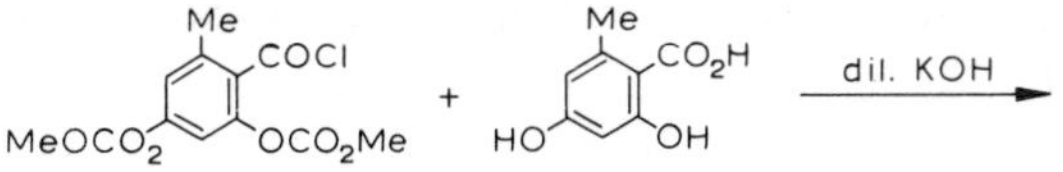

Me COCl MeOCO$_2$ OCO$_2$Me + Me CO$_2$H HO OH —dil. KOH→ Lecanoric acid (III)

A. Robertson and *R. J. Stephenson* (J. chem. Soc., 1932, 1388) protected the carboxyl group as the methyl ester and so synthesised methyl evernate (XX) but the methyl ester function cannot be hydrolysed without cleavage of the depside link, so this route is not suitable for the preparation of depside acids.

The most satisfactory protecting group for the acid substituent was found by starting with the corresponding carbaldehyde which can be oxidised to the acid with potassium permanganate in a final stage. This procedure was introduced by *Fischer (loc. cit.)* and perfected by *Asahina* and his collaborators. The synthesis of imbricaric acid (XXI) is given below as an example (*Asahina* and *T. Yosioka*, Ber., 1937, **70,** 1823):

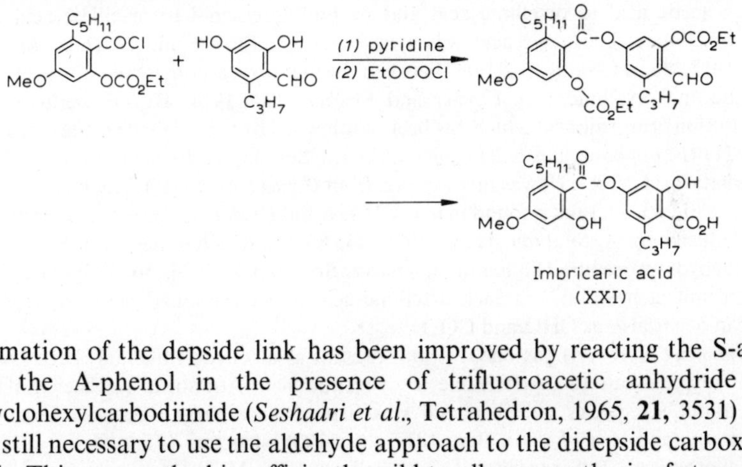

C_5H_{11} COCl MeO OCO$_2$Et + HO OH CHO C_3H_7 —(1) pyridine (2) EtOCOCl→ C_5H_{11} C(=O)–O OCO$_2$Et MeO O CO$_2$Et CHO C_3H_7

→ C_5H_{11} C(=O)–O OH MeO OH CO$_2$H C_3H_7

Imbricaric acid (XXI)

Formation of the depside link has been improved by reacting the S-acid and the A-phenol in the presence of trifluoroacetic anhydride or dicyclohexylcarbodiimide (*Seshadri et al.*, Tetrahedron, 1965, **21,** 3531) but it is still necessary to use the aldehyde approach to the didepside carboxylic acids. This new method is sufficiently mild to allow a synthesis of atranorin (VI, R = H) where the S-component is a formyldihydroxy-*o*-toluic acid.

Seshadri (J. chem. Soc., 1959, 1658) has demonstrated that the depside link is sufficiently strong to allow of the introduction of nuclear substituents for the transformation of one depside into another. Lecanoric acid (III) has been converted into diploschistesic acid (XVII) by successive application of the Gattermann and Dakin reactions:

Lecanoric acid (III) $\xrightarrow[AlCl_3]{HCN,}$ [structure: Me, C(=O)–O, OH, HO, OH, CHO, Me, CO_2H] $\xrightarrow[NaOH]{H_2O_2,}$ Diploschistesic acid (XVII)

The chlorination of depsides has also been examined (*Seshadri et al.*, Indian J. Chem., 1964, **2,** 478).

(*iv*) *Individual depsides*

Table 2 gives references to the structural elucidation and synthesis of depsides. The numbering of formulae is that suggested by *Seshadri* and by *Culberson* (Phytochemistry, 1966, **5,** 815) and which is used in Chemical Abstracts. Beilstein's Handbook uses a different system. Descriptions and indications of the sources of individual depsides are given only for those which are the most abundant or of historical importance.

Lecanoric acid (III) was first isolated from *Lecanora* species by *E. Schunck* (Ann., 1842, **41,** 157) and was shown to be identical with parmelialic acid and glabratic acid isolated from *Parmelia* species by *W. Zopf.* It occurs as *erythrin,* the ester with erythritol and as the 4-*O*-methyl derivative **evernic acid** first isolated by *J. Stenhouse* (*ibid.*, 1848, **68,** 55) from *Evernia vulgaris* Krober (= *E. prunastri.* Ach.). *O. Hesse* (J. pr. Chem., 1900, [ii], **62,** 463) demonstrated the hydrolysis of lecanoric acid by bioling acetic acid to orsellinic acid and by boiling ethanol to orsellinic acid and ethyl orsellinate. Lecanoric acid was synthesised (*E. Fischer* and *H. O. L. Fischer,* Ber., 1913, **46,** 1138) by reaction between bismethoxycarbonylorsellinic acid and chloride and orsellinic acid. *Fischer* and *Fischer* (*ibid.*, 1914, **47,** 505) deduced the constitution of evernic acid which has been synthesised by the aldehyde route discussed on p. 211 (*F. Fujikawa* and *K. Ishiguro,* J. pharm. Soc. Japan, 1936, **56,** 149).

Barbatic acid (V, R = H) was first isolated from *Usnea barbata* (L.) Ach. by *Stenhouse* (Ann., 1880, **203,** 302) and is found in many *Usnea* and *Cladonia* species. Its constitution was deduced by *A. St. Pfau* (Helv., 1928, **11,** 864) and it has been synthesised by the aldehyde procedure (*Fujikawa,* J. pharm. Soc. Japan, 1936, **56,** 182). The extra carbon unit at position 3 in each orsellinic acid portion is found in other depsides and can occur also as CHO and CO_2H.

Atranorin (VI, R = H) has one C-1 unit present as aldehyde which has been shown by ^{14}C experiments to be derivable from formic acid. Atranorin has been isolated from about 90 species of lichen and was first obtained from *Lecanora atra* Ach. (*E. Paterno* and *A. Oglialoro,* Ber., 1877, **10,** 1100); structure (*Pfau,* Helv., 1926, **9,** 650); synthesis (*T. R. Seshadri et al.*, Tetrahedron, 1965, **21,** 3531).

Squamatic acid (XIX) has one C-1 unit present as carboxylic acid. It was first isolated from *Cladonia squamosa* Hoffm. var. *ventricosa* (*Hesse,* J. pr. Chem., 1900, [ii], **62,** 430); structure (see p. 209); synthesis of the dimethyl ester (*Y. Asahina* and *Y. Sakurai,* Ber., 1937, **70,** 64).

Gyrophoric acid (IV) was first isolated in 1849 (*Stenhouse,* Ann., 1849, **70,** 218) from *Gyrophora pustulata* Ach. (= *Umbilicaria pustulata* L.); structure (*Asahina* and *M. Watanabe,* Ber., 1930, **63,** 3044); synthesis (*Asahina* and *Fujikawa, ibid.*, 1932,

Compound	m.p.	R	R^1	OMe for OH at	Other substituents	Ref.
Lecanoric acid[a]	175°	Me	Me	—	—	1
Evernic acid	170°	Me	Me	4	—	1
Tumidulin	174°	Me	Me	Me ester	3,5-Cl_2	2
Diploschistesic acid	174°	Me	Me	—	3-OH	1
Sphaerophorin	137°	Me	C_7H_{15}	4	—	1
Divaricatic acid	137°	C_3H_7	C_3H_7	4	—	1
Stenosporic acid	112°	C_3H_7	C_5H_{11}	4	—	3
Imbricaric acid	126°	C_5H_{11}	C_3H_7	4	—	1
Anziaic acid	124°	C_5H_{11}	C_5H_{11}	—	—	1
Perlatoric acid	108°	C_5H_{11}	C_5H_{11}	4	—	1
Planaic acid	110°	C_5H_{11}	C_5H_{11}	2,2′,4	—	4
Glomelliferic acid	144°	$CH_2\cdot CO\cdot C_5H_{11}$	C_5H_{11}	4	—	1
Miriquidiic acid	140°	$CH_2\cdot CH_2\cdot COEt$	C_5H_{11}	4	—	13
Olivetoric acid	151°	$CH_2\cdot CO\cdot C_5H_{11}$	C_5H_{11}	—	—	1
Confluentic acid	157°	$CH_2\cdot CO\cdot C_5H_{11}$	C_5H_{11}	2′,4	—	5
Microphillic acid	116°	$CH_2\cdot CO\cdot C_5H_{11}$	$CH_2\cdot CO\cdot C_5H_{11}$	4	—	1
Arthoniaic acid	167°	$CH_2\cdot CO\cdot C_5H_{11}$	$CH_2\cdot CO\cdot C_5H_{11}$	2′	—	14
Obtusaic acid	209°	Me	Me	4	3-Me	1
Barbatic acid	187°	Me	Me	4	3,3′-$(Me)_2$	1
Diffractaic acid	190°	Me	Me	2′,4	3,3′-$(Me)_2$	1
Atranorin	196°	Me	Me	Me ester	3-CHO-3′-Me	1
Chloroatranorin	208.5°	Me	Me	Me ester	5-Cl-3-CHO-3′-Me	
Baeomycesic acid	223°	Me	Me	4	3-CHO-3′-Me	1
Squamatic acid	228°	Me	Me	4	3-CO_2H-3′-Me	1

[a] *Erythrin*, m.p. 148°, $[\alpha]_D + 10.6°$, is the erythritol ester of lecanoric acid.[1]

TABLE 2 *(continued)*

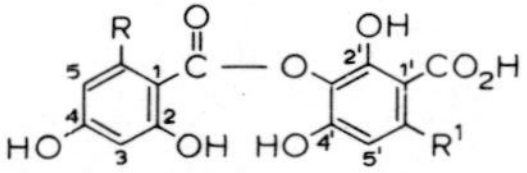

Compound	*m.p.*	*R*	R^1	*OMe for OH at*	*Other substituents*	*Ref.*
Sekikaic acid	147°	C_3H_7	C_3H_7	4,4′		6
Scrobiculin	135.5°	C_3H_7	C_3H_7	4, Me ester		7
Ramalinolic acid	164°	C_3H_7	C_5H_{11}	4		1
Merochlorophaeic acid	164°	C_3H_7	C_5H_{11}	2,4		8
Homosekikaic acid	137.5°	C_3H_7	C_5H_{11}	4,4′		1
Boninic acid	134,5°	C_3H_7	C_5H_{11}	2,4,4′		1
Cryptochlorophaeic acid	182°	C_5H_{11}	C_5H_{11}	2		8
4-*O*-Methylcryptochlorophaeic acid	141.5°	C_5H_{11}	C_5H_{11}	2,4		9

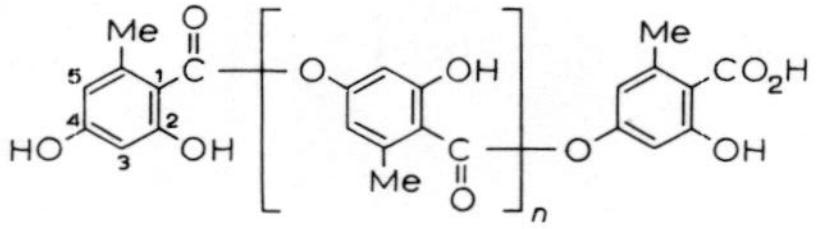

Compound	*m.p.*	*n*	*OMe for OH at*	*Other substituents*	
Gyrophoric acid	220°	1	—	—	1
Methyl gyrophorate	288°	1	Me ester	—	10
Umbilicaric acid	203° (decomp.)	1	2	—	1
Tenuiorin	238° (decomp.)	1	4, Me ester	—	1
Hiascic acid	190° (decomp.)	1	—	5-OH	1
Aphthosin	300° (decomp.)	2	4	—	11

Barbatolic acid	207° (decomp.)	[structure]	1
Nephroarctin, $R_1 = H$, $R_2 = Me$	192°	[structure]	12
Phenarctin, $R_1 = CO_2Me$, $R_2 = H$	167°		15

References
1 *Y. Asahina* and *S. Shibata*, "Chemistry of Lichen Substances", A. Asher, Amsterdam, 1971 (Original Japanese Edition 1949; First English translation 1954).
2 *G. Bendz et al.*, Acta chem. Scand., 1965, **19,** 1188; *S. Huneck*, Ber., 1966, **99,** 1106.
3 *D. F. Culberson*, Phytochem., 1970, **9,** 841.
4 *Huneck*, Z. Naturforsch., 1965, **20b,** 1119.
5 *Huneck*, Ber., 1962, **95,** 328.
6 *T. R. Seshadri* and *G. B. V. Subramanian*, J. Indian chem. Soc., 1963, **40,** 7.
7 *Culberson*, Phytochem., 1967, **6,** 719.
8 *S. Shibata* and *H.-C. Chiang*, *ibid.*, 1965, **4,** 133.
9 *Culberson* and *H. Kristinsson*, Bryologist, 1969, **72,** 431.
10 *U. Sankawa* and *Shibata*, Phytochem., 1970, **9,** 2061.
11 *F. W. Bachelor* and *G. G. King*, *ibid.*, 1970, **9,** 2587.
12 *M. Numo et al.*, Chem. Comm., 1969, 78.
13 *Huneck et al.*, Z. Naturforsch., 1971, **26b,** 1357.
14 *Idem*, *ibid.*, 1970, **25b,** 49.
15 *T. Bruun*, Acta chem. Scand., 1969, **23,** 3601.

65, 983). It is an example of a tridepside. The tetradepside *aphthosin* (see Table 2) was isolated from *Peltigera aphthosa* L. and its great insolubility led to the suggestion that tetradepsides may have been present in, but not extracted from, lichens previously examined.

Sekikaic acid (XXII), structure (*Asahina* and *S. Nonomura*, Ber., 1933, **66,** 30), synthesis (*Asahina* and *M. Yasue*, *ibid.*, 1935, **68,** 1133), is the typical example of a *meta*-depside. In general when the A-component is a 2,3,4-trihydroxybenzoic acid derivative the depside link is formed on the 3-hydroxyl group, see for example **ramalinolic acid** (XXIII) (*Asahina* and *T. Kusaka*, *ibid.*, 1936, **69,** 450, 1896). *Asahina* and *Kusaka* have shown that monobenzoylation of 2,3,4-trihydroxy-6-pentylbenzoic acid occurs on the 3-hydroxyl function by methylation of the product and then hydrolysis to the known 2,4-dimethyl ether. The *meta*-depside link may represent just the most stable position of acylation since benzoyl groups can migrate to adjacent phenolic groups. Evidence has been presented to show that the stable monobenzoyl derivative of 2,3,4-trihydroxybenzoic acid is the 3-derivative although this point has more recently been contested (*A. Critchlow*, *R. D. Haworth* and *P. L. Pauson*, J. chem. Soc., 1951, 1318; *W. Augstein* and *C. K. Bradsher*, J. org. Chem., 1969, **34,** 1349):

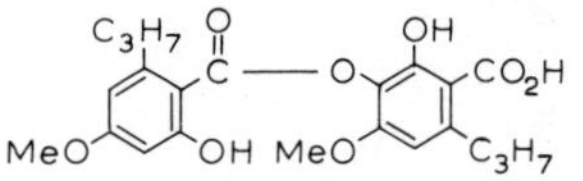

Sekikaic acid
(XXII)

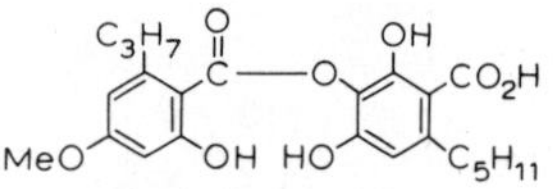

Ramalinolic acid
(XXIII)

(e) Depsidones

(i) Degradation

The lactone ring in depsidones is opened by the action of aqueous alkali and can be closed again by treatment with acetic anhydride (*F. H. Curd* and *A. Robertson*, J. chem. Soc., 1935, 1379). Treatment of the depsidone with dimethyl sulphate and alkali gives a methyl ether acid which can be decarboxylated. The resulting diphenyl ethers have then been degraded to benzoquinones by oxidation and later synthesised by the Ullmann reaction. Fusion of the depsidone with potassium hydroxide and pyrolysis also gives monocyclic phenols useful for structure determination.

(ii) Spectral properties

References to the u.v., i.r., n.m.r. and m spectra of depsidones have been given in the discussion of depsides p. 209. The m.s. fragmentation pattern is not so generally useful in structure determination of depsidones because fragmentation of the diphenyl ether bond is not usually of major importance. Structures have been assigned on the basis of n.m.r. spectra but with the assumption that the related depsides will have the usual structure based on two units of the orsellinic acid type (*S. Shibata* and *H.-C. Chiang*, Chem. Pharm. Bull., Tokyo, 1963, **11,** 926; *S. Huneck, et al.*, Phytochem., 1970, **9,** 2567). The depsidone carbonyl stretching frequency is in the range 1950–1720 cm^{-1}.

(iii) Synthesis

Formation of the diphenyl ether link by oxidation of a depside has been achieved in the synthesis of diploicin (XXV) using manganese dioxide in refluxing chloroform as the reagent (*W. D. Ollis et al.*, Proc. chem. Soc., 1960, 393). In one approach the depside XXIV (R = H), was oxidised and the product methylated, purified and partially demethylated to diploicin (XXV).

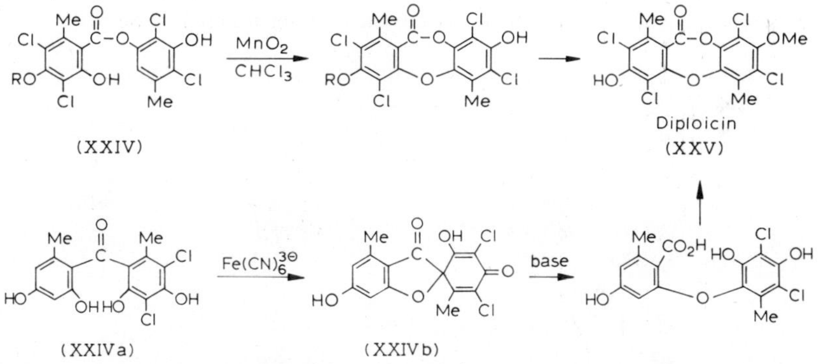

(XXIV) Diploicin (XXV)

(XXIVa) (XXIVb)

In the second approach the benzyl ether (XXIV, R = $PhCH_2$) was oxidised, the product methylated, purified and debenzylated by catalytic reduction to give diploicin and thus establishing the position of the methoxyl group.

Other workers (*J. B. Hendrickson, M. V. J. Ramsay* and *T. R. Kelley*, J. Amer. chem. Soc., 1972, **94,** 6834) have argued that in general it is difficult to form a seven-membered ring, so they employed an oxidative coupling reaction for the synthesis of diploicin which forms the diphenyl ether link initially in a five-membered ring. This is a longer synthetic route than the ones described previously but it gives higher yields. The benzophenone XXIVa was oxidised with potassium ferricyanide to give the dienone XXVIb. Only one dienone was formed and its structure was assigned from spectral data. Hydrolysis of this dienone afforded the ring-opened form of a depsidone from which the latter was readily obtained by reaction with acetic anhydride. This procedure represents a general synthesis of depsidones. The oxidation step gives acceptable yields of one dienone only when one benzene ring of the benzophenone derivative is brominated or chlorinated. The halogen atoms have a directing effect on the ring closure and appear in the dienone ring of the oxidation product.

(iv) Individual depsidones

Depsidones of natural occurrence are listed in Table 3. Evidence for the strùcture of important compounds is given below. In many cases the evidence is not complete in itself and to a small degree the structures rely on the tacit assumption that depsidones are related biosynthetically to depsides. A much greater variation in the oxidation state of the acetate derived methyl group is found in the depsidones (*e.g.* norstictinic acid, psoromic acid, salazinic acid) than the depsides. The diphenyl ether link can be closed on the 5-position as in grayanic acid or on the 3-position of the A-portion as in variolic acid.

Lobaric acid (XXVI) isolated from *Usnea barbata* (L.) Ach. (*Hesse*, Ber., 1877, **10,** 1324), *Lecanora badia* Pers. (2.4% yield) and many *Stereocaulon* species (*Asahina* and *M. Hiraiwa*, *ibid*., 1935, **68,** 1705). Preparation of derivatives (see Scheme 2) indicates the presence of a ketone, phenol and carboxylic acid function in lobaric acid:

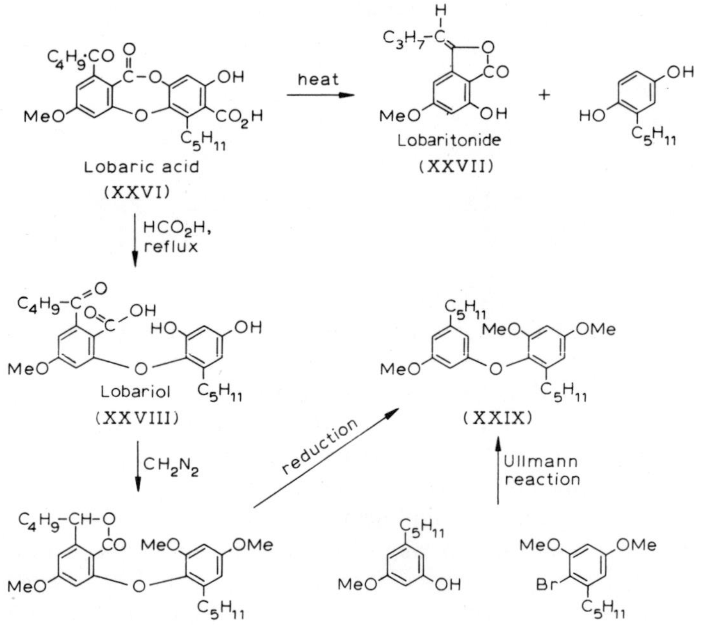

Scheme 2

The depside link is hydrolysed by gentle treatment with alkali to give a dicarboxylic acid which is easily decarboxylated to lobariol (XXVIII), also obtained by the action of formic acid (*Asahina* and *S. Nonomura*, *ibid*., 1935, **68,** 1698). Potash fusion gives resorcinol and *n*-valeric acid and destructive distillation gives the enol lactone lobaritonide (XXVII) which has been converted to 3-hydroxy-5-methoxyvalerophenone and also synthesised (*Asahina* and *Yasue*, *ibid*., 1936, **69,** 643; 1937, **70,** 206; *H. Nogami*,

J. pharm. Soc. Japan, 1941, **61,** 46; C.A., 1941, **35,** 4764). Thus the structure of the S-part of lobaric acid is decided. Lobariol (XXVIII) was converted by methylation, decarboxylation and Wolff–Kishner reduction to the diphenyl ether XXIX, which has been synthesised as shown (*Asahina* and *Yasue, loc. cit.)*. The structure of lobaric acid was proposed on the basis of these experimental results although there is no confirmation of the position of the free carboxyl group other than the observation that since lobaric acid gives a violet coloration with ferric chloride it is probably a 2-hydroxybenzoic acid.

Physodic acid (XXX) first obtained by *Hesse* (Ber., 1897, **30,** 1983) has been isolated from *Parmelia* species, particularly *P. physodes* Ach. (5% yield). Structural evidence (see Scheme 3) was assembled by *Asahina* and *Nogami* (*ibid*., 1934, **67,** 805; 1935, **68,** 77). Physodic acid is very readily converted by base to an isomer which is given the structure X I. Methylation and oxidation of physodic acid to yield *O*-methylolivetonide (XXXII) revealed the structure of the S-portion of the molecule and the nature of this isomerisation. The structure of the A-portion was revealed by degradation to the diphenyl ether XXXIII, which has been synthesised by the Ullmann reaction. Oxidation of XXXIII gives benzoquinones derived from the two benzene rings (Scheme 3).

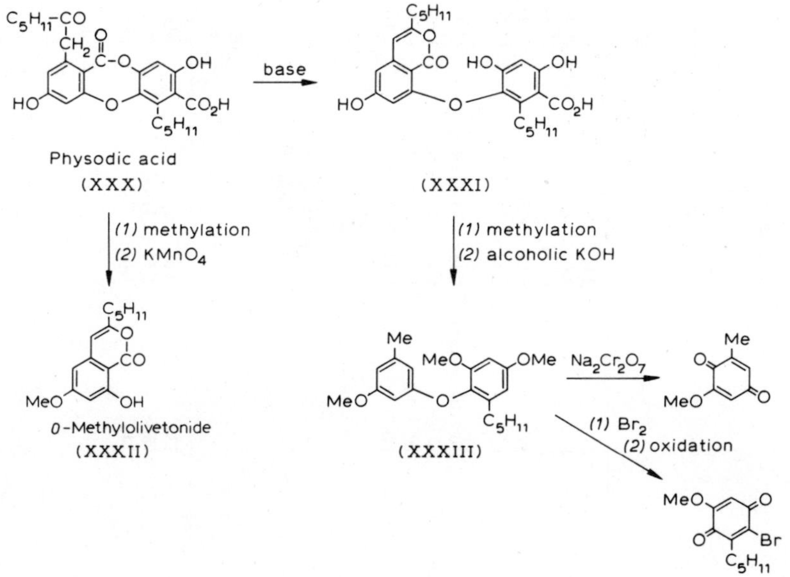

Physodic acid (XXX) → base → (XXXI)

(XXX) → (1) methylation (2) $KMnO_4$ → *O*-Methylolivetonide (XXXII)

(XXXI) → (1) methylation (2) alcoholic KOH → (XXXIII)

(XXXIII) → $Na_2Cr_2O_7$; (1) Br_2 (2) oxidation

Scheme 3

Alectoronic acid (XXXIV, R = H) (see Scheme 4), isolated from *Alectoria japonica* Tuck. and *A. sarmentosa* Ach., gives a methylation product identical with that from α-collatolic acid. α-**Collatolic acid,** *lecanorolic acid* (XXXIV, R = Me) is obtained from *Lecanora atra* Ach. in 1.5% yield (*W. Zopf*, Ann., 1897, **295,** 257). On treatment with

TABLE 3

DEPSIDONES OF NATURAL OCCURRENCE

Compound	*m.p.*	*Structure* 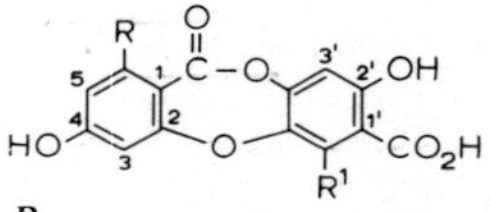 *R*	*R*[1]	*OMe for OH at*	*Other subst.*	*Ref.*
Gangaleoidin	213°	Me	Me	2′, Me ester	3,5-Cl_2	13
Grayanic acid	186°	Me	C_7H_{15}	4		1
Colensoinic acid	173°	C_5H_{11}	C_5H_{11}	4		2
Lobaric acid	192°	$CO \cdot C_4H_9$	C_5H_{11}	4		3
Loxodin	132°	$CO \cdot C_4H_9$	C_5H_{11}	4, Me ester		4
Physodic acid	205° (decomp.)	$CH_2 \cdot CO \cdot C_5H_{11}$	C_5H_{11}	—		3
4-*O*-Methylphysodic acid	151°	$CH_2 \cdot CO \cdot C_5H_{11}$	C_5H_{11}	4		5
Alectoronic acid	193°	$CH_2 \cdot CO \cdot C_5H_{11}$	$CH_2 \cdot CO \cdot C_5H_{11}$	—		3
α-Collatolic acid	124°	$CH_2 \cdot CO \cdot C_5H_{11}$	$CH_2 \cdot CO \cdot C_5H_{11}$	4		3

Compound	*m.p.*	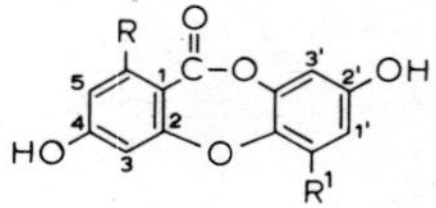 *R*	*R*[1]	*OMe for OH at*	*Other subst.*	*Ref.*
Diploicin	232°	Me	Me	2′	1′,3,3′,5-Cl_4	3
Norlobaridone	186°	$CO \cdot C_4H_9$	C_5H_{11}	—	—	6
Vicanicin	248°	Me	Me	2′	3,4-Cl_2-3′-Me	7
Nidulin[a]		Me	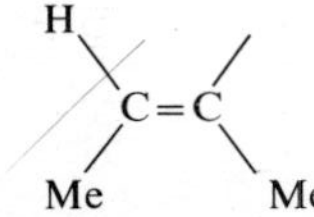 H(Me)C=C(Me)–	2′	1′,3,5-Cl_3	10

[a] A ... of the f... not present in a lichen.

TABLE 3 *(continued)*

Compound	*m.p.*	*Structure* *R*	*R*[1]	*OMe for OH at*	*Other subst.*	*Ref.*
Variolaric acid	296°	$CH_2-O-C=O$		—	—	3,14

Compound	*m.p.*	*R*	*R*[1]	*OMe for OH at*	*Other subst.*	*Ref.*
Norstictinic acid	283° (decomp.)	HOCH – O – CO		—	3-CHO	3
Stictinic acid	268° (decomp.)	HOCH – O – CO		4	3-CHO	3
Psoronic acid	265°	CO_2H	H	2′	3-CHO	3
Pannarin	217°	Me	H	2′	3-CHO, 5-Cl	12
Salazinic acid	260° (decomp.)					3

TABLE 3 *(continued)*

Compound	*m.p.*	*Structure* *R*	*R*¹	*OMe for OH at*	*Ref.*
Hypoprotocetraric acid	250°	Me	Me		5,8
4-*O*-Methylhypoprotocetraric acid	229° (decomp.)	Me	Me	4	11
Virensic acid	245–246°	CHO	Me		9
Nataic acid	226° (decomp.)	Me	H	4	11
Protocetraric acid	250° (decomp.)	CHO	CH_2OH		
Physodalic acid	260° (decomp.)	CHO	CH_2OCOMe		
Fumarprotocetraric acid	250°	CHO	$CH_2OCO{\cdot}CH{=}CH{\cdot}CO_2H$-*trans*		3

References

1 *S. Shibata* and *H.-C. Chiang*, Chem. Pharm. Bull., Tokyo, 1963, **11,** 926.
2 *C. H. Fox, E. Klein* and *S. Huneck*, Phytochem., 1970, **9,** 2567.
3 *Y. Asahina* and *S. Shibata*, "Chemistry of Lichen Substances", A. Asher, Amsterdam, 1971 (Original Japanese Edition, 1949; first English translation, 1954).
4 *T. Komita* and *S. Kurokawa*, Phytochem., 1970, **9,** 1139.
5 *C. F. Culberson, ibid.*, 1966, 5, 815.
6 *G. E. Gream* and *N. V. Riggs*, Austral. J. Chem., 1960, **13,** 285.
7 *S. Neelakantan, T. R. Seshadri* and *S. S. Subramanian*, Tetrahedron, 1962, **13,** 597.
8 *S. Huneck* and *J.-M. Lehn*, Z. Naturforsch., 1966, **21b,** 299.
9 *K. Aghoramurthy, K. G. Sarma* and *T. R. Seshadri*, Tetrahedron, 1961, **12,** 173.
10 *B. W. Bycroft* and *J. C. Roberts*, J. org. Chem., 1963, **28,** 1429; *W. F. Beach* and *J. H. Richards, ibid.*, 1963, **28,** 2746.
11 *T. M. Cresp et al.*, Austral. J. Chem., 1972, **25,** 2167.
12 *D. A. Jackman, M. V. Sargent* and *J. A. Elix*, J. chem. Soc., Perkin I, 1975, 1979.
13 *Sargent, P. Vogel* and *Elix, ibid.*, p. 1986.
14 *N. M. Rana, Sargent* and *Elix, ibid.*, p. 1992.

base α-collatolic acid yields **β-collatolic acid** (XXXV) in a reaction like the isomerisation of physodic acid. Treatment with formic acid gives the di-isocoumarin derivative **collatolone** (XXXVII). Methylation of β-collatolic acid followed by alkali treatment yields the diphenyl ether XXXVI, which has been synthesised. Destructive distillation of α-collatolic acid yields the *O*-methylolivetonide (XXXII). This evidence enabled *Asahina et al.*, (Ber., 1933, **66,** 649) to put forward the structure of α-collatolic acid (Scheme 4).

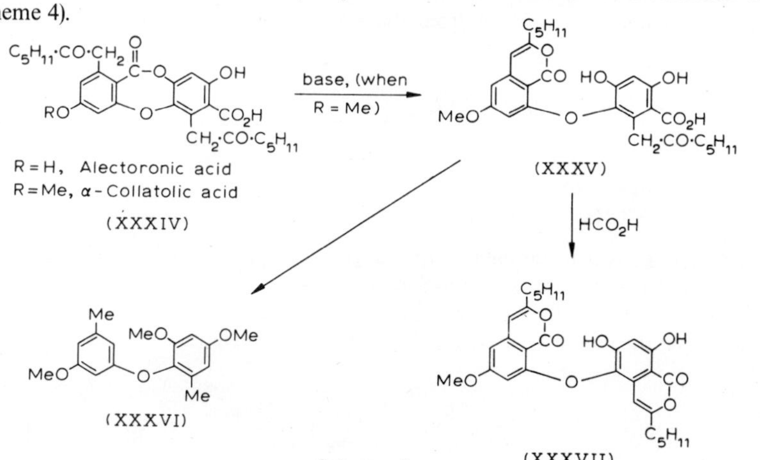

R = H, Alectoronic acid
R = Me, α - Collatolic acid
(XXXIV)

base, (when R = Me)

(XXXV)

HCO_2H

(XXXVI)

(XXXVII)

Scheme 4

Diploicin (XXXVIII), *catolechin*, was first isolated by *Zopf* (Ann., 1904, **336,** 46) from *Diploicia canescens* Dicks. (= *Buellia canescens* [Dicks.] DeNot.). The structure was deduced by *T. J. Nolan et al.* (Sci. Proc. roy. Dublin Soc., 1936, **21,** 333; 1948, **24,** 319) (see Scheme 5). It is one of the small group of depsidones in which the A-acid has been decarboxylated. On heating with hydrobromic acid the lactonic link of diploicin is cleaved with simultaneous decarboxylation. The diphenyl ether link is cleaved with aluminium chloride in benzene to give a depside which is hydrolysed to known fragments (Scheme 5).

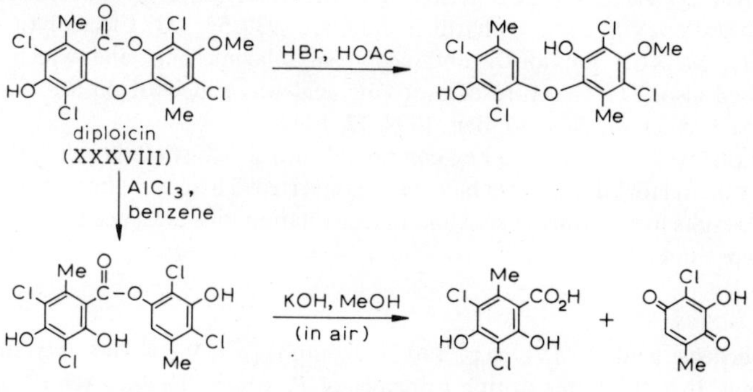

diploicin
(XXXVIII)

HBr, HOAc

$AlCl_3$, benzene

KOH, MeOH (in air)

Scheme 5

The synthesis of diploicin was discussed on p. 217.

Vicanicin (XXXIX) was noted by *Zopf* (Ann., 1905, **340,** 300) as a constituent of *Teloschistes flavicans* and isolated in a state of purity by *Seshadri et al.* (Tetrahedron, 1962, **18,** 597) who chromatographed the crude material over magnesium carbonate. The structure was established by methylation, opening of the depside link by methanol and oxidation of the diphenyl ether to known fragments. The position of the methoxyl group in vicanicin is assumed from the similarity of its u.v. spectrum to that of diploicin:

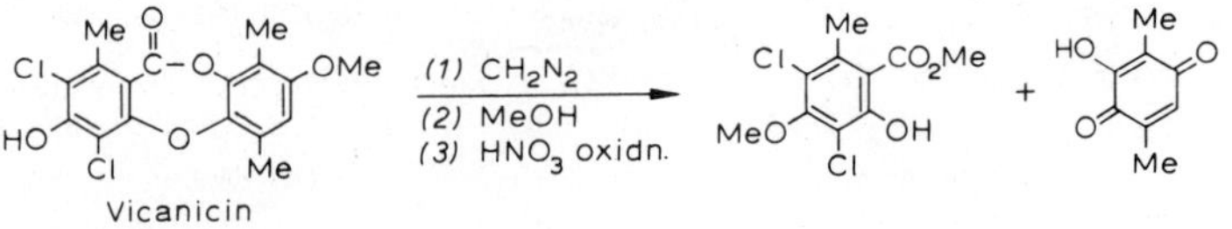

(1) CH_2N_2
(2) MeOH
(3) HNO_3 oxidn.

Vicanicin
(XXXIX)

Nidulin (XL) is a metabolite of the fungus *Aspergillus ustus* (*F. M. Dean, J. C. Roberts* and *A. Robertson*, J. chem. Soc., 1954, 1433). ^{14}C-Tracer studies have shown that the butenyl side chain is derived from isoleucine which may enter as tiglylcoenzyme-A, and the remainder of the molecule from acetate units:

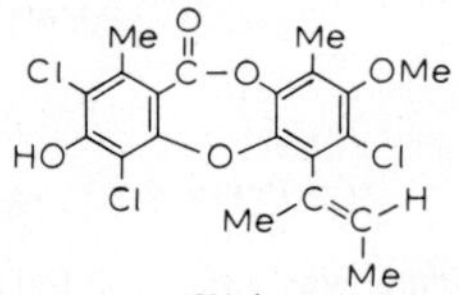

(XL)

Psoromic acid (XLI) (see Scheme 6) was first isolated from *Lecanora crassa* Ach. (*G. Spica*, Ber., 1883, **16,** 427) and is found in many lichens. On fusion with potassium hydroxide, psoromic acid decomposes to orcinol and 3,5-dihydroxy-4-methylbenzoic acid while dry distillation yields atranal. The aldehyde function is reduced to methyl and hydrolysis of the depside link which is accompanied by decarboxylation, followed by methylation yields the acid XLIII. The decarboxylation product of XLIII has been synthesised (*S. Shibata*, J. pharm. Soc. Japan, 1939, **59,** 323). Electrolytic reduction of XLIII gave XLII previously obtained from salazinic acid and which has been synthesised also. The structure of psoromic acid was proposed on the basis of this evidence (*Asahina* and *Shibata*, Ber., 1939, **72,** 1399).

The diphenyl ether XLII is a key compound into which **stictinic acid, protocetraric acid,** and the related depsidones have been converted. These transformations together with the results from potash fusion and dry distillation give evidence for the structures of the depsidones.

(v) Depsones

Picrolichenic acid (XLIV), m.p. 190° (decomp.), $[\alpha]_D$ 0°, is the only member of this group. It is the bitter tasting principle of *Pertusaria amara* (Ach.) Nyl. *Methyl O-methylpicrolichenate*, m.p. 80°, *methyl picrolichenate*, m.p. 102°. Picrolichenic acid

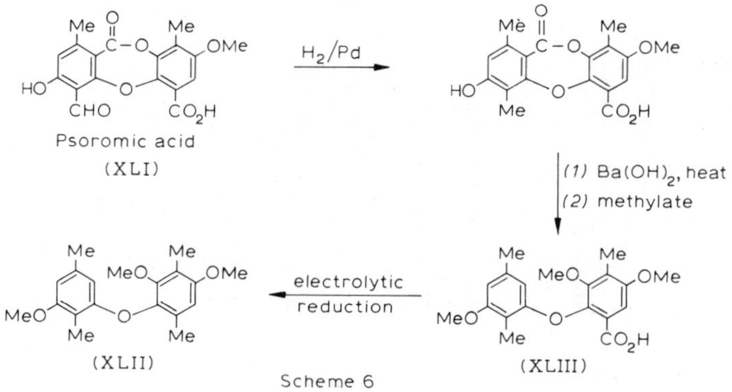

Scheme 6

possesses an unusual carbonyl function with a stretching frequency 1820 cm^{-1}. This is associated with a lactone function which is opened by base but which decarboxylates on acidification to give the phenolic acid XLV (see Scheme 7). The acid decarboxylates on heating to give a phenol which could be transformed into the dibenzofuran derivative XLVI. The latter was identified by conversion to the known 3,7-dimethoxydibenzofuran-1,9-dicarboxylic acid. This evidence led to the proposed structure of picrolichenic acid and the suggestion that it is formed in the lichen by oxidation of the depside XLVII where the normal depsidone formation is prevented by methylation of the necessary hydroxyl group (*C. A. Wachtmeister*, Acta chem. Scand., 1958, **12,** 147):

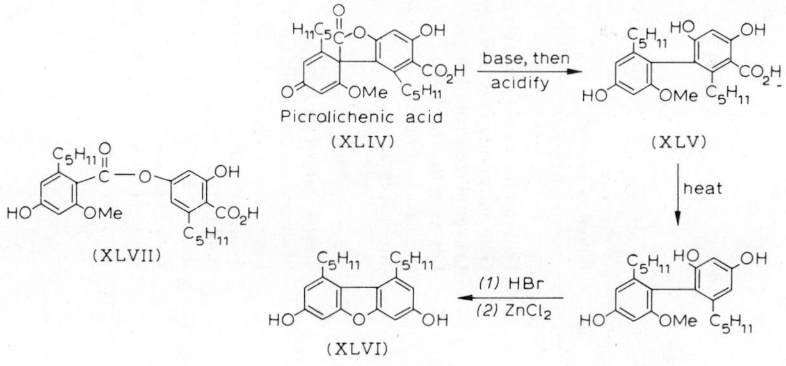

Scheme 7

T. A. Davidson and *A. I. Scott* (J. chem. Soc., 1961, 4075) succeeded in synthesising picrolichenic acid by oxidation of the depside XLVII with manganese dioxide in benzene.

TABLE 4

COMMON HYDROLYSABLE TANNINS

Tannin	*Botanical source*	*Origin*
Gallotannins		
Chinese tannin	Galls on leaves of the shrub *Rhus semialata* Murr.	Himalayas
Turkish tannin	Galls on wood of the tree *Quercus lusitanica* L.	Eastern Mediterranean
Sicilian sumach	Leaves of the shrub *Rhus coriaria* L.	Sicily
Stagshorn sumach	Leaves of the shrub *Rhus typhina* L.	N. America
Tara	Fruit pods of the shrub *Caesalpinia spinosa* (Mol.) Kuntze	Peru
Ellagitannins		
Myrobalans	Unripe nuts of the tree *Terminalia chebula* Retz.	Malaya
Algarobilla	Fruit pods of the shrub *Caesalpinia brevifolia* Baill.	Chile
Divi-Divi	Fruit pods of the shrub *Caesalpinia coriaria* Willd.	S. America
Valonea	Acorn cups of the tree *Quercus aegilops* L.	S. Europe, Asia Minor
Oak bark	Bark of *Quercus* sp.	Europe
Garouille	Root bark of *Q. coccifera* L.	Italy, S. France
Chestnut oak	Bark of *Q. prinus* L.	N. America
Chestnut (Spanish)	Wood of the tree *Castanea sativa* Mill.	France, Italy

2. Hydrolysable tannins

(a) Introduction

Practically all wood and vegetable material contains some form of high molecular weight polyphenolic material, soluble in water, which can be used to tan leather. Commercial tanning materials are obtained from sources rich in tannin. In the process of tanning, hides are soaked in progressively stronger and more acidic solutions of the tannin. Finally they are soaked in a strong solution for several weeks before being subjected to the further stages of leather manufacture. The tannin becomes fixed to the proteins of the skin, probably by hydrogen bonding and this gives the leather increased chemical and biological stability in comparison with fresh skin or hides.

The commercial tanning materials fall into two distinct groups of hydrolysable tannins and condensed tannins depending on their reaction with mineral acid. The hydrolysable tannins, which will be discussed here, are cleaved into a carbohydrate and acid fragments related to gallic and ellagic acids. Table 4 lists the vegetable sources of hydrolysable tannins in which there is a commercial traffic.

The condensed tannins on acid treatment are converted to an amorphous black and insoluble phlobaphene. Hydrolysable tannins are further divided into gallotannins which give no ellagic acid on hydrolysis and ellagitannins which give this acid. The division is important in the manufacture of leather because ellagic acid is responsible for producing a bloom on finished leather. This bloom is a crystalline deposit of ellagic acid. It gives firmness and an aesthetic appeal to the leather but it is objectionable when the goods are to be subsequently dyed.

The hydrolysable and condensed tannins have been reviewed (*E. Haslam*, "Chemistry of Vegetable Tannins", Academic Press, London, 1966).

(b) Isolation and characterisation

(i) Isolation and identification

Tannins are best extracted from plant sources with water. Depsidically bound gallic acid is cleaved by the prolonged action of methanol and ethanol. A few tannins, chebulinic acid, hamamelitannin and aceritannin can be isolated from aqueous extracts of the plant material by crystallisation. Separation of mixtures into phenolic tannins and carboxylic acid tannins is effected by extraction into ethyl acetate from water at pH 6.2 when the phenolic tannin passes into the organic solvent layer and again at pH 2.0 when the carboxylic acid tannin is extracted. Further purification is then effected by fractional precipitation with lead acetate or zinc acetate and by counter-current distribution between ethyl acetate and water (*O. Th. Schmidt et al.*, Ann., 1950, **568,** 165; 1951, **571,** 232). These procedures have been highly successful in separating ellagitannins and the references to individual ellagitannins can be consulted for further examples. In the gallotannin field these processes effect a separation of gross impurities but the separation of the closely related components of gallotannin seems impossible. Chromatography through columns of nylon or cellulose powders has been found useful for purification of gallotannins.

Individual tannins have been characterised by optical rotation, paper chromatography and hydrolysis to a sugar, gallic acid and the acid components discussed on pp. 193 *et seq.* Paper electrophoresis has also been suggested as a separative method and for the characterisation of tannins (*W. Grassmann* and *K. Hannig*, Z. physiol. Chem., 1953, **292,** 32).

(ii) General properties of tannins, determination of molecular weight

Tannins are soluble in water to give solutions with a bitter taste and which darken on exposure to air. These solutions will precipitate alkaloids such as brucine and denature solutions of gelatin. The latter test is usually taken as a qualitative test for tannins but simple galloyl esters of glucose, hamamelitannin and aceritannin which precipitate gelatin do not have the ability to tan leather.

There is considerable evidence that the more complex tannins tend to associate in aqueous solution. For this reason molecular weight determinations made in aqueous solution are unreliable. The association of simple galloyl glucoses has been studied and these have been found to be unassociated in acetone solution (*R. D. Haworth et al.*, J. chem. Soc., 1962, 2944). It is usual therefore to determine the molecular weights of tannins in acetone solution by elevation of boiling point or vapour pressure osmometry methods.

(c) Galloylglucoses and gallotannin

Many simple galloyl derivatives of glucose and other sugars are found in plant extracts. These have little or no tanning action. In the useful gallotannins long chains of depsidically linked gallic acid units are attached to the sugar.

(i) Depsides of gallic acid

The partial acid hydrolysis of Chinese gallotannin yields *m*-digallic (I) and *m*-trigallic acids (II) in significant amounts (*H. G. C. King* and *T. White, ibid.*, 1961, 3231). Fischer showed that the 4-acyl derivatives of gallic acid have no existence but rearrange to the stable 3-acyl derivatives (see p. 198):

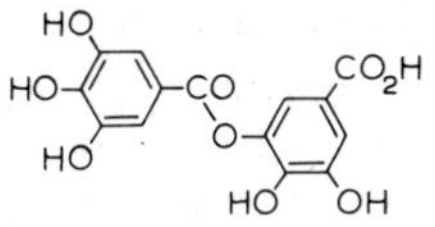

(I) 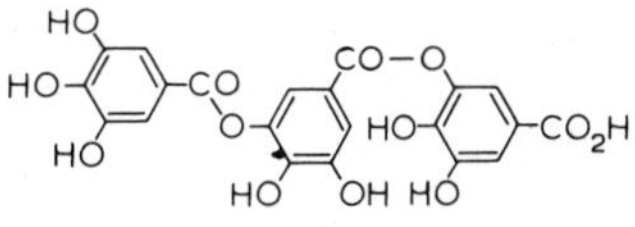 (II)

m-**Digallic acid,** m.p. 268° (decomp.) was first synthesised by condensing tris(*O*-methoxycarbonyl)galloyl chloride with 3,5-bis (*O*-methoxycarbonyl)gallic acid followed by hydrolysis of the ester methoxycarbonyloxy groups by dilute ammonia (*E. Fischer et al.*, Ber., 1913, **46,** 111; 1918, **51,** 45):

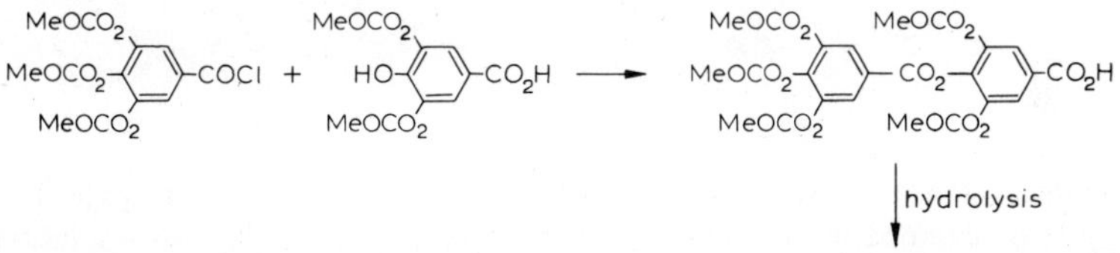

hydrolysis

(I)

Acetylation yields *penta-acetyl-*m*-digallic acid*, m.p. 204° (*methyl ester*, m.p. 167°), which is different from *penta-acetyl-*p*-digallic acid*, m.p. 202° (*methyl ester*, m.p. 192°) prepared by reaction between triacetylgalloyl chloride and 3,5-diacetylgallic acid. Careful hydrolysis of penta-acetyl-*p*-digallic acid yields *m*-digallic acid. Methylation of *m*-digallic acid and then hydrolysis yields 3,4,5-trimethoxybenzoic acid and 3,4-dimethoxybenzoic acid. *m*-Digallic acid has also been obtained by reaction of tri-*O*-benzylgalloyl chloride with diphenylmethyl 3,4-diphenylmethylenedioxy-5-hydroxybenzoate and hydrogenolysis of the protecting groups (*Haworth et al.*, J. chem. Soc., 1965, 6889):

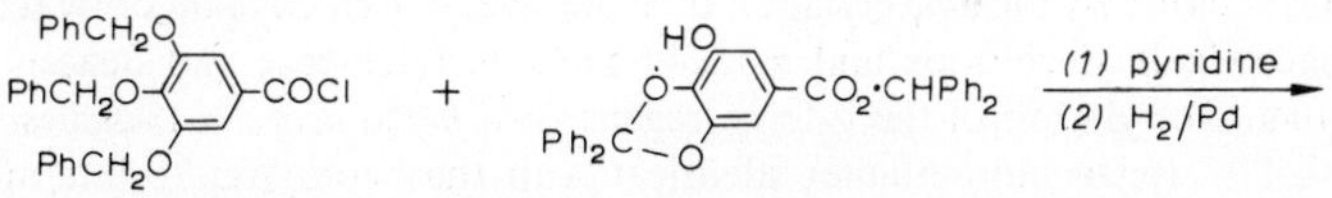

(1) pyridine
(2) H_2/Pd
(I)

m-**Trigallic acid,** m.p. 228°, has been prepared by *Haworth et al., (loc. cit.)*, by an extension of the above method. The depside III was prepared and the acetyl group hydrolysed. The phenol was then condensed with tri-*O*-benzylgalloyl chloride and the product hydrogenolysed to give *m*-trigallic acid:

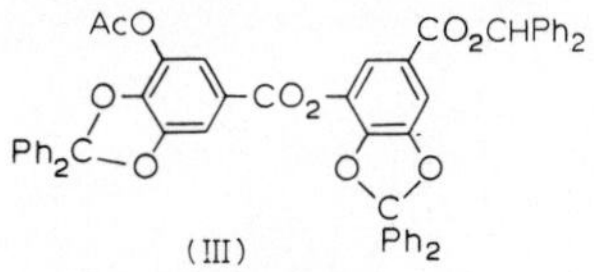

(III)

*Methyl hepta-*O*-methyl-*m*-trigallate*, m.p. 129°.

(ii) Simple gallate esters of glucose

β-D-**Glucogallin** (IV), crystallises from water as the *hydrate* (1.5 H_2O), m.p. 214°, $[\alpha]_D$ −25.6° (water), and was first isolated from the roots of Chinese rhubarb, *Rheum officinale* Baill. (*E. Gibson*, Compt. rend., 1903, **136,** 385). It has been synthesised by *Fischer* and *M. Bergmann* (Ber., 1918, **51,** 1760) who used acetyl as a protecting group and more recently by the condensation of α-2,3,4,6-tetra-*O*-benzylglucose with tri-*O*-benzylgalloyl chloride followed by hydrogenolysis of the protecting groups (*Schmidt* and *H. Schmadel*, Ann., 1961, **649,** 149). This latter synthesis yields both α- and β-glucogallin derivatives depending on the conditions used. β-Glucogallin is slowly hydrolysed by acid and by base and the galloyl group does not migrate from the 1-position.

Hepta-acetyl-β-D-glucogallin, m.p. 125° and 145°, $[\alpha]$ −24.2° (tetrachloroethane):

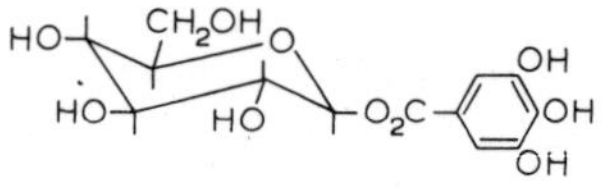

(IV) 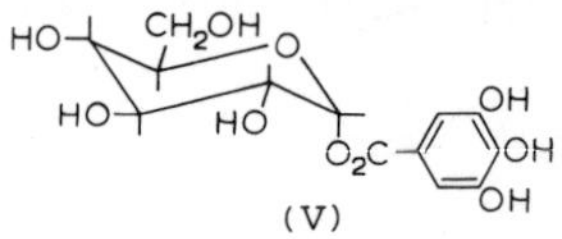 (V)

α-D-**Glucogallin** (V) crystallises from water as the *dihydrate*, m.p. 171°, $[\alpha]_D$ +79.1° (water). It is obtained by reaction of tri-*O*-benzylgalloyl chloride with a solution of α-2,3,4,6-tetra-*O*-benzylglucose in anhydrous pyridine (*Schmidt* and *Schmadel, loc. cit.)* followed by hydrogenolysis of the protecting groups. A previous attempt had been made to synthesise α-glucogallin using acetyl as the protecting group (*Schmidt* and *J. Herok*, Ann., 1954, **587,** 63) but it was later discovered that during removal of the acetyl groups a rearrangement takes place with the formation of 2-galloylglucose (*Schmidt* and *H. Reuss*, *ibid*., 1961, **649,** 137).

Hepta-acetyl-α-D-glucogallin, m.p. 159°, $[\alpha]_D$ +104° (tetrachloroethane).

2-**Galloylglucose** (VIa), α-form crystallised from acetic acid or methanol–ether, m.p. 201° (decomp.), $[\alpha]_D$ +96.8° (water), β-form crystallised from water, m.p. 142° (decomp.), $[\alpha]_D$ +28.9° (water), mutarotating to an equilibrium $[\alpha]_D$ +60.4° (water). This is obtained either by the deacetylation of hepta-acetyl-α-glucogallin or by reaction between triacetylgalloyl chloride and α-1,3,4,6-tetra-acetylglucose and deacetylation of the product. Acetylation of the α-form regenerates *hepta-acetyl-α-D-glucose*, m.p. 165°, $[\alpha]_D$ +113° (tetrachloroethane), identical with the hepta-acetate intermediate in the synthesis from α-1,3,4,6-tetra-acetylglucose.

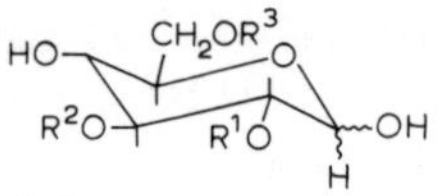

(VI)

(a) R^1 = galloyl, R^2 = R^3 = H
(b) R^2 = galloyl, R^1 = R^3 = H
(c) R^3 = galloyl, R^1 = R^2 = H
(d) R^2 = R^3 = galloyl, R^1 = H

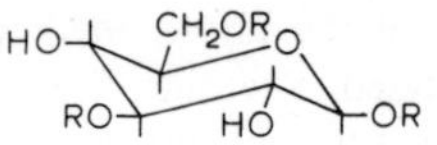

R = galloyl
(VII)

3-**Galloylglucose** (VIb), brittle gum, $[\alpha]_D$ +47.2° (ethanol), *hepta-acetate*, m.p. 149°, $[\alpha]_D$ +15.2° (acetone). Reaction of tri-*O*-benzylgalloyl chloride with 1,2:5,6-di-*O*-isopropylidene-glucose (*C.C.C.* 2nd Edn., Vol. IF, p. 354) gives the 3-tri-*O*-benzylgalloyl derivative from which the protecting groups are removed by successive treatment with dilute acid and hydrogenolysis to yield 3-galloylglucose (*Schmidt* and *A. Schach*, Ann., 1951, **571,** 29).

6-**Galloylglucose** (VIc), decomp. 166°, $[\alpha]_D$ +54.5° → 36.8° (water), is obtained by reaction of 1,2-isopropylidene-3,5-benzylidene-α-D-glucofuranose with tri-*O*-benzylgalloyl chloride and removal of the protecting groups by treatment with acid and hydrogenolysis (*Schmidt* and *Schach, loc. cit.). Hepta-acetate*, m.p. 102°, $[\alpha]_D$ +81.3° (acetone).

3,6-**Digalloylglucose** (VId), decomp. 185°, $[\alpha]_D$ +79° (ethanol), *ennea-acetyl* derivative, m.p. 177°, $[\alpha]_D$ +39.8° (acetone), is a hydrolysis product of chebulinic acid, a natural tannin from myrobalans. It has been synthesised from 3-(tri-*O*-benzylgalloyl)-1,2:5,6-di-*O*-isopropylideneglucose by hydrolysis of the 5,6-*O*-isopropylidene group, reaction with one mole of tri-*O*-benzylgalloyl chloride and then removal of the protecting groups (*Schmidt* and *Schach, loc. cit.).*

β-1,3,6-**Trigalloylglucose** (VII), amorphous, decomp. 190–195°, $[\alpha]_D$ +29.5° (ethanol), gives amorphous acetylation and methylation products. It is characterised by paper chromatography. It is a hydrolysis product of chebulinic acid and terchebin, natural tannins from myrobalans (*Schmidt et al.*, Ann., 1957, **609,** 192; 1967, **706,** 169) and has been synthesised by reaction of tri-*O*-benzylgalloyl chloride with 2,4-di-*O*-benzylglucose and hydrogenolysis of the protecting groups (*Schmidt* and *G. Klinger, ibid.*, 1957, **609,** 199). The β-stereochemistry and conformation of this trigalloylglucose has been established by n.m.r. spectroscopy (*J. C. Jochims, G. Taigel* and *Schmidt, ibid.*, 1968, **717,** 169).

β-1,3,4,6-**Tetragalloylglucose,** m.p. 195°, $[\alpha]_D$ +4.2° (ethanol), *dodecamethyl ether*, m.p. 157°, $[\alpha]_D$ +6.2° (tetrachloroethane), has been isolated from algarobilla (*Schmidt et al., ibid.*, 1969, **729,** 251). The structure was assigned by n.m.r. spectroscopy and after methylation with methyl iodide and silver oxide, the dodecamethyl ether afforded 2-methylglucose on hydrolysis.

2,3,4,6-**Tetragalloylglucose,** amorphous, $[\alpha]_D$ +55.4° (acetone), is a product of the methanolysis of the tannin from Chinese galls, and the leaves of stagshorn and Sicilian sumach (*Haworth et al.*, J. chem. Soc., 1961, 1842). 2,3,4,6-*Tetra*(-O-*benzylgalloyl*)-*glucose*, m.p. 97–98°, has been prepared by the careful hydrolysis of the penta derivative and hydrogenolysis affords tetragalloylglucose (*Haworth et al., loc. cit.; ibid.*, 1962, 2944).

β-1,2,3,4,6-**Pentagalloylglucose,** amorphous, $[\alpha]_D$ +17.7° (acetone), is obtained together with the previously described ester from the methanolysis of gallotannins and has been isolated from myrobalans (*Schmidt et al.*, Ann., 1967, **706,** 186). Reaction of β-glucose with tri-*O*-benzylgalloyl chloride gives the penta-ester which is amorphous and hydrogenolysis of the protecting groups afforded synthetic pentagalloylglucose.

β-Penta(tri-O-methylgalloyl)glucose, m.p. 135° (*Fischer* and *K. Freudenberg*, Ber., 1914, **47,** 2485).

(iii) The gallotannins

The structure of natural gallotannins from Chinese and Turkish galls as a core of pentagalloylglucose to which other gallic acid units are depsidically linked, was put forward by *Fischer* and *Freudenberg* during the period 1914–1919 (*Freudenberg*, "Tannin, Cellulose and Lignin", Verlag Chemie, Berlin, 1933). Much later, as a result of chromatographic evidence, a trisaccharide core was suggested for Chinese gallotannin and a tetrasaccharide core for stagshorn sumach tannin. These results have proved to be erroneous and the structure of the Chinese and Turkish gallotannins has been confirmed as essentially that proposed by *Fischer* and *Freudenberg* (*Haworth et al.*, J. chem. Soc., 1961, 1829, 1842).

N.m.r. spectroscopy has been used to make further refinements to the structures proposed for the gallotannins (*E. Haslam, ibid.*, 1967, 1734). This technique is useful because two adjacent gallate groups attached to glucose adopt a conformation such that each ring falls within the shielding cone of the other. As a result of this the chemical shift of the aromatic protons depends on the position on the sugar to which the gallate group is attached, the 6-position being the most downfield and the 3-position the least downfield. The chemical shift is moved downfield also by a depsidically linked gallate group. Thus the signals for the aromatic protons of gallate groups in a depside chain are downfield from those for single gallate groups attached to glucose. It is possible therefore to make a good estimate of the ratio of depside to ester gallate groups in a sample of gallotannin. A position of attachment for depside chains can be suggested by comparing the spectrum of the tannin with the spectrum of β-pentagalloylglucose where the five distinct signals due to protons on the five aromatic rings have been assigned to specific gallate groups.

Gallotannins are extracted from the natural sources by cold water, extracted into ethyl acetate and the solvent removed. The prolonged action of alcoholic solvents must be avoided as this causes alcoholysis of the gallic acid residues. The crude tannins can be purified by chromatography at 0° over perlon powder and elution with methanol (*Haworth et al., loc. cit.*). Purification by counter-current distribution between aqueous and organic solvent layers is suitable for the removal of gross impurities but it will not effect the separation of isomeric polygalloylated glucoses because these associate in aqueous solution (*idem*, J. chem. Soc., 1962, 2944).

(*1*) *Chinese and Sumach gallotannins.* **Chinese gallotannin** (VIII), white amorphous powder, $[\alpha]_D$ +12.1° (acetone), glucose 12.1%. *Sicilian sumach* and *stagshorn sumach tannins* have similar properties. These are products of *Rhus* species. Chinese tannin is extracted from pathologically damaged leaves of *R. semialata* Murr. galled by an insect. Sicilian and stagshorn sumach tannins are obtained from undamaged leaves.

Chinese gallotannin gives only gallic acid and glucose on hydrolysis in the ratio 9 or 10 to 1. Methylation with diazomethane followed by hydrolysis gave an approximately 1:1 ratio of 3,4,5-tri-*O*-methyl and 3,4-di-*O*-methylgallic acid. *E. Fischer* (Ber., 1919, **52,** 809) suggested that in the substance all the hydroxyl groups in the sugar molecule were esterified with gallic acid and that the tannin had a composition approximating to β-penta-*m*-digalloylglucose. The tannin can be separated into fractions of differing optical rotation by precipitation with alumina which is evidence of it being a complex mixture of galloylated glucoses (*P. Karrer et al.*, Helv., 1923, **6**, 17).

Two new techniques have shed more light on this structural problem. When *Aspergillus niger* is grown on a medium containing tannin, it develops a galloyl esterase which can be purified by chromatography on an ion exchange resin until it shows no carbohydrase activity (*Haworth et al.*, J. chem. Soc., 1961, 1829). This enzyme cleaves gallotannin to gallic acid and glucose only thus confirming the carbohydrate core. The sumach tannins showed the same physical properties as Chinese gallotannin and gave the same proportions of products on hydrolysis and so these gallotannins are considered to be closely related in structure (*idem, ibid.*, 1961, 1842).

Secondly, it was discovered that methanol reacts with tannins to cleave the depside link preferentially. Chinese and sumach tannins were degraded to β-pentagalloylglucose isolated by chromatography over cellulose and identified as its crystalline methylation product (*idem, loc. cit.; ibid.*, 1962, 2944). A small amount of 2,3,4,6-tetragalloylglucose was isolated from tannins derived from old vegetable material but not from the tannin of fresh sumach leaves. Crystalline methyl gallate was also obtained from the methanolysis. The presence of a β-pentagalloylglucose core in the tannins is thus firmly established.

A kinetic investigation into the methanolysis of Chinese gallotannin showed that methyl *m*-digallate is initially formed at a greater rate than methyl gallate, but the didepside is eventually cleaved to methyl gallate. Molecular weight determinations of the tannin gave an average value of 1250 $\pm$60 of which the pentagalloylglucose core accounts for a molecular weight of 940 leaving a residue of relative atomic mass 310. This residue corresponds to only two more galloyl units. Since methanolysis breaks only depsidically linked galloyl groups the two remaining galloyl groups have to be attached as a trigalloyl chain in order to account for the formation of methyl *m*-digallate. The mass spectra fragmentation pattern of fully methylated Chinese gallotannin shows ions at *m/e* 195, 375 and 555 which again indicates the presence of a trigalloyl chain:

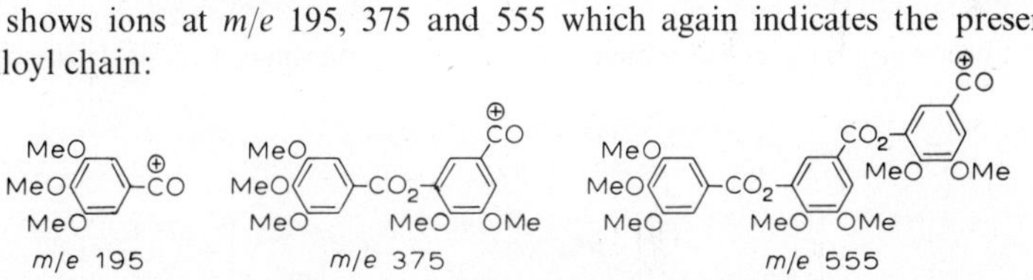

MeO MeO MeO ⊕CO *m/e* 195 — MeO MeO MeO CO_2 MeO OMe ⊕CO *m/e* 375 — MeO MeO MeO CO_2 MeO OMe CO_2 MeO OMe ⊕CO *m/e* 555

A comparison of the n.m.r. spectra of methylated Chinese gallotannin with the spectra of model substances and in particular the position of the aromatic proton resonances suggests that the polygalloyl chain is attached to position 2 of the sugar. Thus the structure of Chinese gallotannin emerges as a mixture of gallate esters of formula VIII:

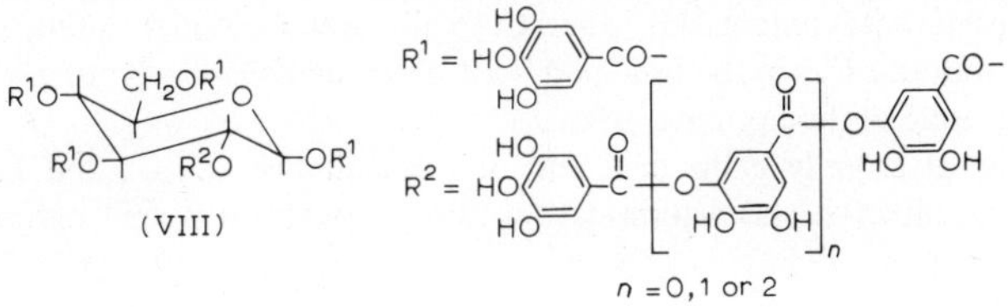

(VIII) — R^1O, CH_2OR^1, R^1O, R^2O, OR^1; R^1 = (HO)$_3$C$_6$H$_2$–CO–; R^2 = (HO)$_3$C$_6$H$_2$–C(=O)–[O–C$_6$H$_2$(OH)$_2$–C(=O)]$_n$–O–C$_6$H$_2$(OH)$_2$–CO–

n = 0, 1 or 2

Old samples of the tannin tend to lose the 1-galloyl group by hydrolysis (*Hàslam et al.*, J. chem. Soc., 1966, 783; 1967, 1734).

Dhava tannin from the leaves of *Anogeissus latifolia* Wall. which is used in S. India has a similar structure to that of Chinese gallotannin (*K. K. Reddy et al.*, Austral. J. Chem., 1964, **17,** 238).

(*2*) *Turkish gallotannin.* **Turkish gallotannin,** amorphous powder, $[\alpha]_D$ +21.7° (acetone), glucose 16.4%, is extracted from the galls produced by a wasp on the young leaf buds of *Quercus lusitanica* L. common in the Levant.

Crude Turkish tannin contains some ellagic acid which is present as a water soluble glucoside. Hydrolysis of the tannin gave gallic acid and glucose in the ratio 5–6 to 1 which led *Fischer* and *Freudenberg* to suggest that the structure approximated to a pentagalloylglucose with extra depsidically linked gallic acid (Ber., 1914, **47,** 2485). *Haworth et al.*, found that Turkish tannin can be obtained free of ellagic acid containing glucosides and that it is hydrolysed solely to gallic acid and glucose (J. chem. Soc., 1961, 1842). Methanolysis under the conditions used for Chinese gallotannin afforded methyl gallate, a non-reducing tetra-*O*-galloylglucose and two reducing tri-*O*-galloylglucoses one of which was crystalline. Thus one sugar hydroxyl group in Turkish tannin must be non-esterified with gallic acid. This is confirmed from the results of methylation of the tannin with methyl iodide and barium oxide in dimethylformamide followed by hydrolysis to give a mixture of 2- and 4-*O*-methylglucose (*Haworth et al., ibid.*, 1962, 3808). Acyl migration can occur during such methylations under alkaline conditions and specifically migration of the acyl group from C-2 to the free hydroxyl on C-4 to give a 2-methoxy derivative where the 4-methoxy derivative would be expected is known for methyl 2,3,6-tri-*O*-benzoylgalactopyranoside (*J. S. D. Bacon et al., ibid.*, 1939, 1248).

The n.m.r. spectrum of Turkish tannin has been interpreted (*Haslam, ibid.*, 1967, 1734) as indicating that the 2-position of glucose is free. The spectrum also indicates the presence of a *m*-trigalloyl chain, on average, which is attached at position 6 and so Turkish tannin is considered to be represented by structure IX. However *β*-1,3,4,6-tetragalloylglucose has been extracted from algarobilla (p. 231) and differs in properties from the tetragalloyl glucose which is obtained by methanolysis of Turkish tannin.

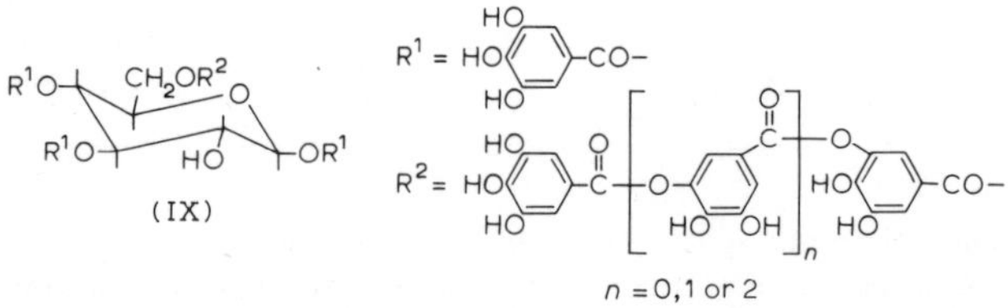

(IX)

n = 0,1 or 2

(*3*) *Tara gallotannin.* **Taratannin** (XI), an amorphous powder, is separated by chromatography on cellulose into fractions A (quinic acid content 29.6%), B, $[\alpha]_D$ −116° (acetone), (quinic acid content 21.0%) and C, $[\alpha]_D$ − 109° (acetone), (quinic acid content 20.2%). It is obtained from the fruit pods of *Caesalpinia spinosa* Kuntze which grows in the Sierra zone of the Peruvian Andes.

Taratannin gives only gallic acid and quinic acid (see C.C.C., 2nd Edn., Vol. II B, p. 139) on hydrolysis and determination of the equivalent weight indicates a tetra-

or penta-galloylated quinic acid structure. Methanolysis afforded 3,4,5-tri-*O*-galloylquinic acid whose structure was confirmed by methylation, hydrolysis and conversion to the crystalline 1-*O*-methylquinide (X). 3,4,5-Tri-*O*-galloylquinic acid has been synthesised and shown to be identical with the methanolysis product from taratannin (*Haslam, Haworth* and *D. A. Lawton*, J. chem. Soc., 1963, 2173). Methyl *m*-digallate and methyl gallate were also identified as products of methanolysis and so structure XI was proposed for taratannin (*Haslam, ≡aworth* and *P. C. Keen, ibid.*, 1962, 3815). This is confirmed by the n.m.r. spectrum which indicates that the depside chain is attached either at positions 3 or 4 where the gallate groups are equatorial and shield each other as is the case for pentagalloylglucose and not at position 5 where the gallate group is axial and therefore out of the shielding cone of the adjacent gallate group (*Haslam, ibid.*, 1967, 1737). The depside chain is thought to be attached at position 3.

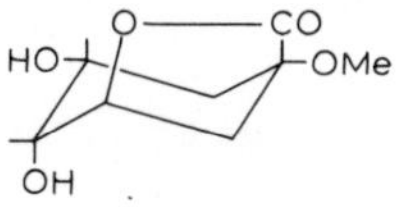

(X) 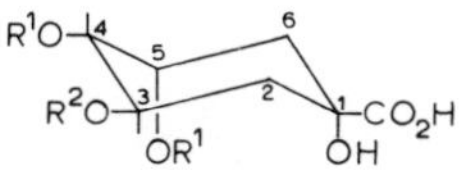 (XI)

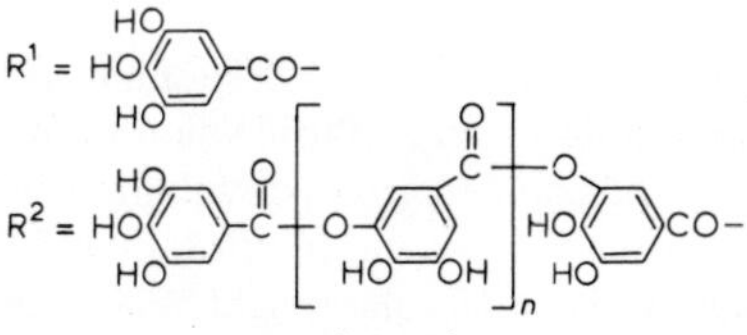

n = 0,1 or 2

3,4,5-*Tri-O-galloylquinic acid*, amorphous powder $[\alpha]_D - 130°$ (water).

(*4*) **Hamamelitannin** (XIIa), crystalline, m.p. 145°, $[\alpha]_D$ +32.6° (water), was first obtained from the bark of witch hazel (*Hamamelis virginiana* L.) and later from chestnut bark (*Castanea sativa* Mill.) and the bark of the American red oak (*Quercus rubra* L.) (*F. Gruttner*, Arch. Pharm., 1898, **236**, 278; *W. Mayer* and *W. Kunz*, Naturwissenschaften, 1959, **46**, 206; *Mayer, Kunz* and *F. Loebich*, Ann., 1965, **688**, 232). It was shown to be a digalloylhexose (*K. Freudenberg*, Ber., 1919, **52**, 177) and the structure of the sugar unit, D-hamamelose, was established as 2-*C*-hydroxymethyl-D-ribose (see C.C.C., 2nd Edn., Vol. IF, p. 538) by *O. Th. Schmidt* (Ann., 1929, **476**, 250; 1934, **515**, 43). The tannin forms a methyl acetal and gives only 3,4,5-tri-*O*-methylgallic acid on methylation and hydrolysis so structure XIIa was proposed. This structure has been confirmed (*Mayer, Kunz* and *Loebich, loc. cit.*) by oxidation of the hexamethyl derivative (XIIb) with periodic acid to yield a triose which afforded 3-tri-*O*-methylgalloylglyceraldehyde osazone identical with a synthetic sample:

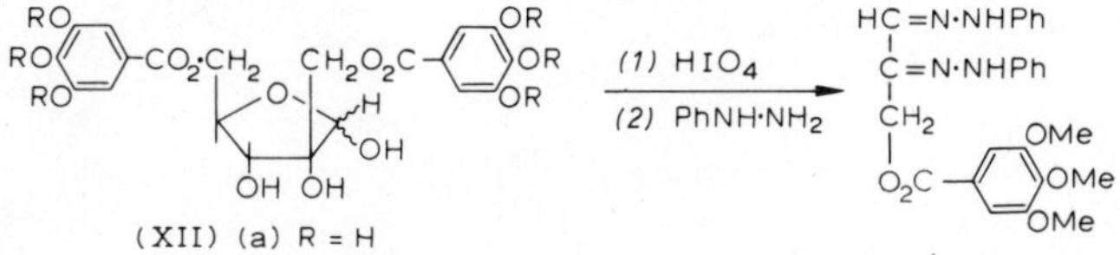

(1) HIO_4
(2) $PhNH{\cdot}NH_2$

(XII) (a) R = H
(b) R = Me

Hamamelitannin does not have the ability to tan leather.

(5) **Aceritannin** (XIII), crystalline, m.p. 164°, $[\alpha]_D$ +20.6° (acetone), *octa-acetate*, m.p. 154°, is isolated from the leaves of *Acer tartaricum* L. (= *A. ginnale*, Maxim.,), native to Korea. Extracts of the leaves are used in N. China for dyeing cotton and silk to a black shade but they have no value as a tannin. Aceritannin was first isolated by *A. G. Perkin* and *Y. Uyeda* (J. chem. Soc., 1922, 66) who showed that hydrolysis gave gallic acid and a non-reducing sugar named aceritol which was subsequently identified as 1,5-anhydro-D-glucitol (C.C.C., 2nd Edn., Vol. I F, pp. 46–48). Methylation of aceritannin with diazomethane gives a hexamethyl derivative which yields tri-*O*-methylgallic acid on hydrolysis. The hexamethyl derivative is unattached by periodic acid and probably does not possess a primary alcoholic group since it does not react with trityl chloride. Structure XIII has therefore been proposed for aceritannin (*N. Kutani*, Chem. Pharm. Bull., Tokyo, 1960, **8**, 72):

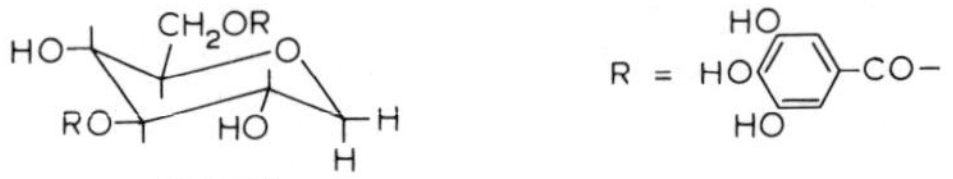

(XIII)

(6) **Gayubatannin,** extracted from the leaves of *Arctostaphylos uva-ursi* L., is used in Spain, admixed with other tannin extracts. Purification of an aqueous extract by chromatography over cellulose and countercurrent distribution afforded a gallotannin, *p*-galloyloxyphenyl-β-D-glucoside and corilagin (*G. Britton* and *E. Haslam*, J. chem. Soc., 1965, 7312). The gallotannin has a molecular weight of 980 ± 30 which corresponds to a mixture of penta- and hexa-galloylglucose. Methanolysis under the conditions used for Chinese gallotannin gave β-pentagalloylglucose. Thus gayuba gallotannin is regarded as the prototype of the complex gallotannins.

(iv) Biosynthesis of gallic acid

Gallic acid is biosynthesised in the higher plants by the shikimic acid route (Scheme 8). In the final stages, two pathways operate. Either (*a*) shikimic acid is dehydrogenated directly to gallic acid or (*b*) it is converted by the usual route to a C_6–C_3 product the side chain of which is then degraded to carboxyl:

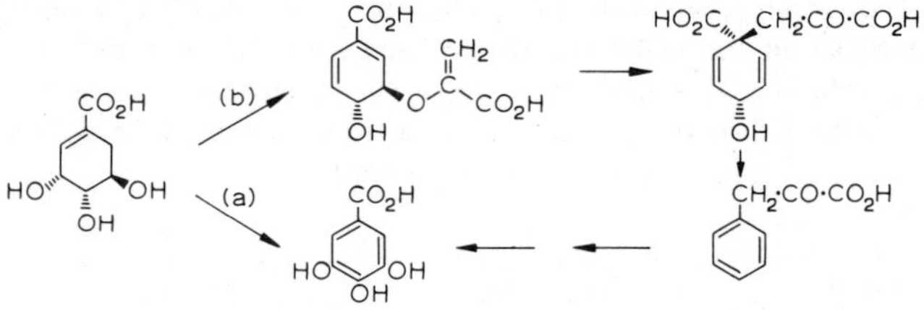

Scheme 8

Phenylalanine has been shown by ^{14}C-labelling techniques to function as a precursor for gallic acid in *Rhus typhina* (*M. H. Zenk*, Z. Naturforsch., 1964, **19b,** 83). However, a more detailed study of the incorporation of shikimic acid and phenylalanine into the tannins of *R. typhina* and gallic acid, quercetin and myrecetin in *Geranium pyrenaicum* Burm. indicates that route (*a*) is the usual one for the biosynthesis of gallic acid. The degradation of C_6–C_3 compounds is a minor pathway (*E. E. Conn* and *T. Swain*, Chem. and Ind., 1961, 592; *P. M. Dewick* and *Haslam*, Biochem. J., 1969, **113,** 537).

(d) Ellagitannins

Acid hydrolysis of ellagitannins gives glucose, usually some gallic acid and an insoluble phenol. Ellagic acid is often present in this insoluble hydrolysis product but early workers failed to notice that other related acids may also be present. These acids which are the building units of the ellagitannins will be described first and the pure tannins which have been isolated from crude extracts will be discussed later. Finally the biogenetic relationship between these acids and gallic acid will be discussed.

(i) Acid components of hydrolysed tannin extracts

Ellagic acid (XIV) recrystallised from pyridine, decomp. ~ 450°, *tetra-acetate*, m.p. 343°, *tetrabenzoate*, m.p. 332°. It is obtained by hydrolysis of some tannin extracts, particularly myrobalans or divi-divi, and by the oxidation of gallic acid with persulphate and acid or by the autoxidation of ethyl gallate in alkaline solution. Treatment of the tetramethyl ether with sodium hydroxide and dimethyl sulphate gives 2,3,4,2′,3′,4′-*hexamethoxy*-6,6′-*diphenic acid* (XV), m.p. 240° (*J. Herzig* and *J. Pollak*, Monatsh., 1908, **29,** 267):

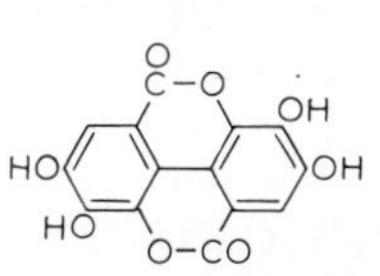

(XIV) 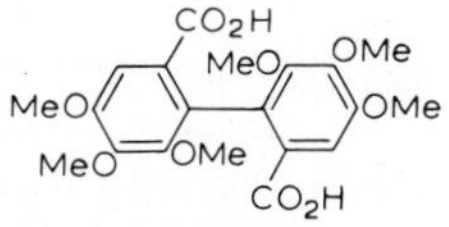 (XV)

Ellagic acid is a very insoluble substance so the way in which it is made soluble in the ellagitannins has provoked discussion. Methylation with diazomethane of the natural ellagitannins corilagin from divi-divi and chebulagic acid from myrobalans followed by hydrolysis gave optically active (+)-hexamethoxydiphenic acid (XV). Thus the ellagic acid is present in these tannins as the diester with glucose of hexahydroxydiphenic acid. The acid is optically active due to restricted rotation about the diphenyl bond (*O. Th. Schmidt et al.*, Ann., 1952, **576,** 75).

(+) and (−)-2,3,4,2′,3′,4′-**Hexahydroxy**-6,6′-**diphenic acids** have been prepared and the various derivatives inter-related as shown below (*Schmidt* and *K. Demmler*, Ann., 1952, **576,** 85; 1954, **586,** 179):

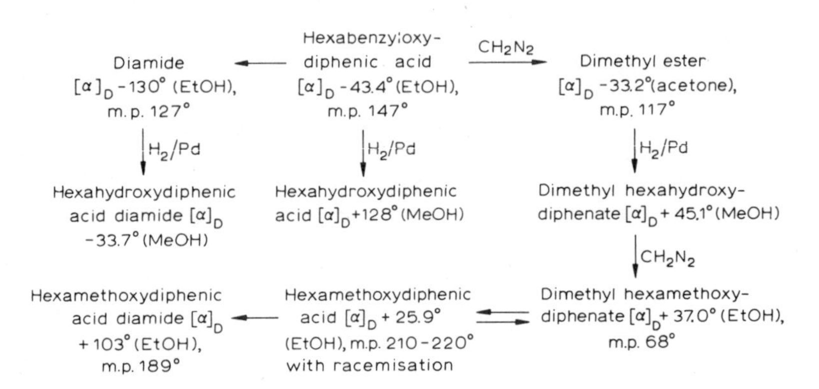

(±)-Hexamethoxydiphenic acid was resolved using quinidine and the hexabenzyloxy compound was resolved using cinchonine. The ethers are not racemised at room temperature and the hexamethyl ether has a half-life of racemisation of 14h in boiling 0.1 *M* sodium hydroxide and 158 min in refluxing acetic acid.

Solutions of hexahydroxydiphenic acid in aqueous methanol slowly deposit ellagic acid. The half-life for racemisation of the hexahydroxydiphenic acid is 281 min at 20° and for the dimethyl ester 215 min at 20°. Thus this acid when present in tannins must be held in one enantiomeric configuration by the glucose molecule to which it is attached. (+)-Hexahydroxydiphenic acid occurs in divi-divi and myrobalans while the (−)-form occurs in valonea Knoppern galls and castalagin from oak and chestnut wood.

In **terchebin,** hexahydroxydiphenic acid is present as a non-aromatic isomer (XVI is one tautomeric form) and can be isolated as the phenazine XVIII after reaction with *o*-phenylenediamine (see Scheme 9).

2,3-[(−)-*Hexahydroxydiphenoyl*]*glucose*, amorphous, $[\alpha]_D$ +55.6° (methanol); *β-enneamethyl* derivative, m.p. 216°, $[\alpha]_D$ −21.3° (acetone) (*Schmidt et al.*, Ann., 1965, **690,** 150) from partial hydrolysis of penduncculagin.

4,6-[(−)-*Hexahydroxydiphenoyl*]*glucose*, prisms from water, $[\alpha]_D$ +45.1° (ethanol), *ennea-acetate*, m.p. 265°, $[\alpha]_D$ + 14° (dioxane) (*idem, ibid.*, 1967, **706,** 164) from brevilagin-2.

3,6-[(+)-*Hexahydroxydiphenoyl*]*glucose*, amorphous, $[\alpha]_D$ small negative (*idem, ibid.*, 1952, **576,** 75), from corilagin.

Hexahydroxydiphenoylglucoses are thought to be formed in nature by the oxidative coupling of a digalloylglucose just as the *in vitro* oxidation of ethyl gallate gives ellagic acid. One particular enantiomer is formed as a result of the chirality and preferred conformation of the digalloylglucose prior to oxidative coupling.

Dehydrohexahydroxydiphenic acid (XVII in one tautomeric form) is the oxidation state of some of the ellagic acid-producing residues in brevilagin-1 and -2 isolated from algarobilla. It is isolated as ellagic acid, in poor yield, when the tannin is hydrolysed by sulphuric acid, as *chloroellagic acid, tetramethyl ether*, m.p. 290°, when the tannin is hydrolysed by concentrated hydrochloric acid followed by methylation of the products and as brevifolincarboxylic acid (XXIII) when the tannin is hydrolysed by water (*idem, ibid.*, 1967, **706,** 131).

This molecule is isolated as the phenazine XIX when the tannin is treated with *o*-phenylenediamine. The phenazinodiphenic acid XVIIIb was synthesised by a standard route from diphenic acid derivatives and shown to be identical with material derived from the phenazine XIX (*idem, ibid.*, 1967, **706,** 180).

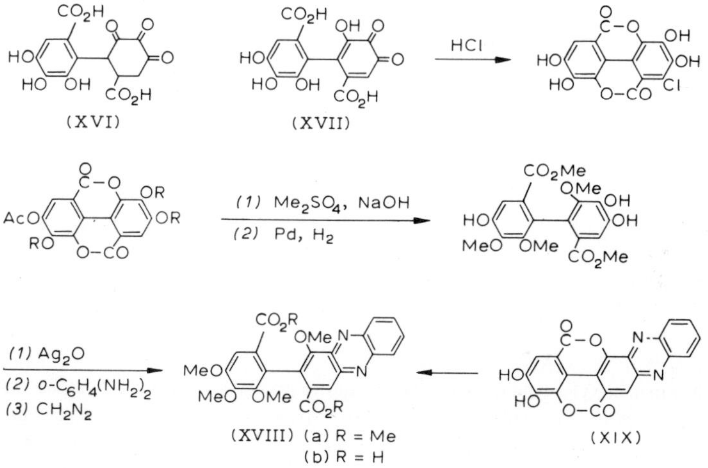

(XVI) (XVII) HCl

(1) Me_2SO_4, NaOH (2) Pd, H_2

(1) Ag_2O (2) o-$C_6H_4(NH_2)_2$ (3) CH_2N_2

(XVIII) (a) R = Me (b) R = H (XIX)

Scheme 9

When the brevilagins and terchebin are methylated, treated with *o*-phenylenediamine and hydrolysed, XVIIIb is isolated in optically active form, the (−)-form from brevilagin and the (+)-form from terchebin (*idem, ibid.*, 1967, **706,** 131, 169).

(+)-**Chebulic acid,** *split acid*, (XXa), amorphous, characterised as its *triethyl ester*, m.p. 188°, $[\alpha]_D$ +24.6° (ethanol). It is isolated from the hydrolysate of myrobalan extract where it is present as part of the tannin chebulinic acid (*K. Freudenberg*, Ber., 1919, **52,** 1238; 1920, **53,** 1728). (−)-*Chebulic acid* has been isolated from algarobilla (*Schmidt et al.*, Ann., 1969, **729,** 249). Chebulic acid was originally purified as the thallium(I) salt but it is more conveniently isolated from myrobalans as the *trimethyl ether*, amorphous, $[\alpha]_D$ +37.4° (water), and characterised as the crystalline *trimethyl ether triamide*, m.p. 257°, $[\alpha]_D$ +48.7° (water) (*Schmidt* and *W. Mayer*, *ibid.*, 1951, **571,** 1; *Mayer, ibid.*, 1951, **571,** 15).

Oxidation of tri-*O*-methylchebulic acid with potassium ferricyanide gave 3,4,5-trimethoxyphthalic acid and pyrolysis gave the isocoumarin XXI thus defining the structure of the aromatic portion of the molecule (*R. D. Haworth* and *L. D. de Silva*, J. chem. Soc., 1951, 3511; 1954, 3611). This aromatic ring was destroyed by oxidation with potassium permanganate to yield a lactonic tricarboxylic acid whose structure was deduced as XXII by further degradation on fusion with potassium hydroxide to acetic, oxalic and succinic acids (*Schmidt* and *Mayer*, *loc. cit.*). The structure

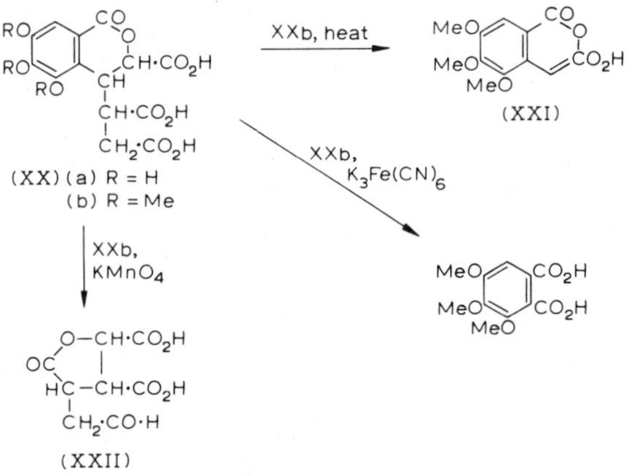

(XX) (a) R = H
(b) R = Me

XXb, heat

(XXI)

XXb, $K_3Fe(CN)_6$

XXb, $KMnO_4$

(XXII)

XXa proposed for chebulic acid has been confirmed by n.m.r. and mass spectra. In addition this evidence indicates a *cis*-arrangement of the substituents on the dihydroisocoumarin ring (*E. Haslam* and *W. Uddin*, J. chem. Soc., C, 1967, 2381; *J. C. Jochims*, *G.' aigel* and *Schmidt*, Ann., 1968, **717,** 169).

Brevifolincarboxylic acid (XXIII) forms yellow prisms which decarboxylate on heating to **brevifolin** (XXIV), yellow crystals; *methyl tri-O-methylbrevifolincarboxylate*, m.p. 164°, *tri-O-methylbrevifolin*, m.p. 213° (*Schmidt* and *K. Bernauer*, *ibid.*, 1954, **588,** 211).

Brevifolincarboxylic acid is obtained by careful hydrolysis of an aqueous extract of algarobilla. Brevifolin separates when an acetone extract of algarobilla is boiled with water. In the early literature this precipitate was mistaken for ellagic acid. Careful hydrolysis of the crystalline tannin algarobin affords (−)-*brevifolincarboxylic acid*, $[\alpha]_D$ −69.1° (methanol) (*Schmidt et al.*, *ibid.*, 1967, **706,** 204):

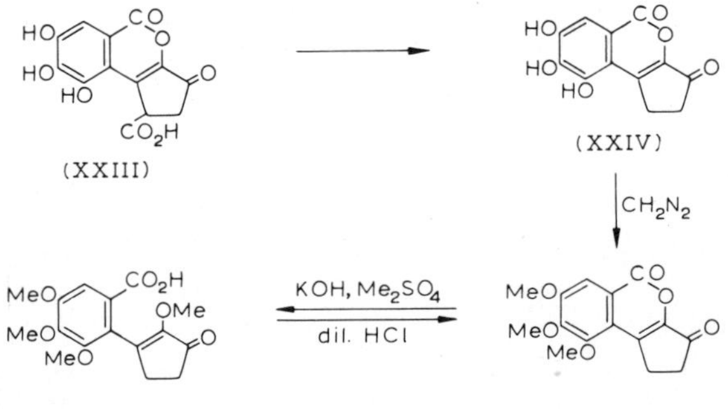

(XXIII)

(XXIV)

CH_2N_2

KOH, Me_2SO_4

dil. HCl

Brevifolin has three phenolic groups methylated with diazomethane, a ketone function and a lactone ring. The lactone ring of tri-*O*-methylbrevifolin is opened by alkali and dimethyl sulphate to give a methoxycarboxylic acid which regenerates tri-*O*-methylbrevifolin on mild treatment with acid. Oxidation of tri-*O*-methylbrevifolin afforded 3,4,5-trimethoxyphthalic acid and succinic acid. This evidence led to structure

XXIV for brevifolin (*Schmidt* and *Bernauer*, *loc. cit.*). Tri-*O*-methylbrevifolin has been synthesised by the Meerwein reaction between the diazonium salt XXV and 1-methoxy-1-cyclopenten-5-one (*idem*, Ann., 1955, **591,** 153) and by cyclisation of the isocoumarin derivative XXVI (*J. Grimshaw* and *R. D. Haworth*, Chem. and Ind., 1955, 199):

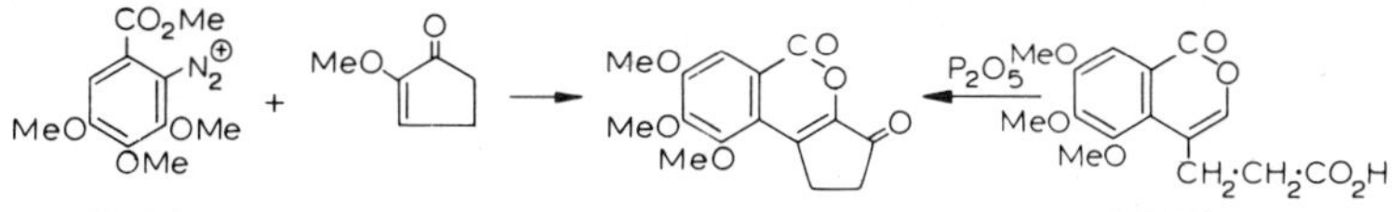

(XXV) (XXVI)

Structure XXIII for brevifolincarboxylic acid was first advanced on the basis of a biogenetic relationship to ellagic acid (*Grimshaw*, *Haworth* and *H. K. Pindred*, J. chem. Soc., 1955, 833; *Schmidt* and *Mayer*, Angew. Chem., 1956, **68,** 103) and subsequently established by synthesis using the Meerwein reaction (*Schmidt* and *R. Eckert*, Ann., 1958, **618,** 71):

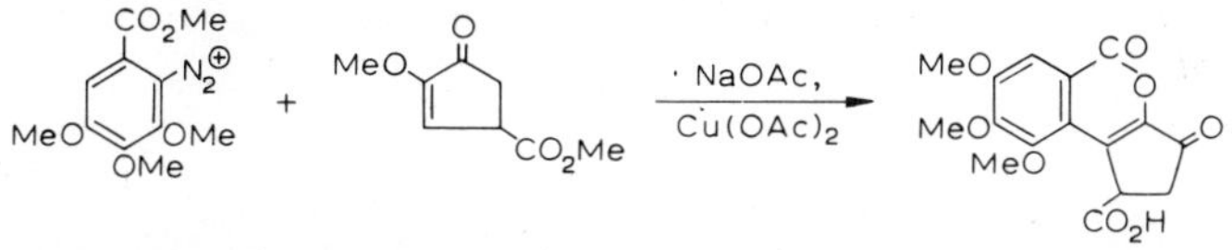

Dehydrodigallic acid (XXVII), darkens 240–260°, *penta-acetate*, m.p. 210°, *hepta-methyl derivative*, m.p. 111°, is obtained from an aqueous extract of chestnut leaves (*Castanea sativa* Mill.) (*Mayer*, Ann., 1952, **578,** 34). Reaction with hot dilute sodium hydroxide gives gallic acid. The structure of dehydrodigallic acid was established by synthesis of the heptamethyl derivative through the Ullmann reaction for the preparation of diphenyl ethers:

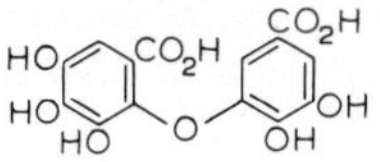

(XXVII) 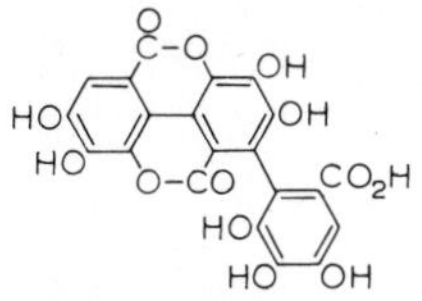 (XXVIII)

Flavogallonic acid (XXVIII) is obtained by hydrolysis of water-soluble tannins and is the stable form of **flavogallol** under these conditions. Solution in sulphuric acid and precipitation by water gives flavogallol which is converted into flavogallonic acid by brief hydrolysis (p. 196). *Methyl hepta*-O-*methyl-flavogallonate*, m.p. 244°. Flavogallol has been obtained by acid hydrolysis of the tannin from the skins of pomegranate (*Punica granatum* L.) (*Schmidt* and *W. Fickert*, Z. Naturforsch. 1958, **13b,** 136) and by hydrolysis of the crystalline tannins castalin and vescalin isolated from the wood of *Castanea sativa* Mill. (chestnut) and *Quercus sessiliflora* Salisb. (European chestnut oak) and from valonea (*Mayer et al.*, Ann., 1967, **707,** 182; 1971, **747,** 51; 1971, **754,** 149).

Valoneic acid dilactone (XXIX) is relatively soluble in acetone from which it crystallises as prisms, *hexa-O-methylvaloneic acid dilactone methyl ester*, m.p. 262°, *hexa-O-methylvaloneic acid dilactone*, m.p. 277°. It is isolated from valonea extract by acid hydrolysis and separated from ellagic acid because of its solubility in acetone (*Schmidt* and *K. Komarck*, *ibid*., 1955, **591,** 156). More recent work on the tannins of valonea indicates that the residual ellagic acid must also contain flavogallonic acid:

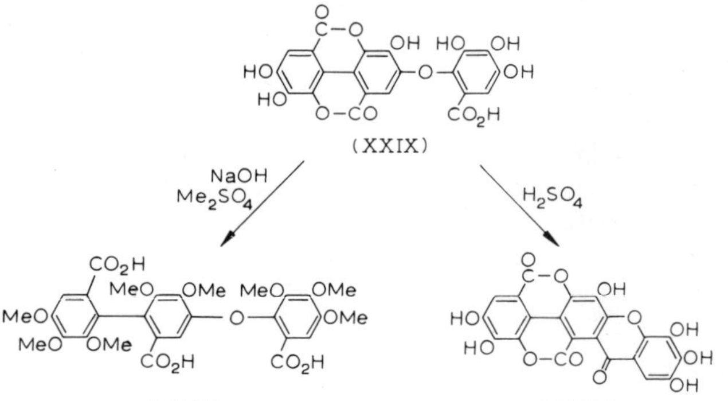

(XXIX)

NaOH, Me_2SO_4 → (XXX)

H_2SO_4 → (XXXI)

Methylation of valoneic acid dilactone with diazomethane gives the hexamethyl ether methyl ester from which the hexamethyl ether can be obtained by hydrolysis. Methylation with dimethyl sulphate and sodium hydroxide gives the octamethyl ether tricarboxylic acid XXX. Valoneic acid is decomposed on boiling with alkali to give ellagic acid and gallic acid thus indicating the general structural features of the molecule. Since valoneic acid dilactone is converted to a xanthone XXXI, structure XXIX was advanced and subsequently established by synthesis. Hydrolysis of tetra-acetylellagic acid gives the diacetyl derivative XXXII which served as starting material for the synthesis of octa-*O*-methylvaloneic acid (XXX) (*Schmidt et al.*, *ibid*., 1957, **602,** 50):

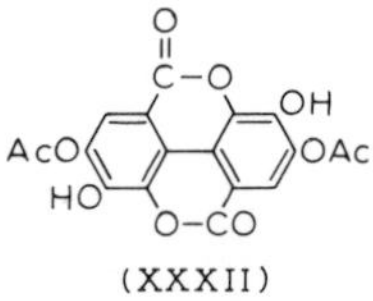

(XXXII)

A water soluble tannin containing valoneic acid has not so far been isolated. However, methylation of the crude tannins from valonea with diazomethane and then hydrolysis has given (−)-octa-*O*-methylvaloneic acid (XXX) and (−)-2,3,4,2′,3′,4′-hexamethoxybiphenyl-6,6′-dicarboxylic acid (hexamethoxydiphenic acid) (*Schmidt* and *H. H. Grunewald*, *ibid*., 1957, **603,** 183). Thus valoneic acid must be bound, in the tannin, to glucose in a non-lactonic form as is ellagic acid.

(ii) The ellagitannins

(*1*) *Myrobalans*. Fruits of the tree *Terminalia chebula* Retz. native to India and

Malaysia, are collected when immature and about 2 cm in diameter. They have a thick hard shell and a kernel. The shell is powdered and sold as myrobalan meal and constitutes an important natural tannin. The crystalline ellagic tannins which have been isolated from this source are given in Table 5 and in addition some simple galloylglucoses have also been isolated. Corilagin and terchebin are isolated from the phenolic components by counter-current distribution.

Chebulinic acid and chebulagic acid are isolated from the carboxylic acid components. Neochebulinic and neochebulagic acids are hydrolysis products of chebulinic and chebulagic acids respectively.

Corilagin and terchebin have also been isolated from the fruits of *Phyllanthus embilica* L. used locally in S. India and Amla tannin (*Y. M. Theresa*, *K. Shastry* and *Y. Nayudamma*, Leather Sci. Madras, 1968, **15**, 337). Chebulinic and chebulagic acids are also obtained from the bark of *Terminalia myriocarpa* Heurck and Muell. (*A. D. Barua*, *J. N. Baruah* and *Shastry*, J. Proc. Inst. Chem. India, 1967, **39**, 111).

Chebulinic acid, (XXXIII), is relatively easy to obtain in a crystalline condition. It gives gallic acid, (+)-chebulic acid (XXa) and 3,6-digalloylglucose on hydrolysis with hot water *(Freudenberg* and *Frank, loc. cit.)*. The reaction with water gives first **neochebulinic acid** (XXXIV). This new acid, unlike chebulinic acid has a free hydroxyl group on the sugar. Methylation with diazomethane then methyl iodide and silver oxide, followed by hydrolysis gives 2-*O*-methylglucose. Neither chebulinic nor neochebulinic acid is decarboxylated by heating with pyridine so the 3,4,5-trihydroxybenzoic acid functions in both molecules are probably esterified. Further hydrolysis of neochebulinic acid gives β-1,3,6-trigalloylglucose. Chebulic acid is thought to be formed in the plant by degradation of a hexahydroxydiphenoyl* residue already bound to glucose as an ester (see p. 239). The particular carboxyl group of the aliphatic part which is esterified to glucose is the one which was originally part of the diphenoyl residue. The orientation with the aromatic carboxyl function at position 2 of glucose follows from an examination of the n.m.r. spectra of chebulic, chebulinic and neochebulinic acids. Neochebulinic acid has the dihydroisocoumarin ring as in chebulic acid but this is absent in chebulinic acid. Chebulinic acid, however, has a free hydroxyl group on the chebulic acid residue. Thus in the rearrangement from chebulinic acid to neochebulinic acid the aromatic ester function of the chebulic acid portion must become detached from glucose and link up to form the dihydroisocoumarin *(Jochims, Taigel* and *Schmidt, loc. cit.)*:

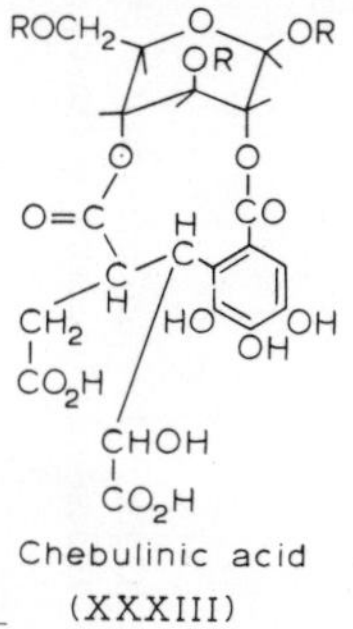

Chebulinic acid
(XXXIII)

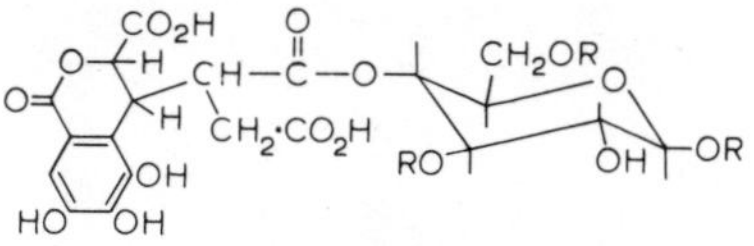

Neochebulinic acid
(XXXIV)

R = HO(HO)(HO)C_6H_2-CO-

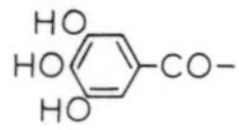

*A convenient trivial name for the 6,6-dicarbonyl-2,2′,3,3′,4,4′-hexahydroxybiphenyl radical.

TABLE 5

ELLAGITANNINS ISOLATED FROM MYROBALAN EXTRACT

Tannin	*Physical properties*[a]	*Hydrolysis products*	*Refs.*
Chebulinic acid[b] (XXXIII)	$+9H_2O$, $[\alpha]_D$ $+65.1°$	Neochebulinic acid	1,2
Neochebulinic acid (XXXIV)	$+6H_2O$, m.p. 193°, $[\alpha]_D$ $+12.4°$	β-1,3,6-Trigalloylglucose, (+)-chebulic acid	2
Chebulagic acid (XXXV)	$+10H_2O$, $[\alpha]_D$ $-57.2°$	Neochebulagic acid	3,4
Neochebulagic acid (XXXVI)	$+9H_2O$, decomp. 196°, $[\alpha]_D$ $-102.3°$	Corilagin, (+)-chebulic acid	4,5
Corilagin (XXXVII)	$+3H_2O$, m.p. 204°, $[\alpha]_D$ $-246°$	(+)-Hexahydroxydiphenic acid, gallic acid, glucose	6,7,8,9
Terchebin (XXXIX)	$+10H_2O$, $[\alpha]_D$ $-53,4°$	β-1,3,6-Trigalloylglucose	10

[a]Optical rotations measured in ethanol.
[b]Classified along with the closely related chebulagic acid but it gives no ellagic acid on hydrolysis.

References to Table 5

1 *Fridolin*, Dissertat. Dorpat, 1884; *K. Freudenberg* and *Th. Frank*, Ann., 1927, **452,** 303.
2 *O. Th. Schmidt et al., ibid.*, 1957, **609,** 192.
3 *Schmidt* and *W. Nieswandt, ibid.*, 1950, **568,** 165.
4 *Schmidt, R. H. Hensler* and *P. Stephen, ibid.*, 1957, **609,** 186.
5 *Schmidt* and *D. M. Schmidt, ibid.*, 1952, **578,** 25.
6 *Idem, ibid.*, 1952, **578,** 31.
7 *O. Th. Schmidt* and *R. Lademann, ibid.*, 1950, **569,** 149.
8 *Schmidt et al., ibid.*, 1952, **576,** 75, 85; 1954, **586,** 179.
9 *O. Th. Schmidt, D. M. Schmidt* and *J. Herok, ibid.*, 1954, **587,** 67.
10 *O. Th. Schmidt et al., ibid.*, 1967, **706,** 169.

The relative stereochemistry of chebulic acid is not fully established and the absolute stereochemistry is not known.

Chebulagic acid (XXXV), on treatment with water gives first **neochebulagic acid** (XXXVI) where a rearrangement has occurred to free the 2-hydroxyl group of glucose. Methylation of neochebulagic acid and hydrolysis gives 2-*O*-methylglucose. Further hydrolysis of neochebulagic acid yields (+)-chebulic acid (XXa) and corilagin:

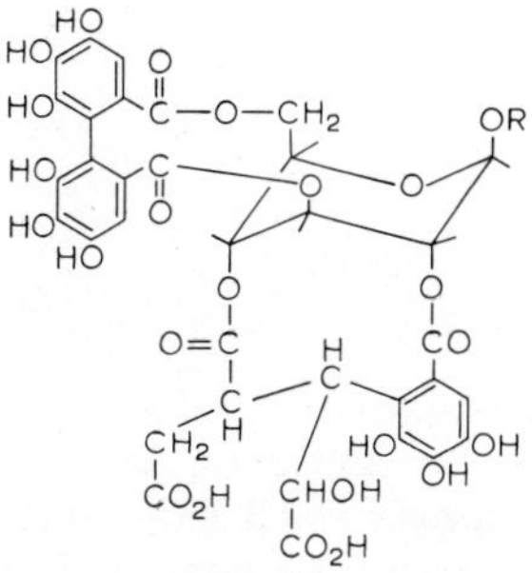

Chebulagic acid
(XXXV)

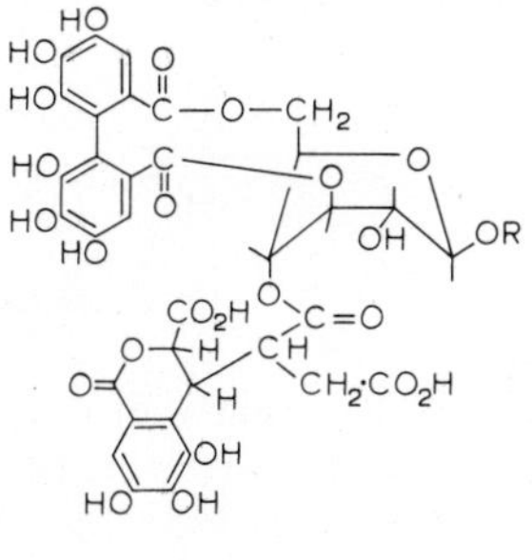

Neochebulagic acid
(XXXVI)

R = HO (3,4,5-trihydroxyphenyl) $-CO-$

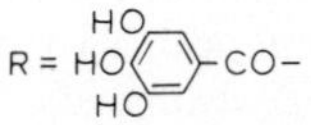

Corilagin (XXXVII), has been isolated from myrobalan and from divi-divi. On acid hydrolysis it gives one mole each of ellagic acid, gallic acid and glucose. Methylation of corilagin with diazomethane and then methyl iodide and silver oxide gives a hendekamethyl derivative which is hydrolysed to 2,4-di-*O*-methylglucose. On stepwise hydrolysis one mole of tri-*O*-methylgallic acid is removed first to give a reducing sugar derivative. From this hydrolysis the ellagic acid residue is isolated as (+)-hexamethoxydiphenic acid. The ellagic acid residue must be present in corilagin esterified to glucose in the form of (+)-hexahydroxydiphenic acid:

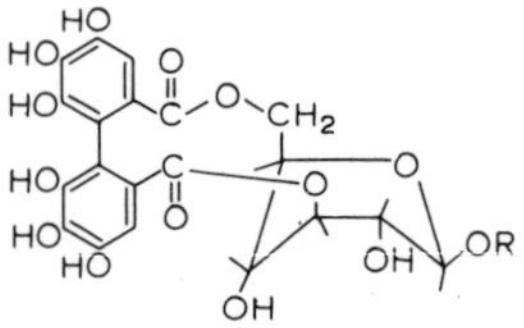

Corilagin
(XXXVII)

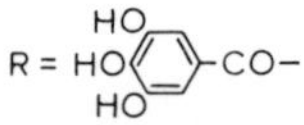

R = HO(HO)(HO)C₆H₂–CO–

The relationship between chebulagic and neochebulagic acids was established by n.m.r. spectroscopy and is like the relationship of chebulinic and neochebulinic acids. The final structural features and the conformations of chebulagic, neochebulagic acids and corilagin were obtained from these spectral data (*J. C. Jochims*, *G. Taigel* and *Schmidt*, Ann., 1968, **717**, 169). However, the absolute stereochemistry of (+)-hexahydroxydiphenic acid and (+)-chebulic acid is not known.

Terchebin (XXXIX) is a yellow crystalline compound. Acid hydrolysis gives ellagic acid, gallic acid and glucose. On treatment with *o*-phenylenediamine, however, terchebin is hydrolysed to β-1,3,6-trigalloylglucose and the phenazine XIX. Further, catalytic hydrogenation of terchebin gives a dihydro derivative which is readily oxidised to terchebin. This evidence suggests that the ellagic acid residue is present in terchebin in the form XXXVIII and similar reactions are known for the corresponding isomer of gallic acid (p. 239):

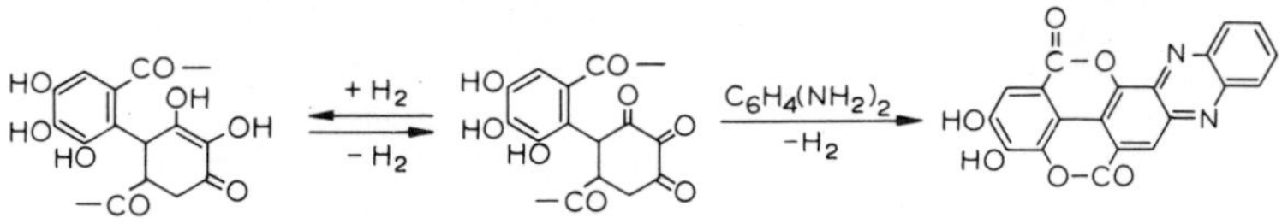

$+H_2$ / $-H_2$ (XXXVIII) $C_6H_4(NH_2)_2$ / $-H_2$ (XIX)

Methylation of terchebin with diazomethane followed by reaction with *o*-phenylenediamine gives the (+)-phenazinedicarboxylic acid XVIIIb (p. 239) so the residue XXXVIII must be held to the glucose in one specific chiral form. This residue is attached to the 2- and 4-positions of glucose and the orientation is deduced from a probable biogenetic relationship between the oxidised diphenic acid residue and the chebulic acid residue in chebulinic acid. Two tautomeric forms of the triketocyclohexane ring are present in solution, only one of which is given in the structure below:

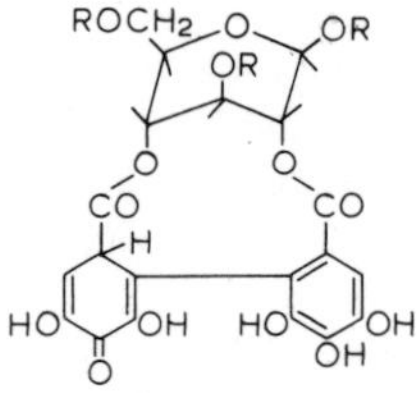

Terchebin
(XXXIX)

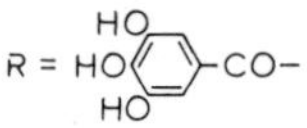

R = HO(HO)(HO)C₆H₂–CO–

(*2*) *Divi-divi* is the short coiled pods of a small tree *Caesalpinia coriaria* Willd. which grows throughout lowland tropical America. The pods are brown in colour and the husk contains up to 50% of tannin which is soluble in water. Separation of the tannin into phenolic and carboxylic acid fractions and further purification as described for myrobalan extract gave corilagin from the phenolic fraction and chebulagic acid from the carboxylic acid fraction (*O. Th. Schmidt* and *R. Lademann*, Ann., 1950, **569,** 165; 1951, **571,** 232).

(*3*) *Algarobilla* is the pods of *Caesalpinia brevifolia* Baill. growing in the dry tropical regions of Chile, south of Atacama. The pod is brittle, rich in tannin and largely soluble in water. The tannins are separated into phenolic and carboxylic acid fractions which can be further fractionated by chromatography over cellulose and elution with acetic acid. Brevilagin-1 and brevilagin-2 are isolated from the phenolic fraction and algarobin from the acidic fraction. The acid function in algarobin is a cyclopentane-1,2-dione group.

TABLE 6

ELLAGITANNINS ISOLATED FROM ALGAROBILLA

Tannin	*Structure*	*Physical properties*	*Ref.*
Brevilagin-1	XL	$+5H_2O$. decomp. 80°, $[\alpha]_D$ +159 (ethanol)	1
Brevilagin-2	XLI	$+6H_2O$. $[\alpha]_D$ +81° (ethanol)	2
Algarobin	XLII	$+3.5H_2O$, decom., 245°, $[\alpha]_D$ −21.4° (ethanol) mutarotating to −42.8°	3

References
1 *O. Th. Schmidt et al.*, Ann., 1967, **706,** 131.
2 *Idem, ibid.*, 1967, **706,** 154.
3 *Idem, ibid.*, 1967, **706,** 204.

Brevilagin-1 (XL) is crystalline and was first isolated by *Schmidt* and *Bernauer (loc. cit.)*. It has one molecule of glucose esterified with two molecules of dehydrohexahydroxydiphenic acid (XVII) which after reaction of the tannin with *o*-phenylenediamine and hydrolysis is isolated as the phenazine XIX (Scheme 9, p. 239). Hydrolysis of the tannin with concentrated hydrochloric acid gives two moles of chloroellagic acid. Methylation of the tannin with diazomethane and then methyl iodide and silver oxide followed by hydrolysis gives 2-*O*-methylglucose so the 1,3,4 and 6 positions of glucose must be esterified. The product of reaction between the tannin and *o*-phenylenediamine undergoes partial hydrolysis with loss of only one dehydrohexahydroxydiphenic acid unit leaving a crystalline glucose ester which by permethylation and hydrolysis gives *β*-1,2,3-tri-*O*-methylglucose. Thus the two dehydrohexahydroxydiphenic acid units in brevilagin-1 must be bound in 4,6 and 1,3 positions to glucose and structure XL follows for the tannin. The conformation has been deduced by n.m.r. spectroscopy (*Jochims et al.*, *ibid.*, 1968, **717,** 169). It is not certain if the dehydroaromatic rings are linked to positions 1 and 4 of glucose as shown or to positions 3 and 6:

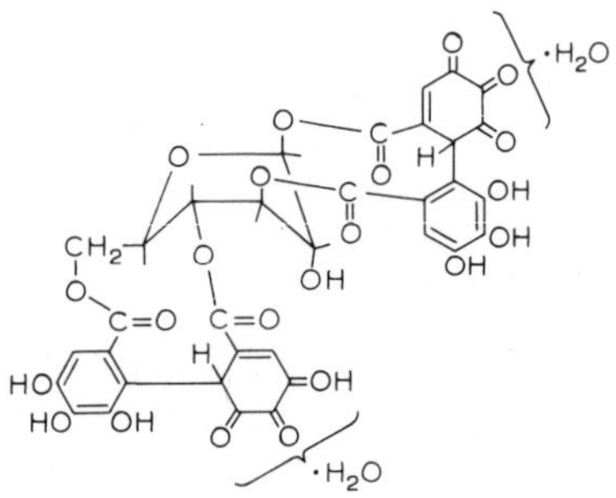

Brevilagin-1
(XL)

Condensation of brevilagin-1 with *o*-phenylenediamine, methylation and hydrolysis yields the (—)-phenazinedicarboxylic acid XVIIIb (p. 239) and methylation of the crystalline glucose 4,6-bound phenazinedicarboxylate obtained by partial hydrolysis also yields (−)-phenazinedicarboxylic acid. The ring planes of both dehydrohexahydroxydiphenic acid molecules in the tannin must be twisted in the same chiral sense.

Brevilagin-2 (XLI) on hydrolysis with concentrated hydrochloric acid gives one mole each of ellagic acid and chloroellagic acid. It contains one hexahydroxydiphenic acid and one dehydrohexahydroxydiphenic acid unit which after reaction with *o*-phenylenediamine, methylation and hydrolysis can be recovered as (−)-hexamethoxydiphenic acid and the (−)-phenazinedicarboxylic acid XVIIIb, respectively. Using the procedures of permethylation and partial hydrolysis the hexahydroxydiphenic acid unit is shown to be linked to the 4,6-positions of glucose and the 2-position is free in the tannin. Structure XLI has been proposed for brevilagin-2 but as with brevilagin-1, it is not certain if the dehydroaromatic ring is linked to position 1 or position 3:

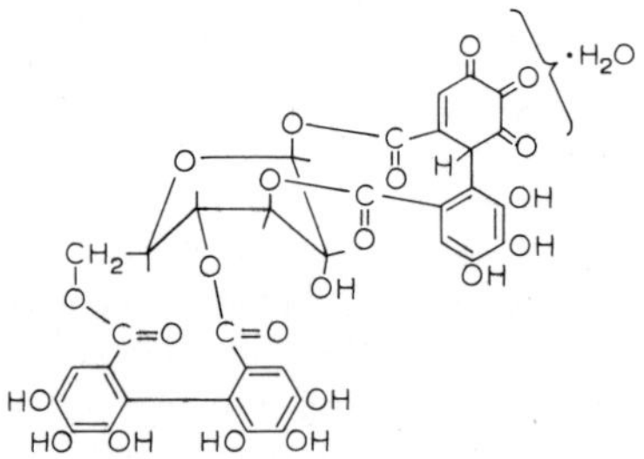

Brevilagin-2
(XLI)

After hydrolysis of either brevilagin-1 or brevilagin-2 by hot water, the dehydrohexahydroxydiphenic acid unit is isolated as a mixture of brevifolincarboxylic acid and brevifolin (XXIV).

Algarobin (XLII) on hydrolysis with hydrochloric acid yields one mole each of glucose and (−)-brevifolincarboxylic acid. It shows mutarotation and gives a colour reaction with aniline phthalate so the 1 and 2 positions of glucose must be free. The rate of hydrolysis is the same as the rate of hydrolysis of 4,6-(hexahydroxydiphenoyl)glucose and faster than that of the 3,6-isomer so algarobin is given structure XLII but it is not certain if the cyclopentanedione ring is attached at position 4, as shown, or position 6 of glucose:

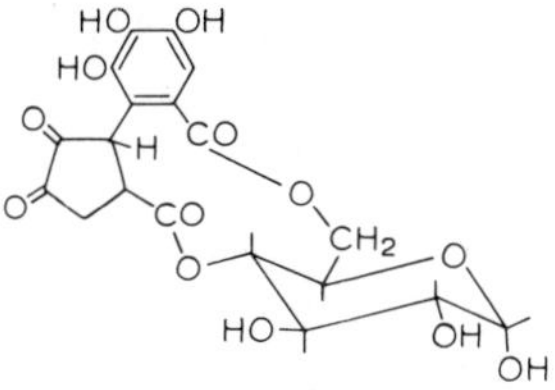

Algarobin
(XLII)

(*4*) *Knopper galls*. Many species of insect produce galls on oak trees but each has a specific point of attack (*E. T. Connold*, "British Oak Galls", Adlard, London, 1908; *A. Darlington*, "Plant Galls", Blandford Press, London, 1968). Knopper galls used in commercial leather production are the result of attack of the wasp *Cynips calicis* Burg. on the immature acorns of oaks of various species. They are common in Hungary. *Scheele* originally isolated gallic acid from the oak apple gall produced by another insect on terminal or axillary buds. Hydrolysis of the crude tannin from Knopper galls yields glucose, ellagic acid (*M. Nierenstein*, J. chem. Soc., 1919, **115,** 1174) and gallic acid.

Pendunculagin (XLIII), $[\alpha]_D$ +106° (methanol), the principal tannin component has been isolated in a crystalline form by chromatography over cellulose powder. On hydrolysis it gives one mole of glucose and two moles of ellagic acid. Methylation and alkaline hydrolysis affords β-methylglucoside and (−)-hexamethoxydiphenic acid so the 1-position of glucose must be free in the tannin. Partial enzymatic hydrolysis of pendunculagin gave 2,3-[(−)-hexahydroxydiphenoyl]glucose identified by permethylation to yield the crystalline nonamethyl derivative which on alkaline hydrolysis gave 4,6-dimethyl-β-methylglucoside. Thus pendunculagin must have structure XLIII (*O. Th. Schmidt*, *L. Wurtele* and *A. Harreus*, Ann., 1965, **690,** 150):

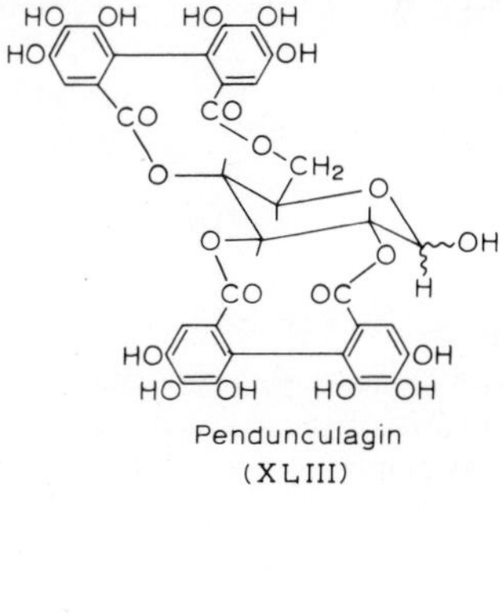

Pendunculagin
(XLIII)

(5) *Oak and Chestnut woods* provide the principal native tanning materials of Italy, France and Germany. Four tannins have been isolated from these materials. **Castalagin,** rhombs from water, decomp. 220°, $[\alpha]_D$ −147.5° (water), pentadecamethyl derivative, m.p. 236°; **vescalagin,** prisms from water, decomp. 200°, $[\alpha]_D$ −109.6° (water), *pentadecamethyl* derivative, m.p. 229°; **castalin,** needles from water, $[\alpha]_D$ +19.3° (water); **vescalin,** amorphous, $[\alpha]_D$ −13.0° (water), are all extracted from wood by water, precipitated with lead acetate and the recovered tannins separated by chromatography over cellulose powder. Castalagin and vescalagin have also been isolated from valonea, the acorn cups of a species of oak (*W. Mayer, W. Bilzer* and *K. Schauerte, ibid.*, 1971, **754,** 149).

Castalagin and castalin are converted to vescalagin and vescalin, respectively, on refluxing their aqueous solutions. On controlled hydrolysis either by acid or by enzymes one mole of ellagic acid is split off so that castalagin is converted to castalin and vescalagin to vescalin (*idem, ibid.*, 1967, **707,** 177).

Castalin and vescalin give flavogallonic acid (XXVIII) (p. 241) and a poor yield of glucose on acid hydrolysis. They must be *C*-glycosides of flavogallonic acid because the n.m.r. spectra indicate only one aromatic proton along with nine phenolic and three aliphatic hydroxyl protons. The n.m.r. signal for the proton at C-1 of glucose is downfield and the protons at C-6 show a large coupling constant so that these three protons of glucose are easily recognised. From this information the remaining proton signals of glucose can be recognised by coupling constant values and double resonance experiments. The position of esterification of glucose by flavogallonic acid can then be deduced from the coupling constant data for CH and OH protons of the glucose chain (*Mayer et al., ibid.*, 1967, **707,** 182; 1971, **747,** 51). Castalin (XLIV) and vescalin (XLV) are formulated as shown on Fischer projection. For the protons attached to glucose and part of the dihydroisocoumarin ring, $J_{1,2}$ is 4.6 Hz in castalin and 2.0 Hz in vescalin. Thus in vescalin these two protons are pseudo-axial and the stereochemistry at C-1 in glucose is as shown. The relationship of the casta- to the vesca-compounds is one of inversion of configuration at C-1 in glucose:

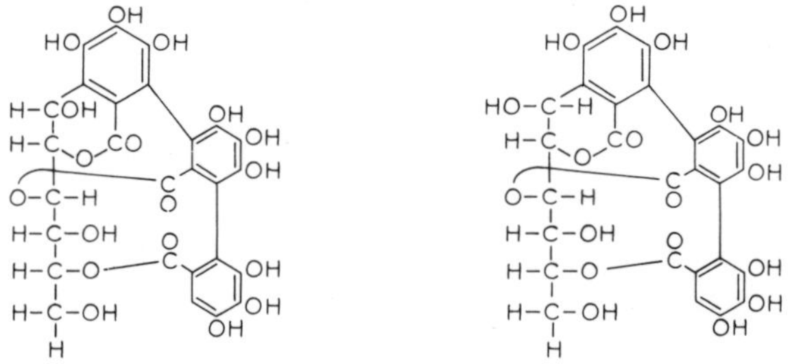

Castalin
(XLIV)

Vescalin
(XLV)

Castalagin (XLVI) and vescalagin (XLVII) on methylation and hydrolysis yield derivatives of castalin and vescalin along with (−)-hexamethoxydiphenic acid. N.m.r. spectral data indicate that the hexahydroxydiphenic acid residues are bound in these compounds to positions 4- and 6- of glucose.

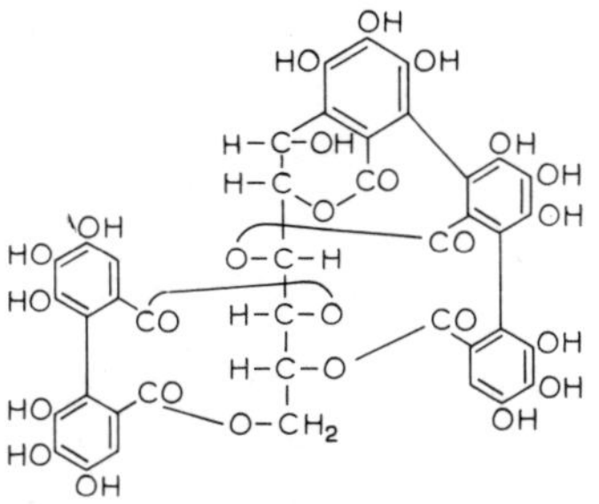

Castalagin
(XLVI)

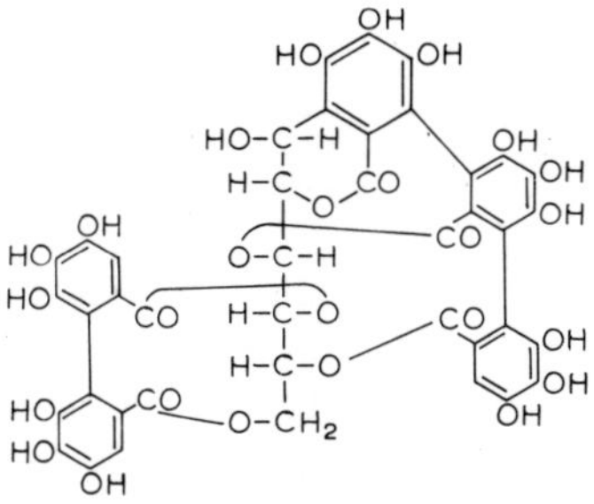

Vescalagin
(XLVII)

(iii) Interrelationship of the ellagitannin acids

When the structure of chebulic and brevifolincarboxylic acids became clear, a biogenetic relationship between these tannin components and hexahydroxydiphenic acid was suggested (*J. Grimshaw, R. D. Haworth* and *H. K. Pindred*, J. chem. Soc., 1955, 833; *Schmidt* and *Mayer*, Ann., 1951, **571,** 1; Angew. Chem., 1956, **68,** 103). The relationship between these acids has been established by the isolation of intermediate oxidation states of hexahydroxydiphenic acid, bound to glucose, and these can be converted *in vitro* either to chebulic acid or to brevifolincarboxylic acid (*Schmidt et al.*, Ann., 1967, **706,** 180, 205) (Scheme 10):

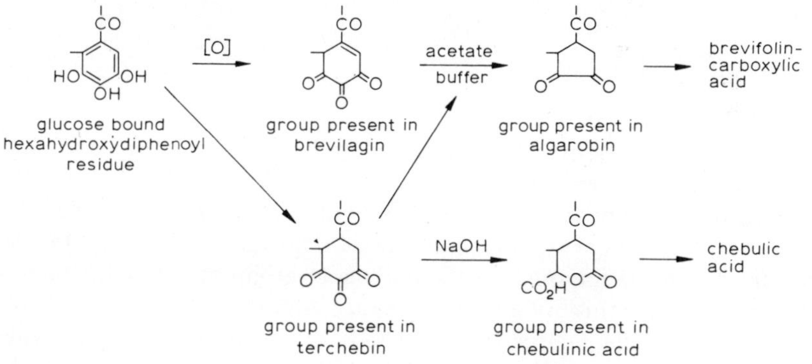

glucose bound hexahydroxydiphenoyl residue → [O] → group present in brevilagin → acetate buffer → group present in algarobin → brevifolin-carboxylic acid

glucose bound hexahydroxydiphenoyl residue → group present in terchebin → NaOH → group present in chebulinic acid → chebulic acid

Scheme 10

Both brevilagin and terchebin give brevifolincarboxylic acid in good yield on heating with an acetate buffer. A reaction analogous to that of brevilagin is the oxidation of 4,6-di-*tert*-butylpyrogallol to di-*tert*-butylcyclopentanedione (XLVIII) (*T. W. Campbell*, J. Amer. chem. Soc., 1951, **73,** 4190):

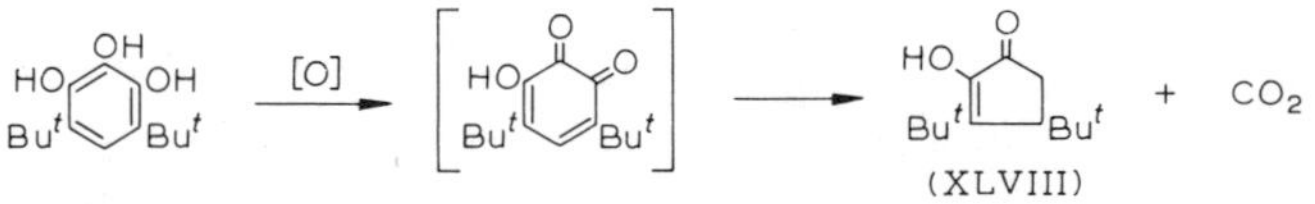

(XLVIII)

The reaction of terchebin must involve the benzilic acid rearrangement, decarboxylation and an oxidation step.

Terchebin gives chebulic acid, identified by paper chromatography, on heating with sodium hydroxide solution and an analogous reaction has been demonstrated for 3,4,5-trioxocyclohexanecarboxylic acid (XLIX):

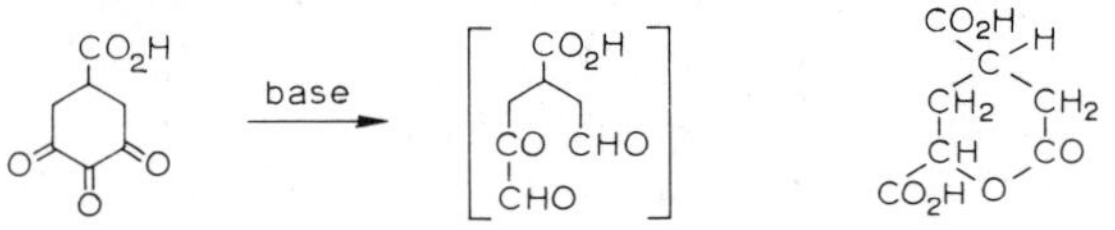

(XLIX)

Brevilagin on treatment with base gives mainly brevifolincarboxylic acid and a small amount of an acid with similar properties to chebulic acid on paper chromatography but which could not be obtained in sufficient quantity for complete identification. This may be chebulic acid or some dehydrochebulic acid (L).

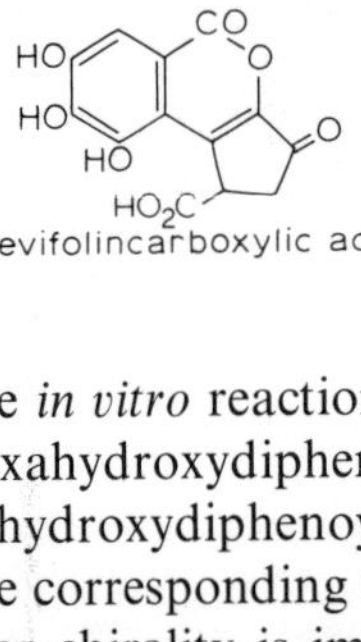

Brevifolincarboxylic acid

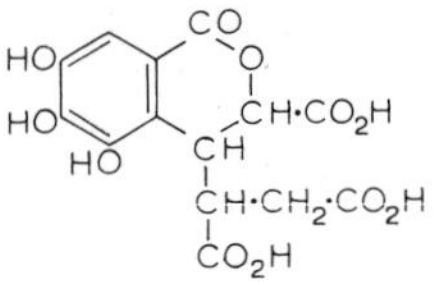

Chebulic acid

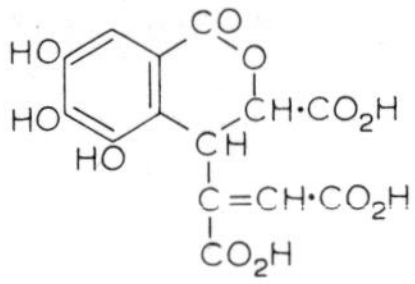

(L)

These *in vitro* reactions are thought to parallel the *in vivo* transformation of hexahydroxydiphenoylglucoses to the various known ellagitannins. The hexahydroxydiphenoyl group itself is thought to arise by oxidative coupling of the corresponding two galloyl groups of a polygalloylglucose. One particular chirality is imposed on the diphenoyl group at this stage because of the necessity for this bridging group to accommodate to the conformation and chirality of the glucose residue. The hexahydroxydiphenic acid which bridges 2,3- or 4,6-positions in glucose is the (−)-enantiomer while that which bridges the 3,6-positions is the (+)-enantiomer. This opposite chirality is maintained in the transformation products of the glucose-bound hexahydroxydiphenoyl group. Hydrolysis products from the brevilagin group which has 1,3- and 4,6-bridges to glucose, and also the crude algarobilla from which these tannins are isolated, have one sign of optical rotation. The corresponding hydrolysis products from chebulagic acid and terchebin which have 2,4- and 3,6-bridges to glucose, and also the crude myrobalans and divi-divi, have the opposite chirality.

3. Lignans

(a) Introduction, biosynthesis

The term lignan was introduced by *R. D. Haworth* (Ann. Reports, 1936, **33,** 266) to describe the groups of naturally occurring compounds which can be formally derived by joining two *n*-propylbenzene derivatives (C_6–C_3 compounds) at the β-carbon atoms of the side chains. The range of structural types found in nature is represented by guaiaretic acid (I), furoguaiacin (II), conidendrin (III) and pinoresinol (V). The lignans have been reviewed (*H. Erdtman*, in "Modern Methods of Plant Analysis", eds. *K. Paech* and *M. V. Tracey*, Springer, Berlin, 1955, Vol. III, p. 428; *W. M. Hearon* and *W. S. MacGregor*, Chem. Reviews, 1955, **55,** 957):

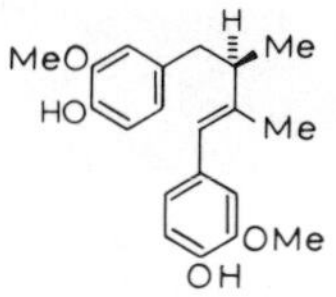

Guaiaretic acid (I) Furoguaiacin (II) Conidendrin (III)

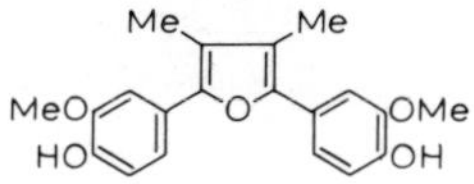

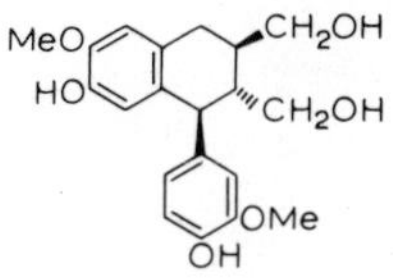

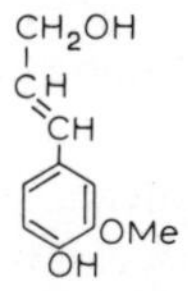

Coniferyl alcohol (IV) Pinoresinol (V) Sinapyl alcohol (VI)

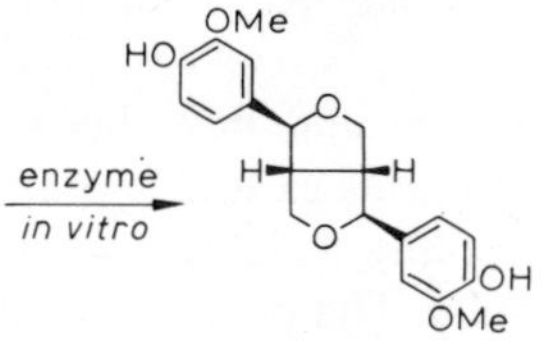

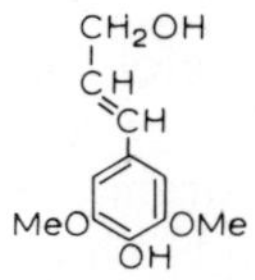

The C_6–C_3 group of compounds is derived in nature by the shikimic acid pathway and the biogenetic route has already been discussed (C.C.C., 2nd Edn., Vol. IIIA, p. 136). In a search for C_6–C_3 compounds which could be oxidised *in vitro* to products of the lignin type, *Erdtman* showed that ferulic acid is oxidised by iron (III) salts to give the dilactone, dehydrodiferulic acid (IX), the structure of which was established by *Cartwright* and *Haworth* (J. chem. Soc., 1944, 535). Coniferyl alcohol (IV) is oxidised by an enzyme from common mushrooms (*Agaricus campestris* L.) to the lignan ($\pm$)-pinoresinol (V) and sinapyl alcohol (VI) is similarly oxidised to syringaresinol (*K. Freudenberg* and *H. Dietrich*, Ber., 1953, **86,** 4, 1157). Oxidation of methyl sinapate with ferric chloride solution gives a compound (IXa) related to the connidendrin type lignans (*A. F. A. Wallis*, Austral. J. Chem., 1973, **26,** 1571). The mechanism of these reactions, and the *in vivo* formation of lignans, is illustrated by the oxidation of coniferyl alcohol. The radical VII is first formed and dimerises to the quinone methide VIII which undergoes further reactions to form the lignan. Lignans are closely related biogenetically to lignin, the polymeric oxidation product of the C_6–C_3 monomer IV. Naturally occurring trimers, tetramers and higher oligomers of IV have not been recorded:

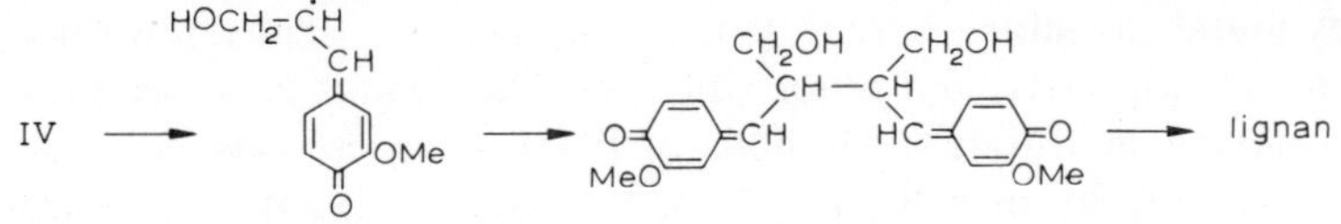

IV → (VII) → (VIII) → lignan

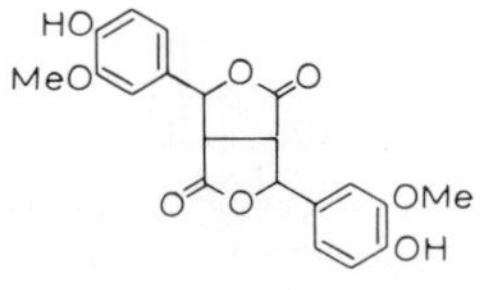

(IX) 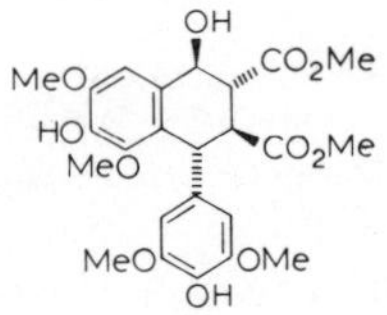 (IXa)

(*b*) *Occurrence and isolation*

The lignans are of wide occurrence and have been obtained from roots, heartwood, foliage, fruits and resinous exudates of plants. Lignans are not volatile with steam. They are usually isolated with a solvent such as ether and separated by chromatography. The phenolic lignans can be chromatographed on silica gel or polyamide powder (*M. K. Seikel, F. D. Hostettler* and *D. B. Johnson*, Tetrahedron, 1968, **24,** 1475). Gas–liquid chromatography can be applied to lignans which contain no hydroxyl functions (*W. A. Jones, M. Beroza* and *E. D. Becker*, J. org. Chem., 1962, **27,** 3232). Lignins containing a number of hydroxyl functions have been separated by counter-current distribution between ether and water (*Freudenberg* and *L. Knof*, Ber., 1957, **90,** 2857) and between ether and aqueous ammonium molybdate (*H. MacLean* and *B. F. MacDonald*, Canad. J. Chem., 1967, **45,** 739).

(c) Individual lignans

(i) 1,4-Diarylbutanes

(−)-**Guaiaretic acid,** (−)-**dihydroguaiaretic acid** and *meso*-**dihydroguaiaretic acid** occur together in the resin of *Guaiacum officinale* L. (*F. E. King* and *J. C. Wilson*, J. chem. Soc., 1964, 4011). The accepted structure of guaiaretic acid (X) was proposed by *G. Schroeter* (Ber., 1918, **51,** 1587) and the absolute configuration was established by chemical relationship with a derivative of L-3,4-dimethoxyphenylalanine (*A. W. Schrecker* and *J. L. Hartwell*, J. Amer. chem. Soc., 1957, **79,** 3827):

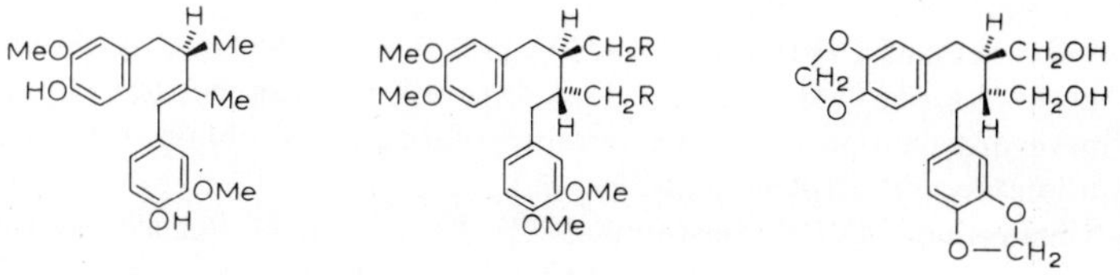

(X) (XI) (XII)

Hydrogenation of di-*O*-methylguaiaretic acid gives both *meso*- and (−)-forms of dihydrodi-*O*-methylguaiaretic acid. A large number of lignans have been related to the (−)-diol (XI, R=OH) which has been converted to dihydrodi-*O*-methylguaiaretic acid (XI, R=H) by reduction of the ditosylate with lithium tetrahydridoaluminate (*idem, ibid.*, 1955, **77,** 432). Thus these compounds are the key to the absolute configuration of many lignans. Methylenedioxy compounds have been related to the corresponding (−)-diol (XII). A link between these two diols to demonstrate their relative stereochemistry has been achieved by the conversion (*Erdtman*, Svensk. kem. Tidskr., 1938, **50,** 161) of (+)-sesamin, related to the diol XII, into (+)-di-*O*-methylpinoresinol (p. 260).

Dehydrogenation of di-*O*-methylguaiaretic acid with iodine and mercuric chloride (*Schroeter, loc. cit.*) gives the naphthalene derivative di-*O*-methyldehydroguaiaretic acid (p. 265).

(±)-Di-*O*-methylguaiaretic acid has been synthesised (*Haworth et al.*, J. chem. Soc., 1934, 1423). The ester, XIII, and the nitrile, XIV, were condensed in the presence of alkali and the product hydrolysed by acid to the amide XV. The ketone obtained from this on further hydrolysis and decarboxylation was treated with methylmagnesium iodide and the resulting alcohol dehydrated to di-*O*-methylguaiaretic acid. Catalytic hydrogenation afforded *meso*-dihydrodi-*O*-methylguaiaretic acid.

(XIII) (XIV) (XV)

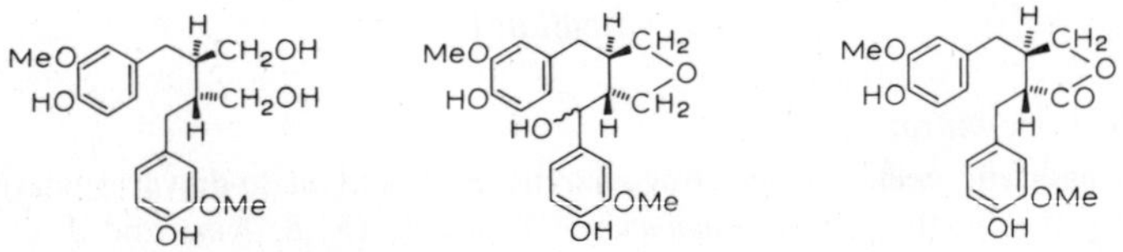

(XVI) (XVII) (XVIII)

(–)-*Guaiaretic acid*, m.p. 100–101°, $[\alpha]_D$ –91° (ethanol); *dimethyl ether*, m.p. 94–94.5°, $[\alpha]_D$ –94° (ethanol); (–)-*di-methyldihydroguaiaretic acid*, m.p. 86–87°, $[\alpha]_D$ –27° (ethanol); meso-*di-O-methyldihydroguaiaretic acid*, m.p. 101–102° (*A. W. Schrecker*, J. Amer. chem. Soc., 1957, **79,** 3823).

Secoisolariciresinol (XVI), m.p. 112–113°, $[\alpha]_D$ –40.8° (acetone), is obtained from *Picea* species. **Liovil** (XVII), m.p. 172–173°, $[\alpha]_D$ –32. 8° (acetone), from *Picea* species gives on hydrogenation the same tetrahydrofuran as is obtained from XVI by dehydration (*Freudenberg* and *Knof, loc. cit.*).

(–)-**Matairesinol** (XVIII) constitutes 50% of the resin of *Podocarpus spicata* R. Br. Reduction of the dimethyl ether with lithium tetrahydridoaluminate (*Haworth* and *L. Wilson*, J. chem. Soc., 1950, 71) gave the (–)-diol XI (R=OH). The structure of matairesinol was confirmed by synthesis of the dimethyl ether (*Haworth* and *D. Woodcock*, *ibid*., 1939, 154). Veratraldehyde and diethyl succinate were condensed in the presence of sodium ethoxide and the product reduced with sodium amalgam when the *meso*-dicarboxylic acid XIX could be isolated. This acid was converted to the (±)-anhydride XX on warming with acetic anhydride. This anhydride, with a *trans*-arrangement of substituents on the ring, is the more stable of the two possible anhydrides. Hydrolysis gave the (±)-acid which was resolved. Reduction of the (+)-anhydride with aluminium amalgam gave (–)-di-*O*-methylmatairesinol. Matairesinol has been synthesised by a similar route using the benzyl ether as a protecting group for the phenol and later removing this group by treatment with acid:

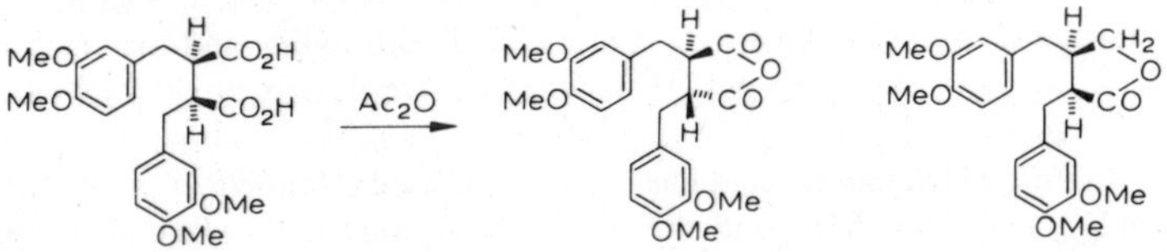

Ac_2O

(XIX) (XX) (XXI)

Concentrated potassium hydroxide converts (–)-di-*O*-methylmatairesinol into an equilibrium mixture from which the starting material and a (+)-rotating isomer XXI can be recovered after acidification. (+)-Di-*O*-methylisomatairesinol (XXI) is reduced by lithium tetrahydridoalı minate (*Haworth* and *Wilson, loc. cit.*). to give the *meso*-form of the diol XI (R=OH). In concentrated alkali the lactone ring is opened and equilibrium established between salts of the corresponding carboxylic acids. Dilute alkali isomerises (+)-di-*O*-methylisomatairesinol to (–)-di-*O*-methylmatairesinol. Here equilibrium is established without opening of the lactone ring and the isomer with *trans* arrangement of benzyl substituents preponderates.

Matairesinol, (XVIII), m.p. 119°, $[\alpha]_D$ −48.6° (acetone); *dimethyl ether*, m.p. 127–128°, $[\alpha]_D$ −35.6° (chloroform) (*Haworth* and *T. Richardson*, J. chem. Soc., 1935, 633); *di-O-methylisomatairesinol*, m.p. 111–112°, $[\alpha]_D$ +78° (chloroform) (*Haworth* and *J. R. Atkinson*, *ibid.*, 1938, 797).

Hydroxymatairesinol (XXII), amorphous, $[\alpha]_D$ −6.3° (alcohol), *dimethyl ether*, m.p. 96–97°, $[\alpha]_D$ +59.8° (tetrahydrofuran) has been obtained from the wood of *Picea excelsa* Link (*Freudenberg* and *Knof*, *loc. cit.*). Catalytic hydrogenation affords matairesinol while acid cyclisation gives α-conidendrin (p. 268). The corresponding ketone, **oxomatairesinol,** m.p. 70–72°, $[\alpha]_D$ +42.6° (tetrahydrofuran) also occurs in the same wood:

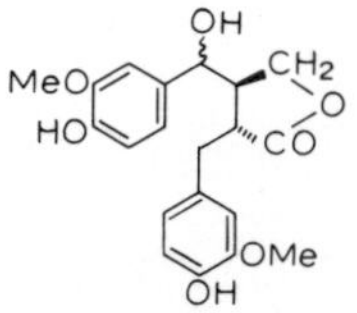

(XXII) 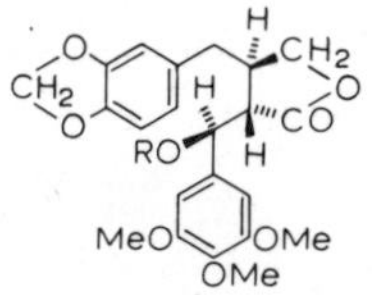 (XXIII) 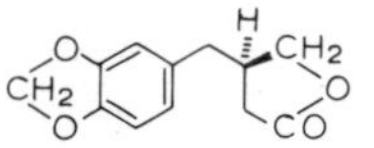 (XXIV) 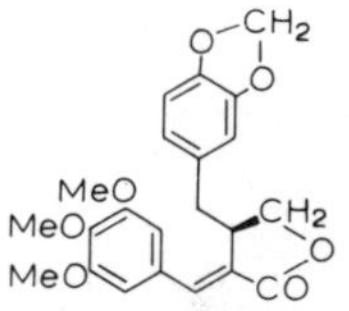 (XXV)

Podorhizol (XXIII, R=H), m.p. 125–126°, $[\alpha]_D$ −51.8° (chloroform), occurs as the *β-glucoside* (XXIII, R=Glu.), m.p. 103–104°, $[\alpha]_D$ −48.6° (methanol), in *Podophyllum emodi* Wall., and *P. peltatum* L. (*M. Kuhn* and *A. von Wartburg*, Helv., 1967, **50,** 1547). Chromic acid oxidation of podorhizol gives a ketone which is converted by base to 3,4,5-tri-*O*-methylbenzoic acid and the lactone XXIV. Reduction of the ketone with sodium tetrahydridoborate gives podorhizol and the epipodorhizol. Epipodorhizol, unlike podorhizol, shows a strong intramolecular hydrogen hond between the hydroxyl and carbonyl functions. Attempted acid hydrolysis of podorhizol glucoside gave (−)-isodesoxypodophyllotoxin (p. 270). These results decide the relative stereochemistry of podorhizol. The absolute stereochemistry was decided by comparison of the molecular rotations of **anhydropodorhizol,** m.p. 78–80°, $[\alpha]_D$ −55.2° (chloroform), obtained from the glucoside by base catalysed elimination, and dihydroanhydropodorhizol with those of savinin (XXIX) and its dihydro derivative. Comparison of the vinyl ^{1}H-n.m.r. line with those in *cis*- and *trans*-3,4,5-tri-*O*-methoxybenzylbutyrolactone established the stereochemistry of anhydropodorhizol about the olefin bond.

Hinokinin, *cubebinolide*, (XXVI), has been isolated from the resin of *Chamaecyparis obtusa* Sieb. et Zucc. (*T. Toshiki* and *T. Ishiguro*, J. pharm. Soc. Japan, 1933, **53,** 11) and is obtained (*E. Mamali*, Gazz., 1935, **65,** 877) by mild oxidation of **cubebin** (XXVII), a constituent of the fruits of *Piper cubeta* L. Catalytic reduction of cubebin (*Ishiguro*, J. pharm. Soc. Japan, 1936, **56,** 444) gives the (−)-diol XII (p. 255). (−)-Hinokinin has been synthesised (*Haworth* and *Woodcock*, J. chem. Soc., 1938, 1985) by a route similar to that described for di-*O*-methylmatairesinol. Concentrated potassium hydroxide affords an equilibrium mixture from which hinokinin and (+)-isohinokinin (XXVIII), can be isolated after acidification (*S. Keimatsu* and *Ishiguro*, J. pharm. Soc. Japan, 1936, **56,** 19). As in the matairesinol series, dilute alkali converts (+)-isohinokinin to (−)-hinokinin:

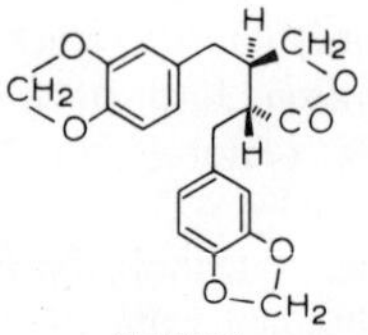

(XXVI) 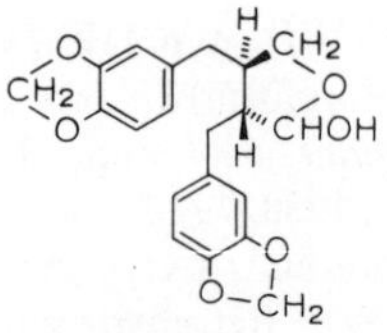 (XXVII)

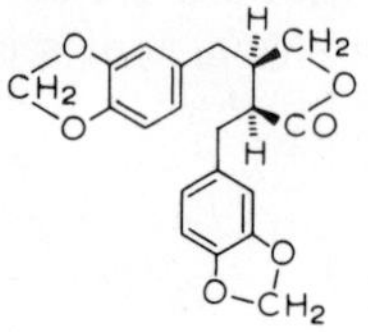

(XXVIII) 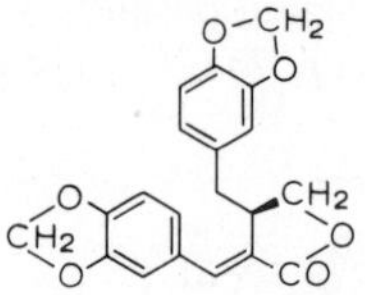 (XXIX)

Hinokinin, (XXVI), m.p. 64–65°, $[\alpha]_D^{17}$ $-34.0°$ (chloroform); *isohinokinin*, (XXVIII), m.p. 116–117°, $[\alpha]_D^{22}$ $+106°$. *Cubebin*, (XXVII), m.p. 128°, $[\alpha]_D$ $-45.7°$ (chloroform) (*E. Mameli*, Gazz., 1907, **37,** II, 483).

Savinin, (XXIX), m.p. 146–147°, $[\alpha]_D$ $-88°$ (chloroform), isolated from the needles of *Juniperus sabina* L. (*Schrecker* and *Hartwell*, J. Amer. chem. Soc., 1954, **76,** 4896) gives (+)-isohinikinin on catalytic hydrogenation. Isomerisation about the olefinic bond occurs in u.v. light. Comparison of the n.m.r. spectra of savinin and isosavinin confirms the stereochemistry indicated for savinin where the vinyl proton is close to the magnetically anisotropic carbonyl group (*K.-T. Wang et al.*, Phytochem. 1967, **6,** 131).

Thujaplicatin (XXX, R=H) and a number of its derivatives have been isolated from an aqueous extract of the wood of *Thuja plicata* Donn. *O*-**Methylthujaplicatin** (XXX, R=Me) affords vanillin and syringaldehyde on oxidation with nitrobenzene. Both lignans are methylated to tri-*O*-methylthujaplicatin which has been synthesised (*H. MacLean* and *K. Murakami*, Canad. J. Chem., 1966, **44,** 1541). Condensation of the lactone XXXI with 3,4,5-trimethoxybenzaldehyde afforded XXXII which was hydrogenated to (±)-tri-*O*-methylisothujaplicatin with the *cis*-arrangement of benzyl groups. This was isomerised by dilute alkali to (±)-tri-*O*-methylthujaplicatin which crystallised from ether as individual (+)- and (−)-rotating crystals. The absolute stereochemistry of the natural (−)-lignans is assumed by analogy with matairesinol and hinokinin:

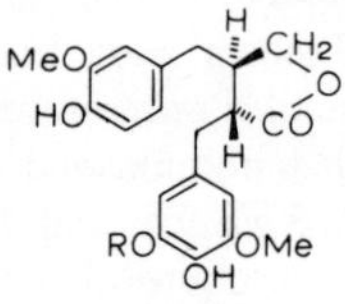

(XXX) 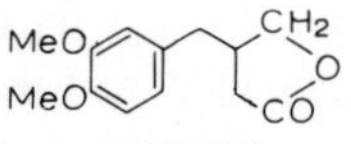 (XXXI) 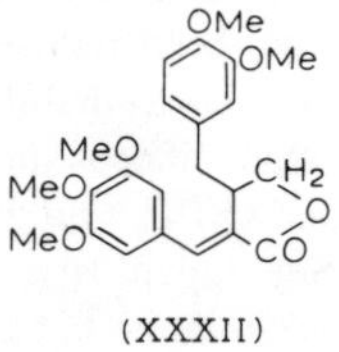 (XXXII)

Thujaplicatin, amorphous; O-*methylthujaplicatin* (XXX, R = Me), m.p. 167–167.5°, $[\alpha]_D$ −48.7° (acetone); *tri-O-methylthujaplicatin*, m.p. 101–102° $[\alpha]_D$ −37.4° (chloroform).

O-**Methylhydroxythujaplicatin** (XXXIII), m.p. 128.5–129.5°, $[\alpha]_D$ −54.8° (chloroform), also occurs in *Thuja plicata* (*MacLean* and *Murakami*, Canad. J. Chem., 1966, **44,** 1827); *tri-O-methylhydroxythujaplicatin*, m.p. 113.5–114.5°, $[\alpha]_D$ −50° (acetone) is unattacked by manganese dioxide and so is unlikely to be a benzyl alcohol, periodate oxidation gives a hydroxyketone XXXIV.

Dihydroxythujaplicatin (XXXV, R = H), amorphous; O-*methyldihydroxythujaplicatin* (XXXV, R = Me), m.p. 95–97° (methanol solvate), $[\alpha]_D$ −97.2° (chloroform), have also been isolated (*MacLean* and *B. F. MacDonald*, *ibid.*, 1967, **45,** 739). Methylation of either gives *tri-O-methyldihydroxythujaplicatin*, m.p. 163.5–164°, $[\alpha]_D$ −105° (chloroform), which is not oxidised by manganese dioxide:

(XXXIII) (XXXIV) (XXXV)

(ii) Tetrahydrofuroguaiacins

Furoguaiacin (XXXVI), occurs in the heartwood of *Guaiacum officinale* L., and has been isolated as its dimethyl and diethyl ethers (*King* and *Wilson*, J. chem. Soc., 1964, 4011). The dimethyl ether is obtained in low yield from selenium dehydrogenation of di-*O*-methylpinoresinol (p. 260) and has been synthesised (*Atkinson* and *Haworth*, *ibid.*, 1938, 1681):

(XXXVI) (XXXVII) (XXXVIII)

Catalytic hydrogenation of the dimethyl ether gave a tetrahydro derivative to which the all *cis* configuration was given (*J. G. Blears* and *Haworth*, *ibid.*, 1953, 1985) and which was later isolated from *Guaiacum officinale* and given the name **di-*O*-methyltetrahydrofuroguaiacin-B** (XXXVII). This hydrofuran was isomerised by acid to *galgravin* (XXXVIII) (*Blears* and *Haworth*, *loc. cit.*) previously isolated from the wood of *Himantandra belgraveana* Diels (*G. K. Hughes* and *E. Ritchie*, Austral. J. Chem., 1954, **7,** 104). The acid catalysed isomerisation involves the benzyl ether centres and yields a less crowded isomer as the product. Catalytic hydrogenation of galgravin affords *meso*-dihydrodi-*O*-methylguaiaretic acid. The stereochemistry of the veratryl substituents of XXXVII and XXXVIII has been confirmed by the ^{1}H-n.m.r. spectra (*King* and *Wilson*, *loc. cit.*; *N. S. Crossley* and *C. Djerassi*, J. chem. Soc., 1962, 1459) which show one doublet for the two methyl groups and one doublet for the two benzyl hydrogens coupled to a vicinal proton. Each lignan must therefore have the two veratryl substituents in an identical environment.

Galbelgin (XXXIX) isolated from *Himantandra baccata* F. M. Bailey (*A. J. Birch et al., ibid.*, 1958, 4471) and **veraguensin** (XL) from *Ococa veraguensis* Mez. (*Crossley* and *Djerassi, loc. cit.*) are both reduced by sodium in liquid ammonia to a dihydro derivative which is cyclised by acid to (−)-galbulin (p. 265). More vigorous reduction of galbelgin gave (−)-di-*O*-methyldihydroguaiaretic acid. The n.m.r. spectrum of veraguensin shows two doublets for the two *C*-methyl groups and two doublets for the two benzyl hydrogens in contrast to the spectrum of galbelgin which shows only one doublet each for these substituents. Thus the stereochemistry shown is deduced for these two lignans:

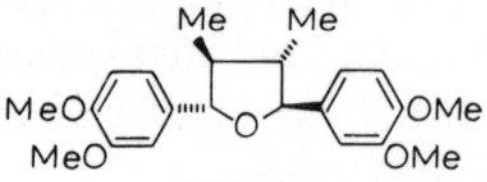

(XXXIX) 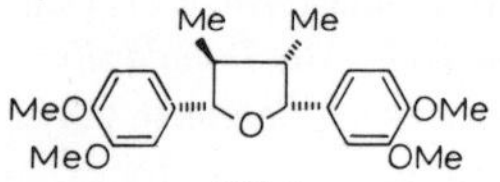 (XL) 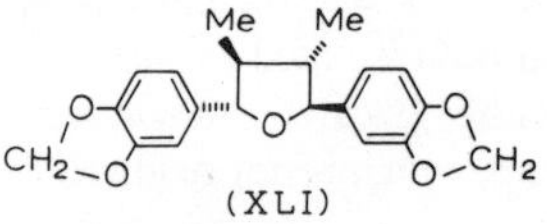 (XLI)

Di-O-methylfuroguaiacin, pale blue fluorescent plates, m.p. 170–171°; *di-O-methyltetrahydrofuroguaiacin-B*, m.p. 132–133°, is dehydrogenated by palladium/charcoal in diphenyl ether to the corresponding furan; *galgravin* (XXXVIII), m.p. 121°, gives di-*O*-methyldehydroguaiaretic acid (p. 265) on treatment with palladium charcoal (*King* and *Wilson, loc. cit.*); *galbelgin* (XXXIX), m.p. 138°, $[\alpha]_D$ −102° (chloroform); *veraguensin* (XL), m.p. 128–129°, $[\alpha]_D$ +34.2° (chloroform).

Galbacin (XLI), m.p. 116°, $[\alpha]_D$ −114° (chloroform), occurs in the wood of *Himantandra baccata* (*Hughes* and *Ritchie, loc. cit.*), treatment with concentrated alkali followed by methylation gave (−)-galbelgin.

(iii) Pinoresinol and lariciresinol types

Pinoresinol (XLII, R=H) comprises up to 10% of the resin exuded by species of pine and fir. Methylation affords (+)-di-*O*-methylpinoresinol (XLII, R=Me) which is the enantiomer (*H. Erdtman*, Ann., 1935, **516,** 162) of (−)-**eudesmin** found in *Eucalyptus* kinos. Catalytic hydrogenation (*Haworth* and *Woodcock*, J. chem. Soc., 1939, 1054) or reduction with sodium in liquid ammonia (*A. J. Birch et al.*, Austral. J. Chem., 1954, **7,** 83) of (+)-di-*O*-methylpinoresinol gives the (−)-diol (XI, R=OH,

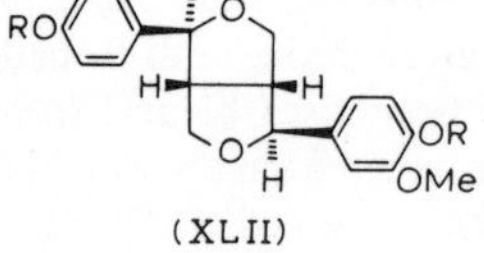

(XLII) 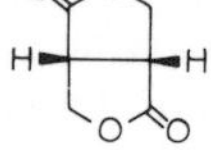 (XLIII) (XLV)

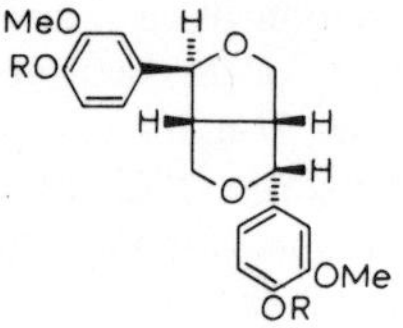

(XLIV)

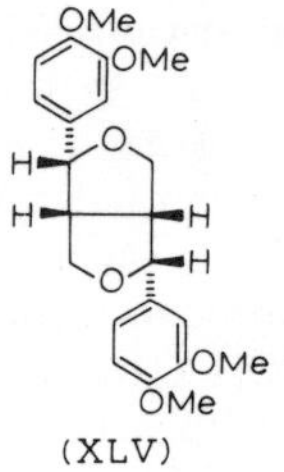

p. 255) and di-*O*-methyllariciresinol (p. 264) can be isolated as an intermediate. Cleavage of dibromodi-*O*-methylpinoresinol with nitric acid gives the (+)-dilactone XLIII and 4-bromo-5-nitroveratrole. Both enantiomers of the dilactone XLIII have been prepared from natural lignans and the (±)-dilactone obtained on mixing the two was identical with a synthetic specimen (*H. Erdtman* and *J. Gripenberg*, Acta chem. Scand., 1947, **1,** 71). The absolute stereochemistry of the ring junction in pinoresinol is decided by these reactions.

Pinoresinol can be partially isomerised on refluxing with ethanolic hydrogen chloride. The products were methylated and separated to give (+)-di-*O*-methylpinoresinol and (+)-di-*O*-methylepipinoresinol (XLIV, R = Me) both of which are converted by acid into an equilibrium mixture of the two (*Gripenberg*, *ibid*., 1949, **3,** 898). This isomerisation involves epimerisation of the benzyl ether centres and theoretically three isomers are possible. The third isomer, (+)-**diaeudesmin** (XLV), has been isolated from *Piper peepuloides* Roxb. (*C. K. Atal et al.*, J. chem. Soc., 1967, 2228). Di-*O*-methylepipinoresinol gave two mononitro derivatives (*Gripenberg*, Acta chem. Scand., 1948, **2,** 82) and therefore must have the structure XLIV (R = Me) with nonequivalent aromatic rings. Pinoresinol gives only one monosubstitution product (*Erdtman*, Svensk. kem. Tidskr., 1936, **48,** 236) and therefore must have equivalent aromatic rings. Since di-*O*-methylpinoresinol is the most stable isomer it is given the structure XLII (R = Me) (*Freudenberg* and *G. S. Sidhu*, Ber., 1961, **94,** 851). (±)-Pinoresinol has been synthesised (*Freudenberg* and *H. Dietrich*, *ibid*., 1953, **86,** 1157) from dehydrodiferulic acid (IX, p. 254) by reduction with lithium tetrahydridoaluminate and ring closure of the resulting alcohol by distillation under reduced pressure.

Sesamin (XLVI) occurs in sesame seed oil. It has been converted into (+)-di-*O*-methylpinoresinol (*Erdtman*, Svensk. kem. Tidskr., 1938, **50,** 161) and hydrogenolysis gives the (−)-diol (XII, p. 255) (*F. von Bruckhausen* and *H. Gerhard*, Ber., 1939, **72,** 830) whose stereochemistry is thus related to that of the guaiaretic acid series of lignans. Epimerisation of (+)-sesamin by acid (*M. Beroza*, J. Amer. chem. Soc., 1956, **78,** 5082) has given all three possible epimers about the benzyl ether groups. Sesamin is the most stable epimer. A second product (+)-**asarinin** (XLVII) also occurs naturally. The third epimer (+)-epiasarinin (XLVIII) is the least stable:

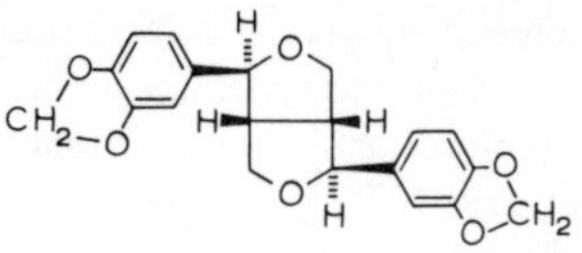

(XLVI) 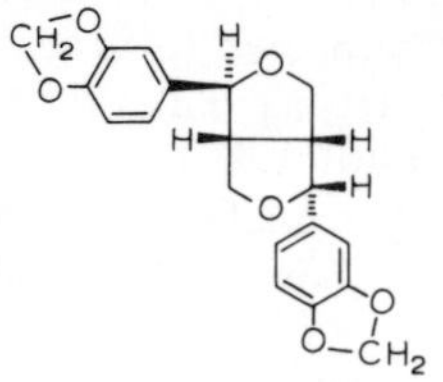 (XLVII) 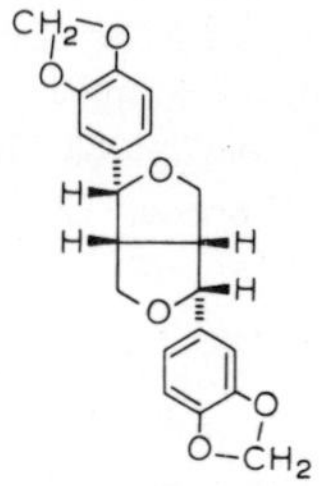 (XLVIII)

Configurations have been assigned (*Freudenberg* and *Sidhu*, *loc. cit.*) on the basis of stability, optical rotation differences and relative behaviour on thin-layer chromatography.

Syringaresinol (LI) was first obtained as the racemate from the enzymic oxidation of syringyl alcohol, and subsequently synthesised (*Freudenberg* and *H. Schraube*, Ber., 1955, **88,** 16) from 4-hydroxy-3,5-dimethoxycinnamic acid. **Liriodendrin,** from the wood of *Liriodendron tulipifera* L., is the diglucoside XLIX (R = glucose) of an optically active syringaresinol (*E. E. Dick*, J. org. Chem., 1958, **23,** 179). Degradation of derivatives with nitric acid gave the (+)-dilactone XLIII also obtained from (+)-pinoresinol:

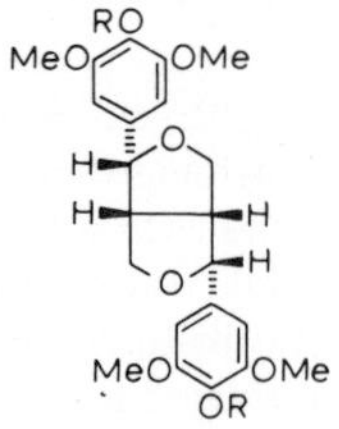

(XLIX) 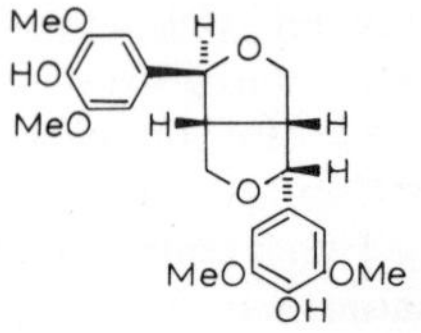 (L) 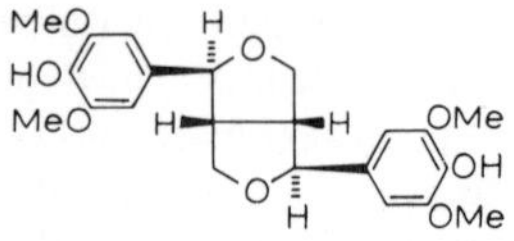 (LI)

The aglucone, lirioresinol-C (XLIX, R = H), was obtained after hydrolysis with emulsin. Acid hydrolysis also caused epimerisation to lirioresinol-A (L) plus lirioresinol-B (LI). **Di-*O*-methyllirioresinol-B** has been isolated from *Eremophila glabra* Ostenf. (*P. R. Jefferies et al.*, Austral. J. Chem., 1961, **14,** 175); it has been shown by X-ray methods to possess a two-fold axis of symmetry (*E. N. Maslen et al., ibid.*, 1962, **15,** 161) and probably has the same configuration as pinoresinol. Liriodendrin and lirioresinol-C (XLIX, R = H) are the least stable to acid and so are assigned the all *endo* configuration. Lirioresinol-A (L) corresponds to epipinoresinol. (±)-**Episyringaresinol,** (±)-lirioresinol-A, has been prepared from (±)-syringaresinol.

The relative stereochemistry assigned to the lignans of the three series described above has been confirmed by ^{1}H-n.m.r. spectroscopy (*K. Weinges*, Ber., 1961, **94,** 2528; *E. D. Becker* and *Beroza*, Tetrahedron Letters, 1962, 157). As expected, the benzyl protons of pinoresinol, sesamin, epiasarinin, and (±)-syringaresinol are equivalent whereas those of isopinoresinol, asarinin and (±)-episyringaresinol are not. The *endo* isomers have the phenyl group close to the methylene protons. Because of this and the magnetic anisotropy of the benzene ring, the position of the methylene proton resonances can be correlated with the stereochemistry of the phenyl groups. These correlations, found in the pinoresinol and sesamin series (*A. J. Birch et al.*, J. chem. Soc., C, 1967, 1968), have been used to assign stereochemistry to related lignans.

(+)-*Pinoresinol* (XLII, R = H), m.p. 120–121°, $[\alpha]_D$ +84,4° (acetone), *diacetate*, m.p. 166–167°, *dimethyl ether*, m.p. 107–108°, $[\alpha]_D$ +64.5° (chloroform). (−)-*Eudesmin*, (−)-*di-O-methylpinoresinol* (enantiomer of XLII, R = Me), m.p. 107°, $[\alpha]_D$ −64.3° (chloroform) (*R. Robinson* and *H. G. Smith*, J. Proc. roy. Soc. N. S. Wales, 1914, **48,** 449). (±)-*Pinoresinol*, m.p. 111°.

(−)-**Epipinoresinol,** *Symplocosigenol*, (enantiomer of XLIV, R = H), m.p. 141–142° *dihydrate*, $[\alpha]_D$ −118.9°, occurs as its *β*-glucoside, **symplocosin,** m.p. 171–171.5°, $[\alpha]_D$ −44.9° (*K. Nishida et al.*, J. Soc. Forestry Japan, 1951, **33,** 407). **Phillyrin,** m.p. 162° and 181°, $[\alpha]_D$ +46.7° (ethanol), occurs in *Forsythia* and *Phillyrea* species and is the *β*-glucoside of *phillygenol*, O-*methylepipinoresinol*, m.p. 134–135°, $[\alpha]_D$ +121.7°

(ethanol), which gives (+)-*di-O-methylepipinoresinol* (XLIV, R = Me), m.p. 125–126°, [α]$_D$ +119° (chloroform), on methylation (*Gripenberg*, Acta chem. Scand., 1949, **3,** 898). Methylation of symplocosin and hydrolysis of the glucosidic link gives an *O*-methylsymplocosigenol, which is probably the enantiomer of phillygenol.

(+)-*Diaeudesmin* (XLV), m.p. 147°, [α]$_D$ +316° (chloroform).

(+)-**Sesamin** (XLVI), m.p. 122–123°, [α]$_D$ +68.1° (chloroform), occurs in sesame oil, (−)-form in *Asarum sieboldii* Miq.; (±)-*sesamin, fagarol*, m.p. 129–130° (*T. Kaku* and *H. Ri*, J. pharm. Soc. Japan, 1937, **57,** 184). (+)-**Asarinin** (XLVII), m.p. 122–123°, [α]$_D$ +118.6° (chloroform); (−)-asarinin occurs in *Asarum sieboldi.* (+)-*Epiasarinin* (XLV I), m.p. 168–171°, [α]$_D$ +385° (chloroform) *(Beroza, loc. cit.).*

Liriodendrin (XLIX, R = glucose), m.p. 269–270°, *octa-acetate*, m.p. 124–125°, [α]$_D$ +7.2° (chloroform). *Lirioresinol-C* (XLIX, R = H), m.po 185–186°, [α]$_D$ +48.9° (chloroform); *dimethyl ether*, m.p. 107–108°, [α]$_D$ +64.5° (chloroform). *Lirioresinol-B* (LI), m.p. 172–177°, [α]$_D$ +62.2° (chloroform); *dimethyl ether*, m.p. 122–123°, [α]$_D$ +45.8° (chloroform) is probably identical with *yangambin* (*R. Haensel et al.*, Z. Naturforsch., 1966, **21b,** 530) isolated from *Piper guineense* Schumach. et Thonn. *Lirioresinol-A* (L), m.p. 210–211°, [α]$_D$ +127° (chloroform); *dimethyl ether*, m.p. 118–120, [α]$_D$ +119° (chloroform) *(Dick, loc. cit.). Acanthoside-D*, m.p. 245–247°, *octa-acetate*, m.p. 110–112°, [α]$_D$ −27.0°, isolated from *Acanthopanax sessiliflorum* Seem may be the diglucoside of a (−)-lirioresinol; acid hydrolysis gives (−)-*lirioresinol-B*, m.p. 177°, [α]$_D$ −90° (*L. A. Elzakova et al.*, Dokl. Akad. Nauk S.S.S.R., 1965, 165, 562).

(±)-**Syringaresinol** (LI), m.p. 174° (*K. Freudenberg* and *H. Dietrich*, Ber., 1953, **86,** 4, 1157) has been isolated from beech wood, *Fagus silvatica* (*H. Nimz* and *H. Gaber*, *ibid.*, 1965, **98,** 538); *dimethyl ether*, m.p. 107–108°. (±)-*Di-O-methylepisyringinol*, m.p. 104–106° (*K. Weinges, ibid.*, 1961, **94,** 2522).

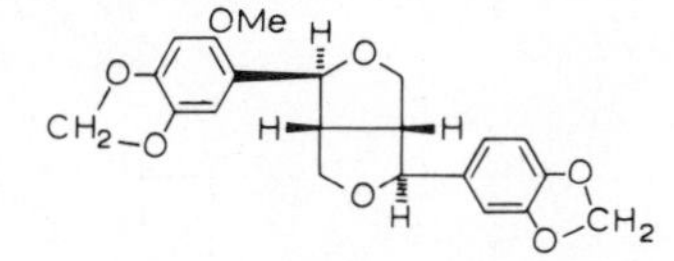

(LII) 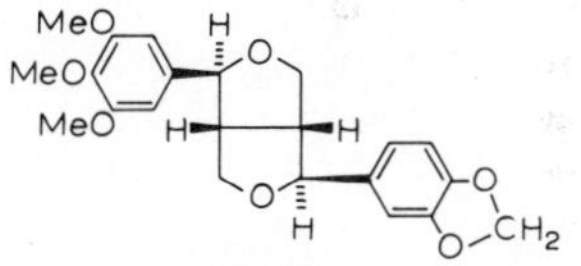 (LIII)

Sesangolin (LII), m.p. 87–88° and 101°, [α]$_D$ +48.5° (chloroform) has been isolated from the seed oil of *Sesamum angolence* Welw. (*W. A. Jones et al.*, J. org. Chem., 1962, **27,** 3232). Nitric acid oxidation gives the dilactone XLIII.

Aschantin (LIII), m.p. 123°, [α]$_D$ +65° (chloroform), occurs in *Piper guineense* Schumach. et Thonn. (*R. Haensel* and *D. Zander*, Arch. Pharm., 1961, **294,** 699).

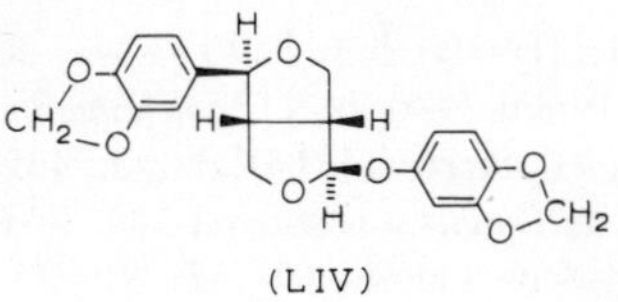

(LIV) 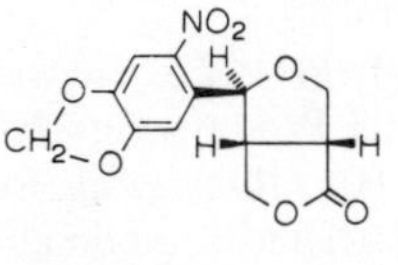 (LV) 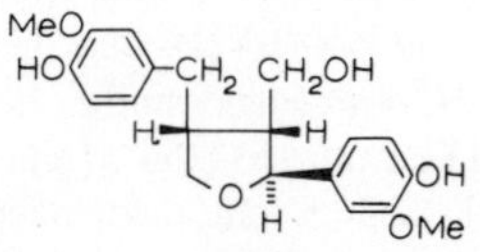 (LVI)

Sesamolin (LIV), m.p. 93–94°, $[\alpha]_D$ +220° (chloroform), occurs in the seed oil of *Sesamum indicum* L. and is responsible for the deep red colour which develops in the aqueous phase when the oil is shaken with hydrochloric acid and furfuraldehyde. Acid hydrolysis of sesamolin gives 3,4-methylenedioxyphenol and a second product which with nitric acid the nitrolactone LV also obtained from the action of nitric acid on bromonitrosesamin (*E. Haslam* and *R. D. Haworth*, J. chem. Soc., 1955, 827; *H. Erdtman* and *Z. Pelchowicz*, Chem. and Ind., 1955, 567; *Beroza*, J. Amer. chem. Soc., 1955, **77,** 3332).

Lariciresinol (LVI), m.p. 167–168°, $[\alpha]_D$ +19.7° (acetone), is isolated from larch resin, *Larix decidua* L.; *dimethyl ether*, m.p. 79–80°, $[\alpha]_D$ +22° (acetone), has been prepared by partial hydrogenation of di-*O*-methylpinoresinol (XLII, R = Me). Lariciresinol is very easily isomerised by acid to the tetralin, isolariciresinol (p. 266) (*Haworth* and *W. Kelly*, J. chem. Soc., 1937, 384):

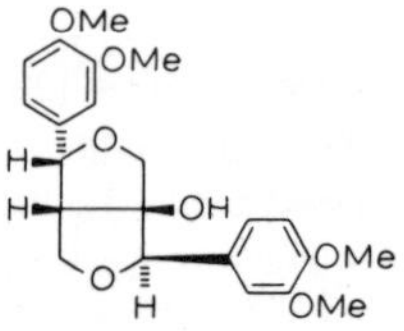

(LVII) 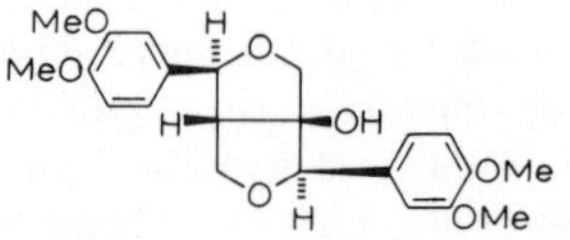 (LVIII)

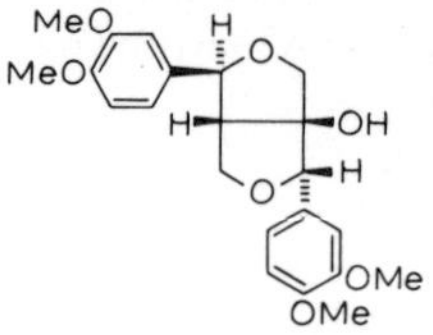

(LIX) 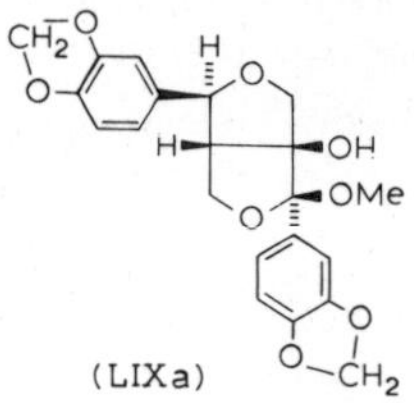 (LIXa) 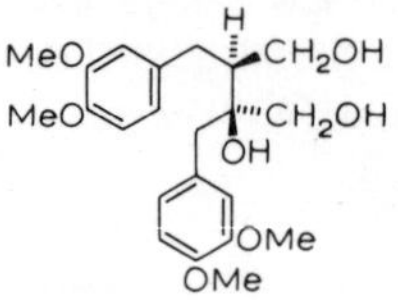 (LX)

Gmelinol (LVII) is isolated from *Gmelina leichhardtii* F. Muell. (*H. Smith*, J. Proc. roy. Soc. N. S. Wales, 1912, **46,** 187). Acid catalysed isomerisation affords first isogmelinol (LVIII) and finally neogmelinol (LIX). Reduction of all three isomers gives the same tetrahydrogmelinol (LX) and this triol is also obtained by reduction of di-*O*-methylolivil (LXI, R = Me) *(Birch et al., loc. cit.)*. The triol has been converted to (−)-di-*O*-methylguaiaretic acid (X, p. 255), (*G. Traverso*, Gazz., 1958, **88,** 851; 1960, **90,** 792). Complete elucidation of stereochemistry in the gmelinol series was effected by comparison of n.m.r. spectra.

Gmelinol (LVII), m.p. 124°, $[\alpha]_D$ +124°; *isogmelinol* (LVIII), m.p. 149°, $[\alpha]_D$ +30° (*R. H.* ≡*arradence* and *F. Lions*, J. Proc. roy. Soc. N. S. Wales, 1940, **74,** 117); *neogmelinol* (LIX), m.p. 163–164°, $[\alpha]_D$ +60° *(Birch et al., loc. cit.)*. **Arboreol** (LIXa), m.p. 135°, $[\alpha]_D$ +84.5 (chloroform) is a related compound from *Gmelina arborea* L. (*A. S. R. Anjaneyulu*, *L. R. Row* and *C. Subrahmanyam*, Tetrahedron Letters, 1972, 2179):

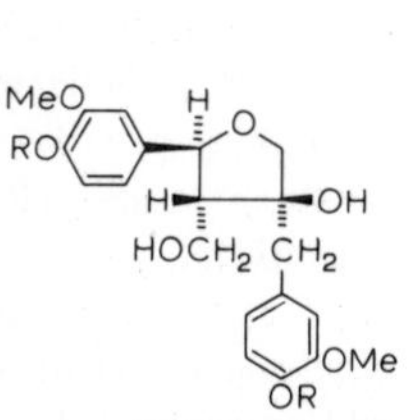

(LXI) 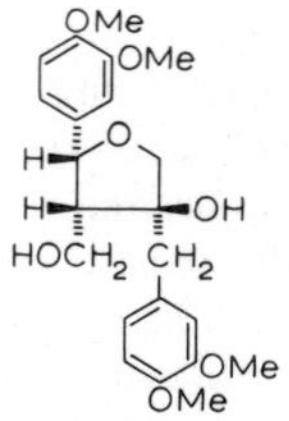 (LXII) 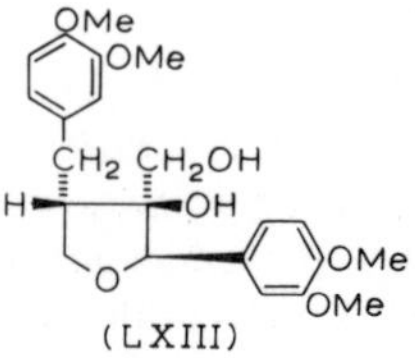 (LXIII)

Olivil (LXI; R=H), m.p. 118–120° (methanol solvate, $[\alpha]_D$ −127°), occurs in the resin of *Olea europa* L., dimethyl ether, m.p. 155–156°, $[\alpha]_D$ −35° (*D. C. Ayres* and *S. E. Mhasalar*, J. chem. Soc., 1965, 3586). Di-*O*-methylolivil and gmelinol give the same triol (LX) on catalytic hydrogenation (*Freudenberg* and *Weinges*, Tetrahedron Letters, 1962, 1077). N.m.r. spectra indicate the structure LXI for olivil (*Birch* and *M. Smith*, J. chem. Soc., 1964, 2705). A stereoisomer of di-*O*-methylolivil, LXII, is obtained by partial reduction of gmelinol along with the structural isomer LXIII, which is the product from partial hydrogenation of isogmelinol. Treatment of olivil with acid affords cyclo-olivil (p. 266).

(iv) Phenyltetralins

Galbulin (LXIV, R=Me) and **galcatin** (LXV) have been isolated from the bark of *Himantandra baccata* Bail. (*G. K. Hughes* and *E. Ritchie*, Austral. J. Chem., 1954, **7,** 104) and galcatin can be converted into galbulin (*Birch et al.*, J. chem. Soc., 1958, 4471). On dehydrogenation, galbulin affords di-*O*-methyldehydroguaiaretic acid (LXVI). Di-*O*-methylisolariciresinol (LXVII, R=R[1]=Me) has been converted into galbulin (*A. W. Schrecker* and *J. L. Hartwell*, J. Amer. chem. Soc., 1955, **77,** 432) by reduction of the ditosylate with lithium tetrahydroaluminate so the stereochemistry at two centres has been related through lariciresinol (p. 264) to that of guaiaretic acid.

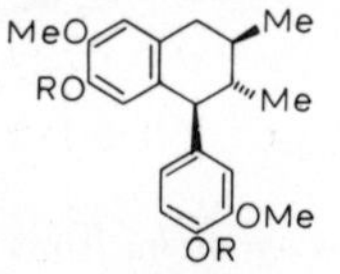

(LXIV) 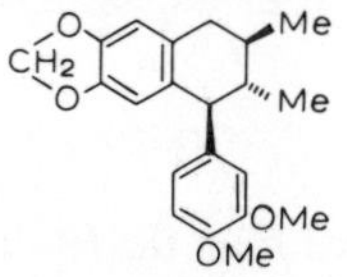 (LXV) 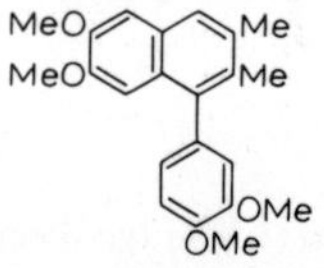 (LXVI)

The phenyl substituent in di-*O*-methylisolariciresinol was assumed (*Haworth* and *Kelly*, J. chem. Soc., 1937, 384) to be equatorial and *trans* to the adjacent hydroxymethyl group and this has been confirmed from the reactions of conidendrin and reterodendrin (p. 268). The relative and absolute stereochemistry of this group is therefore well established:

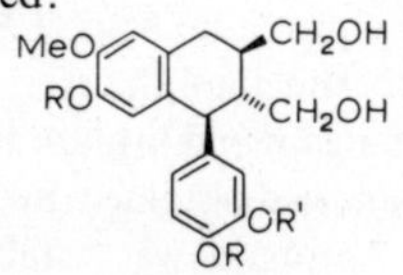

(LXVII) 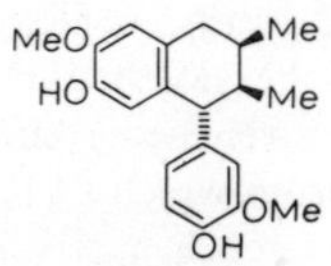 (LXVIII) 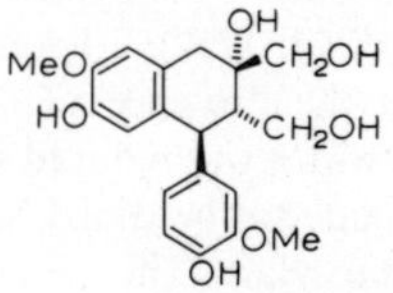 (LXIX)

Galbulin (LXIV, R = Me), m.p. 135°, $[\alpha]_D$ −8.5° (chloroform); *galcatin* (LXV), m.p. 117–118°, $[\alpha]_D$ −8.8° (chloroform) (*Hughes* and *Ritchie, loc. cit.*). **Guaiacin** (LXIV, R = H), isolated from *Guaiacum officinale* L., as the *diethyl ether*, m.p. 114–115°, $[\alpha]_D$ +4° (chloroform), gives a naphthalene derivative on dehydrogenation. The stereochemistry rests on its nonidentity with isoguaiacin and the absolute stereochemistry on a comparison with galbulin (*F. E. King* and *J. C. Wilson*, J. chem. Soc., 1964, 4011). **Isoguaiacin** (LXVIII) was also isolated from *G. officinale* L., as the *dimethyl ether*, m.p. 101–102° and 86–87° (dimorphic), $[\alpha]_D$ −46° (chloroform), and the *diethyl ether*, m.p. 108°, which can be dehydrogenated to the corresponding naphthalene derivatives. The dimethyl ether was identified as the optical enantiomer of a reduction product of di-*O*-methyl-*β*-conidendrin.

Isolariciresinol (LXVII, R = H, R^1 = Me), m.p. 112°, $[\alpha]_D$ +69.4° (acetone), *dimethyl ether*, m.p. 166–167°, $[\alpha]_D$ +20° (chloroform), is readily formed from lariciresinol by the action of acid (*Haworth* and *Kelly, loc. cit.*). **Isotaxiresinol** (LXVII, R = R^1 = H), m.p. 171°, occurs in the heart wood of *Taxus baccata* L. (*King et al.*, J. chem. Soc., 1952, 17). The trimethyl ether is identical with di-*O*-methylisolariciresinol and the triethyl ether gave a benzoylbenzoic acid on permanganate oxidation which was identified by synthesis.

Cyclo-olivil (LXIX), m.p. 167°, $[\alpha]_D$ +61.1° (ethanol), *dimethyl ether*, m.p. 184.5°, $[\alpha]_D$ +35.6° (ethanol) occurs in the resin of the New Zealand olive, *Olea cunninghamii* Maire, (*L. H. Briggs* and *A. G. Frieberg, ibid.*, 1937, 271) and is formed by the action of acid on olivil (p. 265). The stereochemistry follows from that of olivil assuming ring closure to give the tetralin to occur with formation of *trans* adjacent phenyl and hydroxymethyl substituents:

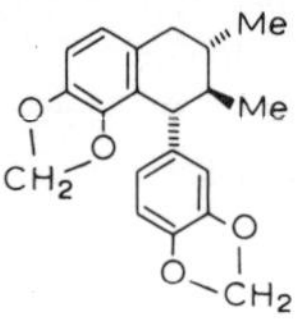

(LXX) 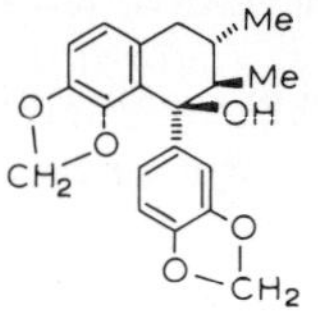 (LXXI) 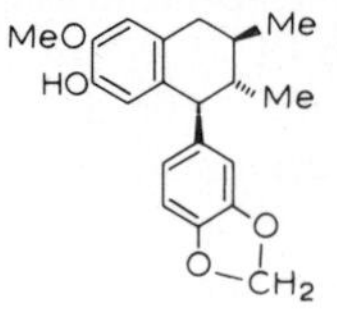 (LXXII)

Otobain (LXX) has been isolated from the fat of *Myristica otoba* fruits along with **hydroxyotobain** (LXXI) and **otobaphenol** (LXXII) (*T. Gilchrist et al.*, J. chem. Soc., 1962, 1780; *N. S. Bhacca* and *R. Stevenson*, J. org. Chem., 1963, **28**, 1638; *Stevenson et al.*, J. chem. Soc., C, 1966, 1775). Dehydrogenation of otobain gives a naphthalene derivative which has been synthesised (*D. Brown* and *Stevenson*, J. org. Chem., 1965, **30**, 1759). The structure LXX was deduced from examination of the n.m.r. spectrum and confirmed by a synthesis of (±)-otobain (*I. Maclean* and *Stevenson*, J. chem. Soc., C, 1966, 1717). 2-Bromo-4,5-methylenedioxyphenylpropiolic acid (LXXIII) cyclised in the presence of dicyclohexylcarbodiimide to the naphthalene derivative LXXIV, which was reduced successively with sodium amalgam and lithium tetrahydridoaluminate to the diol LXXV from which (±)-otobain was obtained by reduction of the ditosylate. The absolute stereochemistry of (−)-otobain was established by comparison of its o.r.d. curve with that of galbulin (*W. Klyne et al.*, *ibid.*, 1966, 893):

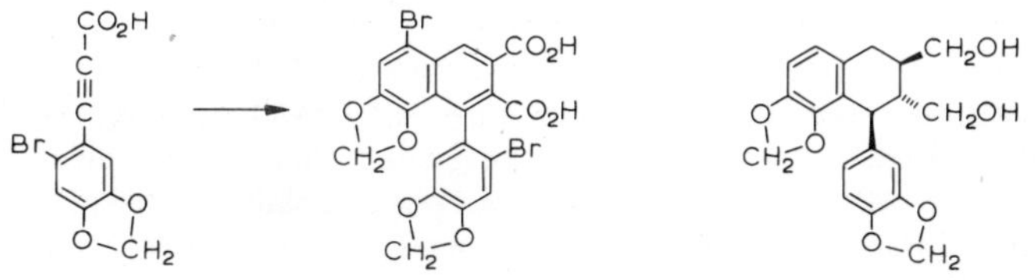

(LXXIII) (LXXIV) (LXXV)

Acid catalysed dehydration of hydroxyotobain and reduction of the resulting olefin gives otobain in low yield. Analysis of the n.m.r. spectrum (*R. Wallace et al.*, J. chem. Soc., 1963, 1445) leads to structure LXXI for hydroxyotobain and the absolute stereochemistry is deduced from the o.r.d. curve. Methylation of otobaphenol gave a compound identical with a synthetic racemate and the ethyl ether was dehydrogenated to a naphthalene derivative which has also been synthesised (*Stevenson et al.*, J. chem. Soc., C, 1966, 1775). The absolute stereochemistry of otobaphenol (LXXII), deduced from the o.r.d. curve *(Klyne et al., loc. cit.)*, is opposite to that of otobain.

Otobain (LXX), m.p. 136–137°, $[\alpha]_D$ −43° (chloroform); *hydroxyotobain* (LXXI), m.p. 116–117°, $[\alpha]_D$ −28° (chloroform); *otobaphenol* (LXXII), m.p. 134–136°, $[\alpha]_D$ +40° (chloroform), *methyl ether, iso-otobain, isogalcatin*, m.p. 105–106°, $[\alpha]_D$ +2° (chloroform), *ethyl ether*, m.p. 106–107°, $[\alpha]_D$ −13° (chloroform).

Lyoniresinol (LXXVI), *dimethoxyisolariciresinol*, m.p. 165–167°, $[\alpha]_D$ +52° (acetone), *dimethyl ether*, m.p. 158–160°, $[\alpha]_D$ +30.1°, occurs as the xyloside, **lyoniside,** m.p. 123.5° and 165° (dimorphous), $[\alpha]_D$ +52° (acetone), in *Sorbus aucuparia* L. (*H. Erdtman et al.*, Acta chem. Scand., 1962, **16,** 518), *Lyonia ovalifolia* Hort. (*M. Yasue* and *Y. Kato*, J. pharm. Soc. Japan, 1960, **80,** 1013), *Alnus glutinosa* L. (*K. Weinges*, Ber., 1961, **94,** 2522), and as the amorphous *rhamnoside* in *Ulmus thomasii* Sarg. (*F. D. Hostettler* and *M. K. Seikel*, Tetrahedron, 1969, **25,** 2325). Lyoniresinol is identical (i.r. spectrum), with the synthetic racemate prepared from the dihydro derivative of (±)-syringaresinol by the action of acid *(Weinges, loc. cit.)*. (±)-**Lyoniresinol,** m.p. 193–194° (anhydrous), *dimethyl ether*, m.p. 164–164.5° has also been isolated from *Ulmus thomasii*. Chromic acid oxidation of (+)-di-*O*-methyllyoniresinol gives *lactone A* (LXXVII), m.p. 193°, and *lactone B* (LXXVIII), m.p. 152°. Lactone A is readily epimerised by base whereas lactone B is not epimerised thus indicating the relative stereochemistry shown (*Y. Kato*, C.A., 1963, **59,** 8668). The absolute stereochemistry (LXXVI) was deduced from molecular rotational difference comparisons *(Kato, loc. cit.*, and *Erdtman, loc. cit.)*.

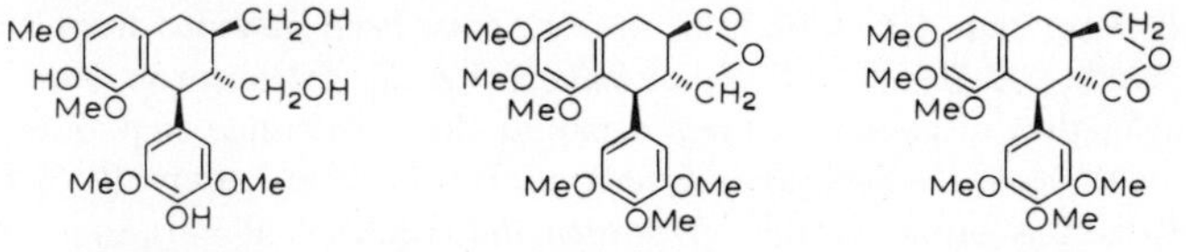

(LXXVI) (LXXVII) (LXXVIII)

O.r.d. measurements have been used both to support this assignment *(Klyne et al., loc. cit.)* and to support the opposite configuration *(F. D. Hostettler* and *M. K. Seikel, loc. cit.)*.

α-**Conidendrin** (LXXIX), m.p. 255–256°, [α]$_D$ −53.7° (acetone), is found in the wood of *Picea excelsa* L., and *Tsuga heterophylla* Sarg. (*I. A. Pearl*, J. org. Chem., 1945, **10,** 219) and has been known for a long time as a crystalline deposit in the liquors from sulphite digestion of wood in pulp manufacture. Dehydrogenation of the *dimethyl ether*, m.p. 179–180°, [α]$_D$ −103.5° (acetone), gives a naphthalene-based lactone which has been synthesised (*R. D. Haworth, T. Richardson* and *G. Sheldrick*, J. chem. Soc., 1935, 636, 1576) from the acid LXXX by the action of formaldehyde. Conidendrin is epimerised by the action of heat or base to *β-conidendrin* (LXXXI), m.p. 210–212°, [α]$_D$ +28° (acetone) (*B. Holmberg*, Ber., 1921, **54,** 2389), *dimethyl ether*, m.p. 156–157°, [α]$_D$ ±0° (*Holmberg* and *M. Sjoberg*, *ibid.*, 1921, **54,** 2406) thus establishing the *trans* lactone ring in α-conidendrin. Conidendrin was transformed by a series of reactions not affecting the asymmetric centres into **reterodendrin** (LXXXII), which is not epimerised by base thus establishing the relative stereochemistry shown (*A. W. Schrecker* and *J. L. Hartwell*, J. Amer. chem. Soc., 1955, **77,** 432). Reduction of di-*O*-methylconidendrin with lithium tetrahydridoaluminate gives di-*O*-methylisolariciresinol (p. 266) and the absolute stereochemistry is thus related to that of guaiaretic acid:

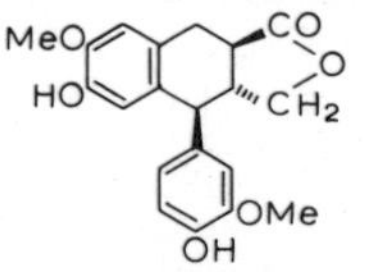

(LXXIX) 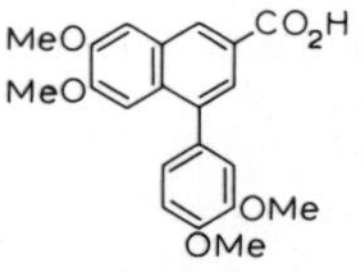 (LXXX) 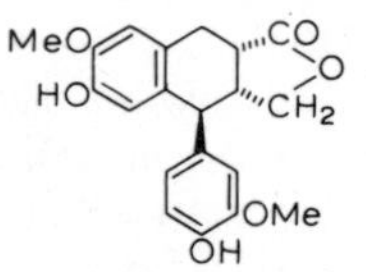 (LXXXI)

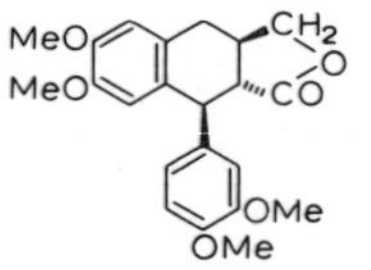

(LXXXII) 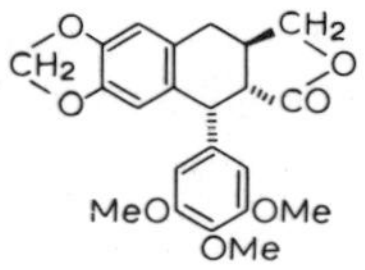 (LXXXIII) 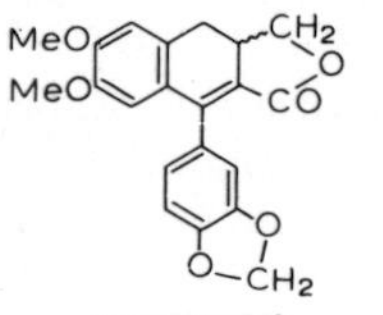 (LXXXIV)

Desoxypodophyllotoxin, *Anthricin, Silicicolin* (LXXXIII), m.p. 168–169°, [α]$_D$ −119° (chloroform), has been isolated from *Juniperus silicicola* L. H. Bailey and prepared by reduction of podophyllotoxin chloride (*Hartwell et al., ibid.*, 1953, **75,** 2138).

Collinusin (LXXXIV), m.p. 196°, [α]$_D$ +132.5°, and **cleistanthin** (LXXXV, R = 3,4-dimethylxylose), m.p. 135–136°, [α]$_D$ −67.2°, have been obtained from *Cleistanthus collinus* Benth. and Hook. f. (*T. R. Govindachari et al.*, Tetrahedron, 1969, **25,** 2815). Dehydrogenation of collinusin gives a naphthalene *derivative*, m.p. 236°, probably identical with *justicidin-B* (*K. Munakata et al.*, Tetrahedron Letters, 1965, 4167; 1967, 3821). Hydrolysis of cleistanthin gives *diphyllin* (LXXXV, R = H), m.p. 291°, which can be methylated to *justicidin-A* (LXXXV, R = Me), m.p. 263°.

Justicidin-C (LXXXVa, R^1 = R^2 = −OMe), m.p. 266°, and **justicidin-D** (LXXXVa, R^1 = R^2 = −OCH_2O-), m.p. 272°, have been isolated from *Justicia procumbeus* L. (*K. Ohoto* and *Munakata, ibid.*, 1970, 923):

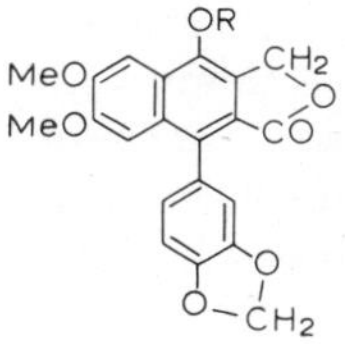

(LXXXV) 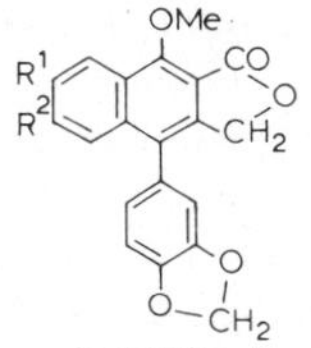 (LXXXVa) 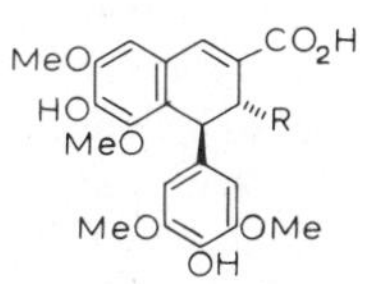 (LXXXVI) 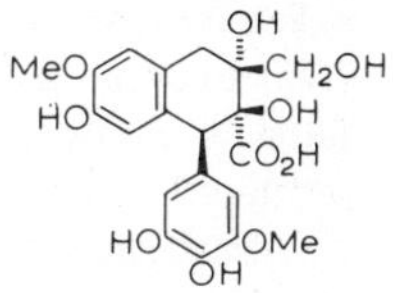 (LXXXVII)

(±)-**Thomasidionic acid** (LXXXVI, R=CO_2H), amorphous, *dimethyl ether methyl ester*, m.p. 142.5–143°, and (±)-**thomastic acid** (LXXXVI, R=CH_2OH), m.p. 232–234°, *dimethyl ether methyl ester*, m.p. 121–122° (*F. D. Hostettler et al.*, Tetrahedron, 1968, **24,** 1475; 1969, **25,** 2325) have been isolated from the heartwood of *Ulmus thomasii* Sarg. Oxidation of di-*O*-methylthomastic acid affords di-*O*-methylthomasidionic acid. Catalytic hydrogenation of methyl di-*O*-methyl thomastate followed by hydrolysis gives a dihydro acid which on lactonisation affords (±)-*lactone-A* derived from (±)-lyoniresinol (p. 267).

Plicatic acid (LXXXVII), amorphous, $[\alpha]_D$ −9.99° (water), *trimethyl ether methyl ester*, m.p. 204–205°, $[\alpha]_D$ −29.9° (chloroform), is the major lignan of the heartwood of *Thuja plicata* Donn. The structure was determined by the application of X-ray methods to the trimethyl ether 4-bromanilide (*J. F. Gardener et al.*, Canad. J. Chem., 1966, **44,** 52). The absolute stereochemistry was deduced by comparison of the o.r.d. curves of the acid and derivatives with those of conidendrin and podophyllotoxin derivatives (*R. J. Swan et al.*, *ibid.*, 1967, **45,** 319).

(v) Lignans of podophyllum resin

The resin derived from *Podophyllum peltatum* L., and *P. emodi* Wall. has long been used in medicine as a purgative. Some of the constituents, particularly podophyllotoxin, have a strong destructive action on sarcoma in mice but toxicity has limited their use in clinical testing (*J. L. Hartwell* and *M. J. Shear*, J. nat. Cancer Inst., 1950, **10,** 1295; C.A., 1951, **45,** 2575). **Podophyllotoxin** (LXXXVIII, R=Me, R′=H) is readily converted into **picropodophyllin** (LXXXIX, R=H) by the action of ammonia or sodium acetate. Both these lactones occur in the resin free and as glucosides. The interconversion was at first thought to involve different sites for closure of the lactone ring, later it was shown to involve epimerisation of the lactone carbonyl (*Hartwell* and *A. W. Schrecker*, J. Amer. chem. Soc., 1951, **73,** 2909). The reaction reversible in *tert*-butanol, catalysed by piperidine, with equilibrium constant 37.0 at 31° (*W. J. Gensler* and *C. D. Gatsonis*, J. org. Chem., 1966, **31,** 3224).

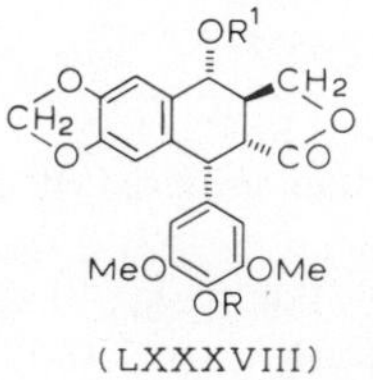

(LXXXVIII) 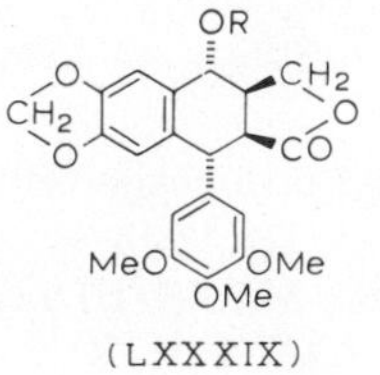 (LXXXIX) 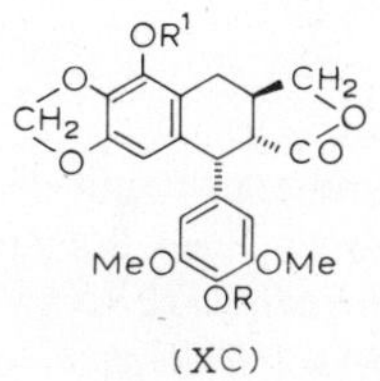 (XC)

α-**Peltatin** (XC, R=R′=H) and β-**peltatin** (XC, R=Me, R′=H), also found in the resin, undergo a parallel epimerisation. Treatment of podophyllotoxin with hydrogen iodide in acetic acid gives podophyllomeronic acid (XCI) which has been synthesised (*A. Robertson* and *R. B. Waters*, J. chem. Soc., 1933, 83). Dehydrogenation gives a naphthalene derivative, XCII, which has also been synthesised (*Haworth* and *J. R. Atkinson, ibid*., 1938, 797). Esters of podophyllotoxin undergo pyrolytic elimination to give α-apopodophyllotoxin (XCIII) more readily than their epimers indicating *cis* arrangement of hydroxyl and adjacent hydrogen in the lignan (*A. W. Schrecker* and *Hartwell*, J. Amer. chem. Soc., 1953, **75**, 5916). Isomers of XCIII can be prepared by migration of the olefinic bond in the tetralin ring. From a study of the catalytic hydrogenation of these isomers, *Schrecker* and *Hartwell (loc. cit.)* deduced the relative stereochemistry of podophyllotoxin. Assignment of absolute stereochemistry rests on an application of the method of molecular rotation differences (*idem, ibid*., 1955, **77,** 433):

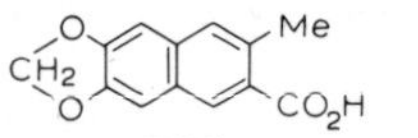

(XCI) 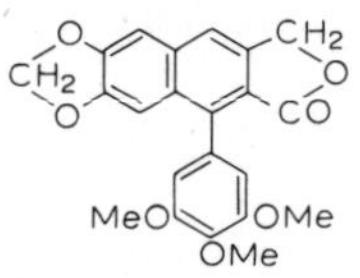 (XCII) 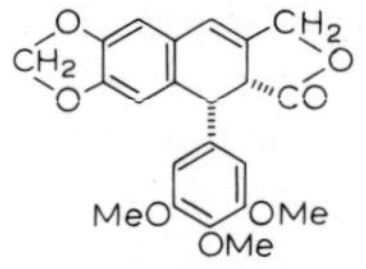 (XCIII)

Podophyllotoxin is converted *via* the chloride into desoxypodophyllotoxin and derivatives of this compound compared with those of conidendrin and reterodendrin (p. 268) whose absolute stereochemistry has been related to that of guaiaretic acid. Hydrogenation of α-apopicropodophyllin gave isodesoxypodophyllotoxin (XCIV) for further comparisons.

Podorhizol, a 1,4-diarylbutane type lignan (p. 255), has also been isolated from podophyllum resin. Its absolute stereochemistry has been deduced by application of the method of molecular rotation differences to that series of lignans and is opposite to the absolute stereochemistry of podophyllotoxin. In agreement with these assignments, acid cyclisation of podorhizol gives (−)-isodesoxypodophyllotoxin while the (+)-enantiomer (XCIV) is obtained from podophyllotoxin:

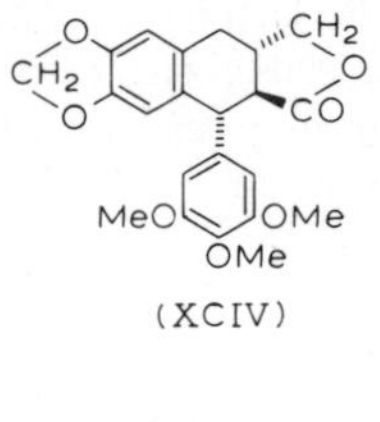

(XCIV) 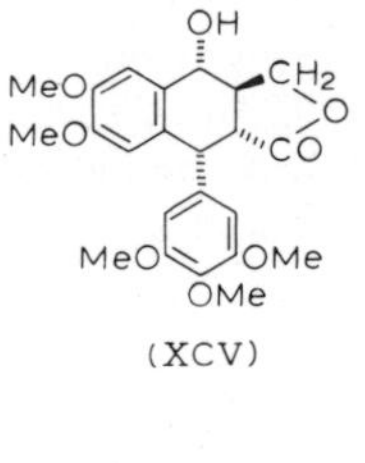 (XCV)

A total synthesis of optically active podophyllotoxin has been achieved (*W. J. Gensler et al*., J. Amer. chem. Soc., 1954, **76,** 315, 5890).

Podophyllotoxin (LXXXVIII, R=Me, R′=H), m.p. 183–184°, $[\alpha]_D$ −132° (chloroform); *demethylpodophyllotoxin* (LXXXVIII, R = R′ = H), m.p. 250–251.6°, $[\alpha]_D$

$-130°$ (chloroform) (*Hartwell et al., ibid.*, 1952, **74,** 280); *demethylpodophyllotoxin β-glucoside* (LXXXVIII, R = H, R′ = glucose), m.p. 165–170°, $[\alpha]_D$ $-75°$ (water) (*A. Stoll et al., ibid.*, 1954, **76,** 5004). *Picropodophyllin* (LXXXIX, R=H), m.p. 231–232°, $[\alpha]_D$ $+9.4°$ (chloroform); *picropodophyllin β-glucoside* (LXXXIX, R = glucose), m.p. 237–238°, $[\alpha]_D$ $-11.5°$ (pyridine) (*Hartwell et al., ibid.*, 1953, **75,** 1308). *α-Peltatin* (XC, R = R′ = H), m.p. 230–232°, $[\alpha]_D$ $-120°$ (chloroform); *β-peltatin* (XC, R = Me, R′ = H), m.p. 231–238°, $[\alpha]_D$ $-119°$ (chloroform) (*Hartwell* and *Detty, ibid.*, 1950, **72,** 246). *α-Peltatin β-glucoside* (XC, R = H, R′ = glucose), m.p. 168–171°, $[\alpha]_D$ $-128.9°$ (methanol); *β-peltatin* ∫*-glucoside* (XC, R = Me, R′ = glucose), m.p. 156–159°, $[\alpha]_D$ $-123°$ (methanol) (*Stoll et al., ibid.*, 1954, **76,** 6431; 1955, **77,** 1710).

Sikkimotoxin (XCV?), m.p. 120°, $[\alpha]_D$ $-91.9°$, has been isolated from the resin of *Podophyllum sikkimensis* (*R. Chatterjee* and *S. C. Chakravarti*, C.A., 1953, **47,** 5920). Structure XCV was proposed; however, the compound of this structure prepared from podophyllotoxin by demethylenation with boron trifluoride at $-60°$ followed by methylation of the resulting phenol, was not identical with sikkimotoxin (*E. Schrier*, Helv., 1964, **47,** 1529).

4. Lignin

Lignin is present in all woody tissue in amounts up to 35% or more so it is an important byproduct of the paper pulp and wood saccharification industries. Its properties and structure are discussed in a number of text-books (*F. E. Brauns*, "The Chemistry of Lignin", Academic Press, New York, 1952; *F. E. Brauns* and *D. A. Brauns*, "The Chemistry of Lignin: Supplement Volume", Academic Press, New York, 1960; *K. Freudenberg* in "Modern Methods of Plant Analysis", eds. *K. Paech* and *M. V. Tracey*, Springer, Berlin, 1955, Vol. III, p. 499; *I. A. Pearl*, "The Chemistry of Lignin", Arnold, London, 1967; *Freudenberg* and *A. C. Neish*, "Constitution and Biosynthesis of Lignin", Molecular-Biologie, Biochemie und Biophysik, Vol. II, Springer, Berlin, 1968; *K. V. Sarkanen* and *C. H. Ludwig*, eds., "Lignins. Occurrence, Formation, Structure and Reactions", Wiley–Interscience, New York, 1971).

(a) Occurrence

Lignin is a component of mature woody tissues. After a certain period in the life of plant cells, lignin begins to be formed. Eventually it permeates the membranous polysaccharides and the spaces between the cells as a coherent ramified mass which strengthens the whole tissue. Its presence brings about the physiological death of the tissue but lignification takes place only with the help of living tissue. The skeleton of lignin is formed by polymerisation of C_6–C_3 units. ^{14}C-Labelled phenylalanine and coniferyl alcohol become incorporated into spruce lignin (*Freudenberg et al.*, Ber., 1969, **102,** 1320).

Lignin is present in all higher plants. It can be divided into three broad classes according to the types of aromatic residue from which it is built (*H. Hibbert et al.*, J. Amer. chem. Soc., 1944, **66,** 32):

(*1*) Softwood (gynosperm) lignins such as that derived from spruce used in the paper industry. These are derived largely from coniferyl alcohol (I).

(*2*) Hardwood (dicotyledonous angiosperm) lignins such as that from aspen or beech. These are derived from coniferyl alcohol and sinapyl alcohol (II).

(*3*) Grass (monocotyledonous angiosperm) lignins such as that from bamboo or maize. These are derived from coniferyl alcohol, syringyl alcohol and 4-hydroxyphenylpropenol (III):

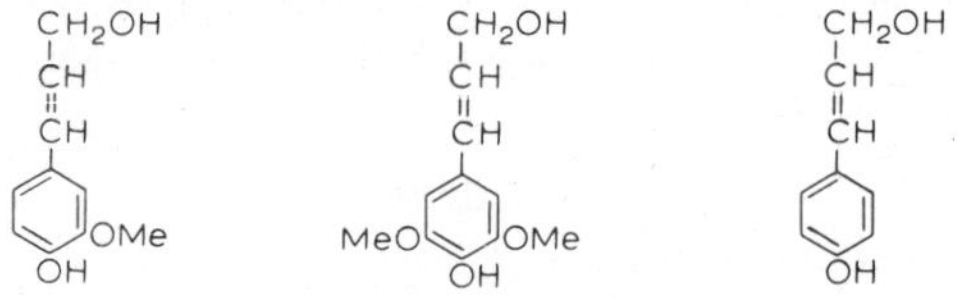

CH_2OH–CH=CH–(ring, OMe, OH) (I) CH_2OH–CH=CH–(ring, MeO, OMe, OH) (II) CH_2OH–CH=CH–(ring, OH) (III)

In the natural state lignin reacts readily with sodium bisulphite or thioglycollic acid to form soluble products. It is not hydrolysed by acids, it is soluble in hot alkali and readily oxidised.

(*b*) *Isolation*

Native lignin. Lignin always suffers some decomposition during isolation. The least degraded material, isolated by mild solvent extraction, is termed native lignin. The wood is first defatted by extraction with ether and the lignin is then extracted by dioxane or acetone. The yield of lignin is usually low. Higher yields have been obtained by first attacking the wood with a wood rotting fungus followed by the mild solvent extraction.

Native lignin is soluble in dilute sodium hydroxide, methanol, dioxane and pyridine, insoluble in water. It is completely soluble in bisulphite solution.

Lignin extracted from cellulose pulp. Some cellulose pulp is manufactured by cooking wood with a solution of sodium hydroxide and sodium sulphide. The lignin passes into solution and can be recovered on acidification. In another pulping process, wood is digested at 125–145° with calcium bisulphite and sulphur dioxide to dissolve the lignin as salts of ligninsulphonic acid and this can be recovered either as the salt or as the free acid.

During these processes the native lignin undergoes extensive modification by reactions of the benzyl ether links. As a model for the sulphite cooking process, the benzyl ether IV has been shown to be converted into an alkylsulphonic acid. A similar reaction occurs with thiols (*Freudenberg et al.*, Ber., 1937, **70,** 500). Sodium sulphide also causes demethylation of phenyl methyl ethers to yield dimethyl sulphide. Some dimethyl sulphoxide is produced commercially by oxidation of this dimethyl sulphide:

(IV)

Lignin residue from cellulose hydrolysis. In the quantitative determination of lignin by the *Klason method*, ground wood is allowed to stand with 66% sulphuric acid to hydrolyse the cellulose and the residual lignin is washed thoroughly with water and dried. The lignin structure is drastically altered in this process by reaction of benzyl ethers and alcohols with aromatic rings under the influence of acid. Gross condensation occurs and the product is no longer soluble in the sulphite cooking process.

Other less vigorous acid treatments have been proposed. The *Willstätter process* uses 40% hydrochloric acid. In the *Freudenberg cuproxam process* wood is first heated with 0.1 *M* sulphuric acid at 100°, then washed and the cellulose removed by solution in ammoniacal copper sulphate. The cellulose will not dissolve without the preliminary acid treatment (see *Freudenberg* in "Modern Methods of Plant Analysis", *loc. cit.*).

The cellulose of wood has also been degraded at room temperature with periodic acid solution to leave a lignin residue (*C. B. Purves et al.*, J. Amer. chem. Soc., 1947, **69,** 1371). This periodate lignin is soluble in bisulphite solution. It must however have undergone extensive degradation of the aromatic rings since periodic acid is known to react with the guaiacol residues which occur in native lignin.

(c) Structure

(i) Evidence from degradation products

(*1*) *Oxidation.* Methylation of spruce lignin with methyl sulphate and alkali followed by oxidation with potassium permanganate affords methoxybenzoic acids. Veratric acid (V), 4,5-dimethoxyisophthalic acid (VI) and the dicarboxylic acids VII and VIII have been isolated. Veratric acid is the most abundant oxidation product and when the lignin is first ethylated and then oxidised, 4-ethoxy-3-methoxybenzoic acid can be isolated (*Freudenberg et al.*, Ber., 1936, **69,** 1415; 1937, **70,** 500; 1938, **71,** 1817; 1962, **95,** 2814):

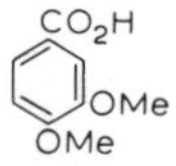

(V) 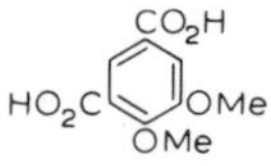 (VI) 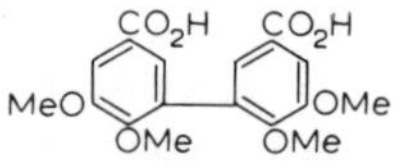 (VII) 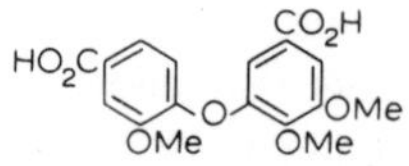 (VIII)

Oxidation of extracted lignins or of raw wood with nitrobenzene and alkali gives vanillin and also syringaldehyde and 4-hydroxybenzaldehyde depending on the type of lignin used (*Freudenberg et al.*, *ibid.*, 1940, **73,** 167; *H. Hibbert et al.*, J. Amer. chem. Soc., 1944, **66,** 32, 37). 5-Carboxyvanillin (IX) and dehydrodivanillin (X) have also been isolated from the oxidation of spruce lignin. Some vanillin is manufactured from softwood lignin by this process. Alkaline copper(II) oxide has also been used to oxidise coniferous lignosulphonates (*I. A. Pearl et al.*, *ibid.*, 1950, **72,** 2309; 1952, **74,** 614; 1954, **76,** 6106):

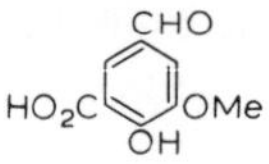

(IX) 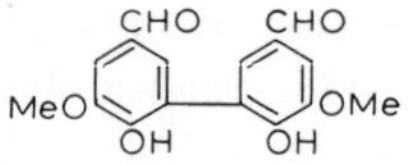 (X)

(*2*) *Reduction.* Hydrogenation of native aspen lignin over copper chromite at high temperature and pressure affords cyclohexane derivatives and over 70% of the lignin was recovered as a mixture of XI, XII and XIII (*E. E. Harris*, *J. D'Ianni* and *H. Adkins*, *ibid.*, 1938, **60,** 1467). Lignins isolated by the soda and sulphite process give a much poorer yield of higher molecular weight products on hydrogenation (*Adkins et al.*, *ibid.*, 1941, **63,** 549):

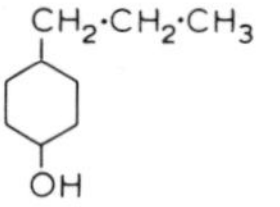

(XI) 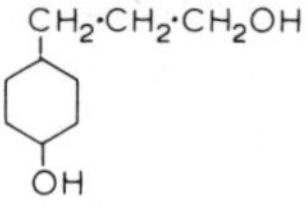 (XII) 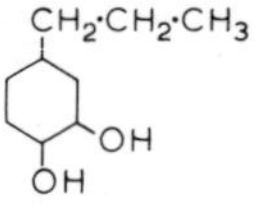 (XIII)

(*3*) *Ethanolysis.* Lignin is decomposed on refluxing with ethanolic hydrogen chloride and some of the products have been identified. Spruce wood decomposition products were separated into a phenolic fraction and this on methylation afforded a crystalline compound, XIV, identified by synthesis. The phenolic fraction from reaction of aspen wood lignin gave XV identified as its 4-nitrobenzoate and also yielded XIV after methylation of the crude products (*Hibbert et al.*, *ibid.*, 1939, **61,** 509, 516):

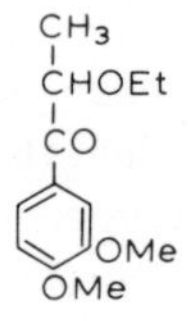

(XIV) 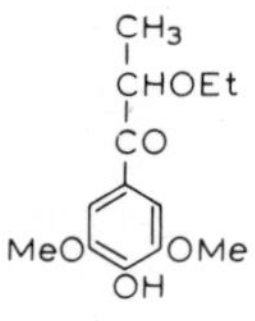 (XV)

Natural lignin contains no ethoxyl groups so the ethoxyl groups in these products must have been derived from the ethanol used as solvent.

(ii) Model for lignin biosynthesis

Enzymic dehydrogenation of coniferyl alcohol can be brought about by the action of enzymes from the juice of the common mushroom (*Agaricus campestris* L.) of from culture filtrates of the wood fungus *Polyporus verisco-lor* (L.) Fr. (*Freudenberg et al.*, Ber., 1958, **91,** 581). The final product is a water insoluble polymer with the same elemental composition as spruce lignin. Methylation and then oxidation of the polymer gave the same range of aromatic acids as were obtained from spruce lignin (*idem, ibid.*, 1962, **95,** 2814). At an intermediate stage during the action of the enzyme dimers of coniferyl alcohol can be isolated from the reaction mixture. Isolated dimers include dehydrodiconiferyl alcohol (XVI), pinoresinol (XVII) and guaiacylglycerol-β-coniferyl ether (XVIII) (*Freudenberg*, Fortschr. Chem. org. Naturstoffe, 1962, **20,** 41):

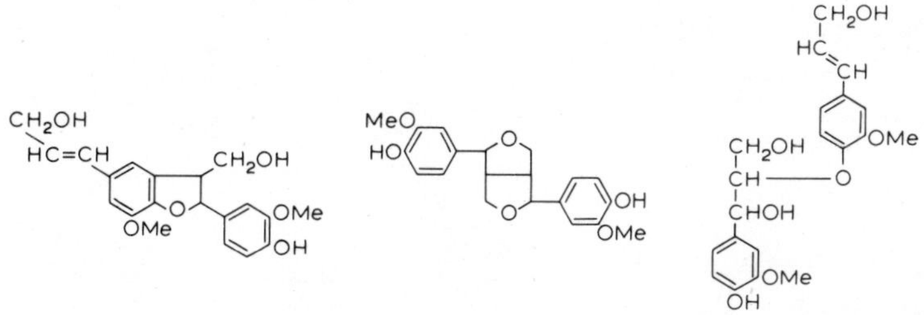

(XVI) (XVII) (XVIII)

The first step in the biosynthesis of lignin from coniferyl alcohol is dehydrogenation to the mesomeric radical XIX which then couples with itself to give dimeric products. Further oxidation then initiates other radical-coupling reactions to build up the lignin polymer. Quinone methides, such as XX, are reactive intermediates in this polymerisation and can themselves react with a phenol without the intervention of an oxidation step and contribute towards the lignin polymerisation reactions. They can also react with an alcohol group from the cellulose constituent of wood and thus bind together the lignin and cellulose macromolecular components of wood:

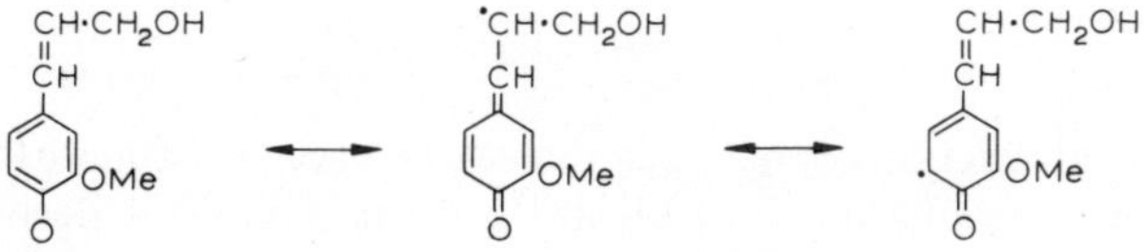

(XIX)

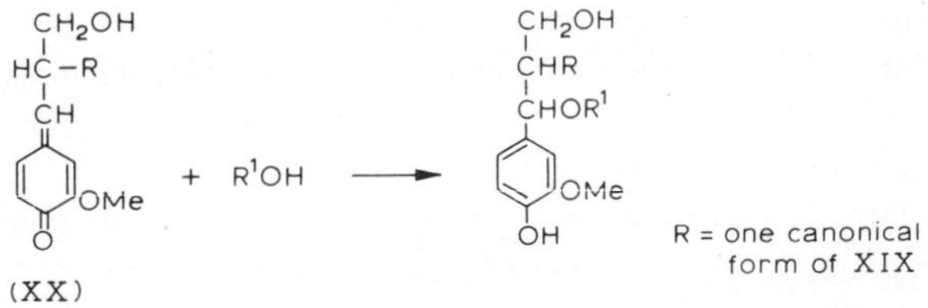

R = one canonical form of XIX

(XX)

(iii) Constitutional model for lignin

Schematic catalogues of the types of units which build up the lignin molecule can be drawn (*Freudenberg*, J. pure and applied Chemistry, 1962, **4,** 9). Scheme 11 is for native spruce wood lignin which is derived mainly from coniferyl alcohol. The lignin of other woods can include coniferyl alcohol (I) and the other C_6–C_3 units II and III. Evidence for the structural units comes from the results of degradation of lignin and the enzymatic dehydrogenation of coniferyl alcohol. N.m.r. spectra provide some further evidence for the occurrence of these units in lignin (*C. H. Ludwig et al.*, J. Amer. chem. Soc., 1964, **86,** 1196):

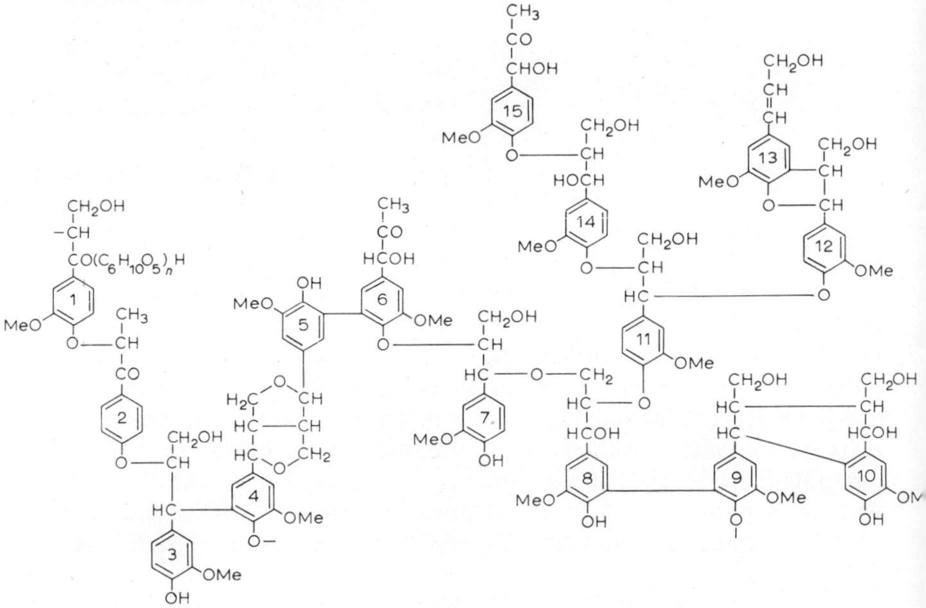

Scheme 11

Some chemical reactions of native lignin can be used to estimate the proportions of these various units which are present. Guaiacol derivatives are oxidised by sodium periodate to the corresponding *o*-quinone with the liberation of methanol. In native spruce lignin about 30% of the guaiacol residues can be so oxidised and so occur in environments like residues 3 and 5 of Scheme 11, while the remaining 70% are thought to be present as guaiacyl ethers such as residues 1 and 13 (*E. Adler et al.*, Acta chem. Scand., 1955, **9,** 319). Oxidation of guaiacol with potassium nitrosodisulphonate gives an *o*-quinone if the *ortho*-position is vacant. Application of this reaction to spruce lignin suggests that 55% of the phenolic units have a free *ortho*-position while the remainder have this position blocked as in unit 8 (*Adler* and *K. Lundquist*, *ibid.*, 1961, **15,** 223):

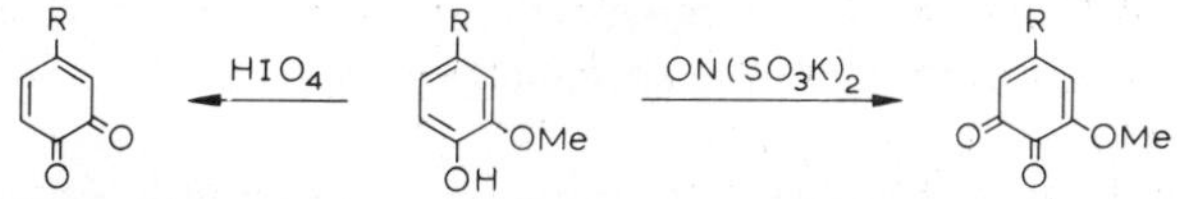

HIO_4 ; $ON(SO_3K)_2$; R ; OMe ; OH

The arylglycerol units, such as 11 and 14, have been shown by a study of model compounds such as XXII to be responsible for the formation of the Hibbert ketones from the alcoholysis of lignin (*Adler et al.*, Ind. eng. Chem., 1957, **49,** 1391):

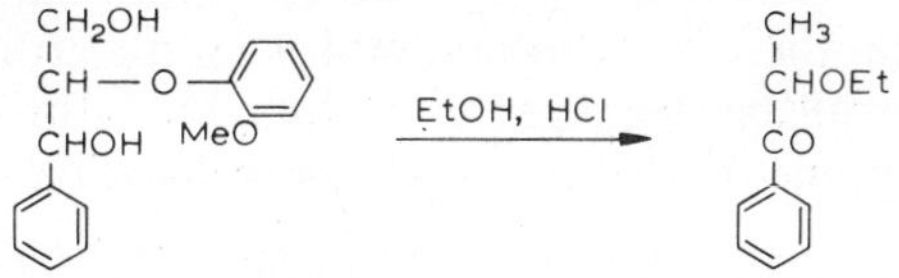

CH_2OH ; CH—O ; CHOH ; MeO ; EtOH, HCl ; CH_3 ; CHOEt ; CO

(XXII)

Molecular weight determinations on native spruce lignin by the ultra centrifuge method give values of 6000–7000 (*G. Meyerhoff*, Naturwiss., 1959, **46,** 143) and similar values have been obtained by viscosity and diffusion methods (*P. R. Gupta*, Pulp Paper Mag. Canada, 1962, **63,** T21).

5. Humic acid

Vegetable and animal remains when left undisturbed are slowly rotted down and form peat or may mix with mineral materials to form soil (*A. Burgess*, Sci. Proc. roy. Dublin Soc., Ser. A, 1960, **1,** 53). The collection of dark coloured organic components of soil and other sedimentary accumulations is called humus. When soil or peat is extracted with sodium hydroxide, much of the organic material is removed into solution and can be reprecipitated by acid. This material is termed humic acid. It is responsible for the fertility of soil where it acts largely as an ion exchanger to retain inorganic ions and as a water-retaining colloid. (For reviews of humic acid see *K. V. Sarkanen* and *C. H. Ludwig* eds., "Lignins. Occurrence, Formation, Structure and Reactions", Wiley–Interscience, New York, 1971, p. 782; *R. D. Haworth*, Soil Science, 1971, **111,** 71.)

Humic acid is formed, in part, by the biological breakdown of cellulose and lignin. It is a very poorly defined amorphous material and in the past the term has also been applied to the blackish products which result from the autoxidation of many polyhydric phenols and benzoquinones. Humic acid isolated from soil, however, always shows an e.s.r. spectrum and this spectrum is not observed for any of the artificial humic acids derived from phenols (*Haworth et al.*, Tetrahedron, 1967, **23,** 1653). The e.s.r. spectrum of humic acid is removed by reduction with sodium dithionite but it is restored again when the specimen is placed in the ambient redox environment.

When humic acid is boiled with dilute acids about 20% of the material is removed as a mixture of polysaccharides, amino acids and metal ions. The residual core still shows an e.s.r. spectrum and contains an extensive aromatic π-system. Distillation with zinc yields hydrocarbons such as pyrene and coronene as well as acridine and benzacridines (*idem, ibid.*, 1967, **23,** 1669). Fusion of humic acid with potassium hydroxide gives a range of phenolic acids and these are also obtained from the artificial humic acids. They probably result by degradation of the material to simple molecules which then recombine to give the products isolated (*idem, ibid.*, 1968, **24,** 5155).

Humic acid is considered to have a central aromatic core to which is attached carbohydrate and α-amino acid residues. The molecular weight has been estimated as 25,000 by sedimentation and viscosity measurements (*E. L. Piret et al.*, Sci. Proc. roy. Dublin Soc., Ser. A, 1960, **1,** 69).

Guide to the Index

This index is constructed in a similar manner to the volume indexes of the first edition of the Chemistry of Carbon Compounds. However, to make the index easier to use, more descriptive entries have been made for the commonly occurring individual, and groups of chemicals.

The indexes cover primarily the chemical compounds mentioned in the text, and also include reactions and techniques, where named, and some sources of chemical compounds such as plant and animal species, oils, etc.

Chemical compounds have been indexed alphabetically under the names used by authors, editing being restricted to ensuring uniformity of entries under the same heading. In view of the alternative nomenclature that can often be used, a limited amount of cross-referencing has been done where it is considered to be helpful, but attention is particularly drawn to Convention 2 below.

For this and the succeeding volumes, the indexing conventions listed below have been adopted.

1. Alphabetisation

(a) The following prefixes have not been counted for alphabetising:

n-	*o-*	*as-*	*meso-*	D	*C-*
sec-	*m-*	*sym-*	*cis-*	DL	*O-*
tert-	*p-*	*gem-*	*trans-*	L	*N-*
	vic-				*S-*
		lin-			*Bz-*
					Py-

Some prefixes and numbering have been omitted in the index, where they do not usefully contribute to the reference.

(b) The following prefixes have been alphabetised:

Allo	Epi	Neo
Anti	Hetero	Nor
Cyclo	Homo	Pseudo
	Iso	

(c) A letter by letter alphabetical sequence is followed for entries, firstly for the main entry, followed by the descriptive entry. The only exception to this sequence is the placing of plural entries in front of the corresponding individual entries to prevent these being overlooked by a strict alphabetical sequence which could lead to a considerable separation of plural from individual entries. Thus "butanes" will come before *n*-butane, "butenes" before 1-butene, and 2-butene, etc.

2. *Cross references*

In view of the many alternative trivial and systematic names for chemical compounds, the indexes should be searched under any alternative names which may be indicated in the main body of the text. Only a limited amount of cross-referencing has been carried out, where it is considered that it would be helpful to the user.

3. *Esters*

In the case of lower alcohols esters are indexed only under the acid, *e.g.* propionic methyl ester, not methyl propionate. Ethyl is normally omitted *e.g.* acetic ester.

4. *Derivatives*

Simple derivatives are not normally indexed if they follow in the same short section of the text.

5. *Collective and plural entries*

In place of "— derivatives" or "— compounds" the plural entry has normally been used. Plural entries have occasionally been used where compounds of the same name but differing numbering appear in the same section of the text.

6. *Main entries*

The main entry of the more common individual compounds is indicated by heavy type. Where entries relate to sections of three pages or more, the page number is followed by "ff".

Index

Acacia blossom oil, 39
Acanthopanax sessiliflorum, 263
Acanthoside-D, 263
Aceritannin, 227, 228, **236**
Aceritol, 236
Acer tartaricum, 236
Acetals, 88, 123
Acetaldehydes, aryl-substituted, 104
Acetaldehyde, 41, 43, 45, 101
Acetaldehyde semicarbazone, 108
Acetaldoxime, 108
Acetamide, 2
2-Acetamidoacetophenone, 136, 137
2-Acetamidobenzaldehyde, 101
4-Acetamidobenzaldehyde, 103
2-Acetamidobenzyl alcohol, 52
2-Acetamidobenzylaniline, 25
2-Acetamidomethylaniline, 25
4-Acetamidophenyl salicylate, 177
Acetanilide, 131
[1-^{14}C]Acetate, 208
Acetic acid, 103, 109, 137
Acetic anhydride, 134
Acetoacetic ester, 92, 113, 168, 191
Acetobromoglucose, 162
Acetomesitylene, 120
Acetone, 43, 101, 130
Acetonedicarboxylic ester, 201
Acetone semicarbazone, 21
Acetonesulphonic acid, 101
Acetonitrile, 9, 134, 169
—, conversion to ketones, 112
Acetonylidenetriphenylphosphorane, 81
Acetonylpyridinium bromide, 138
Acetophenones, 129, 202
—, mass spectra, 121
—, nuclear substituted, 134
—, oxidation, 200
Acetophenone, 13, 16, 41, 42, 43, 82, 106, 107, 109, 115, 116, 117, 119, 120, 121 122, 123, 127, **130,** 159
—, benzylation, 126
—, bromination, 133
Acetophenone, *(continued)*
—, chlorination, 132
—, halogenation, 129
—, nitration, 132
—, oxidation, 131
—, reaction with ethyl acetate, 125
—, — with hydrogen, 134
—, — with hydrogen selenide, 134
—, — with nitrous acid, 131
—, reduction, 124
—, self-condensation, 131
Acetophenone di-*n*-butyl dithioacetal, 134
Acetophenone diethyl acetal, 45, 132
Acetophenone diethyl dithioacetal, 134
Acetophenone dimethyl acetal, 132
Acetophenonehydrazones, 122, 131
Acetophenone oxime, 131
Acetovanillone, **168**
Acetoxime, 59
ω-Acetoxyacetophenone, 133
2-Acetoxybenzaldehyde, 159
3-Acetoxybenzoic acid, 187
4-Acetoxybenzoic acid, 188
2-Acetoxycinnamic acid, 159
2-Acetoxy-4-hydroxybenzoic acid, 190
4-Acetoxy-2-methoxybenzaldehyde, 191
2-Acetoxy-3-methoxybenzoic ester, 188
7-Acetoxy-4-methylcoumarin, 169
α-Acetoxystyrene, 125
Acetylacetone, 101
Acetyl chloride, 45, 89, 125, 132, 177
—, reaction with phenol, 168
—, — with toluene, 134
Acetyl-coenzyme-A, 203, 207
Acetylenes, 79
—, conversion to ketones, 116
Acetylene, 73
Acetylenedicarboxylic acid, 30
Acetylpyruvic acid ester, 101
Acetylsalicylic acid, **177,** 185
Acetylsalicylic anhydride, 177
Acetylsalicyloyl chloride, 177, 180
Acetylsalicyloyl disulphide, 180

Acetylsalicyloylsalicylic acid, 177
Acetylvanillin, 161, 162
Acid chlorides, conversion to aldehydes, 60, 61, 62
—, side-chain reduction, 155
Acridine, 20, 102, 278
Acridone, 179
Acyl azides, 8
Acyl chlorides, reaction with phenols, 166
Acyl halides, reduction to aldehydes, 60
Acylimidazoles, 63
Acylmalonic esters, 62
Acyloins, 76
Acyl phosphonates, 61, 111
Acylpyrazoles, 63
O-Acyl salicyloylamides, 178
Acyltetracarbonylferrates, 108
Agaricus campestris, 253, 275
Alcohols, benzene series, 30
—, conversion to amines, 10
—, from aldehydes, 31, 43, 75
—, from aralkyl halides, 30
—, from carboxylic acids, 32
—, from cyclic ethers, 34
—, from ketones, 31, 43
—, from olefins, 33
—, functional derivatives, 44
—, infrared spectra, 35
—, oxidation, 35, 36
—, — to aldehydes, 56
—, — to ketones, 107
—, phenolic, 141
—, phenyl-substituted, 39, 40
—, physical properties, 35
—, reactions, 35
—, unsaturated, 35
Aldazines, 75
Aldehydes, 131
—, adducts with Grignard reagents, 116
—, as acylating reagents, 82
—, condensation with hippuric acid, 200
—, conversion to alcohols, 31
—, — to carboxamides, 81
—, — to dienes, 79
—, — to epoxides, 83
—, — to ketones, 116
—, — to methylenic olefins, 79
—, — to nitriles, 81
—, decarbonylation, 82
—, deuterated, 67
Aldehydes, *(continued)*
—, from acid chlorides, 61
—, from amides, 63
—, from carboxylic acids, 59
—, from cyclohexenes, 70
—, from 1,2-glycols, 70
—, from Grignard reagents, 66
—, from hydrocarbons, 65
—, from nitroparaffins, 72
—, from olefins, 69
—, from oxazines, 67
—, halogenation, 82
—, isomerisation to ketones, 117
—, oxidation-reduction reaction, 76
—, reactions, 74
—, reaction with alkali, 31
—, — with bromine and cyanide, 72
—, — with diazoalkanes, 82, 83
—, — with *O*-methylhydroxylamine, 6
—, — with primary amines, 74
—, rearrangement to ketones 117
—, reduction with silanes, 44
—, reductive alkylation, 6, 7
—, synthesis, 55
—, α,β-unsaturated, 80
Aldehyde dimethylthioacetal-*S*-oxides, 69
Aldehyde semicarbazones, 64
Aldehyde sulphonic acids, 56
Aldimines, 13, 65
—, α,β-unsaturated, 80
Aldimine hydrochlorides, 64, 151
Aldimine salts, 152
Aldol condensation 126, 159
Aldoximes, 74
—, reduction, 6
Alectoria japonica, 219
Alectoria sarmentosa, 219
Alectoronic acid, **219**, 220
Algarobilla, 226, 231, 234, 238, 239, 240, 247, 252
Algarobin, 240, 247, **248**
Aliphatic aldehydes, aryl-substituted, 104
Alkaloids, 228
Alkanesulphonic acids, 273
Alkoxybenzoic acids, 174
Alkoxybenzoylsulphonohydrazides, 155
Alkoxycarbonyl-alkylidene-phosphoranes, 80
Alkylacetophenones, 134
α-*N*-Alkylamidoacetophenones, 129

Alkyl aminoaryl ketones, 135
N-Alkyl 2-aminobenzylamines, 24
Alkylarenesulphonic acids, 56
Alkyl aryl ketones, 108, 109, 111, 120, 121, 127, 130
—, alkylation, 128
—, conversion to amides, 128
—, oxidation, 123
—, reactions, 122
—, reduction, 123, 124
Alkyl azides, 71, 128
Alkylbenzenes, conversion to ketones, 107
Alkylbenzylamines, 10
C-Alkylbenzylanilines, 92
Alkyl benzyl ketones, 117
Alkyl-9-borabicyclo[3.3.1]nonanes, 72, 119
Alkyl carbonates, 125
Alkyl dichlormethyl ethers, 65
Alkyl halides, 33, 69, 117, 126, 128, 155, 163
—, conversion to ketones, 108
—, reaction with sodium azide, 4
Alkylidenephosphoranes, 110
N-Alkylimines, 91
Alkyl iodides, 118
n-Alkyl ketones, 110
Alkyllithiums, 110
Alkylmagnesium halides, 20, 31, 33
Alkyl (1-methylthio)phosphonate esters, 118
Alkyl nitrites, 125
2-Alkylphenols, 150
4-Alkylphenols, 150
Alkylphenyl ethers, 174
Alkylphenyl ketone anils, 20
Alkylphenylmethanols, 43
Alkyl sulphites, 89
Alloxan, 102
Allyl benzene, 67
Allylbenzyldimethylammonium salts, 15
Allylhydroquinone monobenzoate, 200, 201
Allylic alcohols, 70
Allyl isothiocyanate, 145
2-Allylphenol, 154, 160
Allyl(phenyl)acetaldehyde, 81
Allyl vinyl sulphide, 81
Alnus glutinosa, 267
Aluminium alkoxides, 31, 75, 107
Aluminium hydride, 5
Aluminium isopropoxide, 45
Aluminium methoxide, 139
Aluminium phenoxide, 158
Amarin, 91
Ameliaroside, 168
Amides, 173
—, *N*,*N*-disubstituted, 63
—, from ketones, 128
—, reaction with phenylacetaldehyde, 105
—, reduction, 33, 63
—, — to primary amines, 5
—, α,β-unsaturated, 70
Amidomethylation, 9
Amines, acylation, 11
—, alkylation, 3
—, aralkylation, 2
—, benzoylation, 39
—, by amidomethylation, 9
—, by reductive alkylation, 6
—, condensation with phenol, 141
—, from alcohols, 10
—, from amides, 5
—, from azides, 4
—, from isocyanates, 6
—, from nitriles, 4
—, from organoboranes, 8
—, from organometallic compounds, 8
—, from oximes, 5
—, from quaternary salts, 7
—, from Schiff bases, 6
—, nitration, 14
—, nitrosation, 12
—, preparation by degradative methods, 8
—, reaction with benzaldehyde, 91
—, with oxidising agents, 13
—, reductive cleavage, 12
—, secondary, 3, 4
—, tertiary, 3
Amine hydrochlorides, 4
Amine radical ions, 10
Aminoacetophenones, 134, 135
2-Aminoacetophenone, 109, **135**, 136
3-Aminoacetophenone, 137, 168
4-Aminoacetophenone, 137
ω-Aminoacetophenone, 17
2-Aminoacetophenone oxime, 136
Amino acids, 278
α-Amino acids, benzylidene derivatives, 92
Aminoaryl ketones, 167

Aminobenzaldehydes, 95, 100, 101, 154
2-Aminobenzaldehyde, **101**, 102
3-Aminobenzaldehyde, **102**
4-Aminobenzaldehyde, **102**
Aminobenzoic acids, 51
3-Aminobenzoic acid, 187
2-Aminobenzyl acetate, 52
Aminobenzyl alcohols, 51
2-Aminobenzyl alcohol, **51**, 52
3-Aminobenzyl alcohol, 52
4-Aminobenzyl alcohol, 52
Aminobenzylamines, 24, 26
2-Aminobenzylamine, **24**
2-Aminobenzylaniline, 24, 25
2-Aminobutyrophenone, 137
3-Aminobutyrophenone, 137
4-Aminobutyrophenone, 137
4-Amino-3-hydrazino-5-mercapto-1,2,4-triazole, 73
α-Amino-β-hydroxy acids, 16
3-Amino-2-hydroxybenzoic acid, 182
3-Amino-4-hydroxybenzoic ester, 188
4-Amino-3-hydroxybenzoic ester, 188
α-Aminoketones, 17
β-*tert*-Aminoketone hydrochlorides, 117
Aminonitrobenzaldehydes, 100
4-Amino-3-nitrobenzaldehyde, **103**
2-Amino-4-nitrotoluene, 182
6-Aminopenicillanic acid, 200
β-4-Aminophenylethyl alcohol, 53
1-Amino-1-phenylpropane, **18**
1-Amino-2-phenylpropane, **18**
1-Amino-3-phenylpropane, **19**
2-Amino-1-phenylpropane, **18**
2-Aminopropiophenone, 137
3-Aminopropiophenone, 137
4-Aminopropiophenone, 137
3-Aminosalicylic acid, **182**
4-Aminosalicylic acid, 182
5-Aminosalicylic acid, 182, 183
Amphetamine, 18, 19
Amygdalin, 86
Amyl nitrite, 93, 133
n-Amyl phenyl ketone, 120
Anacyclus pyrethrum, 143
Anchutz's compound, 179
Androsin, 169
Anethole, 160, 188
Anhaline, **143**
Anhalonium sp., 143
Anhydrides, from salicylic acid, 183
1,5-Anhydro-D-glucitol, 236
Anhydropodorhizol, **257**
Anhydrotris-2-aminobenzaldehyde, 101
Anilides, 173
Aniline, 105, 116, 131, 137, 139, 193
—, benzylidene derivatives, 91
—, reaction with benzaldehyde, 91
—, — with benzyl chloride, 20
—, — with formaldehyde, 53
Aniline hydrochloride, 131
Aniline phthalate, 249
2-Anilinobenzaldehyde, 102
Anisaldehyde, 143, 148, 152, **160**
Anise oil, 188
Anisic acid, 160, **188**
Anisic esters, 188
Anisoin, 160
Anisole, 21, 151, 152, 157, 160
Anisonitrile, 188
Anisoyl chloride, 160, 188
Anisylacetone, 168
Anogeissus latifolia, 234
Anthranil, 51, 99, 100, 101, 102
Anthranilic acid, 51, 109, 176, 180, 181
Anthranilic esters, 52
Anthraquinones, 205
Anthricin, 268
Antiarolaldehyde, 165
Antibacterial activity, 208
Antioxidants, 194
Anziaic acid, 213
Aphthosin, 214, 216
Apiolaldehyde, 165
Apionolaldehyde, 165
Apocynin, 149
Apocynol, 149
Apocynum canabinum, 169
α-Apopicropodophyllin, 270
α-Apopodophyllotoxin, 270
Aralkanols, 1
Aralkylalcohols, 1, 55
Aralkylamines 1, 55
—, by reduction, 3
—, nuclear-substituted, 21
—, phenolic, 141
—, physical properties, 11, 16
—, preparation, 1
—, reactions, 11
—, reaction with nitrous acid, 35

Aralkylation, to amines, 2
Aralkylcarbaldehydes, 1, 55
Aralkylcarboxylic acids, conversion to aralkylamines, 8
Aralkylglycidic esters, 104
Aralkyl halides, conversion to alcohols, 30
Aralkylketones, 55
Aralkylmagnesium chlorides, 104
Arboreol, **264**
Arctostaphylos uva-ursi, 236
Arenes, reaction with aziridines, 9
Arenecarbaldehydes, 86
Arenediazonium salts, 175
Argentic picolinate, 107
Aromatic acids, conversion to alcohols, 32
—, from lignins, 275
—, from methyl ketones, 123
Aromatic aldehydes, condensation reactions, 76
—, conversion to carboxamides, 81
—, — to nitriles, 81
—, decarbonylation, 82
—, halogenation, 82
—, oxidation-reduction reaction, 76
—, oxidation to carboxylic acids, 75
—, properties, 73
—, reactions, 74
Aromatic amines, benzoylation, 39
Aromatic hydrocarbons, formylation, 65
—, oxidation to aldehydes, 56
Aromatic ketones, 107
Aroyl chlorides, reduction to aldehydes, 61
Arthoniaic acid, 213
Arylacetic acids, 200
Arylacetylenes, conversion to ketones, 116
3-Arylaldehydes, 70
3-Arylalkan-1-ols, 33
Arylamines, 59
—, conversion to ketones, 108
N-Aryl 2-aminobenzylamines, 24
Aryldialkyl ketones, 115
β-Arylethanols, 34
Arylglycerol units, from lignins, 277
Arylglycidie acids, 115
Arylglycollic acids, 66
Aryl halides, 153
Arylketenes, 139
Arylketones, 70
—, self-condensation, 125
Aryllithiums, 67, 153
Arylmagnesium bromides, 66, 89
Arylmagnesium halides, 173
Arylmethylmethanols, 134
Arylpyruvic acids, 200
Arylthallium bistrifluoroacetate, 38
Aryltrichloromethyl ketones, 129
Aryltrichloromethylmethanols, 66, 78
Asarinin, **261**, 262, **263**
Asarone, 164
Asaronic acid, **198**
Asarum sieboldii, 263
Asarylaldehyde, 164
Aschantin, **263**
Aspergillus sp., 164
Aspergillus niger, 233
Aspergillus ustus, 222, 224
Aspidinol, 170
Aspirin, 171, **177**, 192
Atranal, 224
Atranol, **163**
Atranorin, 204, 208, 211, **212**, 213
Atraric acid, 205, 206, 208
Aubrietin, 145
Aubrietta sp., 145
Auroglaucin, **164**
3-Azetidinols, 129
Azides, 112
—, reduction, 4
—, related to benzylamine, 29
2-Azidobenzaldehyde, **102**
ω-Azidotoluene, 29
Aziridines, 9
Azlactones, 200
Azodicarboxylic acid, 31
Azodicarboxylic esters, 36, 108
Azomethines, 8, 97
Azoxybenzaldehydes, 99
Azoxybenzenedicarbaldehydes, 100
Azoxybenzene-2,2′-dicarbaldehyde dioxime, 100
2-Azoxybenzyl alcohol, 51

Baeomycesic acid, 213
Baeyer-Villiger reaction, 123
Balsams, 39, 42
Barbatic acid, 204, **212**, 213
Barbatolcarboxylic acid, 206
Barbatolic acid, 215
Beckmann rearrangement, 16

Benzacridines, 278
Benzalaniline, 89
Benzalazine, 28, **92**, 93
Benzaldehydes, 35, 56, 59, 154
—, condensation with hippuric acid, 200
—, di-*ortho*-substituted, 57
Benzaldehyde, 7, 12, 13, 16, 25, 36, 39, 41, 42, 44, 45, 46, 47, 48, 49, 55, 58, 61, 71, 72, 73, 75, 77, 83, 84, **86**, 105, 116, 117, 121, 128, 147, 168
—, bisulphite addition compounds, 90
—, chlorination, 82, 87
—, functional derivatives, 88
—, iodination, 82
—, manufacture from toluene, 86
—, nitration, 97
—, oxidation, 87
—, properties, 73
—, reaction with acyl anhydrides, 89
—, — with alkylmagnesium halides, 43
—, — with ammonia, 74
—, — with ammonia and amines, 91
—, — with chlorine, 87
—, — with chloroform, 78
—, — with 1,2- and 1,3-diols, 89
—, — with ethylamine, 19
—, — with hydrogen selenide, 91
—, — with hydrogen sulphide, 90
—, — with sugars, 89
—, — with sulphurous acid, 90
—, sulphonation, 103
Benzaldehyde-d, **87**
Benzaldehyde acetals, 88
Benzaldehyde acetylhydrazone, 92
Benzaldehyde ammonia, **91**
Benzaldehyde benzylhydrazone, 93
Benzaldehyde cyanohydrin, 200
Benzaldehyde cyanohydrin glucoside, 86
Benzaldehyde cyclic acetals, 89
Benzaldehyde diethyl acetal, 89, 91
Benzaldehyde diisopropyl acetal, 89
Benzaldehyde dimethyl acetal, 44, 45, **89**
Benzaldehyde dimethyl dithioacetal, 90
Benzaldehyde dimethylene acetal, 89
Benzaldehyde-2,4-disulphonic acid, **103**
Benzaldehyde ethyl methyl acetal, 89
Benzaldehyde hydrazone, 28, **92**
Benzaldehyde oximes, 93, 94
Benzaldehyde phenylhydrazone, 93
Benzaldehyde sodium dithionite salt, 90
Benzaldehyde-4-sulphonamide, 103
Benzaldehyde-2-sulphonic acid, **103**
Benzaldehyde-3-sulphonic acid, **103**
Benzaldehyde-4-sulphonic acid, **103**
Benzaldehyde trimethylene acetal, 89
Benzaldoximes, 25, 93, 94
Benzaldoxime benzyl ether, 25
Benzaldoxime *O*-methyl ethers, 94
Benzamides, *N*-substituted, 33
Benzamide, 36, 94
Benzanils, 152
Benzcycloheptenone, 202
Benzedrine, 18
Benzene, 34, 130, 132, 133, 138, 186
—, reaction with benzyl alcohol, 39
—, — with carboxylic acid chlorides, 109
—, — with 1,2-epoxypropane, 42
—, — with ethylene oxide, 41
—, reaction with unsaturated amines, 9
2-Benzeneazobenzyl alcohol, 52
Benzeneboronic acid, 147
Benzenediazonium chloride, 29
Benzenesulphenic ester, 49
Benzenesulphinic acid, 49
Benzenethiol, 180
Benzilic acid rearrangement, 252
Benzoates, electrolytic reduction, 33
Benzoate esters, 89
1,3-Benzodioxane-2,4-dione, 178
Benzofurans, 156
Benzoic acids, conversion to alcohols, 32
Benzoic acid, 32, 45, 47, 48, 86, 87, 109, 130, 147, 173, 174
—, electrolytic reduction, 59
—, sulphonation, 191
Benzoic anhydride, 87
Benzoin, 76
Benzoin reaction, 75, 76
Benzonitriles, 169
Benzonitrile, 10, 88, 93, 94
Benzophenones, 217
Benzophenone, 14, 93
Benzoquinones, 216, 219
—, autoxidation, 278
Benzoquinone, 201
1,4-Benzoquinone-2-carboxylic ester, 198
1,4-Benzoquinonemonoxime-2-carboxylic acid, 182
Benzotrichloride, 87
Benzoxazinethiols, 52

Benzoxazinone, 52
Benzoxazolone, 178, 179
Benzoylacetone, 101, 125
Benzoylbenzoic acid, 266
Benzoyl bromide, 61, 87
Benzoyl chloride, 32, 38, 43, 60, 87, 132 133, 148, 169
Benzoyl cyanide, 133
2-β-D-6′-Benzoylglucopyranoside, 149
O-Benzoylhydroxamic acid, 17
1-Benzoyl-2-hydroxyisopropyl-cyclopropanes, 132
Benzoylium ions, 35
Benzoylmethanesulphonic acid, 132
Benzoyl peroxide, 13
Benzoylvanillin, 149
N-Benzylacetamide, 2
Benzyl acetate, 12, 19, 39, **45**
Benzylacetoacetic ester, 138
Benzylacetone, 138
N-Benzylacetoxime, 25
O-Benzylacetoxime, 25
Benzyl alcohols, 154, 156
—, alkyl-substituted, 50
—, conversion to halides, 37
—, electrolytic oxidation, 36
—, from carboxylic acid derivatives, 32
—, halogen-substituted, 49
—, nuclear-substituted, 49
—, oxidation, 35, 36
—, — to aldehydes, 56
—, reaction with thallium trifluoroacetate, 38
—, α-substituted, 37
—, thio analogues, 48
Benzyl alcohol, 12, 29, 30, 32, 33, 35, **39**, 45, 47, 48, 86, 126
—, ethers, 44
—, inorganic esters, 46
—, oxidation, 36
—, reaction with phosgene, 46
—, — with thionyl chloride, 46
—, — with urea, 20
[α-^{2}H]Benzyl alcohol, 16
Benzylamines, 10, 21, 58, 141, 142
—, acylation, 11
—, diazotisation, 59
—, halogen-substituted, 21
—, nitrosation, 12
—, nuclear substituted, 26
Benzylamines, *(continued)*
—, oxidation, 13
—, — to nitriles, 13
Benzylamine, 2, 7, 8, 9, **16**, 19, 20, 21, 35, 58, 59, 93
—, carbonic acid derivatives, 20
—, nitration, 24
—, oxidation, 13
—, physical properties, 11
—, reaction with α,β-epoxy acids, 16
—, — with sodium hydride, 13
—, — with sulphur, 13
[α-^{2}H]Benzylamine, 16
α-Benzylamino-β-hydroxycarboxylic acids, 16
Benzylaminomalonic ester, 30
Benzylammonium polysulphides, 13
Benzylammonium thiocyanate, 21
Benzylanilines, 13, 59
Benzylaniline, 20, 29, 92
Benzyl azide, **29**, 30
N-Benzylbenzaldoxime, 16, 25, 28
Benzyl benzoate, 39, **45**, 46, 76
Benzylbenzyls, 39
Benzylbenzyl alcohols, 39
Benzyl bromide, 29, 47, 138
Benzyl carbamate, 7, **20**
Benzylcarbamic esters, 20
Benzyl chlorides, 21, 95
Benzyl chloride, 16, 19, 20, 28, 29, 39, 47, 87
—, reaction with alkoxides, 44
—, — with ammonia, 2
—, — with aniline, 20
—, — with carboxylic acids, 46
—, — with hydroxylamine, 25
—, — with phenol, 45
—, — with sodium acetate, 45
—, — with sodium azide, 29
—, — with sodium benzoate, 45
—, — with sodium sulphide, 49
—, — with sodium thiosulphate, 46
—, — with thiophenol, 49
—, — with thiourea, 21
Benzyl chloroformate, 20, **46**
Benzyl chloromethyl ether, 44
Benzyl chloromethyl ketone, 138
α-Benzylcinnamic acid, 80
Benzyl cyanide, 17, 144
Benzyldialkylamines, 11
Benzyldiazotates, 29

Benzyl dichloromethyl sulphide, **48**
Benzyl dihydrogen phosphate, **47**
Benzyl dihydrogen phosphite, **47**
Benzyldimethylmethanol, 43
Benzyldimethylsulphonium bromide, 38
3-Benzyl-1,3-diphenyltriazene, 29
Benzyl esters, 45
Benzyl ethers, 33, 37, 38, 44, 72, 217
Benzyl ether links, lignins, 273
α-Benzylethylcarbamide, 18
Benzyl ethyl chloride, 8
Benzyl ethyl ether, **45**
Benzyl ethyl ketone, 137
Benzylethylmethylmethanol, 43
Benzyl ethyl sulphide, 48
Benzyl ethyl sulphone, 48
Benzyl halides, 51, 69
—, conversion to benzaldehydes, 57
—, oxidation, 155
—, — with nitrates, 58
—, reaction with potassium hydrogen sulphide, 48
—, — with thiourea, 48
Benzyl hydrogen sulphate, **46**
Benzylhydrazine, **28**
N-Benzylhydroxylamine, **25**
O-Benzylhydroxylamine, **25**, 28
Benzyl hyponitrite, **46**
Benzylic alcohols, 10
—, oxidation, 13
Benzylic bromides, 58
Benzylideneacetone, 118, 138
Benzylideneamines, 58, 59
Benzylideneamine hydrochloride, 91
Benzylideneanilines, 13
Benzylideneaniline, 6, 20, 30, **91**
Benzylidenebenzylhydrazine, 28
Benzylidene bis-dimethylamine, 75
Benzylidene chloride, 86, 139
Benzylidene diacetate, **90**
Benzylidene diacyl esters, 89
Benzylidenedibenzylhydrazine, 29
Benzylidene dichloride, 87, 89, 90, 158
Benzylidene dihalides, 58, 95
Benzylidene dimethyl ether, 89
Benzylidene dipropionate, 90
Benzylidene-ethylamine, 91
Benzylidene-glycerols, 89
Benzylideneglycine, 92
Benzylidenehydrazine, 92
Benzylidene-imine, 30
Benzylidene-isobutylamine, 91
Benzylidene-methylamine, **91**
Benzylidene-4-nitroaniline, 91
N-Benzylidene-*o*-phenylenediamine, 92
Benzylidenephenylhydrazine, 93
Benzylidenepropylamine, 91
Benzylidenetoluidines, 92
Benzyl iodide, 46
Benzyl isonitrile, 16
Benzyl isothiocyanate, **21**
Benzyllithium, 44
Benzylmagnesium chloride, 31, 42, 43
Benzylmalonic ester, 70
Benzylmethylaminobenzaldehyde, 100, 103
Benzyl methyl ether, 38, **44**
N-Benzyl-*N*-methylformamide, 19
Benzyl methyl ketone, 18, 42, 107, 109, **137**
Benzyl methyl ketoxime, 18
Benzylmethylmethanol, 42
Benzyl methyl sulphide, 38, **48**
Benzyl methyl sulphone, **48**
Benzyl methyl sulphoxide, **48**
L-Benzyl-3-methyltriazene, 29
Benzyl mustard oil, 21
Benzyl nitrate, **47**
Benzyl nitrite, **47**
Benzylnitroamine, 29
N-Benzyl-*N*-nitrosocarbamic ester, 29
N-Benzyl-*N*-nitrosohydrazine, 29
Benzylnitrosohydroxylamine, 25
2-Benzyloxy-3,4-dimethoxybenzoic acid, 193
3-Benzyloxy-4,5-dihydroxybenzoic acid, 198
4-Benzyloxy-2,3-dimethoxybenzoic acid, 193
4-Benzyloxy-3,5-dimethoxybenzoic acid, 197
Benzyl-oxygen bond, cleavage, 37
3-Benzyloxy-4-methoxybenzaldehyde, 162
4-Benzyloxy-3-methoxybenzaldehyde, 162
2-Benzyloxy-4-methoxybenzoic acid, 191
2-Benzyloxy-6-methoxybenzoic acid, 191
3-Benzyloxy-4-methoxybenzoic acid, 189
4-Benzyloxy-2-methoxybenzoic acid, 191
4-Benzyloxy-3-methoxybenzoic acid, 189
2-(4-Benzyloxy-3-methoxyphenyl)-ethylamine, 144
Benzylphenols, 39, 45
1-Benzyl-2-phenylbenzimidazole, 92

Benzyl phenyl ether, **45**
N-Benzyl-*N*-phenylhydrazine, 28
N-Benzyl-*N'*-phenylhydrazine, 28, 93
Benzylphenylhydroxylamine, 92
Benzylphenylmethanol, 45
Benzylphenylnitrosoamine, 28
Benzyl phenyl sulphide, **49**
Benzyl phenyl sulphone, 49
Benzyl phenyl sulphoxide, 49
1-Benzyl-3-phenyltriazene, 29
Benzylphosphonic ester, **47**
N-Benzylphthalimides, 2
Benzyl propyl ketone, 138
Benzyl salicylate, **46**
4-Benzylsemicarbazide, **21**
Benzylsulphamic acid, 25
Benzyl sulphides, oxidation, 38
Benzylsulphinic acid, 48
Benzyl sulphones, 38
Benzyl sulphoxides, 38
S-Benzyl thioformate, 48
N-Benzylthiophthalimide, 49
Benzyl thiosulphate, 48
Benzylthiosulphuric acid, **46**
N-Benzylthiourea, 21
S-Benzylthiourea hydrochloride, 21
S-Benzylthiuronium chloride, 21
Benzyl 4-tolyl sulphide, **49**
Benzyl tosylate, **46**
Benzyl trichloroacetate, **46**
Benzyltrichloromethyl sulphide, **48**
Benzyltrimethylammonium hydroxide, 19
Benzyltrimethylammonium iodide, 19
Benzyltrimethylammonium salts, 14, 38
N-Benzylurea, **21**
Benzyne, 19
Betula lenta, 175
1,2-Bisazoxyethanes, 93
Bis(*n*-butylsulphonyl)phenylmethane, 90
Bis(2-carboxyphenyl) disulphide, 181
Bischler-Napieralski reaction, 17
Bis(α-chlorobenzyl) oxalate, 90
Bis(1,3-diphenylimidazolinylidene-2), 83
α,α-Bis(ethylsulphonyl)ethylbenzene, 134
Bisethylsulphonylphenylmethane, 90
Bis(hydroxybenzyl)amines, 142
3,5-Bis(*O*-methoxycarbonyl)gallic acid, 229
Bismethoxycarbonylorsellinic acid, 212
Bis(*O*-methoxycarbonyl)orsellinic acid chloride, 210
Bis(3-methyl-2-butyl)borane, 63
Bis(methylsulphonyl)phenylmethane, 90
Bismuth gallate, 194
Bismuth iodosubgallate, 194
Bis-β-phenylethyl ether, 41
Bis(trismethylthio)acetophenones, 133
Bistriphenylsilyl chromate, 69
Bitter almond oil, 86
Boninic acid, 214
9-Borabicyclo[3.3.1]nonane, 69
Boron trifluoride etherate, 5
Bourbonal, 162
Bouveault-Blanc reduction, 41
Brevifolin, **240**, 248
Brevifolincarboxylic acid, 239, **240**, 241, 248, 249, 251, 252
Brevilagins, 238, 239, 251, 252
Brevilagin-1, **247**, 248
Brevilagin-2, 247, **248**
Brittle gum, 230
Bromoacetic ester, 202
Bromoacetone, 119
2-Bromoacetophenone, **135**
3-Bromoacetophenone, 129
4-Bromoacetophenone, 135
ω-Bromoacetophenone, **133**
ω-Bromoacetophenone oxime, 133
Bromoalkanes, 2
N-Bromoamides, 108
4-Bromoanisole, 188
Bromobenzaldehydes, 96
4-Bromobenzaldehyde, 95
Bromobenzene, 2
4-Bromobenzonitrile, 24
Bromobenzyl alcohol, 51
Bromobenzylamines, 26
4-Bromobenzylamine, 24
2-Bromobutane, 187
5-Bromo-3,4-dihydroxybenzoic acid, 194
6-Bromo-2,3-dihydroxybenzoic acid, 188
5-Bromo-2,3-dimethoxybenzoic acid, 189
5-Bromo-3,4-dimethoxybenzoic acid, 197
6-Bromo-2,3-dimethoxybenzoic acid, 189
Bromoform, 150
3-Bromo-4-hydroxybenzaldehyde, 161
5-Bromo-2-hydroxybenzaldehyde, 158
5-Bromo-2-hydroxybenzyl alcohol, 148
5-Bromo-4-hydroxy-3-methoxybenzaldehyde, 161

6-Bromo-4-hydroxy-3-methoxybenzaldehyde, 162
6-Bromo-2-hydroxy-3-methoxybenzoic acid, 188
α-Bromoketones, 119
1-Bromomethyl-2,2-dimethyloxirane, 132
2-Bromo-4,5-methylenedioxyphenylpropiolic acid, 266
Bromonitrobenzaldehydes, 98
Bromonitrosesamin, 264
4-Bromo-5-nitroveratrole, 261
6-Bromopiperonal, 162
5-Bromo-β-resorcylic acid, 190
3-Bromosalicylic acid, **181**
4-Bromosalicylic acid, 181
5-Bromosalicylic acid, 176, 181
N-Bromosuccinimide, 37
5-Bromovanillin, 164
Brucine, 228
Buellia canescens, 223
tert-Butanol, 60
4-*tert*-Butylacetophenone, 110, 120
tert-Butylamine borane, 32
4-*tert*-Butylbenzaldehyde, 84
6-*tert*-Butylbenzaldehyde, 84
tert-Butylbenzene, 110
tert-Butyl hypochlorite, 59, 181
N-Butylimines, 130
Butyllithium, 68, 113, 118, 153, 173, 181, 191, 199
tert-Butyl phenyl ketone, 120
Butyl phosphorodichloridite, 179
n-Butylstilbene, 115
Butyraldehyde, 43
Butyrophenone, 116, 120

Caesalpinia brevifolia, 226, 247
Caesalpinia coriaria, 226, 247
Caesalpinia spinosa, 226, 234
Calcium hydridotrimethoxyborate, 31
Callatolone, **223**
Canadian hemp, 149
Candicine chloride, **144**
Cannizzarro reaction, 31, 46, 50, 51, 76, 148, 149, 160
Caprophenone, 120
Capsaicin, 144
Capsicum sp., 144
Caraway oil, 88
Carbaldehydes, 211
—, aliphatic, 1
—, aromatic, 1, 55
—, phenolic, 141
Carbamates, 112
Carbamide derivatives, 52
Carbamoyl chloride, 173, 193
Carbon disulphide, 24, 137, 166
Carbonium ion stabilisation, 146
Carbon oxysulphide, 180
Carbonyl chloride, 24
Carbonyl compounds, α,β-unsaturated, 70
3,4-Carbonyldioxybenzoic acid, 190
3,4-Carbonyldioxybenzoyl chloride, 190
Carbonyl-forming oxidation reactions, 108
Carbonylsalicylamide, 178
Carboxamides, from aldehydes, 81
2-Carboxydiphenyl sulphide, 181
Carboxylic acids, 173
—, conversion to alcohols, 33
—, — to aldehydes, 59
—, — to ketones, 109
—, esters, 43, 131
—, — conversion to methyl ketones, 111
—, — reduction, 32, 33
—, from lignins, 275
—, phenolic, 141
—, reaction with aralkyl halides, 30
—, reduction, 32
—, α,β-unsurated, 80
Carboxylic acid amides, 43, 128
Carboxylic acid anhydrides, 111
—, reaction with benzene, 109
—, with Grignard compounds, 109
Carboxylic acid chlorides, 110, 111
—, reaction with benzene, 109
—,— with Grignard compounds, 109
Carboxylic acid tannins, 227
2-Carboxyphenyl dihydrogen phosphate, 180
(2-Carboxyphenylthio)acetic acid, **181**
5-Carboxyvanillin, 274
Carnation oil, 41
o-Carvacrotic acid, 182
—, macrocyclic anhydrides, 184
Cascara sagrada, 197
Cassia flower oil, 160
Castalagin, 238, **250**
Castalin, 241, **250**
Castanea sativa, 226, 235, 241

Castoreum oil, 130
Catechol, 157, 161, 167
—, acylation, 168
—, carbonation, 188
—, derivatives, 161, 188
—, reaction with ammonium carbonate, 188
Catechol diacetate, 168
Catechol dichloromethylene ether, 173
Catolechin, 223
Cellulose, 161
—, as constituent of wood, 275
—, biological breakdown, 278
—, hydrolysis, 273
Cellulose pulp, 272
Chamaecyparis obtusa, 257
Chebulagic acid, 237, 243, 244, **245**, 246, 247, 252
Chebulic acid, **239**, 243, 244, 245, 251, 252
—, stereochemistry, 245
Chebulinic acid, 227, 231, 239, **243**, 244, 246
Chelation, 155, 156
—, hydroxybenzaldehydes, 158
Chemotaxonomy, 204
Chestnut oak, 226
Chestnut (Spanish), 226
Chinese tannin, 226
Chinese gallotannin, 228, **232**, 233, 234, 236
Chione glabra, 168
Chloral, 46
—, reaction with arylmagnesium bromides, 66
Chloramine, 8, 25, 103
Chloramine-T, 68
Chloratranol, 163
Chloroacetic acid, 181
Chloroacetic acid *N*-hydroxymethylamide, 9
Chloroacetic ester, 105
2-Chloroacetophenone, **135**
4-Chloroacetophenone, 127, 135
ω-Chloroacetophenone, **132**
Chloroacetyl chloride, 132
N-Chloroamines, 10
6-Chloro-2-aminobenzoic acid, 181
Chloroatranorin, 208, 213
Chlorobenzaldehydes, 96
2-Chlorobenzaldehyde, 103
3-Chlorobenzaldehyde, 82, 87
4-Chlorobenzaldehyde, 77, 95
4-Chlorobenzaldoximes, 94
2-Chlorobenzoic acid, 178; 180
4-Chlorobenzoic acid, 50
4-Chlorobenzoic ester, 50
1-Chlorobenzotriazole, 36
2-Chlorobenzotrichloride, 180
Chlorobenzyl alcohol, 50, **51**
Chlorobenzylamines, 26
4-Chlorobenzylamine, 21
4-Chlorobenzyl chloride, 21
α-Chlorobenzyl methyl ether, 45
Chloroellagic acid, 238, 247, 248
β-Chloroethylamines, 8
α-Chloroethyl methyl ether, 119
Chloroform, 186
—, reaction with benzaldehyde, 78
—, — with phenols, 150
Chloroformic ester, 20, 94
2-Chloro-3-hydroxybenzoic acid, 187
3-Chloro-2-hydroxy-4,6-dimethoxybenzoic acid, 199
Chloromethoxymethane, 177
Chloromethylene dibenzoate, 152
N-Chloromethylphthalimide, 142
Chloronitrobenzaldehydes, 98
2-Chloro-4-oxo-4*H*-benzo[1.3.2]dioxaphosphorin, 179
2-Chloro-4-oxo-4*H*-benzo[1.3.2]dioxaphosphorin-2-oxide, 180
α-Chlorophenylacetyl chloride, 139
α-Chloro-α-phenylpropionaldehyde, 132
α-Chloro-α-phenylpropionyl dichloride, 139
4-Chlorosalicylic acid, 181
5-Chlorosalicylic acid, 176, 181
6-Chlorosalicylic acid, 181
α-Chlorostyrene, 130
N-Chlorosuccinimide, 37
Chlorosulphonic acid, 65
Chlorotrimethylsilane, 124
N-Chlorourea, 132
Chromatography, lichen substances, 205
—, lignans, 254
Cicuta virosa, 88
Cinchonine, 238
Cinnamaldehyde, 42, 80
Cinnamaldehyde diethyl acetal, 106
Cinnamamide, 105

Cinnamic acids, 105, 201
Cinnamonitrile, 106
Cinnamon oil, 106
Cinnamoylamide, 143
Cinnamyl alcohol, 35
Cinnolones, 136
Cistus oil, 130
Citronellol, 41
Cladonia sp., 212
Cladonia grayi, 204
Cladonia squamosa, 212
Claisen rearrangement, 154, 200
Clathrate adducts, 186
Cleistanthin, **268**
Cleistanthus collinus, 268
Clemmensen reaction, 124, 156, 202
Clover oil, 161
Cockroach oötheca, 189
Colensoinic acid, 220
α-Collatolic acid, **219**, 220, 223
β-Collatolic acid, **223**
Collinusin, **268**
Column chromatography, 205
π-Complexes, 166
Condensation reactions, aldehydes, 76
Confluentic acid, 213
Conidendrin, 253, 265, 269, 270
α-Conidendrin, 257, 268
β-Conidendrin, 268
Coniferin, 162
Coniferyl alcohol, 253, 254, 272, 275, 276
—, ^{14}C labelled, 271
Copper benzoate, 173
Copper complex, salicoylamide, 178
—, salicylaldehyde, 156
—, salicylaldehyde-imine, 159
Copper phthalate, 187
Copper toluates, 173
Copper-*o*-toluate, 183
Corilagin, 236, 238, 243, 244, **245**, 246, 247
Coronene, 278
Cotton effect, 11, 187
Coumaranone, 200
Coumarins, 169
Coumarin, 159, 175, 201
Cresols, 188
—, carboxylation, 183
—, oxidation to hydroxybenzoic acids, 174
o-Cresol, 151, 176
—, oxidation, 158
—, reaction with formaldehyde, 142
p-Cresol, reaction with carbon dioxide, 171
Cresotic acids, 183
—, cyclic anhydrides, 183
Crotonic acids, 202
Crotonic ester, 191
Cryptochlorophaeic acid, 214
Cubebin, **257**
Cubebinolide, 257
Cumene, peroxidation, 130
Cuminal, 88
Cuminium cyminium, 88
Curcuma oil, 138
Curcumone, 138
Cyanhydrin-β-D-glucosides, 159
Cyanides, 134
Cyanoacetic ester, 136
2-Cyanobenzyl methyl ether, 38
Cyanogen bromide, 151
3-Cyano-2-methyl-2-quinolone, 136
Cyclic acetals, 89
Cyclic anhydrides, from salicylic acids, 183, 184
Cyclisation reactions, 2-hydroxybenzaldehydes, 156
Cyclohexadienones, 150
Cyclohexane derivatives, 37, 274
Cyclohexenes, 70, 135
Cyclohexenylmethanol, 124
β-Cyclohexylaminovinylphosphonic esters, 80
β-Cyclohexylethanol, 41
Cyclohexylmethanol, 124
1,5-Cyclo-octadiene, 69
Cyclo-octatetraene, 104
Cyclo-olivil, 265, **266**
Cyclopentane-1,2-diones, 247
Cymene, 88
Cynips calicis, 249

Dakin reaction, 156, 211
Damascenine, 187
Darzen's glycidic ester condensation, 71
Darzens reaction, 106, 127
Decanophenone, 118
Dehydrochebulic acid, 252
Dehydrodiconiferyl alcohol, 275
Dehydrodiferulic acid, 253, 261

Dehydrodigallic acid, **241**
Dehydrodivanillin, 274
Dehydrohexahydroxydiphenic acid, **238,** 247, 248
2,6-Dehydroxy-4-methylisophthalic acid, 206
Demethylpodophyllotoxin, 270
Demethylpodophyllotoxin-β-glucoside, 271
Depsides, 191, 203, 204, 208
—, biosynthesis, 205, 207
—, chlorination, 212
—, degradation, 208, 209
—, gallic acid, 228
—, hydrolysis, 206
—, hydroxylated, 210
—, *O*-methylated, 210
—, naturally occurring, 213
—, sources, 203, 212
—, spectral properties, 209
—, synthesis, 212
Depside link, hydrolysis, 205
Depsidones, 203, 204, 208, **216**
—, carbonyl stretching frequency, 217
—, naturally occurring, 218, 220
—, spectral properties, 210, 216
—, synthesis, 217
Depsones, 203, 224
Dermatol, 194
Derric acid, **198**
Desaurins, 137
Desoxypodophyllotoxin, **268,** 270
Dhava tannin, 234
2,3-Diacetoxybenzoic acid, 188
2,4-Diacetoxybenzoic acid, 190
3,4-Diacetoxybenzoic acid, 189
3,5-Diacetoxybenzoyl chloride, 163
N,N′-Diacetyl-4,6-di(aminomethyl)-1,3-xylene, 9
3,5-Diacetylgallic acid, 229
Diaeudesmin, **261,** 263
2,6-Dialkoxybenzoic acids, 191
Dialkoxycarbonium tetrafluoroborates, 10
N,N-Dialkylacetamides, 12
Dialkylamines, 141
4-Dialkylaminobenzaldehydes, 103
N,N-Dialkylbenzylamine *N*-oxides, 13
Dialkyl ketones, aryl derivatives, 137
Diamond Black F, 183
Dianhydropodorhizol, 257
Dianilinoethane, 62, 64, 83
1,4-Diarylbutanes, 255
Diazaphospholidine oxide, 81
ω-Diazoacetophenone, 133
Diazoalkanes, reaction with aldehydes, 82, 83
—, — with ketones, 127
2-Diazobenzaldehyde, 99
Diazo compounds, related to benzylamine, 29
Diazoethane, 83
Diazomethane, 62, 82, 83, 94, 127, 132, 138, 155, 165, 192, 193, 199, 232, 237, 240, 242, 243, 245, 246, 247
—, reaction with ketones, 118
Diazonium chloride, 102
Diazonium reaction, 154
Diazonium salts, 108, 146, 241
Diazo reaction, 167, 187
Dibenzofurans, 205, 225
N,N′-Dibenzoylhydrazine, 131
Dibenzoylmethane, 101
Dibenzylamines, 142
Dibenzylamine, 2, 13, **20,** 29, 31, 93
1,4-Dibenzylbenzene, 45
Dibenzyl carbonate, **46**
Dibenzyl chlorophosphonate, 47
Dibenzyldimethylammonium bromide, 15
Dibenzyldiphenyltetrazene, 28
Dibenzyl disulphide, 48, **49**
Dibenzyl dithioacetal, 48
Dibenzyl ether, 39, **45,** 46
2,4-Di-*O*-benzylglucose, 231
Dibenzyl hydrogen phosphate, **47**
N,N′-Dibenzylhydrazine, **28,** 93
Dibenzyl hydrogen phosphite, 47
N,N-Dibenzylhydroxylamine, 13, 25
O,N-Dibenzylhydroxylamine, **28**
Dibenzylideneglycerol, 89
Dibenzylidenehydrazine, 92
N,N-Dibenzylidenephenylenediamines, 92
Dibenzylnitrosamine, 28
2,4-Dibenzyloxybenzoic acid, 191
2,6-Dibenzyloxybenzoic acid, 191
2-(3,4-Dibenzyloxyphenyl)ethylamine, 144
2,4-Dibenzylphenol, 45
3,3-Dibenzyl-1-phenyltriazene, 29
Dibenzyl succinate, **46**
Dibenzyl sulphide, **49**
Dibenzyl sulphite, **46**

Dibenzyl sulphone, **49**
Dibenzyl sulphoxide, 49
N,*N'*-Dibenzylurea, 21
Diborane, 4, 5, 6, 8, 25
ω,ω-Dibromoacetophenone, 133
Dibromobenzaldehydes, 96
3,5-Dibromobenzoic acid, 191
Dibromodi-*O*-methylpinoresinol, 261
3,5-Dibromo-2-hydroxybenzaldehyde, 158
4,6-Dibromo-3-hydroxybenzaldehyde, 160
3,5-Dibromo-4-hydroxybenzoic acid, 188
3,5-Dibromo-2-hydroxybenzyl alcohol, 148
5,6-Dibromo-4-hydroxy-3-methoxybenzaldehyde, 162
4,5-Dibromomethylenedioxybenzene, 162, 190
3,5-Dibromosalicylic acid, 176, 181
Di-*n*-butylamine, 130
1,4-Di-*tert*-butylbenzene, 110
Di-*tert*-butylcyclopentanedione, 251
Di-*tert*-butyl peroxide, 118
4,6-Di-*tert*-butylpyrogallol, 251
6,6-Dicarbonylhexahydroxybiphenyl radicals, 243
Di-*o*-carvacrotide, 184, 185
3,4-Dichloroacetophenones, 134
ω,ω-Dichloroacetophenone, 132
Dichloroacetyl chloride, 133
Dichlorobenzaldehydes, 96
o-Dichlorobenzene, 152
2,4-Dichlorobenzoic acid, 181
Dichlorocarbenes, 16, 150
β,β-Dichloro-4-dimethylaminostyrene, 103
Dichlorodimethyl ether, 37, 44
Dichlorodiphenylmethane, 197
Dichloromethane, 152
4,5-Dichloromethylenedioxybenzene, 190
3,4-Dichloromethylenedioxybenzoyl chloride, 190
Dichloromethylenetriphenylphosphorane, 103
Dichloromethyllithium, 132
Dichloromethyl methyl ether, 88, 152
Dichloromethyl methyl sulphide, 152
3,5-Dichlorosalicylic acid, 176, **181**
Dicobaltoctacarbonyl, 106
D-*m*-cresotide, 184
Di-*p*-cresotide, 184
Dicyclohexylcarbodiimide, 48, 211, 266
Didepsides, 233
Didepsidecarboxylic acids, 211
Diels-Alder addition, 70
Dienes, from aldehydes, 79, 81
1,3-Dienes, 70
Dienones, 147, 217
3,4-Diethoxybenzoic acid, 189
Diethylaluminium chloride, 34
5-Diethylaminoacetamidosalicylic ester, 182
4-Diethylaminobenzaldehyde, 100, **103**
N,*N*-Diethylbenzamide, 5, 19
N,*N*-Diethylbenzylamine, 19
Diethyl cadmium, 196
Diethylphenylmethanol, 43
Diffractaic acid, 209, 213
Digallic acid, 203
m-Digallic acid, 228, **229**, 233
Digalloylglucoses, 238
3,6-Digalloylglucose, **231**, 243
Digalloylhexose, 235
Dihydric phenols, carbonation, 172
Dihydrocoumarin, 201
Dihydroflavoglaucin, 164
Dihydrogallic acid, 197
Dihydroguaiaretic acid, **255**
meso-Dihydroguaiaretic acid, **255**
Dihydroisocoumarins, 250
Dihydroiscoumarin, 243
3,4-Dihydroisoquinoline, 17
Dihydro-1,3-oxazines, for ketone synthesis, 113
Dihydroquinazolines, 25
3,4-Dihydroquinazoline, 24
Dihydro-1,3-oxazines, 67
Dihydroshikimic acid, 197
2,3-Dihydroxyacetophenone, **169**
2,4-Dihydroxyacetophenone, **169**
2,5-Dihydroxyacetophenone, **169**
2,6-Dihydroxyacetophenone, **169**, 191
3,4-Dihydroxyacetophenone, **168**
3,5-Dihydroxyacetophenone, **169**
Dihydroxybenzaldehydes, 161
2,3-Dihydroxybenzaldehyde, **161**
2,4-Dihydroxybenzaldehyde, **163**
2,5-Dihydroxybenzaldehyde, **163**
2,6-Dihydroxybenzaldehyde, **163**
3,4-Dihydroxybenzaldehyde, **161**
3,5-Dihydroxybenzaldehyde, **163**
Dihydroxybenzoic acids, 175
2,3-Dihydroxybenzoic acid, 161, **188**

2,4-Dihydroxybenzoic acid, **190**
2,5-Dihydroxybenzoic acid, 192
2,6-Dihydroxybenzoic acid, 190, **191**
3,4-Dihydroxybenzoic acid, 188, 189
3,5-Dihydroxybenzoic acid, **191**, 198
2,4-Dihydroxybenzoyl chloride, 190
2,5-Dihydroxybenzyl alcohol, **149**
3,4-Dihydroxybenzyl alcohol, **149**
3,4-Dihydroxybenzylamine, **144**
2,2′-Dihydroxydibenzyl ether, 146
1,2-Dihydroxy-1,2-diphenylethane, 71
4,5-Dihydroxy-1,8-diphenyloctane, 71
2,4-Dihydroxy-3-formylbenzoic acid, 163
2,5-Dihydroxy-3-methoxybenzaldehyde, 165
3,4-Dihydroxy-5-methoxybenzaldehyde, 164
4,5-Dihydroxy-3-methoxybenzaldehyde, 164
2,3-Dihydroxy-4-methoxybenzoic acid, 193
2,5-Dihydroxy-4-methoxybenzoic acid, 198
3,5-Dihydroxy-4-methylbenzoic acid, 224
4,5-Dihydroxy-3-methoxybenzoic acid, 197
3,5-Dihydroxy-4-methoxybenzoic ester, 197
4,5-Dihydroxy-3-methoxycarbonyloxybenzoic acid, 198
2,6-Dihydroxy-4-methoxy-3-methylbutyrophenone, 170
3,3′-Dihydroxymethylazobenzene, 51
2,6-Dihydroxy-4-methylbenzaldehyde, **163**
4,6-Dihydroxy-2-methylbenzaldehyde, **163**
2,6-Dihydroxy-4-methylbenzoic acid, 192
4,6-Dihydroxy-2-methylbenzoic acid, 191
2,2′-Dihydroxymethylhydrazobenzene, 51
4,6-Dihydroxy-2-pentylbenzaldehyde, 163
Dihydroxyphenylacetic acids, 199
2,5-Dihydroxyphenylacetic acid, 200
3,4-Dihydroxyphenylacetic acid, 200
3,5-Dihydroxyphenylacetic acid, 201
Dihydroxyphenyl alcohols, 149
Dihydroxyphenylalkylamines, 144
2-(3,4-Dihydroxyphenyl)ethylamine, **144**
Dihydroxyphenyl ketones, 168
3-(2,4-Dihydroxyphenyl)propionic acid, 201
3-(3,4-Dihydroxyphenyl)propionic acid, 201
2,4-Dihydroxypropiophenone, **170**
Dihydroxythujaplicatin, **259**
Diiodomethane, 162
3,5-Diiodosalicylic acid, 182
Diisobutylaluminium hydride, 89
Di-*O*-isopropylideneglucose, 230, 231
β-Diketones, 125
—, for ketone synthesis, 113
Dill-apiol, 165
Dill oil, 165
Dimercaprol injection, 46
2,3-Dimethoxyacetophenone, 169
2,4-Dimethoxyacetophenone, 169
2,6-Dimethoxyacetophenone, 170
2,3-Dimethoxybenzaldehyde, 161
2,6-Dimethoxybenzaldehyde, 153, 163
3,4-Dimethoxybenzaldehyde, 162
3,5-Dimethoxybenzaldehyde, 163
3,4-Dimethoxybenzaldoxime, 144
1,3-Dimethoxybenzene, 163, 169
1,4-Dimethoxybenzene, 193
2,3-Dimethoxybenzoic acid, 188
2,4-Dimethoxybenzoic acid, 191
2,5-Dimethoxybenzoic acid, 193
2,6-Dimethoxybenzoic acid, 191
3,4-Dimethoxybenzoic acid, 189, 229
3,5-Dimethoxybenzoic acid, 169, 191
2,4-Dimethoxybenzoic ester, 190
3,5-Dimethoxybenzoyltoluene-4-sulphonohydrazide, 163
3,4-Dimethoxybenzyl alcohol, 149
3,7-Dimethoxydibenzofuran-1,9-dicarboxylic acid, 225
Dimethoxyisolariciresinol, 267
4,5-Dimethoxyisophthalic acid, 273
2,6-Dimethoxy-4-methylbenzoic acid, 192
2,5-Dimethoxy-3,4-methylenedioxybenzaldehyde, 165
5,6-Dimethoxy-3,4-methylenedioxybenzaldehyde, 165
2,3-Dimethoxy-5-nitrobenzoic acid, 188
3,5-Dimethoxyphenol, 165
2,6-Dimethoxyphenyl acetate, 171
3,4-Dimethoxyphenylacetic acid, 200
3,4-Dimethoxyphenylalanine, 255
2-(3,4-Dimethoxyphenyl)ethylamine, 144
5-(3,4-Dimethoxyphenyl)pentanoic acid, 202
3,5-Dimethoxytoluene, 163
N,*N*-Dimethylacetamide, 19
Dimethylacetamide dimethyl acetal, 41
Dimethylacetophenones, 120
Dimethylamine, 74, 105, 142

4-Dimethylaminoacetophenone, 137
Dimethylaminobenzaldehydes, 100
3-Dimethylaminobenzaldehyde, 102
4-Dimethylaminobenzaldehyde, **102,** 103
N,*N*-Dimethyl-1-amino-3-phenylpropane, 19
β-Dimethylaminostyrene, 105
Dimethylaniline, 47, 102, 103, 137, 177
Dimethylbenzaldehydes, 84
2,4-Dimethylbenzaldehyde, 84
Dimethylbenzyl alcohols, 50
Dimethylbenzylamines, 26
N,*N*-Dimethylbenzylamine, 10, 11, **19**
N-(2,4-Dimethylbenzyl)formamide, 9
Dimethylcadmium, 169
Di-*O*-methylconidendrin, 268
Di-*O*-methyl-*β*-conidendrin, 266
Di-*O*-methyldehydroguaiaretic acid, 255, 260
Di-*O*-methyldihydroguaiaretic acid, 256, 260, 265
Dimethyl disulphide, 69
Di-*O*-methylepisyringinol, 263
Dimethylformamide, 65, 103, 152
Di-*O*-methylfuroguaiacin, 260
3,4-Di-*O*-methylgallic acid, 232
2,4-Di-*O*-methylglucose, 245
Di-*O*-methylguaiaretic acid, 255, 264
N,*N*-Dimethylhydrazine, 122
N,*N'*-Dimethylhydrazine, 105
Di-*O*-methylisolariciresinol, 265, 266, 268
Di-*O*-methylisomatairesinol, 256, 257
Di-*O*-methyllaricireresinol, 261
Di-*O*-methyllirioresinol-B, 262
Di-*O*-methyllyoniresinol, 267
Di-*O*-methylmatairesinol, 256
N,*N*-Dimethylmethanesulphonamide, 111
N,*N*-Dimethyl-4-methoxyphenylthioacetamide, 143
4,6-Dimethyl-*β*-methylglucoside, 249
Di-*O*-methylolivil, 264, 265
2,4-Dimethylphenol, 151
Dimethyl(phenyl)acetaldehyde, 138
N,*N*-Dimethyl-*β*-phenylethylamine, 20
α,α-Dimethyl-α'-phenylethylene glycol, 106
Dimethyl-*β*-phenylethylmethanol, 43
Dimethylphenylmethanol, 42
N,*N*-Dimethyl-*β*-phenylpropionamide, 5
Di-*O*-methylpinoresinol, 255, 259, 260, 261, 262, 264
Di-*O*-methylpipinoresinol, 261, 263
2,4-Dimethylquinazoline, 137
Dimethylsalicylic acids, 182
3,5-Dimethylsalicylic acid, 183
—, macrocyclic anhydrides, 184
Dimethyl sulphide, 37, 273
Dimethyl sulphone, 111
Dimethylsulphonium methylide, 83, 127
Dimethyl sulphoxide, 58, 108, 111, 273
Di-*O*-methyltetrahydrofuroguaiacin-B, 259, 260
2,4-Dimethylthiazole, 68
Dimethylthioformamide, 152
Di-*O*-methylthomasidonic acid, 269
Di-*O*-methylthomastic acid, 269
N,*N*-Dimethyl-2-*o*-tolylethylamine, 7
Dimethylzinc, 138
3,5-Dinitroacetophenone, 135
3,5-Dinitroanisaldehyde, 160
Dinitrobenzaldehydes, 98
2,4-Dinitrobenzaldehyde, 76, **97**
2,6-Dinitrobenzaldehyde, 99
2,4-Dinitrobenzylaniline, 97
3,5-Dinitro-2-hydroxybenzaldehyde, 158
3,5-Dinitrosalicylic acid, 176
Dinitrostilbenedisulphonic acid, 103
2,4-Dinitrotoluene, 97
1,2-Diols, conversion to aldehydes, 71
Diphenic acids, 239
2,5-(Diphenyl)benzoquinones, 205
1,4-Diphenylbuta-1,3-diene, 80
1,1-Diphenyl-1-butene, 116
1,3-Diphenylcyclobutane-2,4-dione, 139
Diphenylcyclopropenone, 139
Diphenyl diselenide 2,2'-dicarboxylic acid, **181**
Diphenyl ethers, 216, 217, 241
α,*β*-Diphenylethyldimethylamine, 15
Diphenylformamidine, 152
Diphenylketene, 139
Diphenylmethane, 39, 45, 142
Diphenylmethyl 3,4-diphenylmethylenedioxy-5-hydroxybenzoate, 229
α-Diphenylnitrone, **92**
Diphenylphosphine, 119
1,3-Diphenylpropene, 105
Diphyllin, 268
Diploicia canescens, 223
Diploicin, 217, 220, **223**, 224
Diplosal, **177**

Diploschistesic acid, 209, 211, 212, 213
Dipyridine-chromium(VI) oxide, 35
Directed aldol condensation, 126
Disalicylaldehyde, 159
Disalicylic acid, **178**
Disalicylide, 183, 184, **185**
Disiamylborane, 63
Disilanes, 111
Disulphides, 49
1,3-Dithianes, 68, 114
Dithioacetals, 123, 124
Di-*o*-thymolides, 183
Di-*o*-thymotide, 184, 185
Di-*o*-tolyl ether, 178
Divaric acid, 212
Divaricatic acid, 213
Divaricatinic acid, 206
Divi-divi, 226, 237, 238, 245, **247**, 253
Duff reaction, 153
Dypnone, 131

Electrolytic reduction, carboxylic acids, 32
Ellagic acid, 194, 195, 227, 234, **237**, 238 240, 241, 242, 245, 246, 248, 249, 250
Ellagitannins, 226, 227, 228, **237**, 242, 252
—, from algarobilla, 247
—, from myrobalan extract, 244
—, methylation, 237
Ellagitannin acids, interrelationship, 251
Emde reduction, 12
Emulsin, 147, 262
Enamines, 130
Ephedrine, 19
Epiasarinin, 261, 262, 263
Epipinoresinol, **262**
Epipodorhizol, 257
Episyringaresinol, **262**
Epoxides, 107, 118
—, conversion to aldehydes, 70
—, — to ketones, 114
—, from aldehydes, 83
—, reduction, 34
α,β-Epoxy acids, reaction with benzylamine, 16
1,2-Epoxyalkanes, 34
α,β-Epoxy-esters, 71
Epoxynitriles, 72, 115
1,2-Epoxypropane, 42
Eremophila glabra, 262
Ergot of rye, 143
Erythrin, 212, 213
Erythritol, 212
Esters, reduction to aldehydes, 62
Ester mesylates, 61
Etard reaction, 56
Ethanethiol, 62, 134
4-Ethoxybenzaldehyde, 160
α-Ethoxybenzyl acetate, 89
Ethoxycarbonyl-α-benzaldoxime, 94
Ethoxycarbonyl chloride, 179
1-Ethoxy,*N*,*N*-dimethylvinylamine, 112
3-Ethoxy-4-hydroxybenzaldehyde, 162
4-Ethoxy-3-hydroxybenzaldehyde, 162
3-Ethoxy-4-methoxybenzoic acid, 189
4-Ethoxy-3-methoxybenzoic acid, 189, 273
α-Ethoxyphenylacetic acid, 73, 89
6-Ethoxy-3-propenylphenol, 162
3-Ethoxypropionitrile, 201
α-Ethoxystyrene, 132
Ethoxyvinyl esters, 79
Ethyl acetate, 64
o-Ethylacetophenone, 109
Ethylamine, 19
N-Ethylbenzamide, 5
Ethylbenzene, 107, 124, 135
—, oxidation, 130
o-Ethylbenzoic acid, 109
N-Ethylbenzylamine, 5, **19**
Ethylene, 34
Ethylene bis(salicylidene-imine), 159
Ethylene chlorohydrin, 31, 41
Ethylenediamine, 159
Ethylene glycol, 89
Ethylene glycol cyclic acetal, 105
Ethylene oxide, 42
—, reaction with benzene, 41
Ethyl gallate, 197
Ethyl hypochlorite, 132
Ethylmagnesium bromide, 137
Ethylmagnesium iodide, 42, 43
Ethylmethylamine, 19
N-Ethyl-*n*-methylbenzylamine, 19
Ethyl methyl ketone, 43
Ethyl nitrite, 99
2-Ethyl-2-phenylbutanal, 107
Ethylphenylmethanol, 42
Ethyl salicylate, 177
Ethylvanillin, 162
Eucalyptus sp., 260
Eucalyptus oils, 88

Eudesmin, **260**, 262
Eugenol, 161
Everinic acid, **192**, 206
Evernia vulgaris, 212
Evernic acid, **212**, 213
Evernic ester, 210
Everninaldehyde, 163
Evodia belahe, 143

Fagarol, 263
Fagus silvatica, 263
Fatty acid amides, 143
Ferulic acid, 253
Fission reactions, ketones, 128
Flavellagic acid, 194, 195
Flavogallol, 195, **196**, **241**
Flavogallonic acid, **241**, 242, 250
Flavoglaucin, **164**
Flavone, 168
Fluorene, 195
Fluorenylidene ketones, 127
Fluorenylidenetriphenylphosphorane, 127
Fluorenylidenetris(*n*-butyl)phosphorane, 127
4-Fluoroanisole, 181
Fluorobenzaldehydes, 56, 96
5-Fluoro-2-methoxybenzoic acid, 181
2-Fluorophenol, 181
4-Fluorophenol, 181
3-Fluorosalicylic acid, **181**
4-Fluorosalicylic acid, 181
5-Fluorosalicylic acid, 181
Fluorotoluenes, oxidation, 56
Formaldehyde, 31, 43, 44, 149
—, reaction with aniline, 53
—, — with *o*-cresol, 142
—, — with phenols, 141, 145, 148
Formaldoxime, 59
Formamides, 18, 119
2-Formamidoacetophenone, 136
Formanilide, 88, 94, 151
[^{14}C]Formate, 208
Formic acid, 7, 24, 65, 88, 103
Formylation reactions, 65
2-Formylbenzenesulphonic acid, 103
Formyldihydroxy-*o*-toluic acid, 211
Formylfluoride, 65
trans-Formylolefins, 80
Forsythia sp., 262
Freudenberg cuproxam process, 273
Friedel-Crafts alkylation, 34
Friedel-Crafts reaction, 109, 110, 130, 132, 133, 134, 166, 173, 193
Friedländer synthesis, 101, 136
Fries reaction, 166
Fries rearrangement, 167, 169, 171
Fumarprotocetraric acid, 222
Furfuraldehyde, 264
Furoguaiacin, 253, **259**
Furoic ester, 199

Galbacin, **260**
Galbelgin, **260**
Galbulin, 260, **265**, 266
Galcatin, **265**, 266
Galgravin, 259, 260
Gallacetophenone, **170**
Gallaldehyde, **164**
Gallamide, 194
Gallic acid, 164, 171, **193**, 194, 197, 198, 227, 228, 232, 233, 234, 236, 237, 241, 242, 243, 244, 245, 246, 249
—, acyl derivatives, 229
—, biosynthesis, 236
—, depsides, 228
—, derivatives, 197
—, ferric complex, 194
—, oxidation, 195, 237
Gallic esters, 194, 197
—, autoxidation, 195
Gall nuts, 193
Gallocyanine mordant dyes, 194
Galloflavin, 195, **196**
Gallotannins, 203, 226, 227, 228, **231**
—, complex, 236
—, extraction, 232
—, methanolysis, 231
—, structures, 232
Galloylated glucoses, 233
Galloyl esterase, 233
Galloylglucoses, 228, 243
2-Galloylglucose, **230**
3-Galloylglucose, **230**, 231
6-Galloylglucose, **231**
p-Galloyloxyphenyl-β-D-glucoside, 236
Gangaleoidin, 221
Garouille, 226
Gattermann aldehyde synthesis, 65
Gattermann-Koch aldehyde synthesis, 65

Gattermann reaction, 151, 160, 163, 164, 165, 211
Gaultheria procumbens, 175
Gaultherin, 175, **177**
Gayuba gallotannin, 236
Gayubatannin, **236**
Geijera sp., 170
Gelatin, 228
Gentisaldehyde, **163**
Gentisic acid, **192,** 193
Gentisyl alcohol, **149**
Geraniol, 41
Geranium oil, 41
Geranium pyrenaicum, 237
Ginger, 170
Glabratic acid, 212
Glomelliferic acid, 213
Gloriosa superba, 191
Glucoaubrietin, 145
4-*O*-β-D-Glucodisoprotocatechuic acid, 189
α-D-Glucogallin, **230**
β-D-Glucogallin, **230**
Glucoses, polygalloyl, 232
Glucose, 177, 232, 233, 234, 237, 243, 244, 245, 246, 247, 249, 250, 251
—, esterification by flavogallonic acid, 250
—, gallate esters, 230
—, galloyl esters, 228
β-Glucose, 231
β-Glucosidase, 189
Glucosides, 145
Glucosyringic acid, 197
Glucovanillin, 162
Glutaric anhydride, 202
Glycerol, 89
Glycidic esters, 71, 105, 127
Glycine, 179
Glycols, from aldehydes, 75
1,2-Glycols, conversion to aldehydes, 10
—, — to ketones, 114
—, oxidation, 69
—, trisubstituted, 115
Gmelina arborea, 264
Gmelina leichhardtii, 264
Gmelinol, **264,** 265
Grass lignins, 272
Grayanic acid, 204, 218, 220
Grignard reaction, 41, 43, 193
Grignard reagents, 8, 30, 92, 111, 122, 153, 188
Grignard reagents, *(continued)*
—, adducts with aldehydes, 74, 116
—, aldehyde synthesis, 66
—, formylation, 66
—, reaction with carbonyl chlorides, 109
—, — with nitriles, 112
Griseofulvin, 199
Grundmann synthesis, aldehydes, 61
Guaiacol, 149, 153, 161, 169, 188, 273, 276, 277
Guaiacum officinale, 255, 259, 266
Guaiacyl ethers, 277
Guaiacylglycerol-β-coniferyl ether, 275
Guaiaretic acid, 253, **255,** 256, 261, 265, 268, 270
Gynosperm lignins, 272
Gyrophora pustulata, 212
Gyrophoric acid, 204, 207, 208, **212,** 214

Haematommic acid, 206
Halle-Bauer reaction, 128
Halogenoacetophenones, 132, 133, 134
Halogenoacetyl halides, 132
Halogenobenzenes, 134
Halogenobenzaldehydes, 95
Halogenobenzoylacetoacetic acid esters, 134
Halogenobenzoylmalonic esters, 134
α-Halogeno esters, 71, 127
Halogenohydrins, conversion to aldehydes, 70
—, — to ketones, 114
α-Halogeno ketones, 119
Hamamelis virginiana, 235
Hamamelitannin, 227, 228, **235**
D-Hamamelose, 235
Hardwood lignins, 272
Helicin, 148, 158
Helicin tetra-acetate, 158
Heliotropin, **162**
Helix conformations, 186
Hemlock oil, 88
Hepta-acetyl-α-D-glucogallin, 230
Hepta-acetyl-β-D-glucogallin, 230
Hepta-acetyl-α-D-glucose, 230
Heptamethoxydiphenic acid dimethyl ester, 195
Hepta-*O*-methylflavogallonic esters, 241
Hepta-*O*-methyl-*m*-trigallic ester, 229
Hexa-*m*-cresotide, 184

Hexagalloylglucose, 236
Hexahydroxy-6,6′-diphenic acids, **237**
Hexahydroxydiphenic acid, 238, 244, 245, 246, 248, 250, 251, 252
Hexahydroxydiphenoylglucoses, 252
2,3-Hexahydroxydiphenoylglucose, 238, 249
3,6-Hexahydroxydiphenoylglucose, 238
4,6-Hexahydroxydiphenoylglucose, 238, 249
Hexahydroxydiphenoyl residues, 243
Hexamethoxybiphenyl-6,6′- dicarboxylic acid, 242
Hexamethoxydiphenic acid, 237, 238, 242, 245, 248, 249, 250
Hexamethylenetetramine, 2, 16, 57, 88, 103, 153
Hexamethylphosphoramide, 66, 75, 124
Hexamethylphosphoric triamide, 10
Hexamethylphosphorous triamide, 83
Hexa-*O*-methylvalonic acid dilactone, 242
n-Hexane, 186
Hexasalicylide, 183, 184
n-Hexyl phenyl ketone, 120
Hiascic acid, 214
Hibbert ketones, 277
Himantandra baccata, 260, 265
Himantandra belgraveana, 259
Hinokinin, **257**, 258
Hippuric acid, 200
Histamine, 143
Hoesch reaction, 166, 169, 170
Hofmann degradation, amides, 8
Hofmann isonitrile synthesis, 16
Hofmann reaction, 105, 178
Homogentisic acid, 200, 201
Homopiperonylic acid, 200
Homosekikaic acid, 214
Hordenine, **143**
Hydratropic alcohol, **42**
Hydratropic aldehyde, 106
Hydrazine, benzyl derivatives, 28
2-Hydrazobenzyl alcohol, 51
Hydrazoic acid, 128
Hydridodiisobutylaluminium, 59, 60
Hydrido(tri-*n*-butylphosphine)copper, 60
Hydroanisoin, 160
Hydrobenzamide, 20, 74, **91**
Hydrocaffeic acid, 201
Hydrocarbons, from alcohols, 75
—, from ketones, 124, 128
—, reaction with carbon monoxide, 65
—, — with dichloromethyl ethers, 65
Hydrocinnamaldehyde, 106
Hydrocinnamdimethylamide, 19
Hydrocinnamic acid, 42
Hydrocinnamonitrile, 19
Hydrocinnamyl alcohol, 42
Hydrogen cyanide, 165
—, adducts with aldehydes, 74
—, reaction with olefins, 10
—, — with phenols, 151
Hydroformylation reaction, 106
Hydrolysable tannins, 203, **227**
Hydroquinone, derivatives, 163, 192
Hydroquinone monoallyl ether, 200
Humic acid, **277**, 278
—, conversion to phenolic compounds, 278
—, from phenols, 278
—, structure, 278
Humus, 277
Hydroxyacetophenones, 167, 174, 199
2-Hydroxyacetophenone, 167, **168**
3-Hydroxyacetophenone, 167, **168**
4-Hydroxyacetophenone, **168**
Hydroxyaralkylamines, 141
Hydroxyaryl alcohols, 145, 147
Hydroxyarylcarbaldehydes, 150
Hydroxyaryl ketones, 168
Hydroxybenzaldehydes, 150, 151, 152, 153, 154, 158, 160
—, methods of formation, 157
2-Hydroxybenzaldehydes, 150, 152, 155, 156, **158**
3-Hydroxybenzaldehyde, 148, 154, 156, **159**, 160
4-Hydroxybenzaldehyde, 150, 152, **160**, 274
4-Hydroxybenzaldoxime, 142
Hydroxybenzoic acids, 171, 174, 189
—, dissociation constants, 175
—, from organometallic compounds, 173
—, properties and reactions, 174
2-Hydroxybenzoic acid, 173, **175**
3-Hydroxybenzoic acid, 160, **187**, 189
4-Hydroxybenzoic acid, 172, 173, 175, 176, **188**
4-Hydroxybenzoic ester, 148

2-Hydroxybenzoyl chloride, 178
3-(3′-Hydroxybenzoyloxy)benzoic acid, 187
4-(4′-Hydroxybenzoyloxy)benzoic acid, 188
2-Hydroxybenzylacetone, 168
3-Hydroxybenzylacetone, 168
4-Hydroxybenzylacetone, 168
Hydroxybenzyl alcohols, 155
2-Hydroxybenzyl alcohol, 145, **147**
3-Hydroxybenzyl alcohol, **148**
4-Hydroxybenzyl alcohol, 145, 147, **148**
Hydroxybenzylamines, 141
2-Hydroxybenzylamine, **142**
4-Hydroxybenzylamine, **142**
2-Hydroxybenzylaniline, 146
2-Hydroxybenzyl-β-D-glucoside, 148
Hydroxybenzyl isothiocyanates, 145
p-Hydroxybenzyl isothiocyanate, **145**
Hydroxybenzylphthalimides, 142
1-Hydroxybutane-1,3,4-tricarboxylic acid, 197
Hydroxycarboxylic acids, 156
2-Hydroxychalcone, 159, 168
4-Hydroxycinnamoylamine, 143
2-Hydroxycyclohexanecarboxylic acids, 176
4-Hydroxy-3,5-diacetoxybenzoic acid, 198
2-Hydroxy-3,4-dimethoxyacetophenone, 170
2-Hydroxy-4,6-dimethoxyacetophenone, 170
4-Hydroxy-3,5-dimethoxyacetophenone, 171
2-Hydroxy-3,4-dimethoxybenzaldehyde, 164
2-Hydroxy-4,6-dimethoxybenzaldehyde, 165
3-Hydroxy-4,5-dimethoxybenzaldehyde, 164
4-Hydroxy-2,6-dimethoxybenzaldehyde, 165
4-Hydroxy-3,5-dimethoxybenzaldehyde, 153
6-Hydroxy-2,4-dimethoxybenzaldehyde, 165
2-Hydroxy-1,3-dimethoxybenzene, 149
2-Hydroxy-3,4-dimethoxybenzoic acid, 193
4-Hydroxy-3,5-dimethoxybenzoic acid, 197
5-Hydroxy-3,4-dimethoxybenzoic acid, 197
4-Hydroxy-3,5-dimethoxybenzyl alcohol, 149
4-Hydroxy-3,5-dimethoxycinnamic acid, 262
2-Hydroxy-4,6-dimethoxy-3-methylacetophenone, 170
4-Hydroxy-3,5-dimethoxytoluene, 149
5-Hydroxy-3,4-diphenylmethylenedioxybenzoic ester, 197
2-Hydroxydithiobenzoic acid, **180**
β-Hydroxy esters, 79
2-Hydroxyformanilide, 101
2-Hydroxyhippuric acid, **179**
Hydroxyiminoacetophenone, **133**
Hydroxyiminobenzaldehydes, 99
2-Hydroxyiminobenzaldehyde, 99
2-Hydroxyiminobenzaldoxime, 100
2-Hydroxyindazole, 102
3-Hydroxy-4-iodobenzoic acid, 187
4-Hydroxyisophthalic acid, 172, 176
Hydroxyketones, 166
Hydroxylamines, *N*-substituted, 74
Hydroxylamine, benzyl derivatives, 25
Hydroxylamine *O*-sulphonic acid, conversion to amines, 8
2-Hydroxylaminoacetophenone, 135
3-Hydroxylaminoacetophenone, 135
2-Hydroxylaminobenzyl alcohol, 51
4-Hydroxymandelonitrile, 143
Hydroxymatairesinol, **257**
2-Hydroxy-3-methoxyacetophenone, 169
2-Hydroxy-4-methoxyacetophenone, 169
2-Hydroxy-5-methoxyacetophenone, 169
3-Hydroxy-4-methoxyacetophenone, 169
4-Hydroxy-2-methoxyacetophenone, 169
4-Hydroxy-3-methoxyacetophenone, 168
6-Hydroxy-2-methoxyacetophenone, 170
2-Hydroxy-3-methoxybenzaldehyde, 161
2-Hydroxy-4-methoxybenzaldehyde, 163
2-Hydroxy-5-methoxybenzaldehyde, 163
3-Hydroxy-4-methoxybenzaldehyde, 162
4-Hydroxy-2-methoxybenzaldehyde, 163
4-Hydroxy-3-methoxybenzaldehyde, 161
2-Hydroxy-3-methoxybenzoic acid, 188
2-Hydroxy-4-methoxybenzoic acid, 190, 191, 198
2-Hydroxy-5-methoxybenzoic acid, 193
2-Hydroxy-6-methoxybenzoic acid, 191
3-Hydroxy-4-methoxybenzoic acid, 189

3-Hydroxy-5-methoxybenzoic acid, 191
4-Hydroxy-2-methoxybenzoic acid, 191
4-Hydroxy-3-methoxybenzoic acid, 189
5-Hydroxy-2-methoxybenzoic acid, 193
4-Hydroxy-3-methoxybenzylamine, **144**
3-Hydroxy-4-methoxybenzyl alcohol, 149
4-Hydroxy-3-methoxybenzyl alcohol, 149
2-Hydroxy-5-methoxycarbonylbenzoic acid, 193
6-Hydroxy-4-methoxycarbonylbenzoic acid, 192
2-Hydroxy-4-methoxy-3-methylacetophenone, 169
6-Hydroxy-4-methoxy-2-methylbenzaldehyde, 163
2-Hydroxy-6-methoxy-4-mehylbenzoic acid, 192
4-Hydroxy-6-methoxy-2-methylbenzoic acid, 192
6-Hydroxy-4-methoxy-2-methylbenzoic acid, 192
3-Hydroxy-4-methoxyphenylacetic acid, 200
4-Hydroxy-3-methoxyphenylacetic acid, 200
1-(4-Hydroxy-3-methoxyphenyl)butan-3-one, **170**
1-(4-Hydroxy-3-methoxyphenyl)ethanol, 149
3-(3′-Hydroxy-4-methoxyphenyl)-propionic acid, 201
3-(4-Hydroxy-3-methoxyphenyl)propionic acid, 201
3-Hydroxy-5-methoxyvalerophenone, 218
3-Hydroxy-2-methylbenzoic acid, 187
3-Hydroxy-4-methylbenzoic acid, 187
3-Hydroxy-5-methylbenzoic acid, 187
4-Hydroxy-2-methylbenzoic acid, 188
4-Hydroxy-3-methylbenzoic acid, 188
5-Hydroxy-2-methylbenzoic acid, 187
2-Hydroxymethyl-β-D-(6-benzoyl)-glucoside, 148
2-Hydroxy-3-methylbenzyldimethylamine, **142**
1-(4-Hydroxy-3-methylphenyl)butan-3-ol, 149
2-Hydroxymethylphenyl-β-D-glucoside, 148
2-*C*-Hydroxymethyl-D-ribose, 235
Hydroxyotobain, **266,** 267
Hydroxyphenylacetaldehydes, 154, 156, 160
Hydroxyphenylacetic acids, 199
2-Hydroxyphenylacetic acid, 200
3-Hydroxyphenylacetic acid, 200
4-Hydroxyphenylacetic acid, 145, 200
4-Hydroxyphenylacetone, **168**
Hydroxyphenylalkylamines, 142
1-(1′-Hydroxyphenyl)butan-3-one, 168
Hydroxyphenylbutyric acids, 202
2-(4-Hydroxyphenyl)ethanol, **149**
2-(4-Hydroxyphenyl)ethylamine, **143**
2-(4-Hydroxyphenyl)ethyl bromide, 147
2-(4-Hydroxyphenyl)ethyldimethylamine, **143**
2-(4-Hydroxyphenyl)ethylmethylamine, 143
2-(4-Hydroxyphenyl)ethyltrimethylammonium chloride, **144**
Hydroxyphenyl ketones, 165
4-Hydroxyphenylpropenol, 272
Hydroxyphenylpropionic acids, 201
3-(2-Hydroxyphenyl)propionic acid, 201
3-(3-Hydroxyphenyl)propionic acid, 201
3-(4-Hydroxyphenyl)propionic acid, 201
2-Hydroxyphenylpropiophenone, **168**
4-Hydroxyphenylpropiophenone, **168**
Hydroxyphenylstyrenes, 154
2-Hydroxythiobenzoic acid, **180**
Hydroxytoluenes, side-chain oxidation, 154
6-Hydroxy-2,3,4-trimethoxybenzaldehyde 165
α-Hydroxy-*o*-xylene, 50
Hypnone, 130
Hypoprotocetraric acid, 222

Illicium sp., 189
Imbricaric acid, 211, 213
Imidazoline, 91
Imine-cobalt complexes, 81
Indazoles, 100
Indigo, 95
Indole, 103
Intramolecular hydrogen bonding, 155, 156
2-Iodoacetophenone, **135**
4-Iodoacetophenone, 135
ω-Iodoacetophenone, **133**
Iodobenzaldehydes, 96
3-Iodobenzaldehyde, 82
4-Iodobenzoyl chloride, 134

4-Iodobenzyl acetates, 51
Iodobenzyl alcohol, 51
Iodobenzylamines, 26
2-Iodo-1,4-dimethoxybenzene, 193
Iodohydrins, 107
Iodomethane, 163
Iodonium nitrate, 36
2-Iodo-1-phenylethanol, 105
3-Iodosalicylic acid, **182**
4-Iodosalicylic acid, 182
5-Iodosalicylic acid, 182
4-Iodosoacetophenone, 135
Iodosobenzaldehydes, 95
Iodoxybenzaldehydes, 95
Iris pallida, 168
Iron octaphenylporphyrazine, 130
Iron pentacarbonyl, 58, 67, 108
Isobutyrophenone, 120
Isocoumarins, 239, 241
Isocyanates, reduction, 6
Isocyanides, 67
Isodesoxypodophyllotoxin, 257, 270
Isoeugenol, 154, 161
Isoeverinic acid, **192**
Isogalcatin, 267
Isogalloflavin, 196
Isogmelinol, 264, 265
Isoguaiacin, **266**
Isohydrocumic acid, 182
Isohinokinin, 257, 258
Isolariciresinol, 264, **266**
Isoleucine, 224
Isonitrosoacetophenone, 133
Isonitroso(hydroxyimino) compounds, 125
Iso-otobain, 267
Isopinoresinol, 262
Isopropenyl acetate, 125
2-Isopropenyloxazine, 114
Isopropenyl stearate, 111
4-Isopropylbenzaldehyde, 84, **88**
Isopropylbenzene, 88
—, peroxidation, 130
4-Isopropylbenzyl alcohols, 50
4-Isopropylbenzylamines, 26
4-Isopropylbenzyl chloride, 88
1,2-Isopropylidene-3,5-benzylidene-α-D-glucofuranose, 231
Isopropylmagnesium iodide, 43
2-Isopropyloxazine, 114
4-Isopropylphenylmagnesium bromide, 88
4-Isopropylsalicylic acid, 182
4-Isopropyltoluene, 88
Isoquinolines, synthesis, 17
Isoquinoline alkaloids, 144
Isosavanin, 258
Isotaxiresinol, **266**
Isothiocyanates, 145
—, reduction, 6
Isovalerophenone, 120
Isovanillic acid, 189
Isovanillin, 162

Jasmine oil, 39, 45
Juniperus sabina, 258
Juniperus silicicola, 268
Justicia procumbeus, 268
Justicidin-A, 268
Justicidin-B, 268
Justicidin-C, 268
Justicidin-D, 268

Ketazines, 75
Ketenes, aryl, 139
Ketene, reaction with salicylic acid, 177
Ketene acetals, 79
Ketene-*N*,*O*-acetals, 68
Ketene diethylacetal, 201
Ketene dithioacetals, 79
Ketimines, 166
α-Ketoaldehydes, 83
Ketones, alkylation, 117, 126
—, aminomethylation, 126
—, aryl-substituted, 115
—, benzene series, 107
—, α-carboxymethylation, 126
—, conversion to alcohols, 31
—, — to aldehydes, 71
—, — to amides, 128
—, — to aromatic hydrocarbons, 124
—, — to β-oxo-acids, 125
—, fission reactions, 128
—, from acetylenes, 116
—, from aldehydes, 116, 117
—, from alkylbenzenes, 107
—, from alkyl halides, 108
—, from arylamines, 108
—, from arylglycidic acids, 115
—, from α-branched aldehydes, 106
—, from carboxylic acids, 109
—, from dihydro-1,3-oxazines, 113

Ketones, *(continued)*
—, from β-diketones, 113
—, from epoxides, 114
—, from esters, 111
—, from 1,2-glycols, 114
—, from nitriles, 112
—, from olefins, 115
—, from oximes, 122
—, from β-oxo-esters, 113
—, from phenyl esters, 166
—, from styrenes, 116
—, halogenation, 129
—, hydroxymethylation, 45
—, phenolic, 141
—, reaction with diazoalkanes, 127
—, — with diazomethane, 118
—, — with *O*-methylhydroxylamine, 6
—, — with ylids, 127
—, reduction with silanes, 44
—, reductive alkylation, 6, 7
—, α,β-unsaturated, 125
Ketoximes, reduction, 6
Klason method, for lignin determination, 273
Knopper galls, 249
Kolbe reaction, 172, 173, 181, 188, 190, 192, 193
Kolbe-Schmitt reaction, 171, 176, 182, 188
Kröhnke aldehyde synthesis, 57

Labdanum oil, 130
Labelled acetate, 208
Lactams, reaction with phenylacetaldehyde, 105
Lactic acid-bacteria, 149
β-Lactones, 208
Lariciresinol, 260, **264**, 265, 266
Larix decidua, 264
Lecanora sp., 212
Lecanora atra, 212, 219
Lecanora badia, 218
Lecanora crassa, 224
Lecanoric acid, 204, 208, 210, 211, **212**
—, erythritol ester, 213
Lecanorolic acid, 219
Leukart reaction, 7, 122, 143
Lichen substances, isolation, 205
Lignans, 203, **253**
—, biosynthesis, 253
—, from podophyllum resin, 269
Lignans, *(continued)*
—, *in vivo* formation, 254
—, isolation, 254
—, occurrence, 254
—, stereochemistry, 258
Lignins, 161, **271**
—, biosynthesis, 275
—, classes, 272
—, ethanolysis, 274
—, isolation, 272
—, occurrence, 271
—, oxidation, 274
—, reduction, 274
Lignin, 160, 203, 254
—, alcoholysis, 277
—, benzyl ether links, 273
—, biological breakdown, 278
—, chemical reactions, 276
—, constitutional model, 276
—, degradation, 276
—, degradation products, 273
—, determination, 273
—, extracted from cellulose pulp, 272
—, molecular weight determinations, 277
—, polymerisation reactions, 275
—, residues from cellulose hydrolysis, 273
—, structure, 273
Ligninsulphonic acid, 272
Lignosulphonates, 274
Limnanthes douglasii, 145
Limnanthin, **145**
Liovil, **256**
Liriodendrin, **262**, 263
Liriodendron tulipifera, 262
Lirioresinols, 262, 263
1-Lithioaldimines, 67
Lithium amide, 126
Lithium aryls, 173
Lithium benzyloxide, 126
Lithium bromide, 32
Lithium bromoacetate, 126
Lithium cyanotrihydridoborate, 7, 19, 31
Lithium dimethyl cuprate, 134
Lithium diisopropylamide, 126
Lithium ethoxyhydridoaluminates, 63
Lithium isopropoxide, 31, 41
Lithium tetrahydridoaluminate, 4, 6, 7, 17, 18, 19, 20, 31, 32, 33, 34, 42, 45, 52, 60, 63, 64, 75, 123, 124, 143, 144, 147, 255, 256, 261, 265, 268

Lithium tri-*tert*-butoxyhydridoaluminate, 31, 60, 62, 72, 97
Lithium triethoxyhydridoaluminate, 64
Lobaric acid, **218**, 219, 220
Lobariol, 218, 219
Lobaritonide, 218
Lophine, 91
Lohpophora williamsii, 144
Lossen rearrangement, 145
Loxodin, 220
Lyonia ovalifolia, 267
Lyoniresinol, **267**, 269
Lyoniside, **267**

McFadyen-Stevens aldehyde synthesis, 64
McFadyen-Stevens reaction, 158, 160
Macrocyclic anhydrides, salicylic acid, 183, 184
Macrocyclic ester formation, 175
Macrocyclic lactones, 176
Magnesium bromohydrosulphide, 134
Magnesium methyl carbonate, 125
Ma Huang, 19
Malathion, 148
Malonic acid, 101
Malonic ester, 30, 92, 171, 183, 202
[1-^{14}C]Malonic ester, 207
Malonyl-coenzyme-A, 203, 207
Malonyl esters, 112
Mandelic acid, 133
Mannich reaction, 117, 127
Matairesinol, **256**, 257
Meerwein-Ponndorf reduction, 123
Meerwein reaction, 241
Melilotic acid, 201
Melilotus officinalis, 201
Mercaptoacetic acid, 181
2-Mercaptobenzoic acid, **180**, 181
6-Mercapto-3-phenyl-*s*-triazolo [4,3-b]-*s*-tetrazine, 73
Merochlorophaeic acid, 214
Mescaline, **144**
Mesityl oxide, 138
Mesotan, 177
α-Mesyloxy ketones, 119
Methacrylic ester, 47
Methanethiol, 90
Methanolamine hydrochlorides, 91
Methazonic acid, 101
Methedrine, 18
Methoxyacetic acid, 159
3-Methoxyacetonitrile, 160
Methoxyacetophenones, 143
2-Methoxyacetophenones, 167, 168
3-Methoxyacetophenone, 168, 169
4-Methoxyacetophenone, 121, 168
Methoxybenzaldehydes, 156
2-Methoxybenzaldehyde, 159
3-Methoxybenzaldehyde, 160
4-Methoxybenzaldehyde, 151, **160**
Methoxybenzenes, 157
Methoxybenzoic acids, 156, 273
2-Methoxybenzoic acid, 174, **178**
3-Methoxybenzoic acid, 187, 200
4-Methoxybenzoic acid, 188
2-Methoxybenzonitrile, 178
3-Methoxybenzonitrile, 142
2-Methoxybenzoyl chloride, 178
3-Methoxybenzoyl chloride, 160, 168
4-Methoxybenzyl alcohol, 146, **148**
3-Methoxybenzylamine, **142**, 145
4-Methoxybenzylamine, 143
2-Methoxybenzyl bromide, 142
α-Methoxybenzyl chloride, 89
2-Methoxybenzyldimethylamine, **142**
3-Methoxybenzyl isothiocyanate, **145**
4-Methoxybenzyl isothiocyanate, 145
3-Methoxybenzylthiourea, 145
Methoxycarbonyl chloride, 177, 192, 193 198, 202
Methoxycarbonylsalicylic acid, 177, 178
Methoxycarbonylsalicyloyl chloride, 179
Methoxycarboxylic acids, 240
1-Methoxy-1-cyclopenten-5-one, 241
2-Methoxy-4,6-dihydroxybenzaldehyde, 165
7-Methoxy-5,6-dimethylphthalide, 206
4-Methoxyhydratropaldehyde, **160**
3-Methoxy-2-methylaminobenzoic ester, 187
N-Methoxymethyl-*N*-methylbenzylamine, 19
Methoxymethyl salicylate, 177
7-Methoxy-1-oxotetrahydronaphthalene, 202
Methoxyphenols, 156, 157
3-Methoxyphenol, 169
4-Methoxyphenol, 193
3-Methoxyphenylacetaldehyde, 160
2-Methoxyphenylacetaldoxime, 160

4-Methoxyphenylacetaldoxime, 160
Methoxyphenylacetic acids, 199, 200
2-Methoxyphenylacetic acid, 200
4-Methoxyphenylacetic acid, 200
α-Methoxyphenylacetic acids, 78, 89
2-(4′-Methoxyphenyl)butan-3-one, **168**
2-(4′-Methoxyphenyl)but-2-ene, 168
4-(2-Methoxyphenyl)butyric acid, 202
4-(3-Methoxyphenyl)butyric acid, 202
4-(4-Methoxyphenyl)butyric acid, 202
1-(4-Methoxyphenyl)ethanol, 147, **148**
2-(4-Methoxyphenyl)ethanols, 147
1-(4-Methoxyphenyl)ethylamine, **143**
2-(4-Methoxyphenyl)ethylamine, 143
2-(2-Methoxyphenyl)ethyl bromide, 202
2-(4-Methoxyphenyl)ethyldimethylamine 143
3-(3-Methoxyphenyl)propylmagnesium iodide, 202
8-Methoxyquinoline, 187
Methyl acetimidate hydrochloride, 132
2-Methylacetophenone, 120
3-Methylacetophenone, 120
4-Methylacetophenone, 120, **134**
Methylamine, 18, 19
4-Methylaminobenzaldehyde, 100, **102**
2-Methylamino-1-phenylpropane, **18**
4-Methyl-2-(2′-aminophenyl)quinoline, 136
Methylaniline, 102
C-Methylanthranil, 135
N-Methylarylamines, 6
Methylbenzaldehydes, 84
2-Methylbenzaldehyde, 88
3-Methylbenzaldehyde, 88
4-Methylbenzaldehyde, 88
N-Methylbenzaldoxime, 94
Methylbenzenes, oxidation to aldehydes, 55, 56
4-Methylbenzo-1,2,3-triazine-3-oxide, 136
S-Methyl benzoylthioformate, 134
Methylbenzyl alcohols, 50
2-Methylbenzyl alcohol, 31
Methylbenzylamines, 26
4-Methylbenzylamine, 88
N-Methylbenzylamine, **19**, 57
Methyl *N*-benzylcarbamate, 20
4-Methylbenzyl chloride, 88
2-Methylbenzyldimethylamines, 14
2-Methylbenzylhydryldimethylamine, 15
2-Methylbut-2-ene, 83
β-Methylcinnamaldehyde, 126
4-*O*-Methylcryptochlorophaeic acid, 214
Methyl cyanide, 122
Methyl *m*-digallate, 233, 235
O-Methyldihydroxythujaplicatin, 259
1-Methyl-3,4-diphenylpyrrole, 105
Methyleneaniline, 30
3,4-Methylenedioxybenzaldehyde, **162**
3,4-Methylenedioxybenzoic acid, 189
3,4-Methylenedioxynitrobenzene, 162
3,4-Methylenedioxyphenol, 264
3,4-Methylenedioxyphenylacetic acid, 200
Methylenetriphenylphosphosphorane, 138
Methyl evernate, 210
N-Methylformanilide, 65, 151, 153, 163
Methyl formimidate hydrochloride, 132
Methyl gallate, 197, 233, 234
2-*O*-Methylglucose, 231, 234, 243, 245, 247
4-*O*-Methylglucose, 234
β-Methylglucoside, 249
Methyl gyrophorate, 214
O-Methylhydroxythujaplicatin, **259**
4-*O*-Methylhypoprotocetraric acid, 222
3-Methylindazole-2-sulphonic acid, 136
Methylisopropylsalicylic acid, 182
Methyllithium, 109
Methyl ketones, 110, 119, 130
—, oxidation to aromatic acids, 123
—, α,β-unsaturated, 81
Methylmagnesium bromide, 115
Methylmagnesium carbonate, 172
Methylmagnesium iodide, 29, 41, 42, 43, 134, 149, 169, 255
2-Methylmercaptobenzoic acid, 181
Methyl methacrylate, 47
Methyl methylthiomethyl sulphoxide, 69, 78
4-Methylmorpholine, 47
Methyl nitrite, 133
O-Methylolivetonide, 219, 223
O-Methyloximes, reduction, 6
2-Methylpentenal, 183
Methylphenols, 141
Methylphenylacetaldehyde, 138
Methylphenylacetamide, 42
2-Methyl-1-phenylbutan-2-ol, **43**
2-Methyl-4-phenylbutan-2-ol, **43**
N-Methyl-*N*-phenylcarbamoyl chloride, 66
2-Methyl-1-phenyl-1,2-epoxypropane, 138

α-Methyl-β-phenylethylamine, 11
N-Methyl-β-phenylethylamine, **20**
N-Methyl-*N*-(α-phenylethyl)aniline, 19
Methyl-β-phenylethylmethanol, 43
Methylphenylglyoxal, 123
Methylphenylketene, 139
Methylphenylmethanol, 41, 42, 123, 135
2-Methyl-2-phenylpentan-4-one, 138
2-Methyl-2-phenylpropanal, 106
2-Methyl-3-phenylpropanal, 107
2-Methyl-1-phenylpropane-1,2-diol, 138
2-Methyl-1-phenylpropan-1-ol, **43**
2-Methyl-1-phenylpropan-2-ol, **43**
2-Methyl-2-phenylpropan-1-ol, **43**
2-Methyl-3-phenylpropionaldehyde, 70, 107
Methylphenyl-*n*-propylmethanol, 43
4-Methyl-5-phenylthiazole, 18
O-Methylphloroglucinol, 165
2-Methylphloroglucinol 1-methyl ether, 170
4-*O*-Methylphysodic acid, 220
O-Methylpicrolichenic ester, 224
O-Methylpipinoresinol, 262
2-Methylquinazoline, 101
1-*O*-Methylquinides, 235
4-Methyl-2-quinolone, 137
Methyl salicylate, 175, 176, **177**, 178, 180
—, reaction with ammonia, 178
—, silicic ester, 185
Methyl salicylate vicianoside, 177
Methylsalicylic acids, 182, 183
3-Methylsalicylic acids, macrocyclic anhydrides, 184
6-Methylsalicylic acid, **183**
O-Methylsimplocosigenol, 263
Methyl sinapate, 254
α-Methylstyrene, 18, 42, 115, 126, 130
α-Methylstyrene epoxide, 115
Methyl sulphate, 165
ω-Methylsulphinylacetophenone, 133
Methylsulphinyl carbanions, 111
Methylthiomethylphosphonic ester, 118
O-Methylthujaplicatin, **258**, 259
Methyl *o*-tolyl ketone, 120
Methyl 2,3,6-tri-*O*-benzoylgalactopyranoside, 234
Methyl vinyl sulphides, 118
Microphillic acid, 213
Miriquidiic acid, 213
Monohydroxyaryl alcohols, 147
Monohydroxybenzaldehydes, 156
Monoketones, benzene series, 107
Monotropitin, 175, **177**
Monotropitin hexa-acetate, 177
Morpholine, 129, 199
Mustard oils, 145
Myrecetin, 237
Myristica otoba, 266
Myrobalans, 226, 231, 237, 238, 239, 242, 245, 253
Myrobalan extract, 239, 244
Myrobalans meal, 243

Naphthalenelithium, 2
Naphthalene-1,3,7-trisulphonic acid, 187
2-Naphthol-6,8-disulphonic acid, 188
1-Naphthol-4-sulphonic acid, 183
2-Naphthylamine, 183
2-Naphthylamine-4,8-disulphonic acid, 187
Nataic acid, 222
Native lignin, 272
Neochebulagic acid, 243, 244, **245**, 246
Neochebulinic acid, **243**, 244, 246
Neogmelinol, 264
Nephroarctin, 215
Nidulin, 220, **224**
Nigella damascena, 187
Nitriles, 74
—, catalytic reduction, 6
—, conversion to aldehydes, 64
—, — to ketones, 112
—, from aldehydes, 81
—, hydrogenation, 4
—, reaction with organometallic reagents, 112
—, — with olefins, 10
—, — with phenols, 166
—, reduction, 4, 64
—, — to primary amines, 5
—, α,β-unsaturated, 70
Nitrilium salts, 10
Nitroacetophenones, 127, 135
2-Nitroacetophenone, **135**
3-Nitroacetophenone, 132, 135
4-Nitroacetophenone, 83, 121, 135
ω-Nitroacetophenone, **133**
Nitroalkanes, 57
3-Nitroanisaldehyde, 160
Nitrobenzaldehydes, 98, 100

Nitrobenzaldehydes, *(continued)*
—, reduction, 99
2-Nitrobenzaldehyde, **95**, 101
3-Nitrobenzaldehyde, 51, **95**, 102, 160
4-Nitrobenzaldehyde, 77, 83, **97**
2-Nitrobenzaldehyde diacetate, 100
4-Nitrobenzaldehyde-2-sulphonic acid, **103**
Nitrobenzaldoximes, 100
Nitrobenzene, 155
4-Nitrobenzenediazonium salts, 146
Nitrobenzene-3-sulphonic acid, 153
Nitrobenzoic acids, 51
4-Nitrobenzoic acid, 51
4-Nitrobenzoyl chloride, 97
Nitrobenzoylmalonic ester, 135
2-Nitrobenzyl acetate, 52
4-Nitrobenzyl acetates, 51
Nitrobenzyl alcohols, **51**, 52
3-Nitrobenzyl alcohol, 51
4-Nitrobenzyl alcohol, 51
Nitrobenzylaldehydes, 51
Nitrobenzylamines, 24, 26
3-Nitrobenzylamine, 24
4-Nitrobenzylamine, 24
4-Nitrobenzylaniline, 100
2-Nitrobenzyl bromide, 95
Nitrobenzyl chlorides, 24
4-Nitrobenzyl chloride, 58
2-Nitrobenzylidene diacetate, 95
4-Nitrobenzylidene diacetate, 97
Nitrobenzyl nitrates, 51
N-Nitrobenzylphthalimides, 24
Nitro compounds, conversion to ketones, 117
—, reduction to aralkylamines, 3
—, — to primary amines, 5
Nitro-2-hydroxybenzaldehydes, 158
β-Nitro-2-methoxystyrene, 160
β-Nitro-4-methoxystyrene, 143
Nitrones, 57, 74
Nitronic esters, 72
2-Nitro-4-nitrosophenol, 76
Nitroparaffins, 72
4-Nitrophenylacetylene, 135
β-4-Nitrophenylethyl alcohol, 53
2-(4-Nitrophenyl)ethyldimethylamine, 143
4-Nitrophenylpropiolic acid, 135
4-Nitrophenyl salicylate, 177
3-Nitrophthalic acid, 182
6-Nitropiperonal, 162
6-Nitropiperonylic acid, 190
Nitropropane, 57, 81, 88
3-Nitroquinoline, 101
5-Nitro-*β*-resorcylic acid, 190
Nitrosalicylic acids, 176
3-Nitrosalicylic acid, **182**
4-Nitrosalicylic acid, 182
5-Nitrosalicylic acid, 182
6-Nitrosalicylic acid, 182
Nitrosobenzaldehydes, 99
2-Nitrosobenzaldehyde, **99**
Nitrosobenzoic acids, 99
2-Nitrosobenzoic acid, 95
2-Nitrosobenzyl alcohol, 51
N-Nitroso-*N*-benzylurea, 21
N-Nitrosodibenzylamines, 12
4-Nitrosodimethylaniline, 57, 97, 103
N-Nitrosodimethylamine, 19
Nitroso-*N*-methylanthranilic acid, 182
N-Nitrosopiperidine, 115
5-Nitrososalicylic acid, **182**
β-Nitrostyrenes, 155, 160
ω-Nitrostyrene, 144
4-Nitrostyrene oxide, 83
Nitrotoluenes, oxidation, 51, 95
2-Nitrotoluene, 95
4-Nitrotoluene, 77, 97, 102
2-Nitrovanillin, 162
2-Nitroveratraldehyde, 162
Norephedrin, 106
Norlobaridone, 220
Norstictinic acid, 218, 221

Oak bark, 226
Obtusaic acid, 213
Ococa veraguensis, 260
Octacarbonyldicobalt, 70
Octa-*O*-methylvaloneic acid, 242
1-Octene, 118
Oenanthophenone, 120
Oil of wintergreen, 175, 176, 177
Olea cunninghamii, 266
Olea europa, 265
Olefins, conversion to alcohols, 33
—, — to aldehydes, 69
—, — to ketones, 115
—, cycloaddition of carbonyl compounds, 83
—, *trans*-formyl, 80
—, from aldehydes, 79
—, reaction with nitriles, 10

Olivetolaldehyde, 163
Olivetolcarboxylic acid, 206
Olivetonic acid, 206
Olivetonide, 206
Olivetoric acid, 209, 213
Olivil, **265**, 266
Oötheca, 189
Orange blossom oil, 41
Orcinol, 163, 191, 192, 224
Orcylaldehyde, **163**, 191
Organoboranes, 119
—, conversion to amines, 8
Organocadmium compounds, 109, 110
Organocopper compounds, 110
Organolithium compounds, 67, 109, 111
Organomercury compounds, 110
Organometallic compounds, 173
—, adducts with aldehydes, 74
—, conversion to amines, 8
Organometallic reagents, 153
—, aldehyde synthesis, 66
Organopalladium compounds, 70
Orsellinic acid, **191**, 192, 205, 206, 207, 208, 211, 212
—, tritium labelled, 208
Orsellinic esters, 192
Otobain, **266**, 267
Otobaphenol, **266**, 267
Orthobenzoic ester, 89
Orthocaine, 188
Orthoform, 188
Orthoformic ester, 88, 89, 104, 123, 132, 152, 153, 165
Orthoform-old, 188
Orthophenylacetic ester, 139
Orthosilicates, 89
Oxalic acid, 239
Oxaloacetic ester, 101
Oxalyl chloride, 90, 178
Oxazines, 67
Oxazolidines, 60
Oxazolines, 60, 66
Oxidation reactions, carbonyl-forming, 108
Oximes, conversion to aralkylamines, 5
Oxiranes, 83, 127
Oxo-acids, 202
α-Oxo-acids, from methyl ketones, 123
β-Oxo-acids, 125
β-Oxo-esters, 125
β-Oxo-esters, *(continued)*
—, conversion to ketones, 113
Oxomatairesinol, **257**
Oxonium sulphates, 196
α-Oxophosphonate esters, 62
β-Oxosulphoxides, 111, 133, 134

Paeonal, 169
Pannarin, 221
Papaver somniferum, 160, 188
Paper chromatography, 205
Paraorsellinic acid, **192**
Parmelialic acid, 212
Parmelia sp., 212, 219
Parmelia physodes, 219
Parmelia tinctorum, 208
Parsley-apiol, 165
Parsley seed oil, 165
Paterno-Büchi reaction, 83
α-Peltatin, **270**, 271
β-Peltatin, **270**, 271
α-Peltatin β-glucoside, 271
β-Peltatin β-glucoside, 271
Peltigera aphthosa, 216
Pendunculagin, 238, **249**
Penicillin-X, 200
Penicillium sp., 183, 189, 192
Penicillium chrysogenum, 200
Penicillium divergens, 149
Penicillium griseofulvum, 183
Penta-acetyl-*m*-digallic acid, 229
Penta-acetyl-*p*-digallic acid, 229
β-Penta-*m*-digalloylglucose, 232
Pentagalloylglucose, **231**, 233, 234, 235, 236
Pentan-2-one, 196
β-Penta(tri-*O*-methylgalloyl)glucose, 231
Pentyl alcohol, 176
Pentyl nitrite, 154
5-Pentylresorcinol, 163
Peptide synthesis, 46, 147, 148
Peracetic acid, 48
Perbenzoic acid, 87
Perlatoric acid, 213
Pertusaria amara, 224
Peru balsam, 42, 45
Pervitin, 18
Phenacetyl chloride, 16
Phenacyl bromide, 129, 132
Phenacyl chloride, **132**, 137
Phenarctin, 215
Phenazines, 239, 246, 247

Phenazinedicarboxylic acids, 246, 248
Phenazinediphenic acids, 239
β-Phenethyl alcohol, 41
Phenetol, 160
Phenetsal, 177
Phenols, autoxidation, 278
—, carbonation, 171
—, condensation with formaldehyde, 141, 142, 145
—, conversion to hydroxyketones, 166
—, formylation, 151, 161
—, from depsidones, 216
—, reaction with acyl chlorides, 166
—, — with chloroform, 150
—, — with formaldehyde, 153
—, — with formic acid derivatives, 152
—, — with hexamethylenetetramine, 153
—, — with hydrogen cyanide, 151
—, — with nitriles, 166
—, — with orthoformic ester, 152
—, side-chain oxidation, 154
Phenol, 42, 130, 142, 157, 176, 177, 178, 181
—, carbonation, 172
—, conversion to salicylic acid, 171
—, Kolbe reaction, 172
—, reaction with acetyl chloride, 168
—, — with benzyl alcohol, 39
—, — with benzyl chloride, 45
—, — with carbon dioxide, 171, 176
—, — with carbon oxysulphide, 180
—, — with carbon tetrachloride, 173
—, — with formaldehyde, 147, 148
Phenolcarboxylic acids, 203
—, from the hydrolysis of depsides, 206
Phenol-formaldehyde-amine condensations, 141
Phenol-formaldehyde condensation, 145
Phenolic aldehydes, 174
Phenolic acids, 278
Phenolic aralkylamines, 141
Phenolic tannins, 227
Phenol ketones, 166, 167
Phenol monocarboxylic acids, 171
1-Phenolpropan-2-ol, **42**
2-Phenoxybenzoic acid, 178
Phenylacetaldehyde, 41, 55, 59, 61, 67, 70, **104**, 154
Phenylacetaldehyde dimethyl acetal, 105
Phenylacetaldehyde oximes, 155
Phenylacetamide, 20
Phenyl acetate, 168, 210
Phenylacetic acid, 17, 20, 78, 109, 137, 159, 200
—, triethyl ortho-ester, 139
Phenylacetic ester, 41, 43, 112, 129
α-Phenylacetoacetonitrile, 137
Phenylacetone, 112, 118, 119, 137
Phenylacetyl chloride, 138
Phenylacyl bromide, 119
Phenylalanine, 17, 236, 237
—, ^{14}C-labelled, 271
α-Phenylaldehydes, 68
Phenylalkylamines, physical properties, 16, 22
N-Phenylbenzaldoxime, 92
Phenyl benzoate, 210
2-Phenylbenzo-1,3-dioxin, 147
N-Phenylbenzylamine, 6, **20**
9-Phenyl-9-borabicyclo[3.3.1]nonane, 119
4-Phenylbutanal, 71
1-Phenylbutan-1-ol, **43**
1-Phenylbutan-3-ol, **43**
2-Phenylbutan-1-ol, **43**
4-Phenylbutan-1-ol, **43**
1-Phenylbutan-2-one, 137
3-Phenylbutan-2-one, 107, 138
4-Phenylbutan-2-one, 118, 138
Phenylbutyric acids, 202
Phenylcarbamates, 94
Phenylcarbene, 75
Phenyl carbonate, 171
2-Phenylchromenylium salts, 159
Phenylcyclopropanes, 76
Phenyldiazomethane, 29, 92
o-Phenylenediamine, 92, 238, 239, 246, 247, 248
Phenyl esters, conversion to ketones, 166
—, photochemical rearrangement, 166
—, reduction to aldehydes, 62
1-Phenylethanol, 20, 41
2-Phenylethanol, 38, 41
α-Phenylethanol, 16, 34
β-Phenylethanol, 34
Phenyl ethers, 151, 153, 173
—, oxidation, 154
Phenylethylacetamide, 43
α-Phenylethyl acetate, 125
β-Phenylethyl acetate, 53
α-Phenylethyl alcohol, 1, **41**

β-Phenylethyl alcohol, 1, **41**
Phenylethylamines, 9, 11
α-Phenylethylamine, 13, **16**
β-Phenylethylamine, 8, **17**, 18, 20, 24, 58
α-Phenylethylaniline, **20**
β-Phenylethyl benzoate, 17
β-Phenylethyl chloride, 8, 20
α-Phenylethyl ethyl ether, **45**
β-Phenylethyl isothiocyanate, 21
β-Phenylethylmagnesium bromide, 43
N-Phenylformimidic ester, 153
Phenylglyoxal, 123, 131, 133
Phenylglyoxylic acid-d, 87
1-Phenylhexan-5-one, 138
3-Phenylhexan-4-one, 138
4-Phenylhexan-3-one, 107
2-Phenylindazole, 52
2-Phenylindole, 131
α-Phenylisobutyraldehyde, 106
Phenyl isocyanate, 94, 173
Phenylisopropylmethanol, 43
Phenyl isothiocyanate, 173
Phenylketene, **139**
Phenylketene diethyl acetal, 139
Phenylketene dimethyl acetal, 139
α-Phenylketones, 114
Phenyllithium, 15, 37, 44, 45, 163
Phenylmagnesium bromide, 18, 29, 41, 42, 43, 67
Phenyl 2-mercaptobenzoate, 181
Phenylmethanethiol, 21, 48, 49
Phenyl methyl ethers, 273
Phenylmethylmalonic acid, 139
Phenylmethylmethanol, 44
1-Phenyl-4-nitroindazole, 99
Phenylnitrosomethane, 25
5-Phenylpentanoic acids, 202
2-Phenylpentan-2-ol, **43**
3-Phenylpentan-3-ol, **43**
1-Phenylpentan-2-one, 138
1-Phenylpentan-3-one, 113, 138
1-Phenylpentan-4-one, 138
2-Phenylpentan-3-one, 138
3-Phenylpentan-2-one, 138
Phenyl β-phenylpropenyl ketone, 131
2-Phenylpropanal, 106
3-Phenylpropanal, 106
1-Phenylpropan-1-ol, **42**
1-Phenylpropan-2-ol, 1
2-Phenylpropan-1-ol, **42**
2-Phenylpropan-2-ol, **42**
3-Phenylpropan-1-ol, 34, 35, **42**
1-Phenylpropan-2-one, 137
3-Phenylpropan-3-one, 119
α-Phenylpropionaldehyde, **106**
β-Phenylpropionaldehyde, **106**
Phenylpropionamides, 144
Phenyl propionate, 168
2-Phenylpropionic acid, 201
3-Phenylpropionic acid, 201
α-Phenylpropionic acid, 139
β-Phenylpropionyl chloride, 138
3-Phenylpropylmagnesium bromide, 43
Phenyl-*n*-propylmethanol, 43
1-Phenylpropyne, 201
Phenyl salicylate, 177
Phenyltetralins, 265
Phenylthioacetmorpholides, 199
Phenyl(trichloromethyl)methanol, 78
2-(1-Phenylvinyl)-4,4,6-trimethyl-1,3-oxazine, 68, 114
Phillygenol, 262, 263
Phillyrea sp., 262
Phillyrin **262**
Phlobaphene, 227
Phloracetophenone, **170**
Phloretic acid, 201
Phloroglucinaldehyde, 165
Phloroglucinol, 151, 157, 165, 167, 170
—, carbonation, 172
—, derivatives, 165, 199
Phloroglucinolcarboxylic acid, 199
Phloroglucinolcarboxylic acid 4-monotosylate, 191
Phosphonate carbanions, 80
Phosphorus ylids, 127
Photo-Fries reaction, 166
Phthalic acid, oxidation, 187
Phyllanthus embilica, 243
Physodalic acid, 222
Physodic acid, **219**, 220, 223
Picea sp., 256
Picea excelsa, 168, 257, 268
Picric acid, 175, 176
Picrin, 168
Picrolichenic acid, **224**, 225
Picrolichenic ester, 224
Picropodophyllin, **269**, 271
Picropodophyllin β-glucoside, 271
Pimelic acid, 176

Pinacols, 75, 124
Pinoresinol, 253, **260**, 261, 262, 275
Piper cubeta, 257
Piper guineense, 263
Piperidine, 101
Piperoic acid, 162
Piperonal, 161, **162**, 189
Piperonyl chloride, 190
Piperonylic acid, 187, **189**, 190
Piperonylidene dichloride, 162
Piper peepuloides, 261
Planaic acid, 213
Plicatic acid, **269**
Podocarpus spicata, 256
Podophyllomeronic acid, 270
Podophyllotoxin, **269**, 270, 271
Podophyllotoxin chloride, 268
Podophyllum emodi, 257
Podophyllum peltatum, 257, 269
Podophyllum resin, 269, 270
Podophyllum sikkimensis, 271
Podorhizol, **257**, **270**
Podorhizol glucoside, 257
Polybenzylbenzenes, 45
Polygalloylglucoses, 232, 252
Polyhydric aldehydes, 155
Polyhydric phenols, autoxidation, 278
Polyphenolic compounds, 227
Polyphenol oxidase, 189
Polyporus veriscolor, 275
Polysaccharides, 271, 278
Polysalicylide, 177, 183
Populin, **148**
Populus sp., 149
Populus alba, 148
Populus balsamifera, 188
Populus tremulata, 148
Populus trichocarpa, 192
Potassium amide, 45
Potassium benzyl oxide, 46
Potassium *tert*-butoxide, 49, 119, 139
Potassium *tert*-butyl hydroperoxide, 58
Potassium 2,6-di-*tert*-butylphenoxide, 119
Primary alcohols, oxidation to aldehydes, 56
Primaverin, 191
Primaverose, 177
Primaveroside, 191, 193
Primula auricula, 169
Primula officinalis, 191, 193
Primulaverin, 193
Propeller conformations, 186
Propionaldoxime, 108
Propionic acid, 170
Propionitrile, 18
Propiophenone, 18, 42, 83, 107, 118, 120, 123
n-Propylbenzene, 107
Propylene oxide, 42
Propyl gallate, 194, 197
n-Propylmagnesium iodide, 43
Proteins, 189
—, bacterial degradation, 143
Protocatechualdehyde, **161**, 162
Protocatechuic acid, 161, **189**
Protocetraric acid, 222, **224**
Prunus, 86
Psoromic acid, 204, 218, 221, **224**
Pulvinic acid, 205
Pummerer rearrangement, 49
Punica granatum, 241
Purpurogallin, 196
Purpurogallincarboxylic acid, 195, **196**
Pyrene, 278
Pyridine, 47, 132
Pyridine-borane, 31
Pyridine dichromate, 35
Pyridine hydrochloride, 160
Pyridinium salts, 57
Pyridinium ylids, 108
Pyrocatechuic acid, 188
Pyrogallol, 157, 170, 194
—, carbonation, 193
—, derivatives, 164, 193
—, oxidation, 195
Pyrogallolaldehyde, **164**
Pyrogallol 1,2-dimethyl ether, 164
Pyrogallolcarboxylic acid, 193
Pyrogallol trimethyl ether, 170
Pyrroles, colour tests, 103

Quaternary ammonium compounds, 14
Quaternary compounds, rearrangement to amines, 15
Quaternary salts, aralkyl halides, 2
—, hydrogenolysis, 7
Quercetin, 237
Quercus sp., 193, 226
Quercus aegilops, 226
Quercus coccifera, 226

Quercus lusitanica, 226, 234
Quercus prinus, 226
Quercus rubra, 235
Quercus sessiliflora, 241
Quinacetophenone, 169
Quinaldine-3-sulphonic acid, 101
Quinic acid, 234
—, galloylated, 235
Quinidine, 238
Quinol, 157, 163, 167, 169
Quinol diacetate, 169
Quinoline derivatives, 101
2-Quinolone, 101
4-Quinolone, 136
o-Quinones, 189, 276
Quinone methides, 254, 275

Ramalinolic acid, 214, **216**
Reductive alkylation, 6
Reformatzky reaction, 79, 202
Reimer-Tiemann reaction, 150, 158, 163, 173
Reissert compounds, 61
Resacetophenone, **169**
Resins, from phenols, formaldehyde and ammonia, 142
Resorcinol, 151, 157, 167, 169, 170, 201
—, carbonation, 172, 190
—, derivatives, 163, 190
Resorcinol bistetrahydropyranyl ether, 191
α-Resorcylaldehyde, **163**
β-Resorcylaldehyde, **163**
γ-Resorcylaldehyde, 163
Resorcylic acids, 190
α-Resorcylic acid, **191**
β-Resorcylic acid, **190,** 191
γ-Resorcylic acid, **191**
Reterodendrin, 265, **268,** 270
Rhamnus purshiana, 197
Rheum officinale, 230
Rhizoninic acid, 206
Rhus sp., 232
Rhus coriaria, 226
Rhus semialata, 193, 226, 232
Rhus typhina, 226, 237
Rissic acid, **198**
Ritter reaction, **10**, 16
Robinia pseudacacia, 197
Rosenmund reaction, 60, 155
Rose oil, 41
Rotenone, 198

Saccharin, 2
Safrole, 162
Salazinic acid, 218, 221, 224
Salicaceae sp., 148
Salicin, 147, **148,** 158, 175
Salicin penta-acetate, 148
Salicin tetra-acetate, 148
Salicoylsalicylic acid, 185
Salicylaldehyde, 147, **158,** 159, 163, 167, 175, 180
—, chelation, 158
—, copper complex, 156
—, manufacture, 150
—, reactions, 158
Salicylaldehyde β-D-glucopyranoside, 158
Salicylaldehyde imine, 156, 159
Salicylaldehyde oxime, 156
Salicylaldoxime, 142, 159, 179
Salicylamine, **142**
Salicylic acid, 147, 158, 171, 172, 173, 174, **175,** 178, 181, 182, 183, 198
—, cyclic anhydrides, 183, 184
—, electrolytic reduction, 59
—, esters, 177
—, ethers, 177
—, high polymers, 183
—, homologues, 182, 183
—, macrocyclic anhydrides, 184
—, manufacture, 176
—, nuclear-substituted, 181
—, oxidation, 192
—, phosphorus derivatives, 178, 179
—, reactions, 176
—, reaction with ketene, 177
—, — with phosphorus compounds, 180
—, reduction, 176
—, self-condensation, 176
—, sulphur derivatives, 180
Salicylic acid *N*-benzenesulphonohydrazide, 158
Salicylic ester, 177
Salicylic methyl ester, 175, 177
Salicylideneacetone, 168
Salicylides, 176, 178
Salicylohydrazide, 179
Salicylonitrile, 178, **179**
Salicyloylamide, **178**
Salicyloylanilide, **179**

Salicyloyl azide, 179
Salicyloyl chloride, **178,** 180
Salicyloyl disulphide, 180
Salicyloylsalicylic acid, 176, **177**
Salicyloyl *p*-toluidide, 179
Saligenin, **147,** 148, 158
Salinigrin, 168
Salirepol, **149**
Salireposide, 149
Salix sp., 149, 168
Salol, 177
Salophen, 177
Savinin, 257, **258**
Schiff bases, 74, 80, 126
—, reduction, 6
Schmidt degradation, carboxylic acids, 8
Schmidt reaction, 105, 128
Sclerotin, 189
Scrobiculin, 214
Secoisolariciresinol, **256**
Secondary alcohols, oxidation to ketones, 107
Sekikaic acid, 214, **216**
Selenoacetophenone, **134**
Selenobenzaldehyde, **91**
Semicarbazide, 64
Sesame seed oil, 261
Sesamin, 255, **261,** 262, **263**
Sesamolin, **264**
Sesamum angolence, 263
Sesamum indicum, 264
Sesangolin, **263**
Shikimic acid, 203, 236, 237, 253
Sicilian sumach, 226
Sicilian sumach tannins, 232
Sikkimotoxin, **271**
Silanes, 44
Silicicolin, 268
Silicon tetraisocyanate, 21
Silver(II) picolinate, 36
Sinalbin, **145**
Sinapis alba, 145
Sinapyl alcohol, 253, 272
Sinigrin, 145
Skatole, 103
Sodamide, 37, 128, 132
Sodium *N*-aroylbenzenesulphonamide, 129
Sodium azide, 16, 102
Sodium benzenesulphonamide, 129
Sodium benzenesulphonochloroamide, 129
Sodium benzyldiazotate, 28
Sodium benzyl oxide, 25, 45, 46
Sodium bis(2-methoxyethoxy)aluminium hydride, 6
Sodium cyanotrihydridoborate, 124
Sodium dihydridobis(2-methoxyethoxy)-aluminate, 31, 32
Sodium dihydridotrithioborate, 124
Sodium hydridobis(2-methoxyethoxy)-aluminate, 62
Sodium phenoxide, conversion to salicylic acid, 171
—, reaction with carbon dioxide, 176
Sodium phenylcarbonate, 172
Sodium tetracarbonylferrate(II), 58
Sodium tetrahydridoborate, 5, 7, 32, 33, 50, 51, 61, 62, 75, 148, 257
Sodium triphenylmethide, 117
Softwood lignins, 272
Soluble aspirin, 177
Sommelet-Hauser rearrangement, 14, 15, 38
Sommelet reaction, 57, 155
Sonn-Müller aldehyde synthesis, 63
Sonn-Müller procedure, 88
Sorbus aucuparia, 267
Sparassis racemosa, 192
Sparassol, 192
Sphaerophorin, 204, 213
Sphaerophorolcarboxylic acid, 206
Sphaerophorus melanocarpus, 204
Spiraea sp., 158
Spiraea ulmaria, 175
Split acid, 239
Squamatic acid, 209, 210, **212,** 213
Stagshorn sumach, 226
Stagshorn sumach tannins, 232
Star anise oil, 160, 168
Stearophenone, 111
Stenosporic acid, 213
Stereocaulon sp., 218
Stevens rearrangement, 14, 15
Stictinic acid, 221, **224**
Stilbenes, 47, 49
Stilbene-2,2′-disulphonic acid, 103
Stirlingia latifolia, 130
Storax, 42
Styrenes, 33
—, conversion to ketones, 116
—, oxidation, 154

Styrene, 41, 42, 92, 105
—, reaction with carbon monoxide and hydrogen, 106
Styrene iodohydrin, 105
Styrene oxide, 34, 42, 82, 104
Succinic acid, 46, 239, 240
Succinic anhydride, 202
Succinic ester, 256
Sugars, cyclic derivatives with benzaldehyde, 89
—, galloyl derivatives, 228
—, photosynthesis from carbon dioxide, 208
Sulphamate esters, 10
Sulphenimides, 49
Sulphinamides, 111
Sulphones, 38
Sulphonium salts, 49
5-Sulphosalicylic acid, **183**
Sulphoxides, 38
Sumach gallotannin, 232
Sumach tannins, 233
Symplocosigenol, 262
Symplocosin, **262**, 263
Syringaldehyde, 155, 161, 258, 274
Syringaresinol, 253, **262**, 263, 267
Syringic acid, **197**
Syringin, 197
Syringyl alcohol, 262, 272

Tannins, algarobilla, 247
—, carboxylic acid, 227
—, condensed, 227
—, determination of molecular weight, 228
—; from knopper galls, 249
—, from oak and chestnut woods, 250
—, hydrolysable, 171, 193, 195, 203, **227**
—, isolation, 227
—, natural, 231
—, paper electrophoresis, 228
—, phenolic, 227
—, properties, 228
—, reaction with methanol, 233
—, valonea, 242
Tannin, 194
—, from Chinese galls, 231
Tannin extracts, hydrolysis, 237
Tanning materials, 227
Tanning process, 227
Tara, 226
Tara gallotannin, 234
Taratannin, **234**
Tartaric acid, 16
Taxus baccata, 266
Teloschistes flavicans, 224
Tenuiorin, 214
Terchebin, 231, **238**, 239, 243, 244, **246**, 251, 252
Terminalia chebula, 226, 242
Terminalia myriocarpa, 243
Tertiary amides, reduction, 63
Tetra-acetylellagic acid, 242
α-1,3,4,6-Tetra-acetylglucose, 230
Tetrabenzaldehyde, 87
α-2,3,4,6-Tetra-*O*-benzylglucose, 230
Tetrabenzylhydrazine, **29**
Tetrabenzyl orthosilicate, 48
Tetrabenzyl pyrophosphate, 47
Tetrabenzyltetrazene, 29
Tetra-*o*-cresotide, 184
Tetra-*p*-cresotide, 184
Tetradepsides, 216
Tetra-*O*-galloylglucose, 234
β-1,3,4,6-Tetragalloylglucose, **231**, 234
2,3,4,6-Tetragalloylglucose, **231**, 233
Tetrahydridoimidazole, 62
Tetrahydrido-2-methylisoquinoline, 7
Tetrahydroacridine, 136
Tetrahydrofurans, 256
Tetrahydrofuroguaiacins, 259
Tetrahydrogmelinol, 264
Tetrahydroimidazoles, 64
Tetrahydroisoquinolines, 17
Tetrahydroquinazolines, 24
Tetrahydroquinazoline-2-thione, 24
Tetrahydroquinazolin-2-one, 24
Tetrahydrosalicylic acid, 176
Tetrahydroxybenzaldehyde, 164, **165**
Tetrahydroxybenzene, derivatives, 165
Tetramethylammonium chloride, 33
Tetramethylbenzaldehyde, 84
Tetramethylbutyl isocyanide, 67
Tetramethyl-1,3-oxazine, 67
Tetrasaccharides, 232
Tetrasalicylide, 183, 184
2,3,4,6-Tetra(tri-*O*-benzylgalloyl)glucose, 231
Thallium trinitrate, 129
Thamnol, 165, 206
Thamnol anil, 206

Thamnolcarboxylic acid, 206
Thamnolic acid, 165
Thiazinethiols, 52
Thiazolidines, 68
Thin-layer chromatography, 205
Thioacetophenone, **134**
Thioamides, 128
—, desulphurisation, 63
Thioanilides, 173
Thiobenzaldehyde, **90**
Thiocarbonyl chloride, 21
Thioesters, reduction, 32
Thioglycollic acid, 272
Thioindigo, 181
Thioindigo dyestuffs, 181
Thiols, 49, 123, 273
Thiol esters, 61
Thiophenol, 49
Thiourea, 21, 48, 60
Thiourea dioxide, 123
Thioxanthone, 181
Thomasidionic acid, **269**
Thomastic acid, **269**
Thuja plicata, 258, 259, 269
Thujaplicatin, **258**, 259
Thymol, reaction with carbon dioxide, 171
o-Thymotic acid, 182, 183
—, macrocyclic anhydrides, 184
Tiglycoenzyme-A, 224
Tolualdehydes, **88**
o-Tolualdehyde, 56, 64
o-Toluanilide, 88
Tolu balsam, 45
Toluenes, from aldehydes, 75
—, oxidation, 95
Toluene, 39, 48, 134
—, chlorination, 87
—, oxidation, 35, 45, 56, 90
—, — to benzaldehyde, 56, 86, 87
—, reaction with acetyl chloride, 134
—, — with carbon monoxide and hydrogen chloride, 88
Toluene-ω-diazotates, **29**
4-Toluenesulphonamide, 103
Toluenesulphonic acids, oxidation, 103
Toluene-2-sulphonic acid, 103
Toluene-4-sulphonic acid, 105, 130
α-Toluenesulphonic acid, 48
4-Toluenesulphonyl chloride, 46, 104, 190
α-Toluenethiol, **48,** 49
α-Toluic aldehyde, 104
p-Toluidine, 179
o-Tolunitrile, 88
β-*p*-Tolylbutyryl chloride, 138
2-Tolylmethanol, 50
4-*p*-Tolylpentan-2-one, 138
3,4,5-Triacetoxybenzoic acid, **198**
3,4,5-Triacetoxybenzoyl chloride, 198
Triacetylgalloyl chloride, 164, 229
Trialkylboranes, 119
1,3,5-Triarylbenzenes, 125
Triarylphosphoranes, 80
Triazenes, related to benzylamine, 29
Triazines, 25, 151
Triazoles, 30
Tribenzaldehyde, 87
Tribenzylamine, 2, **20**
Tribenzyl arsenite, 48
Tribenzyl borate, 48
Tri-*O*-benzylgalloyl chloride, 229, 230, 231
Tribenzylhydrazine, **29**
O,N,N-Tribenzylhydroxylamine, 28
3,4,5-Tribenzyloxybenzoic acid, 198
Tribenzyl phosphate, **47**
Tribenzyl phosphite, **47**
Tribenzyl pyrophosphate, **47**
Tribromobenzaldehydes, 96
2,4,6-Tribromo-3-hydroxybenzaldehyde, 160
2,4,6-Tribromophenol, 158, 174, 176, 188
Tri-*n*-butylhydridostannane, 60
Tributylphosphine, 60
Tri-*o*-carvacrotide, 184, 186
Trichloroacetic acid, 150
Trichloroacetonitrile, 133
Trichloroacetophenone, 133
Trichloroacetyl chloride, 46, 133
Trichlorobenzaldehydes, 96
Trichloroisocyanuric acid, 44
2,4,6-Trichloro-*s*-triazine, 74
Trichocarpin, 192
Tri-*o*-cresotide, 184
Tri-*m*-cresotide, 184
Tri-*p*-cresotide, 184
Tri-*o*-cresyl phosphate, 158
Tridepsides, 216
Tri-3,6-dimethylsalicylide, 184, 186
3,4,5-Triethoxybenzoic acid, 198
Triethyloxonium tetrafluoroborate, 20
Triethyl phosphite, 47, 61, 111

Triethylsilane, 61
Trifluoroacetic anhydride, 211
m-Trigallic acid, 228, **229**
m-Trigalloyl chains, 234
Triacetylgalloyl chloride, 230
Tri-*O*-galloylglucoses, 234
β-1,3,6-Trigalloylglucose, **231**, 243, 244, 246
3,4,5-Tri-*O*-galloylquinic acid, 235
Trihalogenophenols, 148
Trihydric phenols, carbonation, 172
2,3,4-Trihydroxyacetophenone, **170**
2,4,5-Trihydroxyacetophenone, **170**
2,4,6-Trihydroxyacetophenone, **170**
3,4,5-Trihydroxyacetophenone, **170**
2,4,5-Trihydroxybenzaldehyde, 164
2,3,4-Trihydroxybenzaldehyde, **164**
Trihydroxybenzaldehyde, **164**
2,4,6-Trihydroxybenzaldehyde, **165**
3,4,5-Trihydroxybenzaldehyde, **164**
Trihydroxybenzenes, 152
1,2,4-Trihydroxybenzene, 170, 198
1,3,4-Trihydroxybenzene, 164
Trihydroxybenzoic acids, 193
2,3,4-Trihydroxybenzoic acid, **193**, 216
2,3,5-Trihydroxybenzoic acid, **198**
2,3,6-Trihydroxybenzoic acid, **198**
2,4,5-Trihydroxybenzoic acid, **198**
2,4,6-Trihydroxybenzoic acid, **199**
3,4,5-Trihydroxybenzoic acid, 175, **193**
Tris(hydroxybenzyl)amines, 142
3,4,5-Trihydroxycyclohexane-1-carboxylic acid, 197
2,3,6-Trihydroxy-4-methylbezaldehyde, 165
2,3,4-Trihydroxy-6-methylbenzoic acid, 206
3,4,6-Trihydroxy-2-methylbenzoic acid, 206
2,3,4-Trihydroxy-6-pentylbenzoic acid, 216
Trihydroxyphenylacetic acids, 199
Trihydroxyphenyl alcohols, 149
Trihydroxyphenylalkylamines, 144
Trihydroxyphenyl ketones, 168
2,4,6-Trihydroxypropiophenone, **170**
Triisopropylbenzaldehyde, 84
Triketocyclohexane rings, 246
2,3,4-Trimethoxyacetophenone, 170
2,4,6-Trimethoxyacetophenone, 170
3,4,5-Trimethoxyacetophenone, 170
2,3,4-Trimethoxybenzaldehyde, 164
2,4,6-Trimethoxybenzaldehyde, 165
3,4,5-Trimethoxybenzaldehyde, 149, 164, 258
1,2,4-Trimethoxybenzene, 199
1,3,4-Trimethoxybenzene, 164
1,3,5-Trimethoxybenzene, 170
2,3,5-Trimethoxybenzoic acid, 198
2,3,6-Trimethoxybenzoic acid, 199
2,4,5-Trimethoxybenzoic acid, 198
2,4,6-Trimethoxybenzoic acid, 199
3,4,5-Trimethoxybenzoic acid, **198**, 229
3,4,5-Trimethoxybenzoyl chloride, 198
3,4,5-Tri-*O*-methoxybenzylbutyrolactone, 257
3,4,5-Trimethoxybenzyl chloride, 170
2,3,5-Trimethoxybromobenzene, 198
Tri(methoxycarbonyl)galloyl chloride, 164, 229
3,4,5-Trimethoxy-*ω*-nitrostyrene, 144
1-(3,4,5-Trimethoxyphenyl)ethanol, 149
2-(3,4,5-Trimethoxyphenyl)ethylamine, **144**
3,4,5-Trimethoxyphthalic acid, 239, 240
2,4,6-Trimethylacetophenone, 120, 129
Trimethylaluminium, 72
Trimethylbenzaldehydes, 84
2,4,6-Trimethylbenzaldehyde, 73
3,4,5-Tri-*O*-methylbenzoic acid, 257
Trimethylbenzyl alcohols, 50
Trimethylbenzylamines, 26
Tri-*O*-methylbrevifolin, 240, 241
Tri-*O*-methylbrevifolincarboxylic ester, 240
Tri-*O*-methylchebulic acid, 239
Trimethylchlorosilane, 75
Tri-*O*-methyldihydroxythujaplicatin, 259
Trimethylene oxide, 34
Tri-*O*-methylgallic acid, 197, 232, 235, 236, 245
3-Tri-*O*-methylgalloylglyceraldehyde, 235
β-1,2,3-Tri-*O*-methylglucose, 247
Tri-*O*-methylhydroxythujaplicatin, 259
Tri-*O*-methylisogalloflavin, 196
Tri-*O*-methylisothujaplicatin, 258
2,4,6-Trimethylphenylketene, 139
Tri-*O*-methylthujaplicatin, 258, 259
2,4,6-Trinitroacetophenone, 135
2,4,6-Trinitroanisole, 160
2,4,6-Trinitrobenzaldehyde, **97**, 98

β-2,4,6-Trinitrophenylethyl alcohol, 53
2,4,6-Trinitrotoluene, 97
—, reaction with formaldehyde, 53
3,4,5-Trioxocyclohexanecarboxylic acid, 197, 252
1,3,5-Triphenylbenzene, 131
Triphenylcinnamylphosphonium chloride, 79
2,4,5-Triphenylimdiazole, 91
2,3,5-Triphenylisoxazolidine, 92
Triphenylmethyl tetrafluoroborate, 44
Triphenylphosphine, 58
Triphenylphosphine dibromide, 103
Triphenylsilanol, 69
Triphenylstannane, 31
Triphenyltin hydride, 125
Triphenyl trisulphone, 90
2,4,6-Triphenyl-*s*-trithianes, 90
Trisaccharides, 232
Trisalicylide, 183, 184, 185, **186**
3,4,5-Tris(methoxycarbonyloxy)benzoic acid, 198
3,4,5-Tris(methoxycarbonyloxy)benzoyl chloride, 198
Trismethylthioacetophenones, 133
Trithiane, 69
Trithioacetophenone, 134
Tri-*o*-thymotide, 184, **186,** 187
Tritium labelled orsellinic acid, 208
Triton B, 19
Tropolones, 196
Tropylium ions, 35
Tryptophan, 103
Tscherniac-Einhorn reaction, 9
Tsuga heterophylla, 268
Tuber rose oil, 39
Tumidulin, 213
Turkish gallotannin, **234**
Turkish tannin, 226
Tyramine, **143**
[α-^{14}C]Tyramine, 143
Tyramine amides, 143
Tyrosine, 149, 200, 201
—, bacterial degradation, 143
Tyrosol, **149**

Ullman reaction, 196, 216, 218, 219, 241
Ulmus thomasii, 267, 269
Umbilicaria pustulata, 207, 212
Umbilicaric acid, 214
Urea, reaction with benzyl alcohol, 20
—, — with benzylamine, 21
Usnea barbata, 212, 218
Uvitic acid, 196

n-Valeric acid, 218
Valerophenone, 115, 120
Valonea, 226, 241, 250
Valonea tannins, 242
Valoneic acid, 242
Valoneic acid dilactone, **242**
Vanilla beans, 148
Vanilla oil, 160
Vanilla planifolia, 161
Vanillic acid, 161, **189**
Vanillin, 153, 154, **161,** 162, 189, 258, 274
—, manufacture, 150
o-Vanillin, 161, 165, 188
Vanillin-β-D-glucopyranoside, 162
Vanillin oxime, 144, 162
Vanillylideneacetone, 170
Variolaric acid, 221
Variolic acid, 218
Vegetable tannins, 227
Veraguensin, **260**
Veratraldehyde, 149, 162, 189, 256
Veratraldehyde cyanhydrin, 200
Veratric acid, **189,** 273
Veratrine, 189
Veratronitrile, 189
Veratrum sabadilla, 189
Vescalagin, **250**
Vescalin, 241, **250**
Vicanicin, 220, **224**
Vilsmeier formylation procedure, 103
Vilsmeier-Haack aldehyde synthesis, 65
Vilsmeier reaction, 151, 152, 165
Viola cornuta, 177
Violutoside, **177**
2-Vinyloxazine, 67
Virensic acid, 222

Willgerodt reaction, 128, 199
Willstätter process, 273
Wintergreen oil, 175, 176
Wittig olefin synthesis, 119
Wittig reaction, 37, 71, 79
Wolff-Kishner reaction, 75, 124, 219
Wood cellulose, 273

Xanthones, 176, 177, 205, 242
Xanthone-4-carboxylic acid, 178
Xylenes, oxidation, 88
o-Xylene, oxidation, 56
m-Xylene, 21, 88
—, reaction with acetonitrile and formaldehyde, 9
Xylose, 177
Xylosides, 267
o-Xylyl-bromide, 88

Yangambin, 263
Ylids, reaction with ketones, 127

Zanthoxylum clavaherculis, 143
Zieria laevigata, 159
Zierin, 159
Zinc cyanide, 151
Zingerone, 149, **170**
Zingiber officinale, 170